Table of Atomic Weights (Based on Carbon-12)

Element	Symbol	Atomic Weight	Element	Symbol	Atomic Weight
Actinium	Ac	(227)*	Neodymium	Nd	144.2
Aluminum	Al	27.0	Neon	Ne	20.2
Americium	Am	(243)	Neptunium	Np	237.0
Antimony	Sb	121.8	Nickel	Ni	58.7
Argon	Ar	39.9	Niobium	Nb	92.9
Arsenic	As	74.9	Nitrogen	N	14.0
Astatine	At	(210)	Nobelium	No	(254)
Barium	Ba	137.3	Osmium	Os	190.2
Berkelium	Bk	(245)	Oxygen	O	16.0
Beryllium	Be	9.01	Palladium	Pd	106.4
Bismuth	Bi	209.0	Phosphorus	P	31.0
Boron	B	10.8	Platinum	Pt	195.1
Bromine	Br	79.9	Plutonium	Pu	(242)
Cadmium	Cd	112.4	Polonium	Po	(210)
Calcium	Ca	40.1	Potassium	K	39.1
Californium	Cf	(251)	Praseodymium	Pr	140.9
Carbon	C	12.0	Promethium	Pm	(145)
Cerium	Ce	140.1	Protactinium	Pa	231.0
Cesium	Cs	132.9	Radium	Ra	226.0
Chlorine	Cl	35.5	Radon	Rn	(222)
Chromium	Cr	52.0	Rhenium	Re	186.2
Cobalt	Co	58.9	Rhodium	Rh	102.9
Copper	Cu	63.5	Rubidium	Rb	85.5
Curium	Cm	(245)	Ruthenium	Ru	101.1
Dysprosium	Dy	162.5	Samarium	Sm	150.4
Einsteinium	Es	(254)	Scandium	Sc	45.0
Erbium	Er	167.3	Selenium	Se	79.0
Europium	Eu	152.0	Silicon	Si	28.1
Fermium	Fm	(254)	Silver	Ag	107.9
Fluorine	F	19.0	Sodium	Na	23.0
Francium	Fr	(223)	Strontium	Sr	87.6
Gadolinium	Gd	157.3	Sulfur	S	32.1
Gallium	Ga	69.7	Tantalum	Ta	180.9
Germanium	Ge	72.6	Technetium	Tc	98.9
Gold	Au	197.0	Tellurium	Te	127.6
Hafnium	Hf	178.5	Terbium	Tb	158.9
Helium	He	4.00	Thallium	Tl	204.4
Holmium	Ho	164.9	Thorium	Th	232.0
Hydrogen	H	1.008	Thulium	Tm	168.9
Indium	In	114.8	Tin	Sn	118.7
Iodine	I	126.9	Titanium	Ti	47.9
Iridium	Ir	192.2	Tungsten	W	183.8
Iron	Fe	55.8	Unnilennium	Une	(266)
Krypton	Kr	83.8	Unnilhexium	Unh	(263)
Lanthanum	La	138.9	Unnilpentium	Unp	(262)
Lawrencium	Lr	(257)	Unnilquadium	Unq	(261)
Lead	Pb	207.2	Unnilseptium	Uns	(262)
Lithium	Li	6.94	Uranium	U	238.0
Lutetium	Lu	175.0	Vanadium	V	50.9
Magnesium	Mg	24.3	Xenon	Xe	131.3
Manganese	Mn	54.9	Ytterbium	Yb	173.0
Mendelevium	Md	(256)	Yttrium	Y	88.9
Mercury	Hg	200.6	Zinc	Zn	65.4
Molybdenum	Mo	95.9	Zirconium	Zr	91.2

* Parentheses around atomic weight indicate that weight given is that of the most stable known isotope.

The Nature
and Properties of
SOILS

Read

4.1
4.2
4.3
4.4
4.5
4.6
4.8
4.9

6.1
6.2
6.3
6.4
6.5

Reading 11/10/92

7.10 - 7.16

10.1 , 10.3, 10.5, 10.7, 10.8

10.10 to 10.14

11.1 to 11.4
12.1 to 12.4
12.15 to 12.17
13.1 to 13.4

NYLE C. BRADY

CORNELL UNIVERSITY AND
UNITED STATES AGENCY FOR INTERNATIONAL DEVELOPMENT

The Nature and Properties of SOILS

TENTH EDITION

MACMILLAN PUBLISHING COMPANY
NEW YORK
Collier Macmillan Publishers
LONDON

PRINTED IN THE UNITED STATES OF AMERICA

Macmillan Publishing Company
866 Third Avenue, New York, New York 10022

Collier Macmillan Canada, Inc.

LIBRARY OF CONGRESS CATALOGING IN PUBLICATION DATA

Brady, Nyle C.
    The nature and properties of soils/Nyle C. Brady. —10th ed.
        p.   cm.
    Includes bibliographies and index.
    ISBN 0-02-313361-9
    1. Soil science.   2. Soils.   I. Title
S591.B79   1990   89-9863
631.4—dc20   CIP

Printing: 1 2 3 4 5 6 7 8   Year: 0 1 2 3 4 5 6 7 8 9

Preface

Today's headlines focus on two major interactive international problems: widespread hunger and malnutrition, and the deterioration of the quality of the environment resulting from attempts to alleviate this hunger and malnutrition. The quality, management, and conservation of the world's soils are critical elements in each of these problems as well as in their solution.

Soil productivity helps determine how much food and fiber can be provided for the world's ever-increasing human population. Some soils are naturally productive, others are not. Some respond to wise cultural management and can be made more productive. Others will not so respond and could best be left in their native state with natural grass or forest vegetation. In any case, however, without some knowledge of the nature and properties of soils, it is not possible to predict soil quality in a given area or to know how soils should be managed and conserved.

Soils can affect and are affected by environmental deterioration. They can be degraded and even destroyed by excessive soil erosion. Downstream dwellers suffer the consequences of this erosion through the silting of their reservoirs, lakes, and rivers. Soils are also increasingly being used as sites for the disposal of animal, human, and industrial wastes. Knowledge of soil properties is essential to minimize soil erosion and to maximize the safe use of soils as waste disposal recipients.

As were previous editions of this text, the tenth edition is designed to help students understand soils, and to gain a general knowledge of their properties and usefulness. Several organizational changes have been made to make the text more readable and to emphasize properties of soils in the field. Perhaps the most significant of these changes is the early introduction of the concept of soils as natural bodies. The chapters dealing with soil parent materials and with the formation and classification of soils have been moved up front, and the soil classification names are used in subsequent chapters. This will help the students relate the classification scheme to practical soil management.

The chapter concerned with soil acidity and alkalinity has been combined with the one dealing with the liming of soils. Students will now be able to study means of changing soil acidity and alkalinity in the same chapter concerned with the chemical and biological implications of these two important soil properties.

The treatment of soil organic matter also has been brought together in one chapter. This permits consideration of soils dominated by organic matter (peat soils) along with the organic constituents of those soils dominated by mineral matter. Common features as well as pronounced differences are observed.

The text has been thoroughly revised and brought up to date. Recent changes in the classification system, *Soil Taxonomy*, have been incorporated as have been examples of soil contamination from wastes. Sixty eight new illustrations appear in this text Likewise, 42 new tables have been used to show various soil properties and to illustrate how soils can be best managed.

To help both instructors and students, about a dozen study questions have been added for each chapter. These questions focus on practical problems and opportunities, and emphasize the real life nature of soil management problems.

I am most grateful to the reviewers who made valuable suggestions for improvement and change: James L. Allrichs, Purdue University; James Brownell, California State University–Fresno; Terence H. Copper, University of Minnesota–St. Paul; Walter C. Crouse, Clinch Valley College of the University of Virginia; William Guertal, North Carolina State University–Raleigh; Wesley M. Jarrell, Oregon Graduate Center; Melvin L. Northup, Grand Valley State College; Richard M. Vanden Heuvel, University of Illinois–Urbana-Champaign. My colleague Suzanne Chase edited the first draft and made valuable suggestions. I am also indebted to Delores Nelson for some of the word processing of the text and tables. Joyce Torio provided me with valuable reference materials from a variety of sources. But most of all, I am grateful to my dear wife, Martha, who spent hours on the word processor and in making and copying graphs and charts for possible use in the text. Without her the job simply would not have been completed.

N. C. B.

Contents

Read all

Read

3

Soil Formation, Classification, and Survey   47

4

Physical Properties of Mineral Soils   91

8

Soil Reaction: Acidity and Alkalinity   213

9

Organisms of the Soil   253

10

Soil Organic Matter and Organic Soils   279

11

Nitrogen and Sulfur Economy of Soils   315

12

Phosphorus and Potassium   351

13

Micronutrient Elements   381

14

Losses of Soil Moisture and Their Regulation   399

15

Soil Erosion and Its Control   431

16

Fertilizers and Fertilizer Management   471

17

Recycling Nutrients Through Animal Manures and Other Organic Wastes   497

The Nature
and Properties of
SOILS

The Soils Around Us

People are dependent on soils, and, conversely, good soils are dependent on people and the use they make of the land. Soils are the natural bodies in which plants grow. They provide the starting point for successful agriculture.

Soils also have other meanings for humankind. They underlie the foundations of houses and factories and determine whether these foundations are adequate. They are the beds for roads and highways and influence the length of life of these arteries. Soils are used to absorb wastes from sewage systems, wastes from other municipal, industrial, and animal sources. Unfortunately, misused and unprotected soils can be washed into streams and rivers; later they can be deposited in municipal reservoirs, impairing water quality and shortening the period of usefulness of the reservoir. Obviously, soils are as important to city dwellers as to those on the farm.

Most great civilizations have depended on good soils. The ancient dynasties of the Nile were made possible by the food-producing capacity of the fertile soils of the river valley and its irrigation systems. Likewise, the fertile valley soils of the Tigris and Euphrates rivers in Mesopotamia and of the Indus, Yangtze, and Hwang Ho rivers in India and China were the sites of flourishing civilizations. Because fertility was frequently replenished by natural flooding, these valley soils provided continued abundant food supplies. They made possible stable, organized communities and even cities, in contrast to the nomadic, shifting societies they replaced.

Just as good soils helped to build flourishing civilizations, soil destruction or mismanagement was a contributing factor in their downfall. The cutting of timber in the river watersheds resulted in erosion and topsoil loss. Elaborate irrigation and drainage systems, as in the Euphrates and Tigris valleys, often were not maintained. Lack of proper maintenance resulted in the accumulation of harmful salts, and the once productive soils became barren and useless. The proud cities of the river valleys fell into ruin, and their inhabitants migrated elsewhere.

History provides lessons that modern nations have not always heeded. The wasteful use of soil resources by early settlers in the United States during the first century of intensive agricultural production (1840–1940) provides such an example. Even today, many do not fully recognize the long-term significance of soils. They are ignorant of what soils are, what they have meant to past generations, and what they mean today and to future generations.

1.1

What Is Soil?—Concepts of Soil

One reason for the lack of concern for soils is the different concepts as to what soils are. For example, to a mining engineer, soil is the debris covering the rocks or minerals that must be quarried. It is a nuisance and must be removed. To a highway engineer, soil may be the material on which a roadbed is to be placed. If its properties are unsuitable, it will need to be removed and replaced with rock and gravel.

To the average homeowner a good soil is rich, dark, and crumbly as opposed to "hard clay," which resists being spaded into a seedbed for a flower or vegetable garden. The homemaker can relate to the soil's stickiness or tendency to cling to the shoe soles and eventually to carpets. Dirt is soil out of place.

The farmer, along with the homeowner, looks upon the soil as a habitat for plants (Figure 1.1). However, the farmer earns a living from the soil and is therefore forced to pay more attention to its characteristics. To the farmer, soil is more than useful—it is indispensable.

A prime reason for studying soils is to obtain a general concept as to what they are and how they can and should be used. Such a concept is essential to understanding how soils can serve the engineer, the farmer, and the home owner. It is also a requisite for determining how best to conserve soils for future generations.

FIGURE 1.1

Good soils mean bumper crops, such as this partially harvested cornfield in Iowa. Farmers are keenly aware of the value of a productive soil.

1.2

Evolution of Modern Concepts of Soil

Knowledge about soils comes from two basic sources: farmer experience based on centuries of trial and error, and scientific investigations of soils and their management.

EXPERIENCE OF THE CULTIVATOR

From the dawn of agriculture, cultivators recognized good soils, being attracted to the fertile soils of river valleys. More than 42 centuries ago, the Chinese used a schematic soil map as a basis for taxation. Homer, in his *Odyssey* (about 1000 B.C.), makes reference to the use of manure on the land. The Bible refers to the "dung hill" and to the beneficial practice of "dunging" around plants. Early Greek and Roman writers described farming systems that involved leguminous plants and the use of ashes and sulfur as soil supplements. Further development and application of these practical principles of soil management were delayed in Europe by the barbarian invasions of Rome and by the feudal Dark Ages. But by the 17th and 18th centuries, there was a blossoming of scientific inquiry. The stage was set for the application of science to the improvement of agricultural systems, including those involving soils.

EARLY SCIENTIFIC INVESTIGATIONS ON SOIL PRODUCTIVITY

In the early 17th century, Flemish chemist Jan Baptista van Helmont conducted his famous five-year willow tree experiment and concluded that 164 pounds of dry matter resulted primarily from the water supplied because the soil lost no weight while producing the tree. John Woodward, an English researcher, found that muddy water produced more plant growth than rainwater or river water, which led him to conclude that the fine earth was the "principle" of growth. Most others concluded that the principle was organic matter or "humus" taken in by the plants from soil. In the early 18th century, Jethro Tull demonstrated the benefits of cultivation. He thought, erroneously, that stirring the soil made it easier for plants to absorb small quantities of fine earth. All of these men drew erroneous conclusions from their observations.

Through a series of field experiments starting in 1834, the French agriculturist J. B. Boussingault proved that air and rain were the primary sources of carbon, hydrogen, and oxygen in plant tissues. His investigations were largely disregarded until 1840, when the eminent German chemist Justus von Liebig reported findings that crop yields were increased by adding "minerals" or inorganic elements to the soil. He proposed that the mineral elements in the soil and in added manures and fertilizers are *essential* for plant growth.

Liebig's research led to the concept that certain factors were essential for plant growth and that if any one of these factors was limiting, plant production would be reduced accordingly (Figure 1.2). This principle, called the *law of the minimum*, is still valid today, and it may be stated in a practical way as follows: *The level of plant production can be no greater than that allowed by the most limiting of the essential plant growth factors.* Soils are studied today to ascertain which of these factors is least optimum and how its limitation to plant growth can be removed.

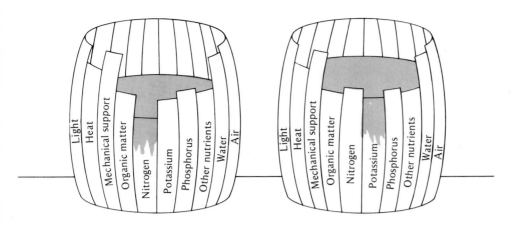

FIGURE 1.2

An illustration of the law of the minimum, which shows how plant growth is constrained by the essential element that is most limiting. The level of water in each barrel above represents the level of crop production. (a) Nitrogen is represented as being the factor that is most limiting. Even though the other elements are present in more than adequate amounts, crop production can be no higher than that allowed by the nitrogen. (b) When nitrogen is added, the level of crop production is raised until it is controlled by the next most limiting factor, in this case potassium.

TESTING LIEBIG'S CONCEPTS

Liebig's work stimulated other investigators. For example, by 1855 J. B. Lawes and J. H. Gilbert at the Rothamsted Experiment Station in England had proved that for cereals, nitrogen did not come from the air, as Liebig had supposed, but from the soil and from nitrogen-containing chemicals added to the soil. Their investigations of phosphorus-containing materials led to the development of acid-treated phosphate rock or "superphosphate," which is still an important source of phosphorus.

Further unraveling of the complexities of nitrogen transformations in soil did not occur until the emergence of soil bacteriology. J. T. Way discovered in 1856 that ammonium compounds were changed to nitrates in soils—a process that, twenty years later, R. Warington demonstrated was biological in nature. In 1890 S. Winogradski isolated the two groups of bacteria responsible for the transformation. Coupled with the discovery in the 1880s that nitrogen-assimilating bacteria grow in nodules of the roots of legumes, these findings provided background information for sound soil and crop management practices.

EARLY RESEARCH IN THE UNITED STATES

The European investigations on soil fertilizers proved to be applicable to the United States. Individuals such as Edmund Ruffin, a Virginia farmer, grasped the concepts relating to nutrient depletion, and in his writings Ruffin was especially critical of those who did not properly care for and replenish their soil. The abundance of open land to the west, however, encouraged the abandonment of the "worn-out" soils in the east rather than the adoption of more realistic management systems.

The establishment of the U.S. Department of Agriculture in 1862 and the first state Agricultural Experiment Station in 1886 accelerated both field and laboratory investigations of soils. Numerous field trials were initiated to test the applicability of findings of the European scientists. For example, C. G. Hopkins of Illinois developed effective soil-management systems based on limestone, rock phosphate, and legumes. Milton Whitney of the U.S. Department of Agriculture emphasized field studies and initiated the first national soil survey systems. F. H. King of Wisconsin studied the movement and storage of water in soils in relation to root penetration and crop growth.

FIELD SOIL INVESTIGATIONS

Except for the field testing of crop response to fertilizer, too little attention was paid to variable characteristics of the soils as found in the field. Likewise, climatic influences on soils were largely ignored. Soils were considered as geological residues on the one hand and reservoirs for plant nutrients on the other.

In 1860 E. W. Hilgard called attention to the relationships among climate, vegetation, rock materials, and the kinds of soils that develop. He conceived of soils not merely as media for plant growth but as dynamic entities subject to study and classification in their field setting. Unfortunately, Hilgard was ahead of his time; many of his concepts had to be rediscovered by others before they were accepted.

In the meantime, a brilliant Russian team of soil scientists led by V. V. Dokuchaev were investigating unique horizontal layers in soils—layers associated with different combinations of climate, vegetation, and underlying soil material. The same sequence of layers was found in widely separated geographical areas that had similar climate and vegetation. The concept of soils as natural bodies was well developed in the Russian studies, as were concepts of soil classification based on field soil characteristics.

The Russian studies began as early as 1870. Unfortunately, they were known to only a few scientists outside Russia until 1914, when they were published in German by K. D. Glinka, a member of the Russian team. The significance of the Russian work was quickly grasped by C. F. Marbut of the U.S. National Soil Survey program. Marbut and his associates developed a nationwide soil classification system based to a great extent on the Russian concepts.

Consideration of soils as natural bodies with distinctive chemical and physical characteristics has led to further modifications in soil classification systems, which will receive attention in later chapters.

1.3

Two Approaches—Pedological and Edaphological

The previous section suggests that two concepts of soil have evolved through two centuries of scientific study. One treats soil as a natural entity, a biochemically weathered and synthesized product of nature. The other treats soil as a natural habitat for plants and justifies soil studies primarily on that basis. These conceptions illustrate the two approaches that can be used in studying soils—that of the *pedologist* and that of the *edaphologist*.

The origin of the soil, its classification, and its description are examined in *pedology* (from the Greek word *pedon*, which means soil or earth). Pedology is the study of the soil as a natural body and does not focus primarily on the soil's immediate practical use. A pedologist studies, examines, and classifies soils as they occur in their natural environment. Pedological findings may be as useful to highway and construction engineers as to farmers.

Edaphology (from the Greek word *edaphos*, which also means soil or ground) is the study of soil from the standpoint of higher plants. Edaphologists consider the various properties of soils in relation to plant production. They are practical and have the production of food and fiber as their ultimate goal. To achieve that goal, edaphologists must determine the reasons for variation in the productivity of soils and find means of conserving and improving productivity.

In this textbook, the dominant viewpoint will be that of the edaphologist. Pedology will be used, however, to the extent that it gives a general understanding of soils as they occur in nature and are classified. Furthermore, because studies of the basic physical, chemical, and biological characteristics of soils contribute equally to edaphology and pedology, these approaches cannot be fully separated. This is illustrated in the following section, which deals with soils as they are found in the field.

1.4

A Field View of Soil

Like the peel of an orange, the soil covers the underlying geological materials of the earth. However, this analogy falls short by not conveying a sense of the enormous variability of soils. Examinations of road cuts from one geographic area to another show marked differences in soil depth, color and mineral makeup. Similar examinations, however, can identify common properties of soils from areas as distant as India, Hawaii, and the continental United States. The common properties will receive attention in the following section; the variations will be treated in succeeding chapters.

Soil Versus Regolith

Unconsolidated material overlying rocks is known as *regolith* (see Figure 1.3). It may be negligibly shallow or tens of meters thick. It may be material that has weathered from the underlying rock or it may have been transported from elsewhere by the action of wind, water, or ice and deposited on bedrock or on other material covering the bedrock. Consequently, the regolith varies greatly in composition from place to place.

The upper 1–2 meters of the regolith differs from the material below. It is higher in organic matter because plant roots concentrate there. Also, plant residues deposited on the surface have been incorporated in the upper regolith by earthworms and other animals and have been further modified by microorganisms. This upper portion of the regolith has also been subject to more weathering than lower portions. Products of this weathering, especially if they have moved vertically, give rise to characteristic layering called *horizons* (see Figure 1.3 and Section 1.5).

This upper and biochemically weathered portion of the regolith[1] is called *the soil*. It is the product of both destructive and synthetic forces. Weathering and microbial decay of organic residues are examples of destructive processes, whereas the formation of new minerals, such as certain clays, and of new stable organic compounds, along with the development of characteristic layer (horizon) patterns, are synthetic in nature.

---

[1] Where the original regolith was relatively uniform in composition, the material below the soil will have a composition similar to the parent material from which the soil was formed.

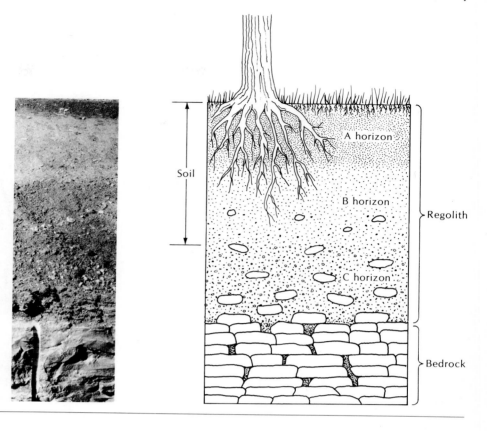

FIGURE 1.3
Relative positions of the regolith, its soil, and the underlying country rock. Sometimes the regolith is so thin that it has been changed entirely to soil; in such a case, soil rests directly on bedrock. [Photo courtesy Tennessee Valley Authority]

THE SOIL VERSUS A SOIL

Characteristics of the soil vary widely from place to place. For example, the soil on steep slopes is generally not as deep and productive as soil on gentle slopes. Soil that has developed from sandstone is more sandy and less inherently productive than soil formed from rocks such as limestone. The properties of a soil that has developed in tropical climates are quite different from those of a soil found in temperate or arctic areas.

Scientists have studied these soil variations and have set up classification systems that recognize a large number of individual soils, each having distinguishing characteristics. Therefore, *a* soil, as distinguished from *the* soil, is merely a well-defined subdivision having recognized characteristics and properties. Thus, a Cecil clay loam, a Marshall silt loam, and a Norfolk sand are examples of *individual soils*, which collectively make up *the soil* covering the world's land areas. The term "soil" is a collective term for all soils, just as "vegetation" is used to designate all plants.

1.5

The Soil Profile and Its Layers (Horizons)

Examination of a vertical section of a soil, as seen in a roadside cut or in the walls of a pit dug in the field, reveals the presence of more or less distinct horizontal layers (Figure 1.4). Such a section is called a *profile*, and the individual layers are known as *horizons*. Every well-developed, undisturbed soil has its own distinctive profile characteristics. These are useful in classifying and surveying soils but are of greatest importance in determining how the soils can best be used.

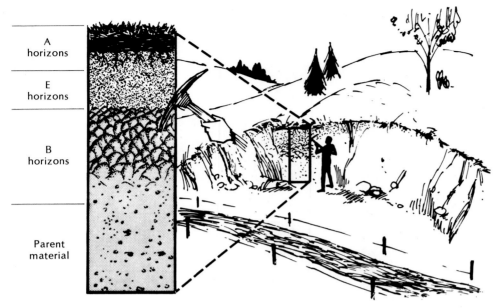

FIGURE 1.4

Field view of a road cut that reveals soil layering and the distinctive character of a *soil profile*. The upper horizons are called the A horizons. They are usually higher in organic matter and darker in color than the underlying horizons. Some constituents such as fine clays have been moved downward from the A horizons by percolating waters. The lower horizons, called B horizons, are sometimes characterized by clay accumulation and by distinctive structures such as shown at the point of the pick. The A and B horizons comprise the *solum*, which is distinct from the parent materials below. The presence and characteristics of different horizons differ sufficiently from soil to soil that it is possible to differentiate one soil from another.

The uppermost layers or horizons of a soil profile are darker in color than the lower horizons. This difference is due to the accumulation of organic matter that results from the decay of plant roots and of other organic residues incorporated into the upper soil layers. Also weathering tends to be more intense in the upper horizon than in the lower layers. Some products of weathering have been leached out of these upper layers, which are collectively termed the A horizons (Figure 1.4).

The underlying layers contain comparatively less organic matter than those nearer the surface. They are characterized, however, by an accumulation of varying amounts of substances such as silicate clays, iron and aluminum oxides, gypsum, and calcium carbonate. These materials may have been washed down from upper layers or they may have been formed in place through the weathering process. These underlying layers are referred to as B horizons (Figure 1.4).

These horizons, both upper and lower, have resulted from the soil-building processes of biochemical breakdown, weathering, and synthesis. They are evidence of the genesis of a natural body distinct from the parent materials from which the body was formed.

Collectively, these horizons make up the *solum* (from the Latin word *solum*, which means soil or land) or the upper part of the profile above the parent material. The solum thus described extends a moderate depth below the surface; a depth of 1–2 meters is representative of temperate region soils, although highly weathered soil of the tropics may be much deeper.

The various layers comprising a soil profile are not always distinct and well defined. Often the transition from one to the other is so gradual that establishing boundaries is difficult. Nevertheless, for a particular soil the various horizons are characteristic, and their properties greatly influence the use that can and should be made of soils.

1.6

Topsoil and Subsoil

When a soil is plowed and cultivated, the natural state of the upper 12–18 centimeters (5–7 inches) is modified. This manipulated part of the soil is referred to as the surface soil or the *topsoil*. This may also be called the *furrow slice* in situations where the soil is turned or "sliced" by the plow. The farmer generally considers "soil" to denote the surface layer, the "topsoil," or the furrow slice.

Topsoil or surface soil is the major zone of root development for crop plants. It contains many of the nutrients available to plants and supplies much of the water necessary for their growth. Through proper cultivation and the incorporation of organic residues, the topsoil can be kept loose and open to assure balanced air and water supplies for plant roots. It can be treated easily with commercial fertilizers and limestone, permitting the soil's fertility, and to a lesser degree its productivity,[2] to be raised or stabilized at levels consistent with economic crop production.

The *subsoil* is comprised of those soil layers underneath the topsoil. It is not seen from the surface and is not commonly disturbed by soil tillage, but there are few land uses that are not influenced by subsoil characteristics. Certainly crop production is affected by root penetration into the subsoil (Figure 1.5) and by the reservoir of moisture and nutrients it represents.

[2] The term *fertility* refers to the inherent capacity of a soil to supply plants with suitable quantities of mineral nutrients (e.g., nitrogen, phosphorus). *Productivity* is related to the ability of a soil to yield economic products (crops) and is the broader term since fertility is only one of a number of factors that determine the magnitude of crop yields.

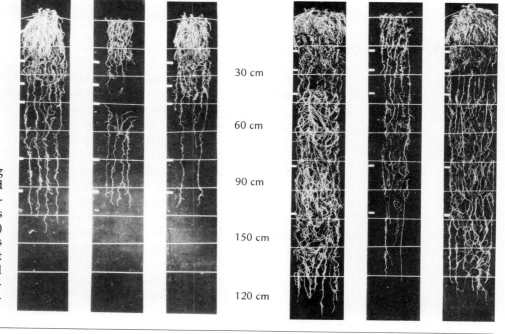

30 cm

60 cm

90 cm

150 cm

120 cm

FIGURE 1.5
Plant roots tell us something about soil characteristics and the treatment the soil has received. The corn crop was grown on an Illinois (Cisne) soil that received no fertilizers or crop residues (left) and that received both fertilizers and crop residues (right). [Courtesy J. B. Fehrenbacher, University of Illinois]

Likewise, downward movement of drainage water is sometimes impeded by impervious subsoils. The consequent wetness is detrimental to the growth of most crop plants. Areas with impervious subsoils should also be avoided in the selection of building sites and the location of roadways.

These observations are of practical significance because the subsoil, unlike the topsoil, is not normally subject to significant human alteration except by drainage. Consequently, land-use decisions often are based more on the nature of the subsoil than on topsoil characteristics.

1.7

Mineral (Inorganic) and Organic Soils

The profile generalizations just described relate to soils that are predominantly *mineral* or *inorganic* in composition. Even in their surface layers, mineral soils are comparatively low in organic matter, which generally ranges from 1 to 6%.

In contrast, soils whose properties are dominated by organic materials are termed *organic* soils. They commonly contain more than 50% organic matter by volume (at least 20% by weight). Typical examples of organic soils are found in wetland areas such as swamps, bogs, or marshes.

When drained and cleared, most organic soils are very productive, especially for high-value crops such as fresh market vegetables. Sizeable areas in Michigan, Wisconsin, and Minnesota illustrate this point. Organic deposits also may be excavated, bagged and sold as organic supplements for home gardens and potted plants. The economic significance of organic soils is considerable in localized regions.

However, since mineral soils occupy such a high proportion of the total land, they will receive the major attention in this text. Organic soils are considered as a unit in Chapter 10. The discussion in other chapters will be concerned primarily with mineral soils.

1.8

Four Major Components of Mineral Soils

Mineral soils consist of four major components: *inorganic or mineral materials, organic matter, water,* and *air*. Figure 1.6 shows the approximate proportions, by volume, of these components in a representative loam surface soil in optimum condition for plant growth. Note that this mineral soil contains about half solids and half pore space (water and air). The solid mineral particles comprise about 45% of the soil volume and organic matter 5%. At optimum moisture for plant growth, the pore space is divided roughly in half, 25% of the volume being water space and 25% air. The proportions of air and water are subject to rapid and great fluctuations. It should be emphasized that these four major soil components occur in a thoroughly mixed condition in soils. The mixture encourages interactions within and between the groups and permits marked variation in the environment for the growth of plants.

The volume composition of subsoils differs somewhat from that of topsoils. Subsoils are lower in organic matter content, are somewhat lower in total pore space, and contain a higher percentage of small pores that are filled much of the time with water rather than air.

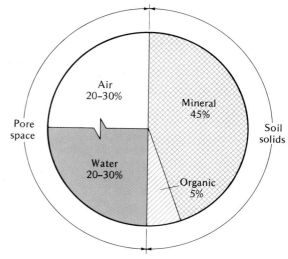

FIGURE 1.6
Volume composition of a loam surface soil when in good condition for plant growth. The air and water in a soil are extremely variable, and their proportions determine in large degree the soil's suitability for plant growth.

1.9

Mineral (Inorganic) Constituents in Soils

The inorganic portion of soils is quite variable in size and composition. It is composed of small rock fragments and minerals[3] of various kinds. The rock fragments are composed of aggregates of minerals and are remnants of massive rocks from which the regolith and, in turn, the soil have been formed by weathering. They are usually quite coarse.

The mineral particles present in soils are extremely variable in size. Some stone and gravel pieces are as large as the smaller rock fragments. *Sands* are somewhat smaller in size [0.05–2 millimeters (mm) in diameter], but each particle can be seen easily with the naked eye. The sand particles do not stick together and feel gritty when rubbed between the fingers. Still smaller in size are the *silt* particles (0.002–0.05 mm), which are powdery when dry and even when wet are not sticky. The smallest mineral particles are the *clays* (<0.002 mm), which form a sticky mass when wet and aggregate into hard clods when dry. The smallest clay particles have colloidal[4] properties and can be seen only with the aid of an electron microscope.

Typical properties of these mineral particles are shown in Table 1.1. It is obvious from this table that both physical properties and the ability of a given soil to supply chemical nutrients are determined to a considerable extent by the proportions of different sized particles present, a property known as *soil texture*. Terms such as sandy loam, silty clay, and clay loam are used to identify the soil texture.

PRIMARY AND SECONDARY MINERALS

Minerals that have persisted with little change in composition since they were extruded in molten lava (e.g., quartz, micas, and feldspars) are known

---

[3] The word "mineral" is used in this book in two ways: (1) as a general term to describe soils dominated by inorganic constituents and (2) as a more specific term to describe distinct minerals found in nature, such as quartz and feldspars. More detailed discussions of the common soil-forming minerals and the rocks in which they are found are given in Chapter 2.

[4] Colloidal systems are two-phase systems in which one phase in very finely divided state is dispersed through a second. Those clay and organic matter particles smaller than about 1 micrometer (μm) in diameter are generally considered to be colloidal in size.

TABLE 1.1

General Properties of Three Major Inorganic Soil Particles.

Range in diameter of particles shown in millimeters (mm).

Property	Sand (0.05–2 mm)	Silt (0.002–0.05 mm)	Clay (<0.002 mm)
1. Means of observation	Naked eye	Microscope	Electron microscope
2. Dominant minerals	Primary	Primary and secondary	Secondary
3. Attraction of particles for each other	Low	Medium	High
4. Attraction of particles for water	Low	Medium	High
5. Ability to hold chemical nutrients and supply them to plants	Very low	Low	High
6. Consistency properties when wet	Loose, gritty	Smooth	Sticky, plastic
7. Consistency properties when dry	Very loose, gritty	Powdery, some clods	Hard clods

as *primary* minerals. They are most prominent in the sand and silt fractions. Other minerals, such as the silicate clays and iron oxides, have been formed by the breakdown and weathering of less resistant minerals as soil formation progressed. These minerals are called *secondary* minerals and tend to dominate the clay and in some cases the silt fraction.

The inorganic fraction of the soil is the original source of most of the mineral elements that Liebig and numerous other scientists have found to be essential for plant growth. Although the bulk of these essential nutrients are held as rigid components of the basic crystalline structure of the minerals, a small but significant portion is in the form of charged ions on the surface of the fine mineral particles (clays). There are important mechanisms that enable plant roots to have access to these surface-held ions, mechanisms of critical importance to rapidly growing plants (see Section 1.14.)

SOIL STRUCTURE

The arrangement of the sand, silt, and clay particles within the soil, termed *soil structure*, is as important as the relative amounts of these particles present (soil texture). The particles may remain relatively independent of each other, but more commonly they are found associated together in aggregates. These aggregates vary from granules to plates, blocks, prisms, and columns. Such structural forms are very important in influencing water and air movement in the soil and are therefore used as criteria in classifying soils.

1.10

Soil Organic Matter

Soil organic matter comprises an accumulation of partially disintegrated and decomposed plant and animal residues and other organic compounds synthesized by the soil microbes as the decay occurs. Such material is continually being broken down and resynthesized by soil microorganisms. Consequently, organic matter is a rather transitory soil constituent, lasting from a few hours to several hundred years. This constituent requires maintenance by the regular addition to the soil of plant and/or animal residues.

The organic matter content of a typically well-drained mineral soil is small, varying from 1 to 6% by weight in the topsoil and even less in the subsoils. The influence of organic matter on soil properties, and consequently on plant growth, is far greater than the low percentage would indicate.

FIGURE 1.7
Soils high in organic matter are darker in color and have greater water-holding capacities than do soils low in organic matter. The same amount of water was applied to each container. As the lower photo shows, the depth of water penetration was less in the soil at the left because of its greater water-holding capacity.

Organic matter binds mineral particles into granules that are largely responsible for the loose, easily managed condition of productive soils and increases the amount of water a soil can hold (Figure 1.7). It is also a major soil source of phosphorus and sulfur and the primary source of nitrogen (three elements essential for plant growth). Finally, organic matter, including plant and animal residues, is the main source of energy for soil organisms. Without it, biochemical activity would come to a near standstill.

In addition to the original plant and animal residues and to their partial breakdown products, soil organic matter includes complex compounds that are relatively resistant to decay. These complex materials, along with some that are synthesized by the soil microorganisms, are collectively known as *humus*. This material, usually black or brown in color, is very fine (colloidal) in nature. For a given mass, its capacity to hold water and nutrient ions greatly exceeds that of clay, its inorganic counterpart. Small amounts of humus thus increase remarkably the soil's capacity to promote plant growth.

1.11

Soil Water—A Dynamic Solution

Two major concepts concerning soil water emphasize the significance of this component of the soil, especially in relation to plant growth.

1. Water is held in the soil pores with varying degrees of tenacity depending on the amount of water present and the size of the pores.
2. Together with its soluble constituents, including nutrient elements (e.g., calcium, potassium, nitrogen, and phosphorus), soil water

makes up the *soil solution*, which is the critical medium for supplying nutrients to growing plants.

When the soil moisture content is optimum for plant growth (Figure 1.6), the water in the large- and intermediate-sized pores can move in the soil and can be used by plants. The movement can be in any direction; downward in response to gravity, upward as water moves to the soil surface to replace that lost by evaporation, and in any direction toward plant roots as they absorb this important liquid. Although some of the soil moisture is removed by the growing plants, some remains in the tiny pores and in thin films around soil particles. The soil solids strongly attract this soil water and consequently compete for it with plant roots.

Thus, not all soil water is *available* to plants. Depending on the soil, as much as one fourth to two thirds of the moisture may remain in the soil after the plants have wilted or died for lack of water.

SOIL SOLUTION

The soil solution contains small but significant quantities of soluble inorganic and organic compounds, some of which contain elements that are essential for plant growth. In Table 1.2 are listed the 17 *essential elements* along with their sources. The soil solids, and particularly the fine organic and inorganic particles, release these elements to the soil solution from which they are taken up by growing plants. Such exchanges, which are critical for higher plants, are dependent on both soil water and the fine soil solids.

One other critical property of the soil solution is its *acidity* or *alkalinity*. Many chemical and biological reactions are dependent on the levels of hydrogen ions (H^+)[5] and hydroxide ions (OH^-) in the soil. These levels influence the solubility, and in turn the availability to plants, of several essential nutrient elements such as iron, manganese, phosphorus, zinc, and molybdenum.

TABLE 1.2
Essential Nutrient Elements[a] and Their Sources

Used in relatively large amounts		Used in relatively small amounts
Mostly from air and water	*From soil solids*	*From soil solids*
Carbon (C)	Nitrogen (N)	Iron (Fe)
Hydrogen (H)	Phosphorus (P)	Manganese (Mn)
Oxygen (O)	Potassium (K)	Boron (B)
	Calcium (Ca)	Molybdenum (Mo)
	Magnesium (Mg)	Copper (Cu)
	Sulfur (S)	Zinc (Zn)
		Chlorine (Cl)
		Cobalt (Co)

[a] Other minor elements, such as sodium, fluorine, iodine, silicon, strontium, and barium, do not seem to be as universally essential as the 17 listed here, although soluble compounds of some will increase the growth of specific plants.

[5] In water solutions hydrogen ions are always hydrated. Consequently, they exist as hydronium ions $[H_3O^+$, often written $H^+(aq)]$ rather than as simple H^+ ions. For reasons of simplicity, however, in this text we shall use the unhydrated formula (H^+) to represent this ion.

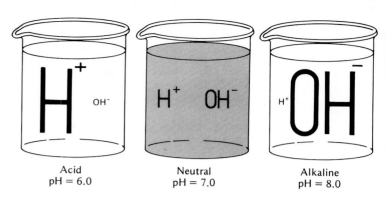

FIGURE 1.8

Diagrammatic representation of acidity, neutrality, and alkalinity. At neutrality the the H⁺ and OH⁻ ions of a solution are balanced, their respective numbers being the same (pH 7). At pH 6, the H^+ ions are dominant, being 10 times greater, whereas the OH^- ions have decreased proportionately, being only one tenth as numerous. The solution therefore is acid at pH 6, there being 100 times more H^+ ions than OH^- ions present. At pH 8, the exact reverse is true; the OH^- ions are 100 times more numerous than the H^+ ions. Hence, the pH 8 solution is alkaline. This mutually inverse relationship must always be kept in mind when pH data are used.

The concentration (activity) of hydrogen (H^+) and hydroxide (OH^-) ions in the soil solution is commonly ascertained by determining its *pH*. Technically the pH is the negative logarithm of the concentration (activity) of H^+ ions ($-\log [H^+]$) in the soil solution. Thus, each unit change in pH represents a tenfold change in the activity of the H^+ and OH^- ions. Figure 1.8 shows very simply the relationship between pH and the concentration of H^+ and OH^- ions. It should be studied carefully along with Figure 1.9, which shows the ranges in pH usually encountered in soils. The pH of the soil will be seen to be of great significance in essentially all aspects of soil science.

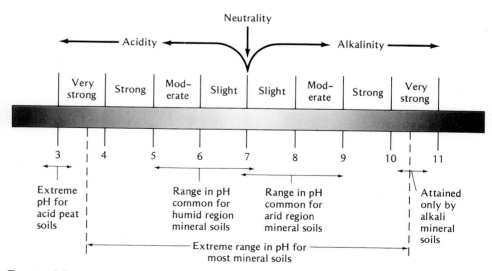

FIGURE 1.9

Extreme range in pH for most mineral soils and the ranges commonly found in humid region and arid region soils. Also indicated are the maximum alkalinity for alkali soils and the minimum pH likely to be encountered in very acid peat soils.

1.12

Soil Air—Another Changeable Constituent

Soil air differs from the atmospheric air in several respects. First, the composition of soil air is quite dynamic and varies greatly from place to place within a given soil. In local pockets, some gases are consumed by plant roots and by microbial reactions while others are released, thereby greatly modifying the composition of the soil air. Second, soil air generally has a higher moisture content than the atmosphere; the relative humidity of soil air approaches 100% when the soil moisture is optimum. Third, carbon dioxide in soil air is often several hundred times higher than the 0.03% commonly found in the atmosphere. Oxygen decreases accordingly and, in extreme cases, may be only 5–10%, or even less, as compared to about 20% for normal atmosphere.

The content and composition of soil air are determined largely by the water content of the soil, since the air occupies those soil pores not filled with water. After a heavy rain or irrigation, large pores are the first to drain and fill with air, followed by the medium-sized pores, and finally the small pores as water is removed by evaporation and plant use.

This drainage sequence explains the tendency for soils with a high proportion of tiny pores to be poorly aerated. In such soils, water dominates. The soil air content is low, and the rate of diffusion of the air into and out of the soil to equilibrate with the atmosphere is slow. The result is high levels of carbon dioxide and low levels of oxygen, unsatisfactory conditions for optimum plant growth and for some soil microbes. This illustrates the relationship between soil physical properties and soil air composition.

1.13

The Soil—A Rich and Varied Biological Laboratory

Soils harbor a diverse population of living organisms, both animals and plants. The entire spectrum ranges in size from rodents and large tree roots to worms and insects down to the tiniest bacteria. Moreover, the number and weight of most organisms vary greatly from soil to soil. For example, depending on conditions, a single gram of soil may contain from a mere hundred thousand to several billion bacteria. In any case, the quantity of living organisms including plant roots is sufficient to influence profoundly the physical, chemical, and especially the biological properties of soils (Figure 1.10).

Activities of soil organisms range widely. They include the largely physical breakdown of plant residues by insects and earthworms as well as the modification and decomposition of these residues by smaller organisms such as bacteria, fungi, and actinomycetes. In addition, the formation of the humus, the most chemically and physically active group of compounds in the soil, is due to soil organisms.

One result of these processes of decay is the release from organic forms of essential plant nutrients, such as nitrogen, phosphorus, and sulfur. Subsequently, still other microorganisms can oxidize, reduce, and otherwise change the state of these and other elements in the soil. These changes have a profound influence on plant growth and otherwise affect soil properties.

FIGURE 1.10
Abundant organic matter, including plant roots, helps create physical conditions favorable for the growth of higher plants as well as microbes (left). In contrast, soils low in organic matter, especially if they are high in silt and clay, are often cloddy (right) and not suitable for optimum plant growth.

1.14

Clay and Humus—The Seat of Soil Activity

The dynamic nature of the finer particles of the soil—clay and humus—has been mentioned. Because of their extremely small particle size, clay and humus have a large surface area per unit weight. Also, they exhibit surface charges that attract negatively and positively charged ions and water.

The attraction (adsorption[6]) of ions such as Ca^{2+}, Mg^{2+}, and K^+ on the surfaces of colloidal clay and humus is not as exciting as is the *exchange* of these ions for other ions in the soil solution. The intimate contact between soil solution ions and adsorbed ions makes such exchanges possible. For example, an H^+ ion released to the soil solution by a plant root can exchange readily with a potassium ion (K^+) adsorbed on the colloidal surface. The K^+ ion is then available in the soil solution for uptake by the roots of crop plants. A simple example of such cation exchange illustrates this point.

$$\boxed{\text{Colloid}}\,K^+ \quad + \quad H^+(aq) \quad \rightarrow \quad \boxed{\text{Colloid}}\,H^+ \quad + \quad K^+(aq)$$
$$\text{(adsorbed)} \quad \text{(in soil solution)} \qquad \text{(adsorbed)} \quad \text{(in soil solution)}$$

This equation illustrates why cation exchange reactions are among the most important in enhancing the growth and development of plants.

Most physical properties of soils are also controlled by clay and humus. For example, their charged surfaces permit them to act as "contact bridges" between larger particles, thus helping to create and maintain the stable aggregates of soil particles that are so desirable in a porous, easily worked soil.

---

[6] *Adsorption* refers to the attraction of ions to the surface of particles in contrast to *absorption*, the process by which ions are taken *into* plant roots.

On a weight basis, humus particles have greater nutrient- and water-holding capacities than do clay particles. However, since clay is generally present in larger amounts, its total contribution to the chemical and physical properties of the soil will usually equal or even exceed that of humus. The most inherently productive agricultural soils contain a balance of these two important soil constituents.

Clay and humus along with other soil solids, soil water, and soil air determine the suitability of soils for all kinds of uses, the most important of which is to sustain plant growth and development. The following section briefly discusses soils as a source of essential mineral elements for plants.

1.15

Interaction of Four Components to Supply Plant Nutrients

In discussing each of the four major soil components—minerals, organic matter, water, and air—emphasis has been placed on their impact on plant growth. Note, however, that the impact of one component is seldom independent of those of the others. Thus soil moisture, which supplies plants with water, simultaneously controls much of the air supply to the plant roots. Likewise, organic matter, by binding mineral particles into clusters, increases the number of large soil pores, thereby influencing the water and air relationships.

ESSENTIAL ELEMENT AVAILABILITY

None of the interactive processes involving the four soil components is more important than the provision of essential nutrient elements to plants. Plants absorb essential nutrients and water directly from the soil solution. However, the level of essential nutrients in the soil solution at any one time is far less than is needed to produce a crop (Figure 1.11). Consequently, the soil solution nutrient levels must be constantly replenished from the inorganic or organic parts of the soil or from fertilizers or manures.

Fortunately, relatively large quantities of these nutrients are associated with the soil solids, both inorganic and organic. By an interesting series of chemical and biochemical processes, nutrients are released from these solid forms to replenish those removed from the soil solution. For example, as was shown in the last section, through ion exchange the ions of essential elements such as calcium (Ca^{2+}) and potassium (K^+) are released from the colloidal surfaces of clay and humus to the soil solution. Nutrient ions are

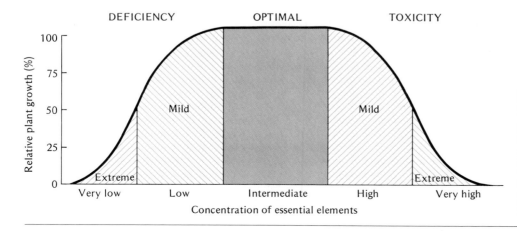

FIGURE 1.11
Relationship between plant growth and concentration in the soil solution of elements that are essential to the plant. Nutrients must be released (or added) to the soil solution in just the right amounts if normal plant growth is to occur.

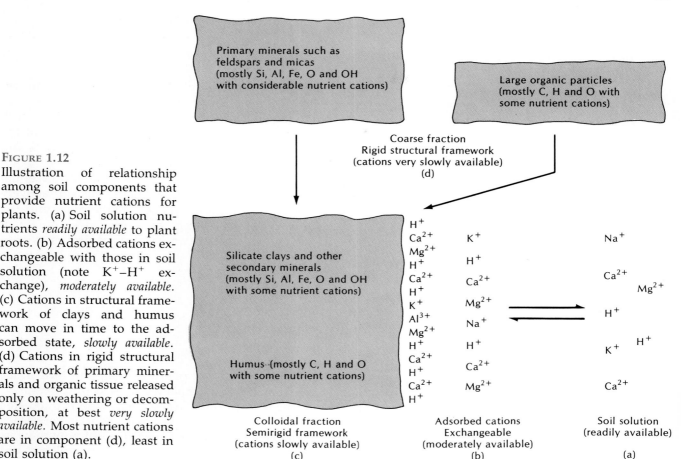

FIGURE 1.12
Illustration of relationship among soil components that provide nutrient cations for plants. (a) Soil solution nutrients *readily available* to plant roots. (b) Adsorbed cations exchangeable with those in soil solution (note K^+–H^+ exchange), *moderately available*. (c) Cations in structural framework of clays and humus can move in time to the adsorbed state, *slowly available*. (d) Cations in rigid structural framework of primary minerals and organic tissue released only on weathering or decomposition, at best *very slowly available*. Most nutrient cations are in component (d), least in soil solution (a).

also released as soil microorganisms decompose organic tissues. Plant roots can readily absorb all of these nutrients from the soil solution, provided there is sufficient oxygen in the soil air to support metabolism of the roots.

The bulk of most nutrient elements, however, is held in the molecular or structural framework of primary and secondary minerals and organic matter. Over a long period of time, these elements may be released for plant use. Thus, the structural framework is a significant source of essential elements in many soils. Figure 1.12 illustrates how the two solid soil components interact with the liquid component (soil solution) to provide essential elements to plants. The quantities of several essential elements found in different forms in typical soils of humid and arid regions are listed in Table 1.3.

TABLE 1.3
Quantities of Six Essential Elements found in Representative Soils in Temperate Regions

Essential element	Humid region soil			Arid region soil		
	In solid framework (kg/ha)	Exchangeable (kg/ha)	In soil solution (kg/ha)	In solid framework (kg/ha)	Exchangeable (kg/ha)	In soil solution (kg/ha)
Ca	8,000	2,250	60–120	20,000	5,625	140–280
Mg	6,000	450	10–20	14,000	900	25–40
K	38,000	190	10–30	45,000	250	15–40
P	900	—	0.05–0.15	1,600	—	0.1–0.2
S	700	—	2–10	1,800	—	6–30
N	3,500	—	7–25	2,500	—	5–20

1.16

Soil and Plant Relations

The absorption of essential elements is determined not only by the availability of soil-held nutrients but by their proximity to the root surfaces (Figure 1.13). As roots penetrate the soil, they come in contact with and absorb ions found in the soil solution as well as those held on the surface of the soil colloids. Also, nutrients in the soil solution can move to the roots and can then be taken up.

It is important to note that plants do not simply absorb the essential nutrients in a passive way. Chemical "carriers" in plant cells can actively transport nutrients from the soil solution across cell membranes into the cells. Likewise, microorganisms in the vicinity of roots can stimulate or inhibit nutrient element uptake. Thus, plant and microbial processes coupled with soil processes ensure the effective use of essential elements for crop production.

1.17

Conclusion

The soils of the world are critical not only to plant, animal, and human life but to the effective use of the Earth's outer crust for myriad purposes.

Soils are studied through the eyes of the edaphologist, who is interested in the soil as a habitat for plants, and of the pedologist, who is concerned with the origin, nature, and classification of soils regardless of how they are used. Both disciplines are important, but the role of the edaphologist will receive primary attention in this text.

A typical temperate region soil in optimum condition for plant growth is made up of about half solid and half nonsolid portions. Variations in the properties of the four soil components—mineral matter, organic matter,

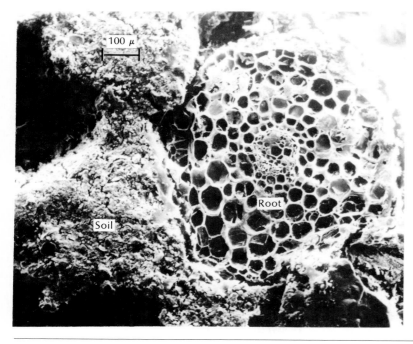

100 μ

Root

Soil

FIGURE 1.13
Scanning electron micrograph of a cross section of a peanut root surrounded by soil. Note the intimacy of contact. [Courtesy Tan and Nopamornbodi (1981)]

water, and air—determine most of the potential usefulness of soils. The interactions among these components are of great significance in providing plants with water, air, and essential nutrients.

Study Questions

1. Give examples in history of the role soils have played in the emergence and destruction of important civilizations.
2. Explain the unique contributions of the two great scientists (a) Liebig and (b) Dokuchaev in enriching our knowledge of soils and their usefulness to humankind.
3. If you were a pedologist, why would your work be of interest to someone involved in crop production? Explain.
4. Compare the activity (concentration) of H^+ ion in a soil at pH 4 with that in a soil at pH 6.
5. Describe soil horizons and explain their importance in determining the properties of soil.
6. What is the "law of the minimum," and why is it important in determining the growth of plants?
7. Explain briefly the role of each of the four major soil components.
8. Which (if any) of the four major soil components is most important and why?
9. What role does organic matter play in determining the effectiveness of the other three soil components (water, air, and mineral matter)?
10. Why is water so important in influencing soil aeration?
11. What unique characteristics of clay and humus make them so important in influencing soil properties?
12. Explain how essential nutrient elements are made available for plant growth.

Reference

TAN, K. H., and O. NOPAMORNBODI. 1981. "Electron microbeam analysis and scanning electron microscopy of soil–root interfaces," *Soil Sci.*, **131**:100–106.

READ
2.1
2.2
2.3

2

Origin, Nature, and Classification of Parent Materials

The influence of weathering, the chemical and physical breakdown of particles, is evident everywhere. Nothing escapes it. It breaks up rocks and minerals, modifies or destroys their physical and chemical characteristics, and carries away the soluble products and even some of the solids as well. The unconsolidated residues, the *regolith,* are left behind (see Section 1.4). But weathering does more than destroy; it synthesizes a soil from the upper layers of this heterogeneous mass. Hence, the upper part of the regolith may be designated the *parent material* of soils.

Parent materials are not always allowed to remain undisturbed on the site of their development. Ice, water, and wind may shift them from place to place until they are allowed to rest long enough for a soil profile to develop. A study of weathering and of the parent materials that result is a necessary introduction to soil formation and classification.

The consideration of the parent materials will begin with a brief review of the kinds of rocks and minerals from which the regolith and, in turn, the soils have formed.

2.1

Classification and Properties of Rocks

The rocks in the Earth's outer surface are commonly classified as *igneous, sedimentary,* or *metamorphic.* Those of igneous origin are formed from molten magma and include such common rocks as *granite* and *diorite* (Figure 2.1).

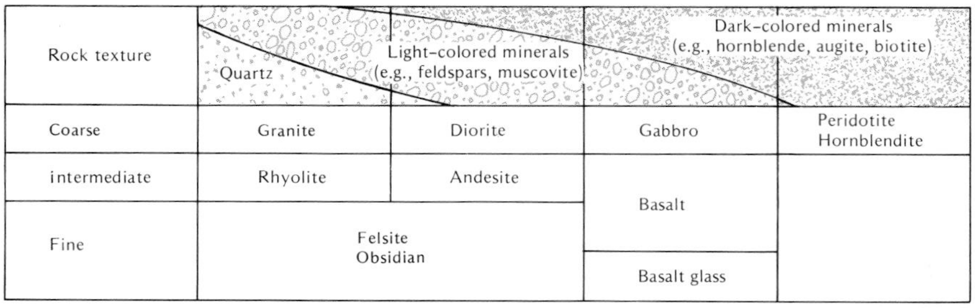

Rock texture	Quartz	Light-colored minerals (e.g., feldspars, muscovite)	Dark-colored minerals (e.g., hornblende, augite, biotite)	
Coarse	Granite	Diorite	Gabbro	Peridotite Hornblendite
intermediate	Rhyolite	Andesite	Basalt	
Fine	Felsite Obsidian		Basalt glass	

FIGURE 2.1

Classification of some igneous rocks in relation to mineralogical composition and the size of mineral grains in the rock (rock texture). Worldwide, light-colored minerals and quartz are generally more prominent than are the dark-colored minerals.

Igneous rock is composed of *primary* minerals[1] such as quartz, the feldspars, and the dark-colored minerals, including biotite, augite, and hornblende. In general, gabbro and basalt, which are high in dark-colored iron- and magnesium-containing minerals, are more easily weathered than are the granites and other lighter-colored rocks.

Sedimentary rocks have resulted from the deposition and recementation of weathering products of other rocks. For example, quartz sand weathered from a granite and deposited on the shore or beach of a prehistoric sea may through geological changes have become cemented into a solid mass called a *sandstone*. Similarly, recemented clays are termed *shale*. Other important sedimentary rocks are listed in Table 2.1 with their dominant minerals. The resistance of a given sedimentary rock to weathering is determined by its particular dominant minerals and by the cementing agent.

Metamorphic rocks are those that have formed by the metamorphism or change in form of other rocks. Igneous and sedimentary masses subjected

TABLE 2.1

Some of the More Important Sedimentary and Metamorphic Rocks and the Minerals Commonly Dominant in Them

	Type of rock	
Dominant mineral	*Sedimentary*	*Metamorphic*
Calcite ($CaCO_3$)	Limestone	Marble
Dolomite ($CaCO_3 \cdot MgCO_3$)	Dolomite	Marble
Quartz (SiO_2)	Sandstone	Quartzite
Clays	Shale	Slate
Variable	Conglomerate[a]	Gneiss[b]
Variable		Schist[b]

[a] Small stones of various mineralogical makeup are cemented into conglomerate.

[b] The minerals present are determined by the original rock, which has been changed by metamorphism. Primary minerals present in the igneous rocks commonly dominate these rocks, although some secondary minerals are also present.

[1] Primary minerals have not been altered chemically since they formed as molten lava solidified. Secondary minerals are recrystallized products of the chemical breakdown and/or alteration of primary minerals.

TABLE 2.2

The More Important Primary and Secondary Minerals Found in Soils Listed in Order of Decreasing Resistance to Weathering Under Conditions Common in Humid Temperate Regions

The primary minerals are also found abundantly in igneous and metamorphic rocks. Secondary minerals are commonly found in sedimentary rocks.

Primary minerals		Secondary minerals		
		Geothite	$FeOOH$	Most resistant
		Hematite	Fe_2O_3	
		Gibbsite	$Al_2O_3 \cdot 3H_2O$	
Quartz	SiO_2			
		Clay minerals	Al silicates	
Muscovite	$KAl_3Si_3O_{10}(OH)_2$			
Microcline	$KAlSi_3O_8$			
Orthoclase	$KAlSi_3O_8$			
Biotite	$KAl(Mg,Fe)_3Si_3O_{10}(OH)_2$			
Albite	$NaAlSi_3O_8$			
Hornblende[a]	$Ca_2Al_2Mg_2Fe_3Si_6O_{22}(OH)_2$			
Augite[a]	$Ca_2(Al,Fe)_4(Mg,Fe)_4Si_6O_{24}$			
Anorthite	$CaAl_2Si_2O_8$			
Olivine	$(Mg,Fe)_2SiO_4$			
		Dolomite	$CaCO_3 \cdot MgCO_3$	
		Calcite	$CaCO_3$	
		Gypsum	$CaSO_4 \cdot 2H_2O$	Least resistant

[a] The given formula is only approximate since the mineral is so variable in composition.

to tremendous pressures and high temperature succumb to metamorphism. Igneous rocks are commonly modified to form *gneisses* and *schists*; those of sedimentary origin, such as sandstone and shale, may be changed to *quartzite* and *slate*, respectively. Some of the common metamorphic rocks are shown in Table 2.1. As was the case for igneous and sedimentary rock, the particular mineral or minerals that dominate a given metamorphic rock will influence its resistance to weathering (see Table 2.2 for a listing of the more common minerals).

2.2

Weathering—A General Case

Weathering is basically a combination of destruction and synthesis (see Figure 2.2). Where the weathering starts, rocks are first broken down physically into smaller rocks and eventually into the individual minerals of which they are composed. Sand particles are commonly made up of individual minerals. Simultaneously, rock fragments and the minerals therein are attacked by chemical forces and changed to new minerals, either by minor modifications (alterations) or by complete chemical changes. Chemical changes are accompanied by a continued decrease in particle size and by the release of soluble constituents, which are subject to loss in drainage waters or recombination into new (secondary) minerals.

The new secondary minerals that are formed are shown (Figure 2.2) in two groups: (a) the silicate clays, resulting from alteration or decomposition and recombination, and (b) the very resistant end products, including iron and aluminum oxide clays. Along with the very resistant primary minerals such as quartz, these two groups dominate soils of temperate regions. In highly weathered soils of tropical regions, the oxides of iron and aluminum predominate.

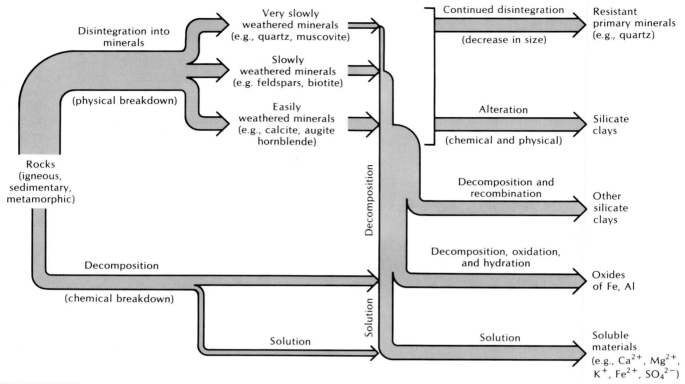

FIGURE 2.2

Pathways of weathering that occur under moderately acid conditions common in humid temperate regions. The disintegration of rocks into minerals is a physical process, whereas decomposition, recombination, and solution are chemical processes. Alteration of minerals involves both physical and chemical processes. Note that resistant primary minerals, new secondary minerals synthesized as weathering occurs, and soluble materials are the products of the weathering. In arid regions the physical processes would predominate, but in humid tropical areas decomposition and recombination would be most prominent.

Two basic processes, mechanical and chemical, are involved in the changes indicated in Figure 2.2. The former is often designated as *disintegration*, the latter as *decomposition*. Both are operative, moving from left to right in the weathering diagram. Disintegration results in a decrease in size of rocks and minerals without appreciably affecting their composition. During decomposition, however, definite chemical changes take place, soluble materials are released, and new minerals are synthesized, some of which are resistant end products. Mechanical and chemical processes may be outlined as follows.

1. Mechanical (disintegration)
   a. Temperature—differential expansion of minerals, frost action, and exfoliation
   b. Erosion and deposition—by water, ice, and wind
   c. Plant and animal influences
2. Chemical (decomposition)
   a. Hydrolysis
   b. Hydration
   c. Acidification
   d. Oxidation
   e. Dissolution

2.3

Mechanical Forces of Weathering (Disintegration)

TEMPERATURE

During the day, rocks become heated; at night they often cool much below the temperature of the air. The warming causes some minerals within a given rock to expand more than others. Likewise, when cooled, some minerals contract more than others. With every temperature change, therefore, differential stresses are set up that eventually must produce cracks, thus encouraging mechanical breakdown.

Because heat is conducted slowly, the outer surface of a rock is often warmer or colder than the inner and more protected portions. This differential heating and cooling sets up lateral stresses that, in time, may cause the surface layers to peel away from the parent mass. This phenomenon is referred to as *exfoliation*, which at times is sharply accelerated by the freezing of included water (Figure 2.3). The force[2] developed by the freezing of water is equivalent to about 1465 metric tons (megagrams) per square meter (Mg/m^2) or 150 tons/ft^2, a pressure that widens cracks in huge boulders and dislodges mineral grains from smaller fragments.

WATER, ICE, AND WIND

Rainwater beats down on the land and then travels oceanward, continually shifting, sorting, and reworking the sediments that it carries. When loaded with sediments, water has a tremendous cutting power, as is amply demonstrated by the gorges, ravines, and valleys around the world. The rounding of sand grains on an ocean beach is further evidence of the abrasion that accompanies water movement.

Ice is an erosive detachment and transporting agency of tremendous capacity. A look at a Greenland or an Alaskan glacier illustrates its power. The abrasive action of glaciers as they move under their own weight disintegrates rocks and minerals alike. Even though they are not so extensive at the present time, glaciers in ages past transported and deposited parent materials over millions of hectares.

Wind, an important carrying agent, also exerts an abrasive action when armed with fine debris. Dust storms of almost continental extent have occurred in the past. Tons of material have been picked up from one area and transferred to another. As dust is transferred and deposited, abrasion of one particle against another occurs. The rounded rock remnants in some arid areas of the west are caused largely by wind action.

PLANTS AND ANIMALS

Simple plants, such as mosses and lichens, grow on exposed rock. They catch dust, which encourages further plant growth, and a thin film of highly organic material accumulates. Roots of higher plants sometimes exert a prying effect on rocks, which results in some disintegration. However, such influences, as well as those exerted by animals, are of little importance in producing parent material when compared to the drastic physical effects of water, ice, wind, and temperature changes.

FIGURE 2.3

An illustration of an important type of concentric weathering called *exfoliation*. A combination of physical and chemical processes stimulate the mechanical breakdown, which produces layers that appear much like the leaves of a cabbage.

---

[2] The international system of measurements called the *SI system* is used in this text. A list of the most commonly used units and their equivalents in the old British system is shown inside the front cover. The comparative SI and British units will both be shown in the text in most cases.

2.4

Chemical Processes of Weathering (Decomposition)

As soon as physical disintegration of rock and minerals begins, chemical decomposition starts. This is especially noticeable in warm and humid regions, where chemical and physical processes are intense and tend to accelerate each other.

Chemical weathering is accelerated by the presence of water (with its omnipresent solutes), oxygen, and the organic and inorganic acids that result from the microbial breakdown of plant residues. These agents commonly act in concert to convert primary minerals (e.g., feldspars and micas) to secondary minerals (e.g., silicate clays) and to soluble forms that carry essential elements that support plant growth.

WATER AND ITS SOLUTIONS

Water and its dissolved salts and acids are perhaps the most pervasive factors in the weathering of minerals. Through *hydrolysis, hydration,* and *dissolution,* water enhances the degradation, alteration, and resynthesis of minerals. A simple example is the action of water on microcline, a potassium-containing feldspar.

$$KAlSi_3O_8 + H_2O \xrightarrow{\text{hydrolysis}} HAlSi_3O_8 + K^+ + OH^-$$
$$\text{(solid)} \quad \text{(liquid)} \qquad\qquad \text{(solid)} \qquad \text{(solution)}$$

$$2\,HAlSi_3O_8 + 11\,H_2O \xrightarrow{\text{hydrolysis}} Al_2O_3 + 6\,H_4SiO_4$$
$$\text{(solid)} \qquad \text{(liquid)} \qquad\qquad \text{(solid)} \quad \text{(solution)}$$

$$Al_2O_3 + 3\,H_2O \xrightarrow{\text{hydration}} Al_2O_3 \cdot 3H_2O$$
$$\qquad\qquad\qquad\qquad \text{(hydrated solid)}$$

Note that these reactions illustrate dissolution and hydration as well as hydrolysis. The potassium released is soluble and is subject to adsorption by soil colloids, to uptake by plants, and to removal in the drainage water. Likewise, the silicic acid (H_4SiO_4) is soluble. It can be removed slowly in drainage water, or it can be recombined with other compounds to form secondary minerals such as the silicate clays. Hydration is evident through the formation of the hydrated aluminum oxide ($Al_2O_3 \cdot 3H_2O$). Iron oxides also are commonly hydrated.

ACID SOLUTION WEATHERING

Weathering is accelerated by the presence of the hydrogen ion in water, such as that provided by carbonic and organic acids. For example, the presence of carbonic acid (H_2CO_3) results in the chemical solution of calcite in limestone, as illustrated in the following reaction.

$$CaCO_3 + H_2CO_3 \rightarrow Ca^{2+} + 2\,HCO_3^-$$
$$\text{Calcite}$$
$$\text{(solid)} \quad \text{(solution)} \qquad\qquad \text{(solution)}$$

Other much stronger acids, such as nitric acid (HNO_3), sulfuric acid (H_2SO_4), and some organic acids are found in soils (Figure 2.4). Also present

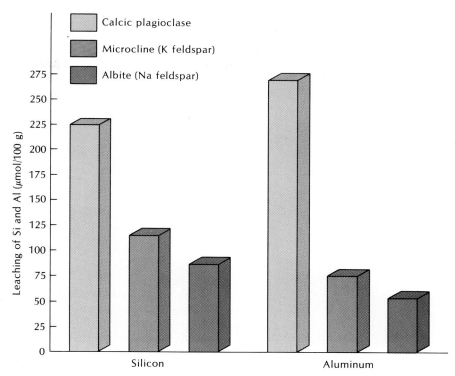

FIGURE 2.4

The leaching of silicon and aluminum by organic acids from three feldspars. Note that the calcic plagioclase released more of both elements than did the potassium and sodium feldspars. [Drawn from data (averages of leaching with eight different organic acids) in E. P. Manley and L. J. Evans, "Dissolution of feldspars by low-molecular-weight aliphatic and aromatic acids," *Soil Science*, **141**:106–12, © by Williams & Wilkins, 1986.]

are the hydrogen ions associated with soil clays. Each of these sources of acidity is available for reaction with other soil minerals.

An example of the reaction in a moist soil of hydrogen ions with soil minerals is that of an acid clay $\left(\boxed{\text{Clay}}\genfrac{}{}{0pt}{}{H^+}{H^+}\right)$ with a feldspar such as anorthite.

$$CaAl_2Si_2O_8 + \boxed{\text{Clay}}\genfrac{}{}{0pt}{}{H^+}{H^+} \rightarrow H_2Al_2Si_2O_8 + Ca^{2+}\boxed{\text{Clay}}$$

Anorthite (solid)   Acid clay (solid)   Acid silicate (solid)   Calcium clay (solid)

The mechanism is simple: The hydrogen ions of the acid clays replace the calcium ions of the fresh minerals that they may closely contact. The unstable acid silicate that results is subject ultimately to dissolution and subsequent reaction to form silicate clays. Thus, "clay begets clay" by a mechanism that depends on the removal of base-forming cations from the weathering mass of minerals by insoluble inorganic acids.

OXIDATION

Concomitant with the action of water in weathering is that of oxidation. It is particularly manifest in rocks that contain iron, an element that is easily oxidized. In some minerals, iron is present in the reduced ferrous (Fe^{2+}) form. If ferrous iron is oxidized to the ferric (Fe^{3+}) state, other ionic adjustments must be made because a three-valent ion is replacing a two-valent one. These adjustments result in a less stable mineral, which is then subject to both disintegration and decomposition.

In other cases, ferrous iron may be released from the mineral and is almost simultaneously oxidized to the ferric form. An example of this is the hydration of olivine and the release of ferrous oxide, which may be immediately oxidized to ferric oxide (geothite).

$$3\ MgFeSiO_4 + 2\ H_2O \rightarrow H_4Mg_3Si_2O_9 + SiO_2 + 3\ FeO$$

| Olivine | Serpentine | Ferrous oxide |

$$4\ FeO + O_2 + 2\ H_2O \rightarrow 4\ FeOOH$$

| Ferrous oxide | Geothite |

When ions such as Fe^{2+} are removed or are oxidized within the minerals, the rigidity of the mineral structure is weakened and the mechanical breakdown is made easier. This provides a favorable environment for further chemical reactions.

INTEGRATED WEATHERING PROCESSES

The chemical weathering processes just described occur simultaneously and are interdependent. For example, hydrolysis of a given primary mineral may release ferrous iron that is quickly oxidized to the ferric (three-valent) form, which, in turn, is hydrated to give a hydrous oxide of iron. Hydrolysis also may release soluble cations (Table 2.3), silicic acid, and aluminum or iron compounds. These substances can be recombined to form secondary silicate minerals such as the silicate clays. Such reactions are illustrated in a general way in Figure 2.5, which shows how a hypothetical primary silicate mineral is changed by chemical weathering. Figures 2.2 and 2.5 should be useful in visualizing how weathering helps produce the specific soil components that are studied in this text.

TABLE 2.3

Comparative Losses of Mineral Constituents as Weathering Takes Place in a Granite and in a Limestone

The losses are compared to that of aluminum, which was considered in these cases to have remained constant during the weathering process.

Granite to clay		Limestone to clay	
Constituent	Comparative loss (%)	Constituent	Comparative loss (%)
CaO	100.0	CaO	99.8
Na_2O	95.0	MgO	99.4
K_2O	83.5	Na_2O	76.0
MgO	74.7	K_2O	57.5
SiO_2	52.5	SiO_2	27.3
Fe_2O_3	14.4	Fe_2O_3	24.9
Al_2O_3	0.0	Al_2O_3	0.0

Data for granite from Merrill (1879), limestone from Diller (1898).

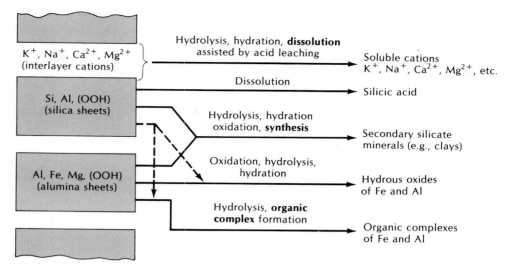

FIGURE 2.5

Illustration of chemical weathering of a hypothetical silicate mineral—K^+, Na^+, Ca^{2+}, Mg^{2+}, Fe, Al, Mg (OOH), Si, Al, (OOH)—showing the various chemical processes responsible for the weathering. Si, with some Al, and associated O^{2-} and OH^- ions are assumed to be in the "silica" sheets; Fe, Al, and Mg and associated O^{2-} and OH^- in the "alumina" sheets; and the K^+, Na^+, Ca^{2+}, and Mg^{2+} ions between these sheets. Silicon is removed by dissolution, forming silicic acid, or is combined with Al, Fe, and Mg in secondary silicate minerals (clays). Iron may be oxidized and, with aluminum, hydrated to form hydrous oxides or complex organic compounds. Note the dominance of water-dependent reactions. Silicate clays and iron and aluminum oxides are secondary minerals formed as chemical weathering proceeds. Soluble metal cations, organic complexes of Fe and Al, and even a soluble silicon compound are products of these reactions and are subject to removal from the soil by leaching.

2.5

Factors Affecting Weathering of Minerals

Weathering is influenced by three major factors: (a) climatic conditions, (b) physical properties, and (c) chemical characteristics of the rocks and minerals. Each will be discussed briefly.

CLIMATIC CONDITIONS

Climatic conditions, more than any other factor, control the rate and nature of weathering. Under arid conditions where physical forces dominate, the size of the particles has been decreased with relatively little change in composition. The original primary minerals are more prominent, whereas minerals that require water for their synthesis are less so. Physical changes due to temperature fluctuation and wind action are accompanied by only modest chemical action. Consequently, the soils of arid regions are remarkably like the parent materials from which they were formed.

In a humid region, however, the forces of weathering are more varied, and most of them are active. Vigorous chemical changes accompany disintegration. New minerals such as silicate clays (Figure 2.6) and oxides of iron and aluminum are more evident. The process is accelerated and intensified

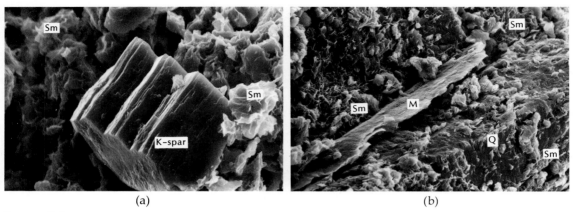

FIGURE 2.6

Scanning electron micrographs illustrating silicate clay formation from weathering of a granite rock in southern California. (a) A potassium feldspar (K-spar) is shown surrounded by smectite (Sm) and a vermiculite (both silicate clays). (b) Mica (M) and quartz (Q) associated with smectite (Sm). [Courtesy J. R. Glasmann]

by the decay of large quantities of organic matter coming from abundant plant growth.

The high year-round temperatures and the luxuriant plant growth in the humid tropics provide optimum conditions for weathering. As a result, in these regions the primary silicate minerals have largely succumbed to intense weathering and only the more highly weathered silicate clays still remain. The more resistant products of chemical weathering, such as the oxides of iron and aluminum, tend to dominate soils of humid tropical regions.

PHYSICAL CHARACTERISTICS

Particle size, hardness, and nature and degree of cementation are among the physical characteristics that influence weathering. Rocks comprised of minerals with large crystals disintegrate more easily than those with fine crystals because of differences in expansion due to temperature changes. Once the rocks have disintegrated into the individual minerals, however, the finer crystals are more subject to chemical change than the larger ones. The much larger surface area of finely divided material presents greater opportunity for chemical attack.

Hardness and cementation influence weathering. For example, a dense quartzite or a sandstone cemented firmly by a slowly weathered mineral resists mechanical breakdown and presents a small total surface area for chemical activity. On the other hand, porous rocks such as volcanic ash or coarse limestone, having a larger surface area for chemical attack, are readily broken down into smaller particles and are more easily decomposed.

CHEMICAL AND STRUCTURAL CHARACTERISTICS

Chemical and crystalline characteristics also influence the ease of chemical removal or breakdown. Some minerals such as gypsum ($CaSO_4 \cdot 2H_2O$) or calcite ($CaCO_3$) can be solubilized in water saturated with carbon dioxide and are relatively easily removed from parent material. Others, such as the ferromagnesium minerals olivine and biotite, are quite easily weathered. They contain readily oxidizable iron, and their various ions are not very tightly packed in the mineral crystal structures. In contrast, more tightly packed crystal units and lack of oxidizable iron give muscovite considerable resistance to weathering.

The stability of soil-forming minerals is dependent on climatic and biotic conditions. Consequently, no listing of minerals based on their resistance to weathering would be valid for all conditions. However, for humid temperate conditions, the following is a general order of weathering resistance of the sand- and silt-sized particles of some common minerals (Barshad, 1955).

Quartz (most resistant) > muscovite and potassium feldspars >

sodium and calcium feldspars > biotite, hornblende, and augite >

olivine > dolomite and calcite > gypsum

This order would be changed slightly depending on the climate and other environmental conditions. In any case it accounts for the absence of gypsum, calcite, and dolomite in soils of humid regions and for the predominance of quartz in the coarser fraction of most soils.

2.6

Geological Classification of Soil Parent Materials

Two groups of inorganic parent materials are recognized: (a) *sedentary* (formed in place) and (b) *transported*. The latter may be subdivided according to the agencies of transportation and deposition, as indicated below and diagrammed in Figure 2.7.

1. Sedentary
   Still at original site   Residual
2. Transported
   a. By gravity   Colluvial
   b. By water   Alluvial
                                    Marine
                                    Lacustrine
   c. By ice   Glacial
   d. By wind   Eolian

These terms properly relate only to the placement of the parent materials. However, they are sometimes applied to the soils that occur on these

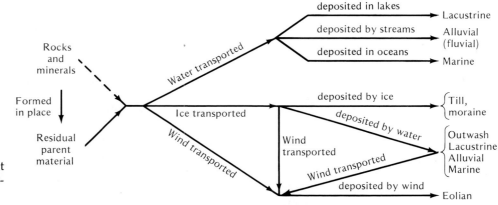

FIGURE 2.7
How various kinds of parent material are formed, transported, and deposited.

deposits—for example, glacial soils, alluvial soils, and residual soils. These groupings are very general because a wide diversity occurs within each soil group.

2.7

Residual Parent Material

Residual parent material develops in place from the underlying rock. Typically, it has experienced long and often intense weathering. In a warm, humid climate, it is likely to be thoroughly oxidized and well leached. Residual material is generally comparatively low in calcium and magnesium because these constituents have been largely leached out.

Red and yellow colors due to oxides of iron are characteristic when weathering has been intense as in hot, humid areas. In cooler and especially drier climates, weathering of residual materials is much less drastic and the oxidation and hydration of the iron may be hardly noticeable. Also, the calcium content is often higher and the colors of the weathering mass are subdued. Large areas of this type of material are found on the Great Plains and in other regions of the western United States.

Residual materials are widely distributed on all continents. In the United States, the physiographic map (Figure 2.8) shows six great eastern and central provinces where residual materials are prominent: (a) the Piedmont Plateau, (b) the Appalachian Mountains and plateaus, (c) the limestone valleys and ridges near the Appalachians, (d) the limestone uplands, (e) the sandstone uplands, and (f) the Great Plains region. The first three groups alone encompass about 10% of the area of the United States. In addition, great expanses of these sedentary accumulations are found west of the Rocky Mountains.

A great variety of soils occupy the regions covered by residual debris because of the marked differences in the nature of the rocks from which these materials have evolved. The varied soils are also a reflection of wide differences in climate and vegetation. As will be seen in Chapter 3, the profile of a well-developed soil is profoundly influenced by climate and associated vegetation.

2.8

Colluvial Debris

Colluvial debris is made up of the fragments of rock detached from the heights above and carried down the slopes mostly by gravity. Frost action has much to do with the development of such deposits. Rock fragment (talus) slopes, cliff rock debris (detritus), and similar heterogeneous materials are good examples. Avalanches are made up largely of such accumulations.

Parent material developed from colluvial accumulation is dependent on the sources of the material. It is frequently coarse and stony because physical rather than chemical weathering has been dominant. Soils developed from colluvial materials are generally not of great agricultural importance because of their small area, inaccessibility, and unfavorable physical and chemical characteristics. However, some useful timber and grazing lands in very mountainous regions have colluvial materials.

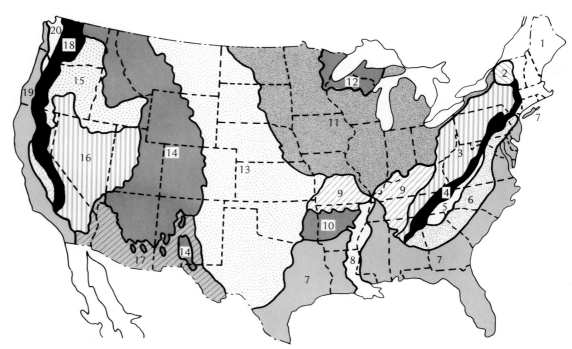

FIGURE 2.8

Generalized physiographic and regolith map of the United States. The regions are as follows (major residual areas italicized).

1. New England: mostly glaciated metamorphic rocks.
2. Adirondacks: glaciated metamorphic and sedimentary rocks.
3. *Appalachian Mountains and plateaus*: shales and sandstones.
4. *Limestone valleys and ridges*: mostly limestone.
5. Blue Ridge mountains: sandstones and shales.
6. *Piedmont Plateau*: metamorphic rocks.
7. Atlantic and Gulf coastal plain: sedimentary rocks with sands, clays, and limestones.
8. Mississippi floodplain and delta: alluvium.
9. *Limestone uplands*: mostly limestone and shale.
10. *Sandstone uplands*: mostly sandstone and shale.
11. Central lowlands: mostly glaciated sedimentary rocks with till and loess, a wind deposit of great agricultural importance (see Figure 2.18).
12. Superior uplands: glaciated metamorphic and sedimentary rocks.
13. *Great Plains region*: sedimentary rocks.
14. *Rocky Mountain region*: sedimentary, metamorphic, and igneous rocks.
15. Northwest intermountain: mostly igneous rocks; loess in Columbia and Snake river basins (see Figure 2.15).
16. Great Basin: gravels, sands, alluvial fans from various rocks; igneous and sedimentary rocks.
17. Southwest arid region: gravel, sand, and other debris of desert and mountain.
18. *Sierra Nevada and Cascade mountains*: igneous and volcanic rocks.
19. *Pacific Coast province*: mostly sedimentary rocks.
20. Puget Sound lowlands: glaciated sedimentary.
21. California central valley: alluvium and outwash.

2.9

Alluvial Stream Deposits

There are three general classes of alluvial deposits: floodplains, alluvial fans, and deltas. Each will be considered in order.

→ Stream channel

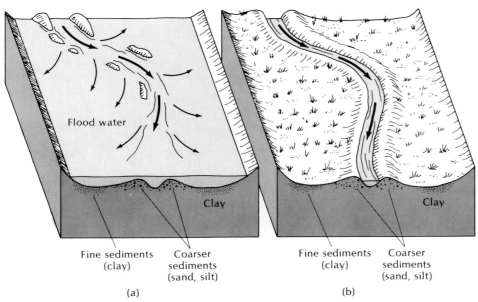

FIGURE 2.9

Illustration of floodplain development. (a) A stream is at flood stage, has overflowed its banks, and is depositing sediment in the floodplain. The coarser particles are being deposited near the stream channel where water is moving most rapidly, while the finer clay particles are being deposited where water movement is slower. (b) After the flood the sediments are in place and vegetation is growing. [Redrawn from *Physical Geology*, 2nd ed., by F. R. Flint and B. J. Skinner, copyright © 1977 John Wiley & Sons, Inc. Reprinted by permission of John Wiley & Sons, Inc.]

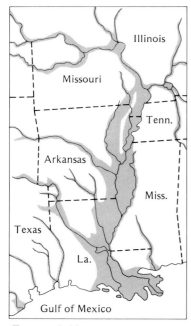

FIGURE 2.10

Floodplain and delta of the Mississippi River. This is the largest continuous area of alluvial soil in the United States.

FLOODPLAINS

Streams commonly overflow their banks and flood the surrounding area. That part of a valley which is inundated during floods is a floodplain. Sediment carried by the swollen stream is deposited during the flood, with the coarser materials being laid down near the river channel and finer materials farther away. Figure 2.9 illustrates the relationship between a stream and its surrounding floodplain. If, over a period of time, there is a change in grade, a stream may cut down through its already well-formed alluvial deposits. This leaves *terraces* above the floodplain on one or both sides. Often two or more terraces of different heights can be detected. These suggest periods when the stream was at each of these elevations.

The floodplain along the Mississippi River is the largest in the United States (Figure 2.10), varying from 20 to 75 miles in width. But throughout the country floodplains large and small provide significant parent materials for other important soil areas. The soils derived from these sediments are generally rich in nutrients, but they may require drainage and protection from overflow.

Equally productive soils are found on the floodplains of other countries. The Nile River valley in Egypt and the Sudan, the Euphrates, Ganges, Indus, Brahmaputra, and Hwang Ho river valleys of Asia, and the Amazon Basin of Brazil are good examples. Some of these floodplain deposits are used for the production of wetland rice, grown by most of the low-income people of these areas.

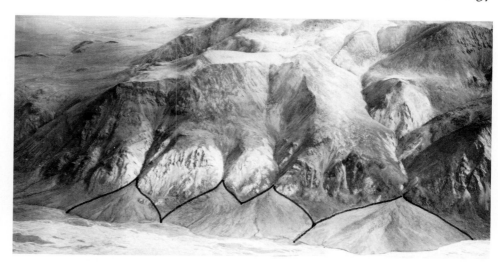

FIGURE 2.11
Characteristically shaped alluvial fans alongside a river valley in Alaska. Although the areas are small and sloping, they can develop into well-drained soils. [Courtesy U.S. Geological Survey]

ALLUVIAL FANS

Streams that leave a narrow valley in an upland area and suddenly descend to a much broader valley below deposit sediment in the shape of a fan (alluvial fans) (Figure 2.11). Fan material generally is gravelly and stony, somewhat porous and well-drained.

Alluvial fan debris is found over wide areas in mountainous and hilly regions. The soils derived from this debris often prove very productive, although they may be quite coarse-textured. In certain glaciated sections, such deposits also occur in large enough areas to be of considerable agricultural importance.

DELTA DEPOSITS

Much of the finer sediment carried by streams is not deposited in the floodplain but is discharged into a lake reservoir or ocean into which the stream flows. Some of the suspended material settles near the mouth of the river, forming a delta. Delta deposits are by no means universal, being found at the mouths of only a few rivers of the world. A delta often is a continuation of a floodplain (its "front," so to speak). It is clayey in nature and is likely to be swampy as well.

Areas of delta sediments that are subject to flood control and drainage become rather important agriculturally. The combination deltas and floodplains of the Mississippi, Ganges, Amazon, Hwang Ho, Po, Tigris, and Euphrates rivers are striking examples. The Nile Valley of Egypt illustrates the fertility and productivity of soils originating from such parent material.

2.10

Marine Sediments

Much of the sediment carried away by streams eventually is deposited in oceans, lakes, seas, and gulfs. The coarser fragments are deposited near the shore and the finer particles at a distance (Figure 2.12). Over long periods of time, these underwater sediments build up and can become quite deep. In some areas, changes in the elevation of the earth's crust have resulted in these marine deposits being raised above sea level. The deposits are then subject to weathering and to soil formation, giving rise in some cases to valuable agricultural soils. Such regions are found along the Atlantic and

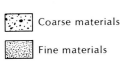
Coarse materials

Fine materials

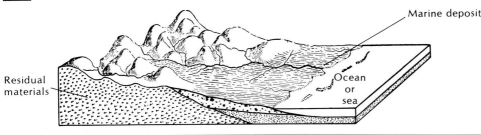

Marine deposit

Residual materials

Ocean or sea

FIGURE 2.12
Marine materials in relation to residual uplands. Note the coarse-textured (sandy) materials near the shoreline and finer materials (silt and clay) farther from the shore.

Gulf coasts of the United States and make up about 10% of the country's land area.

Marine deposits are quite variable in texture. Some are sandy, as is the case of the Atlantic seaboard coastal plain areas. Others are high in clay, as are deposits found in the Atlantic and Gulf coastal flatwoods and in the interior pinelands of Alabama and Mississippi. In any case, these marine sediments have been subjected to soil-forming processes for relatively short periods of time. As a consequence, the properties of the soils that form are heavily influenced by those of the marine parent materials. Fortunately, with proper management and fertilization, soils developed on some marine deposits are quite productive.

2.11

The Pleistocene Ice Age

During the Pleistocene epoch, northern North America, northern and central Europe, and parts of northern Asia were invaded by a succession of great ice sheets. Certain parts of South America and areas in New Zealand and Australia were similarly affected. Antarctica undoubtedly was capped with ice much as it is today. The sea level was about 130 m (425 ft) lower than it is at present.

It is estimated that the Pleistocene ice at its maximum extension covered perhaps 20% of the land area of the world. It is surprising to learn that present-day glaciers, which we consider as mere remnants of the Great Ice Age, occupy an area about one third that occupied by the Pleistocene glaciers. The present volume of ice, however, is much less, since our living glaciers are comparatively thin. Even so, if present-day glaciers were to melt, the world sea level would increase by about 65 m (210 ft).

In North America, the major centers of ice accumulation were in central Labrador and the western Hudson Bay region, with a minor concentration in the Canadian Rockies. From the major centers, great continental glaciers pushed outward in all directions but especially southward. At different times, they covered most of what is now Canada, southern Alaska, and the northern part of the contiguous United States. The southernmost extension was down the Mississippi Valley, where the least resistance was met because of the lower and smoother topography (Figure 2.13).

Europe and central North America apparently sustained four distinct ice invasions over a period of 1–1.5 million years (Table 2.4). In the United States, these invasions are identified consecutively as Nebraskan, Kansan,

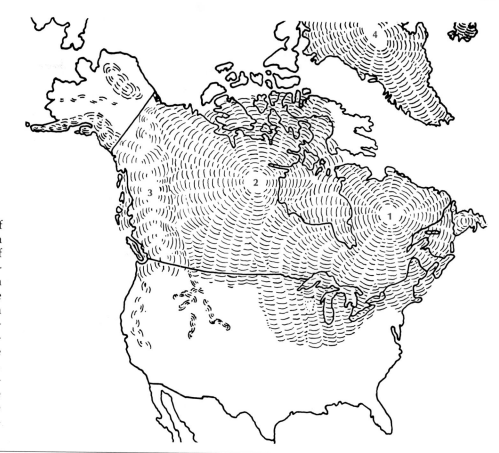

FIGURE 2.13
Maximum development of continental glaciation in North America. The four centers of ice accumulation are numbered. Apparently the eastern and central United States were invaded from the Labradorian (1) and Hudson Bay (2) centers. Note the marked southerly advance of the ice in the upper Mississipi Valley, where topography offered little resistance. To the east the Appalachian highlands more or less blocked the ice invasion.

Illinoian, and Wisconsin. The surface glacial deposits of the eastern United States and Canada are largely of Wisconsin age.

The invasions were separated by long interglacial ice-free intervals that covered a total time period considerably longer than the periods of glaciation. Some of the interglacial intervals evidently were times of warm or semitropical climate in regions that are now definitely temperate. Since the Wisconsin glacial ice disappeared from northern Iowa and central New York only about 12,000 years ago, we now may be enjoying the mildness of another interglacial period.

TABLE 2.4
Nomenclature Used to Identify Periods of Glaciation and Interglaciation in North America During the Pleistocene Period

Period name	Type of period	Approximate date starting
Holocene	Interglacial	10,000 B.C.
Wisconsin	Glacial	100,000
Sangamonian	Interglacial	225,000
Illinoian	Glacial	325,000
Yarmouthian	Interglacial	600,000
Kansan	Glacial	700,000
Aftonian	Interglacial	900,000
Nebraskan	Glacial	1–1.5 million B.C.

As the glacial ice was pushed forward, it conformed to the unevenness of the areas invaded. It rose over hills and mountains with surprising ease. The existing regolith with much of its mantle of soil was swept away, hills were rounded, valleys filled, and, in some cases, the underlying rocks were severely ground and gouged. Thus, the glacier became filled with rock and all kinds of unconsolidated materials and carried great masses of these materials as it pushed ahead (Figure 2.14). Finally, as the ice melted and the glacier retreated, a mantle of glacial debris or *drift* remained. The drift provided a new regolith and fresh parent material for soil formation.

The area covered by glaciers in North America is estimated at 10.4 million km² (4 million mi²), and at least 20% of the United States is influenced by the deposits. An examination of Figures 2.13 and 2.15 indicates the magnitude of the ice invasion at maximum glaciation in this country.

FIGURE 2.14

Tongues of a modern-day glacier in Canada. Note the evidence of transport of materials by the ice and the ''glowing'' appearance of the two major ice lobes. [Photo A-16817-102 courtesy National Air Photo Library, Surveys and Mapping Branch, Canadian Department of Energy, Mines, and Resources]

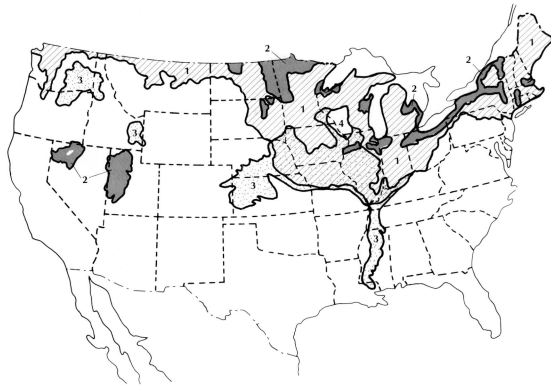

FIGURE 2.15
Areas in the United States covered by the continental ice sheet and the deposits either directly from, or associated with, the glacial ice. (1) Till deposits of various kinds; (2) glacial-lacustrine deposits; (3) the loessial blanket (note that the loess overlies great areas of till in the Midwest); (4) an area, mostly in Wisconsin, that escaped glaciation and is partially loess covered.

2.12

Glacial Till and Associated Deposits

The name *drift* is applied to all material of glacial origin, whether deposited by the ice or associated waters. The materials deposited directly by the ice, called *glacial till*, are heterogeneous mixtures of debris of great diversity, which vary from rocks and boulders to clay. Glacial till is found mostly as irregular deposits called *moraines*, of which there are various kinds (Figure 2.16).

Terminal moraines are hilly ridges that characterize the southernmost extensions of the various ice lobes when the ice margin was stationary long enough to permit an accumulation of debris. Terminal moraines occupy little land area and are of interest only as they help us understand the glaciation process.

As the ice rapidly retreated, a thinner and more level deposition called *ground* moraine was laid down. Ground moraine is by far the most widely distributed of all glacial deposits and is of considerable agricultural significance. As the ice receded to the north, there were periods when the front was relatively stationary again, and hilly *recessional* moraines of limited area were formed.

Special small features associated with the moraine are *kames* (conical hills or short ridges of sand and gravel deposited by ice), *eskers* (long, narrow

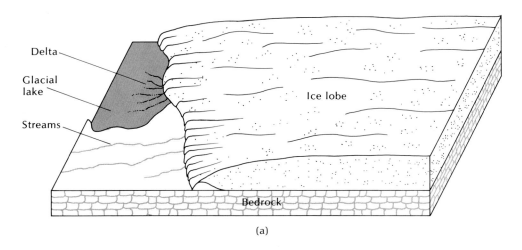

(a)

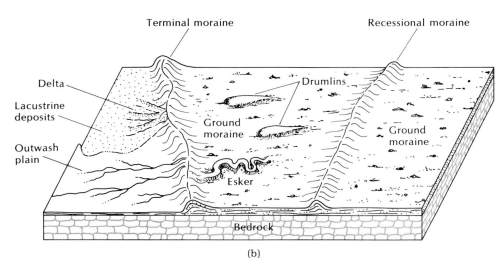

(b)

FIGURE 2.16
Illustration of how several glacial materials were deposited. (a) A glacier ice lobe moving to the left, feeding water and sediments into a glacial lake and streams, and building up glacial till near its front. (b) After the ice retreats, terminal, ground, and recessional moraines are uncovered along with cigar-shaped hills (drumlins), the beds of rivers that flowed under the glacier (eskers), and lacustrine, delta, outwash deposits.

ridges of coarse gravel deposited by ice-walled streams coming from the glacier), and *drumlins* (cigar-shaped hills composed of till and oriented parallel to the direction of the ice movement).

An outstanding feature of glacial till is its variability. Consequently, the soils derived from such material are very heterogeneous. Thus, the term "glacial soil" is of value only in suggesting the mode of deposition of the parent materials. It indicates practically nothing about the characteristics of a soil so designated.

2.13

Glacial Outwash and Lacustrine Sediments

Torrents of water were constantly gushing from the ice lobes of glaciers, especially during the summer. The vast loads of sediment carried by such streams were either dumped immediately or carried to other areas for deposition. As long as the water had ready egress, it flowed away rapidly to deposit its load downstream.

OUTWASH PLAINS

An *outwash* plain is formed by streams that are heavily laden with glacial sediment flowing from the ice (Figure 2.16). Since the sediment is sorted by flowing water, sands and gravels are common. Outwash deposits are found in valleys and on plains where the glacial waters were able to flow away freely. These valley fills are common in the United States.

GLACIAL LAKE (LACUSTRINE) DEPOSITS

When the ice front came to a standstill where there was no ready escape for the water, ponding began and ultimately very large lakes were formed (Figure 2.17). Particularly prominent were those south of the Great Lakes in New York, Ohio, Indiana, and Michigan, and in the Red River Valley of Minnesota (see Figure 2.15). The latter lake, called Glacial Lake Agassiz, was about 1200 km (745 mi) long and 400 km (248 mi) wide at its maximum extension. Somewhat smaller lakes also were formed in the intermountain regions west of the Rockies (e.g., Willamette Valley in Oregon), as well as in the Connecticut River Valley in New England and elsewhere.

The deposits formed in these glacial lakes range from coarse delta materials and beach deposits near the shore, to larger areas of fine silts and clay deposited from the deeper, more still waters at the center of the lake. Areas of inherently fertile soils have developed from these materials as the lakes dried. Extending westward from New England along the Great Lakes to the broad expanse of the Red River Valley, these deposits have produced some of the most important soils of the northern and intermountain regions of the United States.

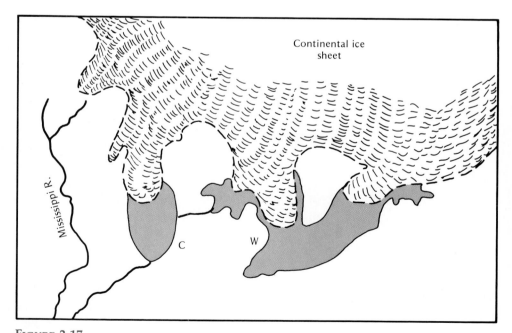

FIGURE 2.17
When conditions of topography were favorable, the Wisconsin ice sheet acted as a great waning dam. A stage in the development of glacial lakes in Chicago (C) and Warren (W) is represented. [After Daly (1934)]

2.14

Glacial-Associated Eolian Deposits

Wind-transported (*eolian*) parent materials are among the most important in the United States, especially in the central part of the country. These materials are associated mostly with glaciation.

ORIGIN AND LOCATION

During the glacial periods, conditions were ideal for wind erosion. The weather was cold and windy, and vegetative cover in areas just to the south of the glacier was sparse or nonexistent. In the winters, winds picked up fine alluvial materials, deposited in previous summers by the ice-fed streams, and moved them southward. Fine particles from glacial till and even residual materials were similarly transported.

These wind-blown materials, comprised primarily of silt with some fine sand and clay, are called *loess* (pronounced "luss"). They covered existing soils and parent materials—in some areas to depths of more than 8 m (26 ft). The thicker deposits are found adjacent to the wider alluvial flood plains.

Loess and other associated eolian deposits are found over wide areas of the Central United States, as shown in Figure 2.18. The deepest deposits are located along the Mississippi and Missouri rivers. Significant deposits are found from eastern Colorado to western Ohio and as far south as northwestern Texas. The materials westward toward the Rocky Mountains are sandy in texture. Extensive loessial deposits also occur in the Palouse region of Washington and Idaho and other areas of the northwestern United States.

Loessial deposits also are found in other countries. For example, deposits reaching 30–100 m (98–328 ft) in depth are found in some 800,000 km^2 (309,000 mi^2) in central and western China. These materials have been wind-blown from the dryland areas of Central Asia and are generally not associated directly with glaciers.

Loess has given rise to soils of considerable diversity. A comparison of the loessial area of the Midwest (Figure 2.18) with the map of soil orders (see

FIGURE 2.18

Approximate distribution of loess in central United States. The soil that has developed therefrom is generally a silt loam, often somewhat sandy. Note especially the extension down the eastern side of the Mississippi River and the irregularities of the northern extensions. Smaller areas of loess occur in Washington, Oregon, and Idaho (see Figure 12.15).

Figure 3.11) shows the presence of loess in six distinct soil regions. This situation indicates the probability of great differences in the fertility and productivity of loess soils. The variety of loess soils also emphasizes the influence of climate as a strong determinant of soil characteristics.

OTHER EOLIAN DEPOSITS

Eolian deposits other than loess exist, such as volcanic ash and sand dunes. Soils from volcanic ash occur in Montana, Oregon, Washington, Idaho, Nebraska, and Kansas. They are light and porous and generally are of less agricultural value than soils developed from loess.

Except where they are stabilized by vegetation, sand dunes are of little agricultural value and may become a menace to agriculture if they are moving. Examples of such deposits are found in the great Sahara and Arabian deserts. Smaller areas occur in the United States in Nebraska, Colorado, New Mexico, Michigan, and along the eastern seaboard.

2.15

Agricultural Significance of Glaciation

The Pleistocene glaciation in the United States has been generally beneficial and especially so for agriculture. The leveling and filling actions of the glaciers have created a smoother topography more suited to farming operations. The same can be said for the glacial lake sediments and loess deposits. Also, the parent materials thus supplied are geologically fresh, and the soils derived from them usually are not drastically leached. "Young" soils generally are higher in available nutrients and, under comparable conditions, superior for crop production.

Glaciation was less beneficial to areas near the centers of ice accumulation. For example, in eastern Canada the moving ice picked up unconsolidated materials (including soils) and transported them to areas in the northern United States. Although these materials generally contributed to the productivity of soils in the United States, the glaciation left large areas of shallow soils in eastern Canada and New England.

In glaciated areas, there is great variability in the nature of the soils that develop. On glacial outwash plains, the soils may be coarse textured and even high in gravel, resulting in low moisture retention and droughtiness. Glacial lacustrine clays are often poorly drained because of their high clay content. Some glacial tills are also quite compact, and the soils developing on them tend to have imperfect drainage (Table 2.5). Soils developed on

TABLE 2.5
Relationship Between Certain Parent Materials in Pennsylvania and Soil Drainage Class

Parent material class	Number of profiles	Percent of profiles in drainage classes	
		Well and moderately well drained	Somewhat poorly, poorly, and very poorly drained
Alluvium	23	81	19
Limestone	23	100	0
Till	52	65	35
Shale	77	74	26
Sandstone	21	100	0

From Petersen et al. (1971).

loess tend to be erosion prone even though they may be quite productive. These differences emphasize the definite influence of parent material on the kinds of soil that develop.

2.16

Conclusion

Together with climate, the nature and properties of parent materials are the most significant factors affecting the kind and quality of the world's soils. Knowledge of these parent materials, their source or origin, mechanisms for weathering, and means of transport is essential if we are to gain an understanding of soils—and especially if we are to classify them properly. These facts should be kept in mind as we turn to a study of soil formation and classification in the next chapter.

Study Questions

1. In general, which rocks weather more rapidly under humid conditions: rocks with light colored minerals such as the feldspars and quartz or those with darker colored minerals such as hornblende? What are the reasons for this difference?
2. What are the comparative actions of (a) physical and (b) chemical weathering forces in arid versus humid regions? Why the difference?
3. Distinguish between primary and secondary minerals and give examples of each.
4. How do silicate clays influence the weathering of primary minerals?
5. Assume that a mineral containing iron and magnesium is being weathered. Illustrate how the processes of oxidation, hydration, and dissolution are effective in this weathering.
6. Would you be more likely to find gypsum ($CaSO_4 \cdot 2H_2O$) in soils of humid regions or of arid regions? Why?
7. Explain how water and temperature interact in the physical weathering of rocks and minerals.
8. Where would you go if you wanted to find soil areas with the greatest chemical weathering? What colors would you expect to find there for soils and why?
9. Give examples of physical and chemical characteristics of rocks and minerals that influence the rate at which they will weather.
10. Would you be more apt to find soils developed from residual soil materials in North Carolina or New York? Why?
11. Why are soils developed from alluvial stream deposits apt to be quite fertile? What about their ease of drainage?
12. Would you expect soils developed on typical marine deposits to be fine (clay-like) or coarse (sand-like)? Explain.
13. Differentiate among glacial till, glacial outwash, and glacial eolian deposits.
14. Was glaciation beneficial or detrimental to agriculture in the United States? Explain.

References

BARSHAD, I. 1955. "Soil development," in F. E. Bear (Ed.), *Chemistry of the Soil* (New York: Reinhold).

DALY, R. A. 1934. *The Changing World of the Ice Age* (New Haven: Yale Univ. Press).

DILLER, J. S. 1898. *Educational Series of Rock Specimens*, Bull. 150, U.S. Geological Survey.

FLINT, F. R., and B. J. SKINNER. 1977. *Physical Geology*, 2nd ed. (New York: Wiley).

MANLEY, E. P., and L. J. EVANS. 1986. "Dissolution of feldspars by low-molecular-weight aliphatic and aromatic acids," *Soil Sci.*, **141**:106–12.

MERRILL, G. P. 1879. *Weathering of Micaceous Gneiss*, Bull. 8, Geol. Soc. Amer., p. 160.

PETERSEN, G. W., R. L. CUNNINGHAM, and R. P. MATELSKI. 1971. "Moisture characteristics of Pennsylvania soils: III. Parent material and drainage relationships," *Soil Sci. Soc. Amer. Proc.*, **35**:115–19.

Soil Formation, Classification, and Survey

Read 3.1 to 3.4

To study and use soils effectively, some sort of classification is necessary. A classification enhances the interchange of information on soils from one area to another. It also helps us predict the suitability of different soil areas for agricultural production and for nonagricultural uses such as roadbeds, building sites, and so on.

To increase our understanding of soils as they occur in the field, three phases must be considered: (a) soil genesis, the evolution of a soil from its parent material; (b) soil classification, the grouping of soils having common properties; and (c) soil survey, the depiction of the geographic distribution of soils and their relationship to landscapes. Soil genesis will receive attention first.

3.1

Factors Affecting Soil Formation

Studies of soils throughout the world have shown that the kinds of soil that develop are largely determined by five major factors:

1. Climate (particularly temperature and precipitation).
2. Living organisms (especially native vegetation, microbes, soil animals, and human beings).
3. Nature of parent material.
4. Topography of the site.
5. Time that parent materials are subjected to soil formation.

In fact, soils often are defined in terms of these factors as "dynamic natural bodies having properties derived from the combined effect of *climate* and *biotic activities*, as modified by *topography*, acting on *parent materials* over periods of *time*."

CLIMATE

Climate is perhaps the most influential of the factors because it determines the nature of the weathering that occurs. For example, temperature and precipitation affect the rates of chemical, physical, and biological processes responsible for profile development. For every 10 °C rise in temperature, the rates of biochemical reaction double. Furthermore, biochemical changes by soil organisms are sensitive to temperature as well as moisture. Temperature and effective moisture influence the organic matter content of soil (see Figure 10.8). The very modest profile development characteristic of cold areas contrasted with the deep-weathered profiles of the humid tropics is further evidence of climatic control (Figure 3.1).

Climate also influences the natural vegetation. In humid regions, plentiful rainfall provides an environment favorable for the growth of trees (Figure 3.2). In contrast, grasslands are the dominant native vegetation in semiarid regions, and shrubs and brush of various kinds in arid areas. Thus, climate also exerts its influence through a second soil-forming factor, the living organisms (compare Figures 3.2 and 3.12).

LIVING ORGANISMS

Soil organisms play a major role in profile differentiation. Organic matter accumulation, profile mixing, nutrient cycling, and structural stability are all

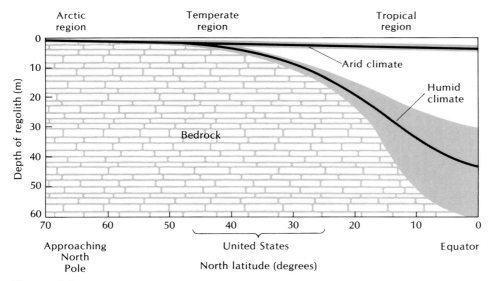

FIGURE 3.1

Illustration of the effects of two climatic variables, temperature and moisture (precipitation), on the depth of weathering as indicated by regolith depth. In cold climates (arctic regions) the regolith is shallow under both humid and arid conditions. At lower latitudes (higher temperatures), the depth of the regolith increases sharply in humid area but is little affected in arid regions. Under humid tropical climates the regolith may be 50 or more meters in depth. While soil depth may be only a fraction of regolith depth, the influence of climatic considerations on weathering is obvious from this diagram.

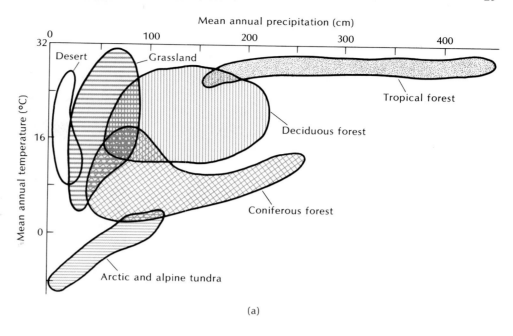

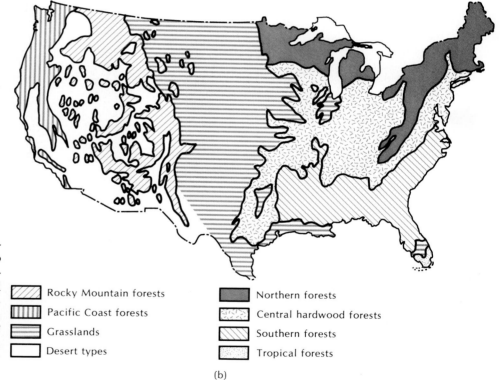

FIGURE 3.2
The effect of climate on vegetation. (a) General relationship among mean annual temperatures and precipitation and types of vegetation. (b) General types of vegetation in the United States. [(a) From NSF (1975); (b) redrawn from a more detailed map of U.S. Geological Survey]

enhanced by the activities of organisms in the soil. Vegetative cover reduces natural soil erosion rates, thereby slowing down the rate of mineral surface removal.

The effect of vegetation on soil formation can be seen by comparing properties of soils formed under grassland and forest vegetation (Figure 3.3). The organic matter content of the "grassland" soils is generally higher than that of soils in forested areas, especially in the subsurface horizons. The

(a)

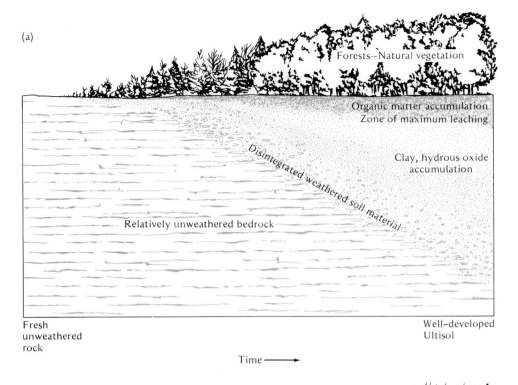

Forests—Natural vegetation

Organic matter accumulation
Zone of maximum leaching

Clay, hydrous oxide
accumulation

Disintegrated weathered soil material

Relatively unweathered bedrock

Fresh
unweathered
rock

Well–developed
Ultisol

Time ⟶

(b)

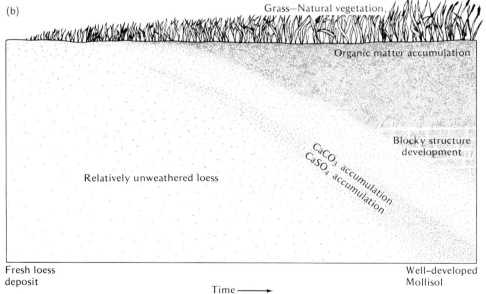

Grass—Natural vegetation

Organic matter accumulation

Blocky structure
development

$CaCO_3$ accumulation
$CaSO_4$ accumulation

Relatively unweathered loess

Fresh loess
deposit

Well–developed
Mollisol

Time ⟶

FIGURE 3.3
How two soil profiles may have developed in climatic areas that encouraged as natural vegetation either (a) forests or (b) grasslands. Organic matter accumulation in the upper horizons occurs in time, the amount and distribution depending on the type of natural vegetation present. Clay and iron oxide accumulate and characteristic structures develop in the lower horizons. The end products differ markedly from the soil materials from which they formed.

higher organic content gives the soil a darker color and higher moisture- and cation-holding capacity as compared to the "forest" soil. Also, structural stability of soil aggregates tends to be encouraged by the grassland vegetation.

The mineral element content in the leaves, limbs, and stems of the natural vegetation strongly influences the characteristics of the soils that develop and especially their acidity. Coniferous trees (e.g., pines, firs) tend to be low in metallic cations such as calcium, magnesium, and potassium. The cycling of nutrients from the litter falling from conifers will be low compared to that of some deciduous trees (e.g., oaks, maples) that are much higher in metallic cations (Figure 3.4). Consequently, soil acidity is more likely to develop

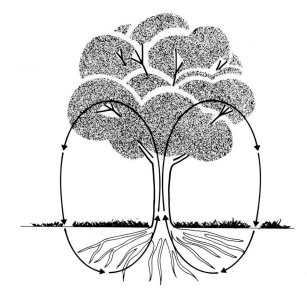

FIGURE 3.4
Nutrient recycling is an important factor in determining the relationship between vegetation and the soil that develops. If residues from the vegetation are low in bases, acid weathering conditions are favored. High base-containing plant residues tend to neutralize acids, thereby favoring only slightly acid to neutral weathering conditions.

under pine vegetation with its low content of base-forming cations than under oak or maple tree vegetation.[1] Also, the removal of base-forming cations by leaching is more rapid under coniferous vegetation.

In addition, an interaction develops among the natural vegetation, other soil organisms, and the characteristics of the soil. Thus, as soils develop under native grasslands, bacteria (*Azotobacter*) that can "fix" atmospheric nitrogen into compounds usable by plants tend to flourish. This provides an opportunity for an increase in the nitrogen content of the soils and, as we shall see in Chapter 10, the organic matter content as well. Other microorganisms attack plant and animal residues, producing slimy organic materials. These along with an abundance of plant roots help bind soil particles into stable aggregates. Living organisms are a critical factor in determining soil character.

Human activities can also significantly influence soil formation. Destroying the natural vegetation (trees, grass) and subsequently tilling the soil for crop production abruptly modifies the soil-forming factors. Likewise, irrigating a soil in an arid area drastically influences the soil-forming factors, as does adding fertilizer and lime to soils of low fertility. Although human activities are pertinent only in recent geological times, they have had significant influences on soil-forming processes in some areas.

PARENT MATERIAL

Geological processes have brought to the Earth's surface numerous *parent materials* in which soils form (Figure 3.5). The nature of the parent material profoundly influences soil characteristics. For example, soil texture (see page 91) is largely influenced by parent materials. In turn, soil texture helps control the downward movement of water, thereby affecting the translocation of fine soil particles and plant nutrients. The chemical and mineralogical compositions of parent material also can influence weathering directly and, simultaneously, can affect the natural vegetation. For example, the presence

[1] Soil acidity is determined primarily by the relative proportion of H^+ ions, which can form acids such as H_2CO_3 and HNO_3, and of metallic cations (e.g., Ca^{2+}, Mg^{2+}, and K^+), which are capable of forming bases such as $Ca(OH)_2$ and KOH. Thus, Ca^{2+}, Mg^{2+}, and K^+ are known as *base-forming cations*. See Section 8.1 for further explanation.

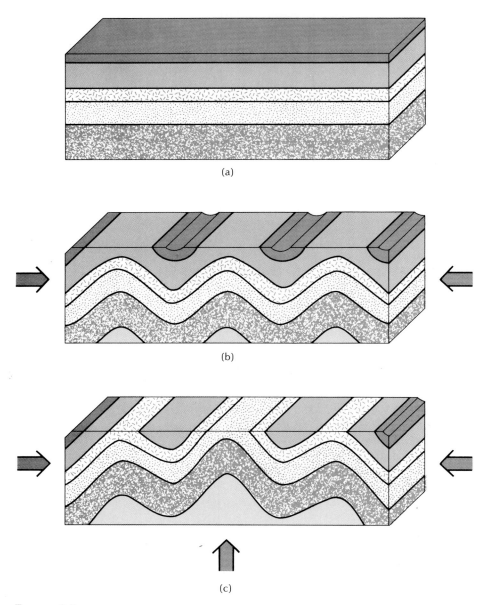

FIGURE 3.5
Diagrams showing how geological processes have brought different rock layers to the surface in a given area. (a) Unaltered layers of sedimentary rock with only the uppermost layer exposed. (b) Lateral geological pressures deform the rock layers. At the same time, erosion removes much of the top layer, exposing part of the first underlying layer. (c) Localized upward pressures further reform the layers, thereby exposing two more underlying layers. As these four rock layers are weathered, they give rise to the parent materials on which different kinds of soils can form.

of limestone in parent material will delay the development of acidity, a process that moist climates encourage. In addition, the leaves of trees found on limestone materials are relatively high in calcium and other base-forming metallic cations. As these high-base leaves are incorporated into the soil and are decomposed, they further delay the process of acidification or, in humid temperate areas, the progress of soil development.

Parent material also influences the quantity and type of clay minerals present in the soil profile. The parent material itself may contain varying amounts and types of clay minerals, perhaps from a previous weathering cycle. Also, the nature of parent material greatly influences the kinds of clays that can develop as the soil evolves. In turn, the nature of the clay minerals present markedly affects the kind of soil that develops.

TOPOGRAPHY

Topography relates to the configuration of the land surface and is described in terms of difference in elevation, slope, and so on. The topography of the land can hasten or delay the work of climatic forces. In smooth, flat country, excess water is removed less rapidly than in rolling areas. Rolling to hilly topography encourages natural erosion of the surface layers, which reduces the possibility of a deep soil (Figure 3.6). On the other hand, if water stands for part or all of the year on a given area, the climate is less effective in regulating soil development.

There is a definite interaction among topography, vegetation, and soil formation. In the grassland–forest transition zones, trees commonly occupy the slight depressions in an otherwise prairie vegetation. This is apparently a moisture effect. As would be expected, the nature of the soil in the depressions is quite different from that in the uplands. Topography, therefore, not only modifies climate and vegetative effects, but often has a major direct effect on soil formation and on the type of soil that forms.

TIME

The length of *time* that materials have been subjected to weathering influences soil formation. Time effects can be seen by comparing soils of a glaciated region with those in a comparable unglaciated area nearby. The effect of time is also apparent when comparing soils on recent Wisconsin glacial material with those on adjacent older deposits of the Illinoian or Kansan age. The influence of parent material is much more apparent in the soils of glaciated regions, where insufficient time has elapsed since the ice retreated to permit the full development of many soils.

Soils located on alluvial or lacustrine materials (see Sections 2.9 and 2.13) generally have not had as much time to develop as have the surrounding upland soils. Also, some coastal-plain parent materials have been uplifted only in recent geological time, and, consequently, the soils thereon have had relatively short exposure to weathering.

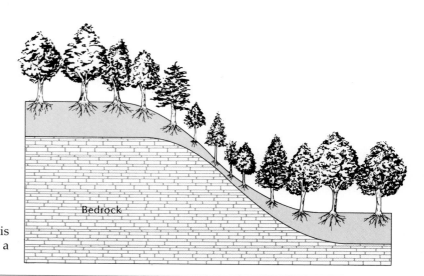

FIGURE 3.6
Topography influences soil properties. In this case depth is constrained by sharp slopes of a hillside.

Bedrock

The interaction of time with other factors affecting soil formation must be emphasized. The time required for the development of a horizon definitely will be influenced by the parent material, the climate, and the vegetation. These factors are highly interdependent in determining the kind of soil that develops.

It should be emphasized that two or more of the factors influencing soil formation are usually active simultaneously and interdependently; thus, climate and parent material influence vegetation, which, in turn, over a period of time, has helped influence the nature of the parent material available. Likewise, there is a relation between some parent materials and topography. The interaction and interdependence of these factors introduces some complexity in evaluating just how a given soil was formed.

3.2

Soil Formation in Action

An examination of a soil pit or a fresh roadcut reveals a soil profile with distinctive horizontal layers, some of which are highly visible. Such layers would not be found if a similar cut were to be made in unconsolidated materials recently moved by a bulldozer or laid down following a recent major volcanic eruption. Obviously then, significant changes have been made as soils develop from relatively unconsolidated parent materials. A study of soil formation (genesis) gives some notion as to how these changes occur and why they can stimulate the development of so many different kinds of soil.

THREE MAJOR PROCESSES

Soil genesis is brought about by a series of processes, the most significant of which are (a) *weathering* and *organic matter breakdown*, by which some soil constituents are modified or destroyed and others are synthesized; (b) *translocation* of inorganic and organic materials up and down the soil profile, the materials being moved mostly by water but also by soil organisms; and (c) *accumulation* of soil materials in horizontal layers (horizons) in the soil profile, either as they are formed in place or translocated from above or below the zone of accumulation.

A SIMPLIFIED EXAMPLE

The role of these major processes can be seen by following the changes that take place as soils form from relatively uniform parent material. When plants begin to grow and their residues are deposited on the surface of the parent materials, soil formation has truly begun. The plant residues are disintegrated and partly decomposed by soil organisms that also synthesize new organic compounds that make up humus. Earthworms, which burrow into and live in the soil, along with other small animals such as ants and termites, mix these organic materials with the underlying mineral matter near the surface of the parent material. This mixture, which comes into being rather quickly, is commonly the first soil horizon developed; it is different in color and composition from the original parent material.

As plant residues decay, organic acids are formed. These acids are carried by percolating waters into the soil, where they stimulate the weathering processes. They solubilize some chemicals that are translocated (leached) from upper to lower horizons or are completely removed from the emerging soil by percolating waters.

As weathering proceeds, some primary minerals are disintegrated and altered to form different kinds of silicate clays. Others are decomposed and the decomposition products are recombined into new minerals such as other silicate clays and hydrous oxides of iron and aluminum (see Figure 2.2).

The newly formed minerals may accumulate in place where they are formed or may move downward and accumulate in lower soil zones. As materials are translocated from one soil zone to another, soil horizons are formed. Upper horizons may be characterized by the removal of specific constituents, while the accumulation of these or other constituents may characterize the lower horizons. In either case, soil horizons are created that are different in character from the original parent material.

TRANSPORT OR ACCUMULATION OF SIMPLE COMPOUNDS

Weathering also produces soluble materials, including positively charged ions (cations, e.g., Ca^{2+}) and negatively charged ions (anions, e.g., SO_4^{2-}) (see Figure 2.2). In humid regions, these materials are moved downward in rainwater and are either taken up by plant roots and recycled back into the soil or removed in the drainage water. In areas with lower rainfall, however, once some of these ions are moved to the subsurface horizon, they are combined to form insoluble compounds such as calcite ($CaCO_3$) and gypsum ($CaSO_4 \cdot 2H_2O$). Layers containing these compounds are common in soils of semiarid and arid areas.

In some soils in arid areas, soluble salts such as sodium chloride (NaCl) and sodium sulfate (Na_2SO_4) also accumulate at or near the soil surface, having been transported there by upward-moving water before it evaporates into the atmosphere (see Section 8.18).

PHYSICAL PROPERTIES

Soil genesis involves change in physical as well as chemical properties of different soil horizons. The best examples of such changes relate to soil structure, that is the grouping or arrangement of the soil particles into aggregates.[2] The organo-mineral layer near the soil surface commonly has a granular-type structure, differentiating it from the subsurface horizons and the original parent material. Likewise, blocky and prismatic-type structures characterize some subsurface horizons, especially those with clay accumulation. As will be discussed in Section 3.3, soil structure is one of the major properties used to characterize soil horizons and, in turn, to classify soils.

SOIL GENESIS IN NATURE

The examples we have considered of weathering, translocation, and accumulation illustrate rather simply how soil horizons come into being and, in turn, how the process of soil genesis proceeds. But before studying specifically the horizons found in soils, two additional points should be emphasized. First, the parent materials from which many soils develop are not uniform, varying considerably according to depth. These differences in the properties of a given parent material commonly existed before soil genesis started. For example, water-deposited materials may vary widely in texture with depth.[3] Consequently, in characterizing soils, consideration must be given not only to the *genetic* horizons and properties that come into

---

[2] Soil structure was discussed briefly in Section 1.9. See Figure 4.8 for a description of the various structural types.

[3] As mentioned in Section 1.9, soil texture is related to the size of the individual soil particles (sand, clay, etc.). See Figure 4.6 for the sand, silt, and clay components of different textural classes.

being during soil genesis but to those layers or properties that may have been *inherited* from the parent material.

Second, it is well to remember that soil genetic processes are still going on at different rates in different soils. Consequently, in some soils, the process of horizon differentiation has only begun whereas in others it is well-advanced. The latter are sometimes referred to as "well-developed soils." This means that the soil profiles we see today are likely quite different from those that existed 1500–2000 years ago and from those that will exist by the year 2200. The dynamic nature of soil genesis must be kept in mind.

3.3

The Soil Profile

The layering or horizon development described in the previous section gradually gives rise to natural bodies called *soils*. Each soil is characterized by a given sequence of these horizons. A vertical exposure of the horizon sequence is termed a *soil profile*. Attention now will be given to the major horizons making up soil profiles and the teminology used to describe them.

THE MASTER HORIZONS AND LAYERS

For convenience in study and description, five *master* soil horizons are recognized. These are designated using the capital letters O, A, E, B, and C. Subordinate layers or distinctions within these master horizons are designated by lowercase letters. A common sequence of horizons within a profile is shown in Figure 3.7.

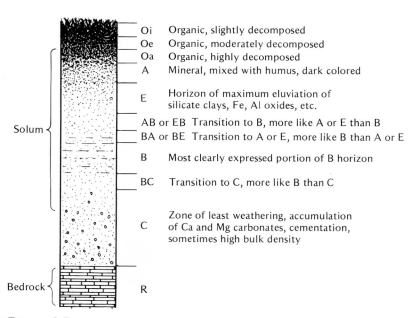

FIGURE 3.7

Hypothetical mineral soil profile showing the major horizons that may be present in a well-drained soil in the temperate humid region. Any particular profile may exhibit only some of these horizons, and the relative depths vary. In addition, however, a soil profile may exhibit more detailed subhorizons than indicated here. The solum includes the A, E, and B horizons plus some cemented layers of the C horizon.

O Horizons (Organic). The O group is comprised of organic horizons that form above the mineral soil. They result from litter derived from dead plants and animals. O horizons usually occur in forested areas and are generally absent in grassland regions. The specific horizons are

Oi: Organic horizon of the original plant and animal residues, only slightly decomposed.
Oe: Organic horizon, residues intermediately decomposed.
Oa: Organic horizon, residues highly decomposed.

A Horizons. The A horizons are the topmost mineral horizons. They contain a strong mixture of partially decomposed (humified) organic matter, which tends to impart a darker color than that of the lower horizons.

E Horizons. E horizons are those of maximum leaching or *eluviation* (from Latin *ex* or *e*, out, and *lavere*, to wash) of clay, iron, and aluminum oxides, which leaves a concentration of resistant minerals, such as quartz, in the sand and silt sizes. An E horizon is generally lighter in color than the A horizon and is found under the A horizon.

B Horizons (Illuvial). The subsurface B horizons include layers in which *illuviation* (from the Latin *il*, in, and *lavere*, to wash) of materials has taken place from above and even from below. In humid regions, the B horizons are the layers of maximum accumulation of materials such as iron and aluminum oxides and silicate clays. In arid and semiarid regions, calcium carbonate, calcium sulfate, and other salts may accumulate in the B horizon.[4]

The B horizons sometimes are referred to incorrectly as the *subsoil*. In soils with shallow A horizons, part of the B horizon may be incorporated into the plow layer and thus become part of the "topsoil." In other soils with deep A horizons, the plow layer or topsoil may include only the upper part of the A horizons, and the subsoil would include the lower part of the A horizon along with the B horizon. This emphasizes the need to differentiate between terms often used in soil management (topsoil, subsoil) and those used to describe the soil profile.

C Horizon. The C horizon is the unconsolidated material underlying the solum (A and B). It may or may not be the same as the parent material from which the solum formed. The C horizon is outside the zones of major biological activities and is generally little affected by the processes that formed the horizons above it. Its upper layers may in time become a part of the solum as weathering and erosion continue.

R Layers. Underlying consolidated rock, with little evidence of weathering.

Transition Horizons. These horizons are transitional between the master horizons (O, A, E, B, and C). They may be dominated by properties of one horizon but have prominent characteristics of another. Both capital letters are used to designate the transition horizons (e.g., AE, EB, BE, and BC), the dominant horizon being listed before the subordinate one. Letter combinations such as E/B are used to designate transition horizons where distinct parts of the horizon have properties of E and other parts have properties of B.

---

[4] In some soils of arid and semiarid regions the accumulation of these calcium compounds takes place in the C horizon (below).

TABLE 3.1
Lowercase Letter Symbols to Designate Subordinate Distinctions Within Master Horizons

Letter	Distinction	Letter	Distinction
a	Organic matter, highly decomposed	o	Accumulation of Fe and Al oxides
b	Buried soil horizon	p	Plowing or other disturbance
c	Concretions or nodules	q	Accumulation of silica
d	Dense unconsolidated materials	r	Weathered or soft bedrock
e	Organic matter, intermediate decomposition	s	Illuvial accumulation of O.M.[a] and Fe and Al oxides
f	Frozen soil	t	Accumulation of silicate clays
g	Strong gleying (mottling)	v	Plinthite, (high iron, red material)
h	Illuvial accumulation of organic matter	w	Distinctive color or structure
i	Organic matter, slightly decomposed	x	Fragipan (high bulk density)
k	Accumulation of carbonates	y	Accumulation of gypsum
m	Cementation or induration	z	Accumulation of soluble salts
n	Accumulation of sodium		

[a] O.M. = organic matter.

SUBORDINATE DISTINCTIONS

Master horizons are further characterized by specific properties such as distinctive color or the accumulation of materials such as clays and salts. These subordinate distinctions are identified by using lower case letters that designate specific characteristics. A list of these subordinate distinctions and their meaning are shown in Table 3.1. By illustration, a Bt horizon is a B horizon characterized by clay accumulation (t); likewise in a Bk horizon carbonates (k) have accumulated.

HORIZONS IN A GIVEN PROFILE

It is not likely that the profile of any one soil will show all of the horizons that collectively are cited in Figure 3.7. The ones most commonly found in well-drained soils, are Oe, or Oa if the land is forested; A or E, or both, depending on circumstances; B or Bw; and C. Conditions of soil genesis will determine which others are present and their clarity of definition.

When a virgin soil is plowed and cultivated, the upper 15–20 cm (6–8 in.) becomes the furrow slice. The cultivation, of course, destroys the original layered condition of this upper portion of the profile, and the furrow slice becomes more or less homogeneous. In some soils, the A and E horizons are deeper than the furrow slice (Figure 3.8). In other cases where the upper

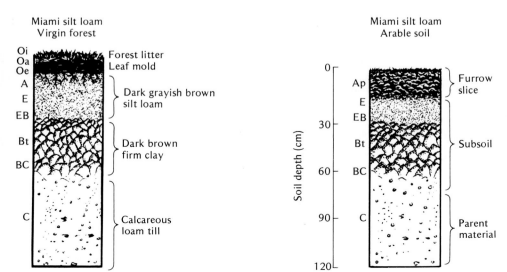

FIGURE 3.8

Generalized profile of the Miami silt loam, one of the Alfisols of the eastern United States, before and after land is plowed and cultivated. The surface layers are mixed by tillage and are termed the Ap (plowed) horizon. If erosion occurs, they may disappear, at least in part, and some of the B horizon will be included in the furrow slice.

horizons are quite thin, the plowline is just at the top of or even down into the B.

In some cultivated land, serious erosion has produced a *truncated* profile. As the surface soil was swept away, the plowline was gradually lowered to maintain a sufficiently thick furrow slice. In many cases the furrow slice is almost entirely within the B zone, and the C horizon is correspondingly near the surface. Many farmers are today cultivating the B horizons. In profile study and description, such a situation requires careful analysis. Comparison to a nearby eroded site can show how much erosion has occurred.

3.4

Concept of Individual Soils

As soil formation proceeds, there is great variability in the soil profile characteristics from one location to another. Differences are found in properties such as soil depth, clay and organic matter content, and, perhaps most importantly, the degree of horizon development. These differences in profile characteristics are noted not only from continent to continent or region to region, but from one part of a given field to another. In fact, notable differences sometimes are found within a matter of a few meters.

To study the soil intelligently at different locations on the Earth's surface and to communicate information about what is found, systems of soil classification have been developed. Through such classification systems *the* soil is perceived as being composed of a large number of individual units or natural bodies called *soils*. Each individual soil has a given range of soil properties that distinguish it from other soils.

PEDON

Examination and study of the soil in the field is necessary to establish the kinds and ranges of properties that are to characterize given soil units. Such field study should be focused on a three-dimensional unit that, although it is small in size, is sufficiently large so that the nature of its horizons can be ascertained and the range of its properties identified. It varies in size from about 1 to 10 m^2 (11 to 110 ft^2) and is called a *pedon* (rhymes with "head on," from the Greek *pedon*, ground) (Figure 3.9).

Because of its very small size, a pedon obviously cannot be used as the basic unit for a workable field soil classification system. However, a group of pedons, termed a *polypedon*, closely associated in the field and similar in properties, is of sufficient size to serve as a basic classification unit, or a *soil individual*. Such grouping conceptually approximates what in the United States has been called a *soil series*. Nearly 17,000 soil series have been characterized in this country. They are the basic units used in classification of the nation's soils.[5]

Two extremes in the concept of soils now have been identified. One extreme is that of a natural body called *a* soil, characterized by a three-dimensional sampling unit (pedon), related groups of which (polypedon) are termed a soil series. At the other extreme is *the* soil, a collection of all these natural bodies that is distinct from water, solid rock, and other natural parts of the earth's crust. These two extremes represent opposite ends of elaborate soil classification schemes that are used to organize knowledge of soils.

---

[5] It is well to point out the conceptual soil *classification* units may not coincide strictly with *mapping* units as used in the field. Thus, as applied in the preparation of field maps, the series designation may include aggregates of polypedons along with some inclusions of others.

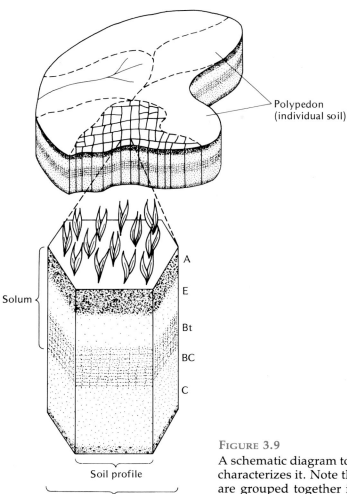

Polypedon
(individual soil)

Solum

A
E
Bt
BC
C

Soil profile

Soil pedon

FIGURE 3.9
A schematic diagram to illustrate the concept of pedon and of the soil profile that characterizes it. Note that several contiguous pedons with similar characteristics are grouped together in a larger area (outlined by broken lines) called a polypedon or individual soil. Several individual soils are present in this landscape.

3.5

Evolution of Soil Classification Systems

Throughout history humans have used some kind of system to name and classify soils. From the time crops were first cultivated, humans noticed differences in soils and classified them, if only in terms of "good" and "bad." The early Chinese, Egyptian, Greek, and Roman civilizations acknowledged differences in soils as media for plant growth. Such recognition is common today as soils are described as being good "cotton," "soybean," "alfalfa," "wheat," or "rice" soils.

Soils also have been classified in terms of the geological parent materials from which they were formed. Terms such as "sandy" or "clayey" soils as well as "limestone" soils and "lake-laid" soils have a geological connotation and are used today.

The concept of soils as natural bodies was first developed by the Russian soil scientist V. V. Dokuchaev and his associates. They noted the relationship among climate, vegetation, and soil characteristics, a concept that Dokuchaev published in 1883. Because of poor international communications, his concept was not promoted in the United States until the early part of the twentieth century. C. F. Marbut of the U.S. Department of Agriculture grasped the concept of soils as natural bodies, and in 1927 he developed a soil classification scheme based on this principle. The scheme was improved in 1935, and more comprehensive schemes followed in 1938 and 1949, the 1949 system serving well for about 25 years.

The Soil Survey Staff of the U.S. Department of Agriculture, in cooperation with soil scientists in other countries, developed a new comprehensive system of soil classification based on soil properties. This system has been in use in the United States since 1965, and it is used, at least to some degree, by scientists in 45 other countries. This system will be used in this text.

3.6

New Comprehensive Classification System—Soil Taxonomy[6]

The comprehensive soil classification system, called *Soil Taxonomy* (Soil Survey Staff, 1975), maintains the natural body concept and has two other major features that make it most useful. First, the system is based on *soil properties* that are easily verified by others. This lessens the likelihood of controversy over the classification of a given soil, which can occur when scientists deal with systems based on soil genesis or presumed genesis.

The second significant feature of *Soil Taxonomy* is the *unique nomenclature* employed, which gives a definite connotation of the major characteristics of the soils in question. Consideration will be given to the nomenclature used after brief reference is made to the major criteria for the system—soil properties.

BASES OF SOIL CLASSIFICATION

Soil Taxonomy is based on the properties of soils as they are found today. Although one of the objectives of the system is to group soils similar in genesis, the specific criteria used to place soils in these groups are those of soil properties. In so doing, soil genesis is not ignored. Since soil properties often are related directly to soil genesis, it is difficult to emphasize soil properties without at least indirectly emphasizing soil genesis as well.

All of the chemical, physical, and biological properties presented in this text are used as criteria for *Soil Taxonomy*. A few examples are the moisture, temperature, color, texture, and structure of the soil. Chemical and mineral properties such as the contents of organic matter, clay, iron and aluminum oxides, silicate clays, salts, the pH, the percentage base saturation,[7] and soil depth are other important criteria for classification. The presence or absence of certain diagnostic soil horizons also determines the place of a soil in the classification system. They will now receive our attention.

DIAGNOSTIC HORIZONS

Diagnostic soil horizons are found in the surface or the subsurface. The diagnostic surface horizons are called *epipedons* (from the Greek *epi*, over, and *pedon*, soil). The epipedon includes the upper part of the soil darkened by organic matter, the upper eluvial horizons, or both. It may include part of the B horizon (see Section 3.3) if the latter is significantly darkened by organic matter. Six epipedons are recognized (Table 3.2), but only four are of

---

[6] Taxonomy is the science or principles of classification. For a review of the achievements and challenges of *Soil Taxonomy*, see SSSA (1984).

[7] The percentage base saturation is the percentage of a soil's total cation adsorption (exchange) capacity that is satisfied by base-forming cations such as Ca^{2+}, Mg^{2+}, and K^+ (see Section 8.3).

TABLE 3.2

Major Features of Diagnostic Horizons in Mineral Soils Used to Differentiate at the Higher Levels of *Soil Taxonomy*

Diagnostic horizon (and designation)	Major feature
Surface Horizons = Epipedons	
Mollic (A)	Thick, dark colored, high base saturation, strong structure
Umbric (A)	Same as Mollic except low base saturation
Ochric (A)	Light colored, low organic content, may be hard and massive when dry
Histic (O)	Very high in organic content, wet during some part of year
Anthropic (A)	Man-modified Mollic-like horizon, high in available P
Plaggen (A)	Man-made sod-like horizon created by years of manuring
Subsurface Horizons	
Argillic (Bt)	Silicate clay accumulation
Natric (Btn)	Argillic, high in sodium, columnar or prismatic structure
Spodic (Bhs)	Organic matter, Fe and Al oxides accumulation
Cambic (B)	Changed or altered by physical movement or by chemical reactions
Agric (A or B)	Organic and clay accumulation just below plow layer resulting from cultivation
Oxic (Bo)	Highly weathered, primarily mixture of Fe, Al oxides and nonsticky-type silicate clays
Duripan (m)	Hard pan, strongly cemented by silica
Fragipan (x)	Brittle pan, usually loamy textured, weakly cemented
Albic (E)	Light colored, clay and Fe and Al oxides mostly removed
Calcic (k)	Accumulation of $CaCO_3$ or $CaCO_3 \cdot MgCO_3$
Gypsic (y)	Accumulation of gypsum
Salic (z)	Accumulation of salts
Kandic	Accumulation of low activity clays
Petrocalcic	Cemented calic horizon
Petrogypsic	Cemented gypsic horizon
Placic	Thin pan cemented with iron alone or with manganese or O.M.[a]
Sombric	Organic matter accumulation
Sulfuric	Highly acid with Jarosite mottles

[a] O.M. = organic matter.

any prominence in the soils of the United States. The other two, called *anthropic* and *plaggen*, are the result of man's intensive use of soils. They are found in parts of Europe and Asia.

Many subsurface horizons characterize different soils in the system. Those that are considered diagnostic horizons are shown along with their major features in Table 3.2. Each of these layers is used as a distinctive property to help place a soil in its proper class in the system.

CATEGORIES OF THE SYSTEM

There are six categories of classification in *Soil Taxonomy*: (a) order (the broadest category), (b) suborder, (c) great group, (d) subgroup, (e) family, and (f) series (the most specific category). These categories may be compared with those used for the classification of plants. The comparison would be as shown in Table 3.3, where white clover (*Trifolium repens*) and Miami silt loam are the examples of plants and soils, respectively.

Just as *Trifolium repens* identifies a specific kind of plant, the Miami series identifies a specific kind of soil. The similarity continues up the classification

TABLE 3.3

Comparison of the Classification of a Common Cultivated Plant, White Clover (*Trifolium repens*), and a Soil, Miami Series

Plant classification			Soil classification	
Phylum	Pterophyta		Order	Alfisols
Class	Angiospermae		Suborder	Udalfs
Subclass	Dicotyledoneae	Increasing specificity	Great Group	Hapludalfs
Order	Rosales		Subgroup	Typic Hapludalfs
Family	Leguminosae		Family	Fine loamy, mixed, mesic
Genus	*Trifolium*		Series	Miami
Species	*repens*		Phase[a]	Miami, eroded phase

[a] Technically not a category in *Soil Taxonomy* but used in field surveying.

scale to the highest categories—phylum for plants and order for soils. With this general background, a brief description of the six soil categories and the nomenclature used in identifying them is presented.

Order. The *order* category is based largely on soil-forming processes as indicated by the presence or absence of major diagnostic horizons. A given order includes soils whose properties suggest that they are not too dissimilar in their genesis. As an example, many soils that developed under grassland vegetation have the same general sequence of horizons and are characterized by a thick, dark epipedon (surface horizon) high in metallic cations. Soils with these properties are thought to have been formed by the same general genetic processes and are included in the same order, Mollisols. There are eleven soil orders in *Soil Taxonomy* (Table 3.4).

TABLE 3.4

Names of Soil Orders in *Soil Toxonomy* with Their Derivation and Major Characteristics

	Formative element		
Name	Derivation	Pronunciation	Major characteristics
Entisols	Nonsense symbol	Re*cent*	Little profile development, ochric epipedon common
Inceptisols	L. *inceptum*, beginning	In*cept*ion	Embryonic soils with few diagnostic features, ochric or umbric epipedon; cambic horizon
Mollisols	L. *mollis*, soft	*Moll*ify	Mollic epipedon, high base saturation, dark soils, some with argillic or natric horizons
Alfisols	Nonsense symbol	Ped*alf*er	Argillic or natric horizon; high to medium base saturation
Ultisols	L. *ultimus*, last	*Ult*imate	Argillic horizon, low base saturation,
Oxisols	Fr. *oxide*, oxide	*Ox*ide	Oxic horizon, no argillic horizon, highly weathered
Vertisols	L. *verto*, turn	In*vert*	High in swelling clays; deep cracks when soil dry
Aridisols	L. *aridus*, dry	*Arid*	Dry soil, ochric epipedon, sometimes argillic or natric horizon
Spodosols	Gk. *Spodos*, wood ash	*Podzol*; odd	Spodic horizon commonly with Fe, Al, and humus accumulation
Histosols	Gk. *Histos*, tissue	*Hist*ology	Peat or bog; >30% organic matter
Andisols[a]	Modified from Ando	*And*esite	From volcanic ejecta, dominated by allophane or Al-humic complexes

[a] Only recently added as a soil order; less detail will be given on these soils.

Suborder. The *suborders* are subdivisions of orders that emphasize properties that suggest genetic homogeneity. Thus, wetness, climatic environment, and vegetation, which help determine the nature of the genetic processes, also will help determine the suborder in which a given soil is found. Some 47 soil suborders are recognized, and 44 are used in the United States.

Great Group. Diagnostic horizons (Table 3.2) are the primary bases for differentiating the *great groups* in a given suborder. Soils in a given great group have the same kind and arrangement of these horizons. More than 230 great groups are recognized, some 187 of which are used in the United States.

Subgroup. The *subgroups* are subdivisions of the great groups. The central concept of a great group makes up one subgroup (Typic). Other subgroups may have characteristics that are intergrades between those of the central concept and soils of other orders, suborders, or great groups. More than 1200 subgroups are recognized, about 1000 of which are found in the United States.

Family. In the *family* category are found soils with a subgroup having similar physical and chemical properties affecting their response to management and especially to the penetration of plant roots (e.g., soil–water–air relationships). Differences in texture, mineralogy, temperature, and soil depth are primary bases for family differentiation. Nearly 6600 families have been recognized.

Series. The *series* category is the most specific unit of the classification system. It is a subdivision of the family, and its differentiating characteristics are based primarily on the kind and arrangement of horizons. Conceptually, it includes only one polypedon; however, in the field, aggregates of polypedons and associated inclusions are included in the soil series mapping units. There are about 16,800 soil series recognized in the United States (Figure 3.10).

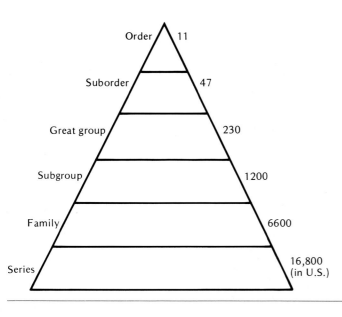

FIGURE 3.10

The categories of *Soil Taxonomy* and approximate number of units in each category. [Data from U.S. Department of Agriculture, personal correspondence]

NOMENCLATURE

A major feature of the system is the nomenclature used to identify different soil classes. The names of the classification units are combinations of syllables, most of which are derived from Latin or Greek and are root words in several modern languages. Since each part of a soil name conveys a concept of soil character or genesis, the name automatically describes the general kind of soil being classified. For example, soils of the order *Aridisols* (from the Latin *aridus*, dry, and *solum*, soil) are characteristically dry soils. Those of the order *Inceptisols* (from the Latin *inceptum*, beginning, and *solum*, soil) are soils with only the beginnings of profile development. Thus, the names of orders are combinations of (a) formative elements, which generally define the characteristics of the soils, and (b) the ending *sol*.

The names of suborders automatically identify the order of which they are a part. For example, soils of the suborder *Aquolls* are the wetter soils (from the Latin *aqua*, water) of the Mollisols order. Likewise, the name of the great group identifies the suborder and order of which it is a part. *Argiaquolls* are Aquolls with clay or argillic (Latin *argilla*, white clay) horizons.

The nomenclature as it relates to the different categories in the classification system might be illustrated as follows.

Mollisols · · · · · · · · order

Aqu*olls* · · · · · · · · · · · · · · suborder

Argiaqu*olls* · · · · · · · · · · · · · · · · · great group

Typic Argiaqu*olls* · subgroup

Note that the three letters *oll* identify each of the lower categories as being in the Mollisols order. Likewise, the suborder name *Aquolls* is included as part of the great group and subgroup name. If one is given only the subgroup name, the great group, suborder, and order to which the soil belongs are automatically known.

Family names in general identify subsets of the subgroup that are similar in texture, mineral composition, and soil temperature at a depth of 50 cm. Thus the name *Typic Argiaquolls, fine, mixed, mesic,* identifies a family in the *Typic Argiaquolls* subgroup with a fine texture, mixed clay mineral content, and mesic (8–15 °C) soil temperature.

Soil series names have local significance since they normally identify the particular locale in which the soil is first found. Thus, series such as Fort Collins, Cecil, Miami, Norfolk, and Ontario identify soils first described near the city or locality named.

In field soil surveying, soil series are sometimes further differentiated on the basis of surface soil texture or other characteristics. These field mapping units are called *soil phases*. Names such as Fort Collins loam, Cecil clay, or Cecil, eroded phase are used to identify such phases. Note, however, that soil phases, practical as they may be in local situations, are not a category in the *Soil Taxonomy* system.

With this brief explanation of the nomenclature of the new system, the order category of the system now will be considered.

3.7

Soil Orders

Eleven orders are recognized. The names of these orders and their major characteristics are shown in Table 3.4. Note that all order names have a common end, *sol* (from the Latin *solum*, soil).

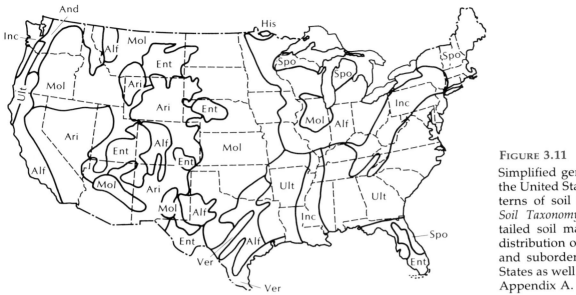

FIGURE 3.11
Simplified general soil map of the United States showing patterns of soil orders based on *Soil Taxonomy*. For more detailed soil maps showing the distribution of both soil orders and suborders for the United States as well as the world, see Appendix A.

Figure 3.11 is a simplified general soil map showing the distribution of soil orders in the United States. The major areas of the country dominated by each order are evident. Profiles of each soil order are shown in color on Plate I. More detailed general soil maps (along with the map legends) for the United States and for the world appear in Appendix A.

ENTISOLS (little if any profile development)

Entisols are weakly developed mineral soils without natural genetic (subsurface) horizons or with only the beginnings of such horizons (Plate I). Most have an ochric epipedon and a few have man-made anthropic or agric epipedons. The extremes of highly productive soils on recent alluvium and infertile soils on barren sands as well as shallow soils on bedrock are included.

Soils of this order are found under a wide variety of environmental conditions in the United States (see Figure 3.11 and Appendix A). For example, in the Rocky Mountain region and in southwest Texas, shallow, medium-textured Entisols (Orthents) over hard rock are common. They are used mostly as rangeland. Sandy Entisols (Psamments) are found in Florida, Alabama, and Georgia and typify the sand hill section of Nebraska. Psamments are used for cropland in humid areas. Some of the citrus-, vegetable-, and peanut-producing areas of the South are typified by Psamments.

Entisols are probably found under even more widely varied environmental conditions outside the United States (see Appendix A). Psamments are typical of the Sahara Desert and Saudi Arabia and dominate parts of southern Africa and central and north-central Australia. Entisols having medium to fine textures (Orthents) are found in northern Quebec and parts of Alaska, Siberia, and Tibet. Orthents are typical of some mountain areas such as the Andes in South America and some of the uplands of an area extending from Turkey eastward to Pakistan.

The agricultural productivity of the Entisols varies greatly depending on their location and properties. With adequate fertilization and a controlled water supply, some Entisols are quite productive; in fact, Entisols developed on alluvial floodplains are among the world's most productive soils. However, restrictions on the depth, clay content, or water balance of most Entisols limit the intensive use of large areas of these soils.

INCEPTISOLS (few diagnostic features)

The horizons of Inceptisols are thought to form quickly and result mostly from the alteration of parent materials (Plate I). Inceptisols have ochric or umbric epipedons and/or cambic subsurface horizons. Subsurface horizons associated with more advanced weathering and/or accumulation of silicate clays, iron and aluminum oxides, and organic matter are absent. Profile development of Inceptisols is more advanced than that of the Entisols order but less so than for the other orders. Many agriculturally useful soils are included along with others for which productivity is limited by factors such as slow drainage of water.

Inceptisols are found in each of the continents (see Appendix A); in the United States they appear in all but arid environments. For example, some Inceptisols called Ochrepts (from the Greek *ochros*, pale), with thin, light-colored surface horizons, extend from southern New York through central and western Pennsylvania, West Virginia, and eastern Ohio. Ochrepts also dominate an area extending from southern Spain through central France to Germany, and are present as well in Chile, North Africa, eastern China, and western Siberia.

Tropepts (Inceptisols formed under continually warm conditions) are found in northwestern Australia, central Africa, southwestern India, and southwestern Brazil. Wet Inceptisols, or Aquepts (from the Latin *aqua*, water), are found in areas along the Amazon and Ganges rivers.

There is considerable variability in the natural productivity of Inceptisols. For instance, those found in the Pacific Northwest are quite fertile and provide some of the world's best wheat lands. In contrast, some of the low-organic Ochrepts in southern New York and northern Pennsylvania are not naturally productive. They have been allowed to reforest following earlier periods of crop production.

MOLLISOLS (dark soils of grasslands, mollic epipedon, base rich)

The Mollisols order includes some of the world's most important agricultural soils. Mollisols are characterized by a *mollic epipedon*, or surface horizon, which is thick, dark, and dominated by base-forming cations such as Ca^{2+} and Mg^{2+} (high base saturation) (Table 3.4 and Plate I). They may have an argillic (clay), natric, albic, or cambic horizon, but not an oxic or spodic one. The surface horizon generally has granular or crumb structures, largely resulting from the organic matter present, that are not hard when the soils are dry. This justifies the use of a name that implies softness (see Table 3.4).

Most of the Mollisols have developed under grass vegetation (Figure 3.12). Grassland soils of the central part of the United States, lying between the Aridisols on the west and the Alfisols on the east, make up the central core of this order. However, a few soils developed under forest vegetation (primarily in depressions) have a mollic epipedon and are included among the Mollisols. The soil temperature and soil moisture characteristic of Mollisols are shown in Figure 3.13.

Mollisols are dominant in the Great Plains states and in Illinois (see Figure 3.11). Those where soil moisture is not limiting are called *Udolls* (from the Latin *udus*, humid). A region extending from North Dakota to southern Texas is characterized by *Ustolls* (from the Latin *ustus*, burnt), which are intermittently dry during the summer. Farther west in parts of Idaho, Utah,

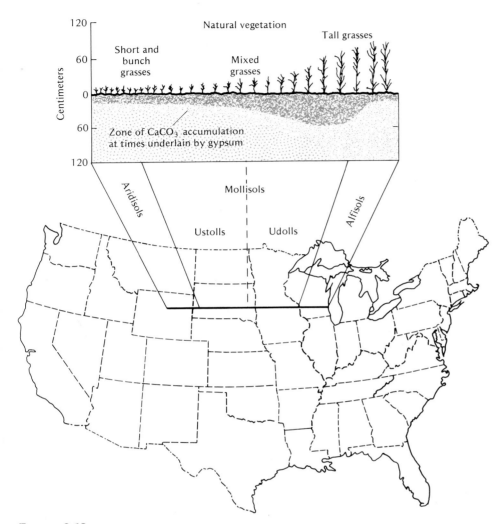

FIGURE 3.12
Correlation between natural grassland vegetation and certain soil orders is graphically shown for a strip of territory in north central United States. The control, of course, is climate. Note the deeper organic matter layer and deeper zone of calcium accumulation as one proceeds from the drier areas in the west toward the more humid region where prairie soils are found. Alfisols develop under grassland vegetation, but more commonly occur under forests and have lighter colored surface horizons.

Washington, Oregon, and California are found sizable areas of *Xerolls* (from the Greek *xeros*, dry), which are the driest of the Mollisols. This soil order characterizes a larger land area in the United States than any other soil order (Table 3.5).

The largest area of Mollisols outside the United States is that stretching from east to west across the heartland of the Soviet Union (mostly Borolls). Other sizable areas are found in Mongolia and northern China and in northern Argentina, Paraguay, and Uruguay.

The high native fertility of Mollisols (especially the Udolls) makes them among the world's more productive soils. When they were first cleared for cultivation, their high native organic matter released sufficient nitrogen and other nutrients to produce high crop yields even without fertilization. Yields on these soils were unsurpassed by other unirrigated areas. Even today, when moderate to heavy fertilization increases the productivity of naturally

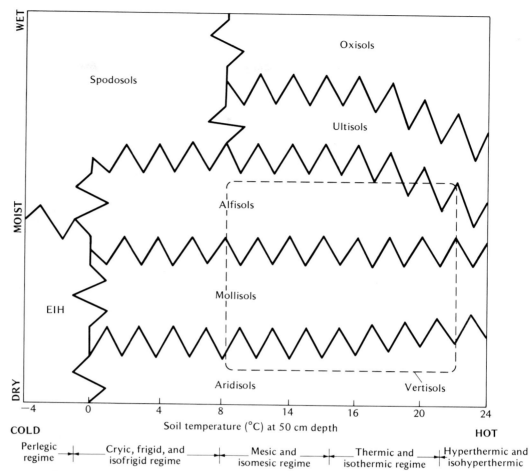

FIGURE 3.13

Diagram showing the general soil moisture and soil temperature regimes that characterize the most extensive soils in each of the seven soil orders. Soils of the other three orders (Entisols, Inceptisols, and Histosols) may be found under any of the soil moisture and temperature conditions (including the area marked EIH). Major areas of Vertisols are found only where clayey materials are in abundance and are most extensive where the soil moisture and temperature conditions approximate those shown inside the box with broken lines. Note that these relationships are only approximate and that less extensive areas of soils in each order may be found outside the indicated ranges. For example, some Ultisols (Ustults) and Oxisols (Ustox) have soil moisture levels for at least part of the year that are much lower than this graph would indicate. (The terms used at the bottom to describe the soil temperature regimes are those used in helping to identify soil families.)

---

infertile soils in more humid areas, Mollisols still rate among the best. Photographs of two Mollisols are shown in Figure 3.14.

ALFISOLS (argillic or natric horizon, medium-high bases)

Alfisols are moist mineral soils having no mollic epipedon or oxic or spodic horizons (Plate I and Figure 3.14). They have gray to brown surface horizons (commonly an ochric epipedon), medium- to high-base[8] status, and contain an illuvial horizon in which silicate clays have accumulated. The cation

---

[8] The term *base* is used to indicate base-forming cations such as Ca^{2+}, Mg^{2+}, K^+, and Na^+. The cation exchange capacity is a measure of the soil's ability to adsorb exchangeable cations.

TABLE 3.5

Approximate Land Areas of Different Soil Orders and Suborders of the United States and the World (in Parentheses), Their Major Land Use and Natural Fertility Status

Order and suborder	Land area (%)	Major land uses	Fertility
Alfisols	13.4 (13.2)		
Aqualfs	1.0	Cropland	High
Boralfs	3.0	Forest	High
Udalfs	5.9	Cropland	High
Ustalfs	2.6	Cropland	High
Xeralfs	0.9	Rangeland	High
Aridisols	11.5 (18.8)		
Argids	8.6	Rangeland	Low
Orthids	2.9	Rangeland	Low
Entisols	7.9 (8.3)		
Aquents	0.2	Wetland	Moderate
Orthents	5.2	Rangeland, forest	Moderate
Psamments	2.2	Cropland	Low
Histosols	0.5 (0.9)	Wetland, cropland	Moderate
Inceptisols	16.3 (8.9)[a]		
Aquepts	11.9	Cropland	Moderate
Ochrepts	4.3	Cropland, forest	Moderate to low
Umbrepts	0.7	Forest	Low
Mollisols	24.6 (8.6)		
Aquolls	1.3	Cropland	High
Borolls	4.9	Cropland	High
Udolls	4.7	Cropland	High
Ustolls	8.8	Cropland, rangeland	High
Xerolls	4.8	Cropland, rangeland	High
Spodosols	5.1 (4.3)		
Aquods	0.7	Forest	Low
Orthods	4.4	Forest	Low
Ultisols	12.9 (5.6)		
Aquults	1.1	Forest	Low to moderate
Humults	0.8	Forest	Low
Udults	10.0	Forest, cropland	Low
Vertisols	1.0 (1.8)		
Uderts	0.4	Cropland	High
Usterts	0.6	Cropland	High
Andisols[b]	1.9	Cropland, forest	Moderate
Oxisols	0.01 (8.9)	Cropland, forest	Low
Miscellaneous land	4.5 (19.7)	Barren land	—

Data from USDA Soil Conservation Service, Soil Survey Staff (personal communication), and McCracken et al. (1985).

[a] Includes Andisols.

[b] Andisol world percentage is included in the Inceptisols; United States percentage is an approximation.

exchange capacity of the clay horizon is more than 35% saturated with base-forming cations (Ca^{2+}, Mg^{2+}, etc.). This horizon is termed *argillic* if only silicate clays are present and *natric* if, in addition to the clay, it is more than 15% saturated with sodium and has prismatic or columnar structure (see Figure 4.8).

The Alfisols appear to be more strongly weathered than the Inceptisols but less so than the Spodosols. They are formed in cool to hot humid areas but also are found in the semiarid tropics. Most often Alfisols are developed under native deciduous forests, although in some cases grass is the native vegetation.

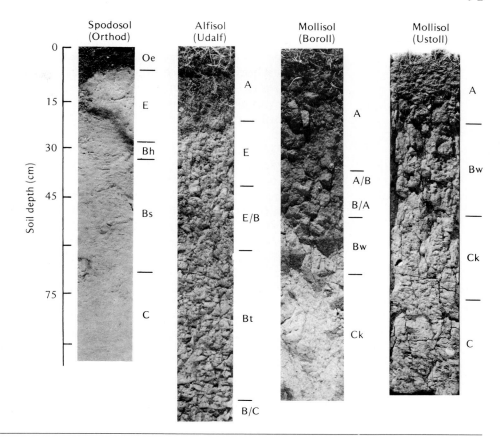

FIGURE 3.14
Monoliths of profiles representing four soil orders. The
suborder names are also
shown (in parentheses). Note
the spodic horizons in the Spodosol characterized by humus
(Bh) and iron (Bs) accumulation. In the Alfisol is found the
illuvial clay horizon (Bt), and
the structural B horizon (Bw) is
indicated. The thick dark surface horizon (mollic epipedon)
characterizes both Mollisols.
Note that the zone of calcium
carbonate accumulation (Ck) is
higher in the Ustoll, which has
developed in a dry climate.
The E/B horizon in the Alfisol
has characteristics of both A
and B horizons.

Alfisols are typified by *Udalfs* (from the Latin *udus*, humid) in Ohio,
Indiana, Michigan, Wisconsin, Minnesota, Pennsylvania, and New York
(see Figure 3.11). They are found in sizable areas of *Xeralfs* (from the Greek
xeros, dry) in central California; some cold *Boralfs* (from the Greek *boreas*,
northern) in the Rockies and Minnesota; *Ustalfs* (from the Latin *ustus*, burnt)
in areas of hot summers, including Texas and New Mexico; and the wet
Aqualfs (from the Latin *aqua*, water), in parts of the midwest.

Alfisols occur in other countries around the world (see Appendix A). A
large area dominated by Boralfs is found in northern Europe stretching from
the Baltic States through western Russia. A second large area is found in
Siberia. Ustalfs are prominent in the southern half of Africa, eastern Brazil,
eastern India, and southeastern Asia. Large areas of Udalfs are found in
central China, England, France and central Europe, and southeastern Australia. Xeralfs are prominent in southwestern Australia, Italy, and central
Spain.

In general, Alfisols are productive soils. Good crop yields are favored by
their medium- to high-base status, generally favorable texture, and location
(except for some Xeralfs) in regions with sufficient rainfall for crop production at least part of the year. In the United States these soils rank favorably
with the Mollisols and Ultisols in their productive capacity.

ULTISOLS (argillic horizon, low bases)

Ultisols are soils with an argillic (clay) horizon and a low-base status (less
than 35% of the exchange capacity satisfied with base-forming metallic
cations). Their mean annual temperatures at 50 cm depth are above 8 °C.

They are more highly weathered and acidic than Alfisols but less acid than Spodosols. Most Ultisols have developed under moist conditions in warm to tropical climates. Except for the wetter members of the order, their subsurface horizons are commonly red or yellow in color, evidence of accumulations of oxides of iron (Plate I). Ultisols still have some weatherable minerals, however, in contrast to the Oxisols. Ultisols are formed on old land surfaces, normally under forest vegetation, although savannah or even swamp vegetation is also common.

Most of the soils of the southeastern part of the United States fall in this order (Figure 3.11 and Appendix A). Udults, the moist, well-drained Ultisols, extend from the east coast (Maryland to Florida) to and beyond the Mississippi River Valley and are the most extensive of soils in the humid southeast. Humults (high in organic matter) are found in the United States in Hawaii, eastern California, Oregon, Washington; Humults are also present in the highlands of some tropical countries. Xerults (Ultisols in Mediterranean-type climates) occur locally in southern Oregon and northern and western California.

Ustults are found in the semiarid areas with a marked dry season. Together with the Ustalfs, the Ustults occupy large areas in Africa and India.

Ultisols are prominent on the east and northeast coasts of Australia (see Appendix A). Large areas of Udults are located in southeastern Asia and in southern China. Important areas also are found in southern Brazil and Paraguay.

Although Ultisols are not naturally as fertile as Alfisols or Mollisols, they respond well to good management. They are located mostly in regions of long growing seasons and of ample moisture for good crop production. The silicate clays of Ultisols are usually of the nonsticky type,[9] which, along with oxides of iron and aluminum, assures ready workability. Where adequate chemical fertilizers are applied, Ultisols are quite productive. In the United States, the well-managed Ultisols compete well with Mollisols and Alfisols as first-class agricultural soils.

Oxisols (oxic horizon, highly weathered)

The Oxisols are the most highly weathered soils in the classification system (Figure 3.13). They have ochric or umbric epipedons, but their most important diagnostic feature is a deep *oxic* subsurface horizon. This horizon is generally very high in clay-size particles dominated by hydrous oxides of iron and aluminum. Weathering and intense leaching have removed a large part of the silica from the silicate minerals in this horizon. Some quartz and 1:1-type silicate clay minerals remain, but the hydrous oxides are dominant.

The clay content of Oxisols is very high, but the clays are of the nonsticky type. The depth of weathering in Oxisols is much greater than for most of the other soils—16 m or more having been observed.

Oxisols occupy old land surfaces and occur mostly in the tropics. Relatively less is known about them than most of the other soil orders. They occur in large geographic areas, however, and millions of people in the tropics depend on them for food and fiber production.

The largest known areas of Oxisols occur in South America and Africa (see Appendix A). Orthox (Oxisols having a short dry season or none) occur in northern Brazil and neighboring countries. An area of Ustox (hot, dry summers) nearly as large appears in Brazil to the south of the Orthox.

---

[9] Silicate clays vary considerably in their chemical and physical properties. Some are sticky when wet and are cloddy and hard when dry. Others are not sticky and are desirable for ease of tillage of the soil. Silicate clay types will be covered in Chapter 7.

Oxisols are found in large areas of central Africa, being located on old land surfaces of this area, and in Malaysia and Indonesia.

The management of Oxisols presents both problems and opportunities. Most of them have not been cleared of their native forest vegetation or have been tilled using only primitive methods. The few instances where modern farming techniques have been used have met with mixed success. Heavy fertilization, especially with phosphorus-rich materials, is required. Deficiencies of micronutrients are common. In some areas, torrential rainfall makes cultural practices that leave the soil bare extremely harmful. For that reason, maintenance of forest cover may well be the wisest land use for some areas dominated by these soils. Such cover can be provided by some plantation crops including oil-palm and rubber, which are ideal for these soils.

Research on crop production on Oxisols suggests that the potential of some of these soils for food and fiber production is far in excess of that currently being realized. In Brazil and central Africa, selected areas of these soils have been demonstrated to be high in productivity when they are properly managed. However, the native forest vegetation should be maintained on most of these soil areas.

VERTISOLS (dark, swelling clays)

The Vertisols order of mineral soils is characterized by a high content (>30%) of sticky or swelling-and-shrinking-type clays to a depth of 1 m, which in dry seasons causes the soils to develop deep, wide cracks (Plate I). A significant amount of material from the upper part of the profile may slough off into the cracks, giving rise to a partial "inversion" of the soil. This accounts for the term *invert*, which is used to characterize this order in a general way.

Vertisols are found mostly in subhumid to semiarid environments and where the average soil temperatures are higher than 8 °C (Figure 3.13). There are several small but significant areas of Vertisols in the United States (Figure 3.11). Two are located in humid areas, one in eastern Mississippi and western Alabama and the other along the southeast coast of Texas. These soils are of the *Uderts* (from the Latin *udus*, humid) suborder because their moist condition prevents cracks from persisting for more than three months of the year.

Two other Vertisol areas are found in east central and southern Texas, where the soils are drier. Since cracks persist for more than three months of the year in these areas, the soils belong to the *Usterts* (from the Latin *ustus*, burnt) suborder, characteristic of areas with hot, dry summers. *Xererts* (Greek *xeros*, dry) are also located in California.

In India, Ethiopia, the Sudan, and northern and eastern Australia, large areas of Vertisols are found (see Appendix A). Smaller areas occur in sub-Saharan Africa and in Mexico, Venezuela, Bolivia, and Paraguay. These latter soils probably are of the Usterts or Xererts suborder, since dry conditions persist long enough for the wide cracks to stay open for periods of three months or longer.

In each of the major Vertisol areas mentioned, some of the soils are used for crop production. Even so, their very fine texture and marked shrinking and swelling characteristics make them less suitable than other soils for crop production and for building foundations and highway bases. They are sticky and plastic when wet and hard when dry. Because they are too sticky when wet and too hard when dry, the period of time during which they can be plowed or otherwise tilled is very short. In areas such as those in India and the Sudan, where slow-moving animals or human power are commonly used to till the soil, the cultivators cannot perform tillage operations on time and are limited to the use of small, near-primitive tillage implements that their animals can pull through the "heavy" soil.

Despite their limitations, Vertisols are widely tilled, especially in India, Ethiopia, and the Sudan. Sorghum, corn, millet, and cotton are crops commonly grown, but the yields are generally low. Recent research shows that these large soil areas can produce greatly increased yields of food crops with improved soil management practices.

ARIDISOLS (dry soils)

Aridisols are dry soils characterized by an ochric epipedon generally light in color and low in organic matter (Plate I). They may have a horizon of accumulation of calcium carbonate (calcic), gypsum (gypsic), soluble salts (salic), or sodium (natric). Except where there is groundwater or irrigation, the soil layers are moist only for short periods during the year. These short moist periods may be sufficient for drought-adapted desert shrubs and annual plants, but not for conventional crop production. If groundwater is present near the soil surface, soluble salts often accumulate to levels that most crop plants cannot tolerate.

Aridisols are found mostly in dry areas. A large area of Aridisols called *Argids* (from the Latin *argilla*, white clay), which have a horizon of clay accumulation, occupies much of the southern parts of California, Nevada, Arizona, and central New Mexico (Figure 3.11). The Argids also extend down into northern Mexico. Smaller areas of *Orthids* (Aridisols without argillic or natric horizons) are found in several western states.

Vast areas of Aridisols are present in the Sahara desert in Africa, the Gobi and Taklamakan deserts in China, and the Turkestan desert of the Soviet Union. Most of the soils of southern and central Australia are Aridisols, as are those of southern Argentina, southwestern Africa, Pakistan, and the Middle East countries.

Without irrigation, Aridisols are not suitable for growing cultivated crops. Some areas are used for sheep or goat grazing, but the production per unit area is low. Where irrigation water is available, Aridisols can be made highly productive. Irrigated valleys in arid areas are among the most productive in the world. However, they must be carefully managed to prevent the accumulation of soluble salts.

SPODOSOLS (spodic horizon, forests, low bases)

Spodosols are mineral soils that have a spodic horizon, a subsurface horizon with an accumulation of organic matter and of oxides of aluminum with or without iron oxides (Plate I and Figure 3.14). This illuvial horizon, which is not very thick, usually occurs under an eluvial horizon, normally an albic horizon, which is light in color and, therefore, has been described as "wood ash."

Spodosols form mostly on coarse-texture, acid, parent materials subject to ready leaching. They occur only in moist to wet areas and are most common where it is cold or temperate (Figure 3.13). Forests are the natural vegetation under which most of these soils have developed.

Species low in metallic ion contents, such as pine trees, seem to encourage the development of Spodosols. As the litter from these low-base species decomposes, strong acidity develops. Percolating water leaches acids down the profile, and the upper horizons succumb to this intense acid leaching. Most minerals except quartz are removed. In the lower horizons, oxides of aluminum and iron as well as organic matter precipitate, thus yielding the interesting Spodosol profiles.

Many of the soils in the northeastern United States, as well as those of northern Michigan and Wisconsin, belong to this order (Figures 3.11 and 3.14). Most of them are Orthods, the "common" Spodosols described above. Some, however, are Aquods because they are seasonally saturated with water and possess characteristics associated with this wetness. Important areas of Aquods occur in Florida.

The area of Spodosols in the Northeast extends up into Canada. Other large areas of this soil are found in Northern Europe and Siberia. Smaller but important areas occur in the southern part of South America and in the cool mountainous areas of temperate regions.

Spodosols are not naturally fertile. When properly fertilized, however, these soils can become quite productive. For example, the productive po-tato" soils of northern Maine are Spodosols, as are some of the vegetable- and fruit-producing soils of Florida, Michigan, and Wisconsin. Even so, the low native fertility of most Spodosols generally makes them uncompetitive for tilled crops. They are covered mostly with forests, the vegetation under which they originally developed.

HISTOSOLS (organic soils)

Histosols are characterized by high organic carbon contents (Plate I). All of them must have at least 12% organic carbon (about 20% organic matter). This minimum carbon content for Histosols rises in soils with appreciable clay content and must be at least 18% (30% organic matter) in soils with 50% or more of clay.

Histosols are typically formed in a water-saturated environment such as that found in a peat bog. They can form in any climate from the equator to permafrost areas so long as water is available. In uncultivated areas, the organic matter retains much of the original plant tissue form. Upon drainage and cultivation of the area, the original plant tissue form tends to disappear as decomposition takes place.

Less progress has been made on the classification of Histosols than of the other orders. However, four suborders have been established and are being tested in the field (see Table 3.6). When artificially drained, these soils are of great practical importance in local areas, being among the most productive, especially for vegetable crops. Their characteristics are given more detailed consideration in Chapter 10.

ANDISOLS [10]

The Andisols order specifically includes soils developed in volcanic ejecta. Since these materials have been deposited in recent geological time, they have not been highly weathered. The colloidal fraction of at least the upper 35 cm of Andisols is dominated by the silicate minerals allophane and imogolite and /or aluminum–humus complexes. The upper layers are char-acteristically dark in color and have low bulk densities.

The minerals in the volcanic materials have weathered in place, producing amorphous or poorly crystallized minerals such as allophane, imogolite, and ferrihydrite, but no downward translocation of these colloids has taken place. Andisols were previously classified as suborders of the Inceptisols order, which implies their minimum profile development.

Andisols are found in areas where significant depths of volcanic ash and other ejecta have accumulated. They occur in some very productive wheat-

---

[10] Since Andisols were only recently established as a soil order, specific information on their classification and utilization will not be presented in this text.

growing soils of Washington, Idaho, and Oregon (see Figure 3.11). Other productive Andisols are found in Japan, Ecuador, Colombia, and Indonesia where volcanic eruptions have deposited pertinent materials.

KEYS TO SOIL ORDERS

A simple key to soil orders in *Soil Taxonomy* is shown in Figure 3.15. This key, as well as a more detailed key in Appendix A, helps illustrate the relationships among the soil orders. The keys should be studied carefully.

The approximate land area for each of the eleven soil orders in the United States and the world is shown in Table 3.5. Note that the percentages of Mollisols and Inceptisols are higher than the others in the United States, whereas the percent of land area on which Aridisols and Alfisols predominate is higher worldwide.

3.8

Soil Suborders, Great Groups, and Subgroups

SUBORDERS

The eleven orders just described are subdivided into about 47 suborders, as shown in Table 3.6. The characteristics used as a basis for subdividing into the suborders are those that give the class the greatest genetic homogeneity. Thus, soils formed under wet conditions generally are identified under

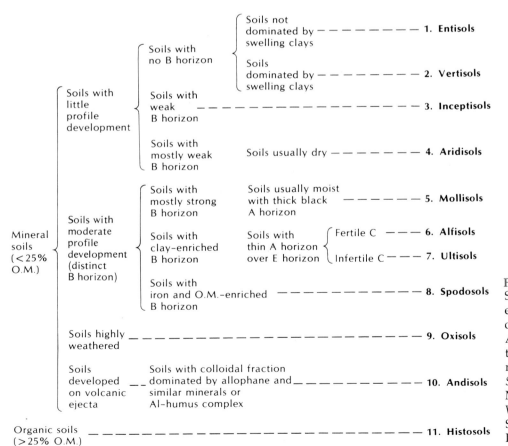

FIGURE 3.15
Simplified key to help differentiate among the eleven soil orders of *Soil Taxonomy* (see Appendix A for a more detailed key). [Reprinted by permission from *Soil Science Simplified*, Second Edition, by M. Harpstead, F. D. Hole, and W. F. Bennett, © 1988 by Iowa State University Press, Ames, Iowa 50010.]

TABLE 3.6
Soil Orders and Suborders in *Soil Taxonomy*
Note that the ending of the suborder names identifies the order in which the soils are found.

Order	Suborder	Order	Suborder	Order	Suborder
Entisols	Aquents	Alfisols	Aqualfs	Aridisols	Argids
	Arents		Boralfs		Orthids
	Fluvents		Udalfs		
	Orthents		Ustalfs	Spodosols	Aquods
	Psamments		Xeralfs		Ferrods
					Humods
Inceptisols	Andepts[a]	Ultisols	Aquults		Orthods
	Aquepts		Humults		
	Ochrepts		Udults	Histosols	Fibrists
	Plaggepts		Ustults		Hemists
	Tropepts		Xerults		Saprists
	Umbrepts				Folists
		Oxisols	Aquox		
Mollisols	Albolls		Perox	Andisols	Aquands
	Borolls		Torrox		Cryands
	Rendolls		Udox		Torrands
	Udolls		Ustox		Udands
	Ustolls				Ustands
	Xerolls				Xerands
		Vertisols	Torrerts		Vitrands
			Uderts		
			Usterts		
			Xererts		

[a] Many Andepts would now be included in the Andisols order.

separate suborders (e.g., Aquents, Aquerts, Aquepts), as are the drier soils (e.g., Ustalfs, Ustults). This arrangement also provides a convenient device for grouping soils outside the classification system (e.g., the *wet* and *dry* soils).

To determine the relationship between suborder names and soil characteristics, refer to Table 3.7. Here the formative elements for suborder names are identified, and their connotations given. Thus, the *Borolls* (from the Greek *boreas*, northern) are cool Mollisols. Likewise, soils in the *Udults* suborder (from the Latin *udus*, humid) are moist Ultisols. Identification of the primary characteristics of each of the other suborders can be made by cross reference to Tables 3.6 and 3.7.

GREAT GROUPS

The *great groups* are subdivisions of suborders. They are defined largely by the presence or absence of diagnostic horizons and the arrangements of those horizons. These horizon designations are included in the list of formative elements for the names of great groups shown in Table 3.8. Note that these formative elements refer to epipedons such as umbric and ochric (see Table 3.2), to subsurface horizons such as argillic and natric, and to pans such as duripan and fragipan.

Remember that the great group names are made up of these formative elements attached as prefixes to the names of suborders in which the great groups occur. Thus, a *Ustoll* with a *natric* horizon (high in sodium) belongs to the *Natrustolls* great group. As might be expected, the number of great groups is high, more than 230 having been identified.

The names of selected great groups from three orders are given in Table 3.9. This list illustrates again the usefulness of *Soil Taxonomy*, especially the nomenclature it employs. The names identify the suborder and order in

TABLE 3.7
Formative Elements in Names of Suborders in *Soil Taxonomy*

Formative element	Derivation	Connotation of formative element
alb	L. *albus*, white	Presence of albic horizon (a bleached eluvial horizon)
and	Modified from Ando	Ando-like
aqu	L. *aqua*, water	Characteristics associated with wetness
ar	L. *arare*, to plow	Mixed horizons
arg	L. *argilla*, white clay	Presence of argillic horizon (a horizon with illuvial clay)
bor	Gk. *boreas*, northern	Cool
cry	Gk. *kruos*, icy cold	Cold
ferr	L. *ferrum*, iron	Presence of iron
fibr	L. *fibra*, fiber	Least decomposed stage
fluv	L. *fluvius*, river	Floodplains
fol	L. *folia*, leaf	Mass of leaves
hem	Gk. *hemi*, half	Intermediate stage of decomposition
hum	L. *humus*, earth	Presence of organic matter
ochr	Gk. base of *ochros*, pale	Presence of ochric epipedon (a light surface)
orth	Gk. *orthos*, true	The common ones
perud	Continuously humid	Of year-round humid climates
plagg	Modified from Ger. *plaggen*, sod	Presence of plaggen epipedon
psamm	Gk. *psammos*, sand	Sand textures
rend	Modified from Rendzina	Rendzina-like
sapr	Gk. *sapros*, rotten	Most decomposed stage
torr	L *torridus*, hot and dry	Usually dry
ud	L. *udus*, humid	Of humid climates
umbr	L. *umbra*, shade	Presence of umbric epipedon (a dark surface)
ust	L. *ustus*, burnt	Of dry climates, usually hot in summer
xer	Gk. *xeros*, dry	Annual dry season

TABLE 3.8
Formative Elements for Names of Great Groups and Their Connotation

These formative elements combined with the appropriate suborder names give the great group names.

Formative element	Connotation	Formative element	Connotation	Formative element	Connotation
acr	Extreme weathering	gibbs	Gibbsite	psamm	Sand texture
agr	Agric horizon	gyps	Gypsic horizon	quartz	High quartz
alb	Albic horizon	gloss	Tongued	rhod	Dark red colors
and	Ando-like	hal	Salty	sal	Salic horizon
arg	Argillic horizon	hapl	Minimum horizon	sider	Free iron oxides
bor	Cool	hum	Humus	sombr	Dark horizon
calc	Calcic horizon	hydr	Water	sphagn	Sphagnum moss
camb	Cambic horizon	kand	Low activity clay	sulf	Sulfur
chrom	High chroma	luv, lu	Illuvial	torr	Usually dry
cry	Cold	med	Temperate climates	trop	Continually warm
dur	Duripan	nadur	See *natr* and *dur*	ud	Humid climates
dystr, dys	Low base saturation	natr	Natric horizon	umbr	Umbric epipedon
eutr, eu	High base saturation	ochr	Ochric epipedon	ust	Dry climate, usually hot in summer
ferr	Iron	pale	Old development		
fluv	Floodplain	pell	Low chroma	verm	Wormy, or mixed by animals
frag	Fragipan	plac	Thin pan		
fragloss	See *frag* and *gloss*	plagg	Plaggen horizon	vitr	Glass
		plinth	Plinthite	xer	Annual dry season

Plate I

Photographs of profiles of ten of the soil orders in *Soil Taxonomy*.

1. Alfisols—an Aeric Ochraqualf from western New York.

2. Aridisols—a Typic Camborthid from western Nevada.

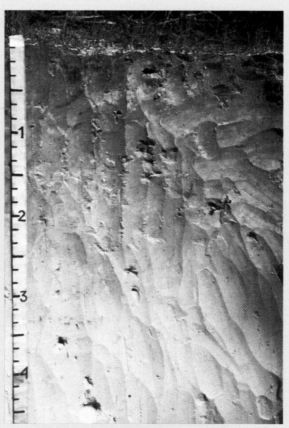

3. Entisols—a Typic Quartzipsamment from eastern Texas.

4. Histosols—a Limnic Medisaprist from southern Michigan.

5. Inceptisols—a Typic Dystrochrept from West Virginia.

6. Mollisols—a Typic Hapludoll from Rio de Janeiro, Brazil.

7. Oxisols—a Tropeptic Haplorthox from central Puerto Rico.

8. Spodosols—a Typic Haplorthod
from northern New York.

9. Ultisols—a Typic Hapludult
from western Arkansas.

10. Vertisols—a Typic Pellustert
from Queensland, Australia.

Plate II

Landsat satellites have provided valuable remote sensing information on the earth's soils and other natural resources.

11. In this photo of an area in Tippecanoe County, Indiana, a false color composite of three spectral bands (two visible, one infrared) is used to show wooded areas (red), cultivated soils developed under prairie vegetation (dark), and cultivated soils developed under forest vegetation (light areas).

12. This photo shows an area along the Chester River in Queen Annes County, Maryland. The modern thematic mapper (TM) scanner used provided reasonably good detail, and a false color composite of three spectral bands showing wooded areas in brown and red and areas with little soil cover in green and blue. Soil differences are obvious in some fields.

Photo 1 courtesy Bill Waltman and D. J. Lathwell, Cornell University; photos 2, 3, 6, and 10 courtesy A. C. Orvedal; photos 4, 5, and 7–9 courtesy Soil Science Society of America; photo 11 courtesy M. F. Baumgardner. Purdue University; photo 12 courtesy Goddard Space Flight Center, NASA.

TABLE 3.9

Examples of Names of Great Groups of Selected Suborders of the Mollisol, Alfisol, and Ultisol Orders

The suborder name is identified as the italicized portion of the great group name.

	Dominant feature of great group		
	Argillic horizon	*Minimum horizon development*	*Old land surfaces*
Mollisols			
1. Aquolls (wet)	Argi*aquolls*	Hapl*aquolls*	—
2. Udolls (moist)	Argi*udolls*	Hapl*udolls*	Pale*udolls*
3. Ustolls (dry)	Argi*ustolls*	Hapl*ustolls*	Pale*ustolls*
4. Xerolls (med.)[a]	Argi*xerolls*	Hapl*oxerolls*	Pal*exerolls*
Alfisols			
1. Aqualfs (wet)	—		Pale*udalfs*
2. Udalfs (moist)	Arg*udalfs*	Hapl*udalfs*	Pale*udalfs*
3. Ustalfs (dry)	—	Hapl*ustalfs*	Pale*ustalfs*
4. Xeralfs (med.)[a]	—	Hapl*oxeralfs*	Pal*exeralfs*
Ultisols			
1. Aqults (wet)	—	—	Pale*aqults*
2. Udults (moist)	—	Hapl*udults*	Pale*udults*
3. Ustults (dry)	—	Hapl*ustults*	Pale*ustults*
4. Xerults (Med.)[a]	—	Hapl*oxerults*	Pal*exerults*

[a] Med. = Mediterranean climate; distinct dry period in summer.

which the great groups are found. Thus, Argiudolls are Mollisols of the Udolls suborder characterized by an argillic horizon. Cross reference to Table 3.8 identifies the specific characteristics separating the great group classes from each other. Careful study of these two tables will show the utility of this classification system.

SUBGROUPS

Subgroups are subdivisions of great groups. There are more than 1200 subgroups recognized, some 1000 of which are in the United States. Subgroups permit characterization of the core concept of a given great group and of gradations from that central concept to other units of the classification system. The subgroup most nearly representing the central concept of a great group is termed *Typic*. Thus, the *Typic Hapludolls* subgroup typifies the Hapludolls great group. A Hapludoll with restricted drainage would be classed as an *Aquic Hapludoll*. One with evidence of intense animal (earthworm) activity would fall in the *Vermic Hapludolls* subgroup.

Some intergrades may have properties in common with other orders or with other great groups. Thus, the *Entic Hapludolls* subgroup intergrades toward the Entisols order. The subgroup concept illustrates very well the flexibility of this classification system.

3.9

Soil Families and Series

FAMILIES

The family category of classification is based on properties important to the growth of plant roots. The criteria used include broad classes of *particle size, mineralogy, temperature,* and depth of the soil penetrable by roots. Table 3.10 gives an example of the classes used (a more complete listing is given in Appendix B). Terms such as loamy, sandy, and clayey are used to identify the broad textural classes. Terms used to describe the mineralogical classes include smectitic, kaolinitic, siliceous, carbonatic, and mixed. For temperature classes, terms such as frigid, mesic, and thermic are used.

TABLE 3.10

Commonly Used Examples of Particle Size, Mineralogy, and Soil Temperature Used to Differentiate Families

Partical-size class	Mineralogical class	Soil temperature class	
			Mean annual temperature (°C)
Fragmental	Carbonatic	Frigid	<8
Sandy	Micaceous	Mesic	8–15
Loamy	Siliceous	Isomesic[b]	8–15
Fine-loamy	Kaolinitic	Thermic	15–22
Loamy skeletal[a]	Smectitic	Hypothermic	>22
Clayey	Oxidic		
	Mixed		

[a] Skeletal refers to presence of up to 35% rock fragments.
[b] "Iso" prefix refers to soils where the difference between summer and winter temperatures are less than 5 °C; in other soil temperature classes, ths difference is greater than 5 °C.

Thus, a Typic Argiudoll from Iowa, loamy in texture, having a mixture of clay minerals and with annual soil temperatures (at 50 cm depth) between 8 and 15 °C, is classed in the *Typic Argiudolls loamy, mixed, mesic* family. In contrast, a sandy-textured Typic Cryorthod, high in quartz and located in a cold area in eastern Canada, is classed in the *Typic Cryorthods sandy, siliceous, frigid* family.

The terms *shallow* and *micro* are sometimes used at the family level to indicate unusual soil depths. About 6600 families have been identified in the United States.

SERIES

The families are subdivided into the lowest category of the system called *series.* One series may be differentiated from another in a family through differences in one or more but not necessarily all of the soil properties. Series are, of course, established on the basis of profile characteristics. This requires a careful study of the various horizons as to number, order, thickness, texture, structure, color, organic content, and reaction (acid, neutral, or alkaline). Features such as hardpan at a certain distance below the surface, a distinct zone of calcium carbonate accumulation at a certain depth, or striking color characteristics greatly aid in series identification.

In the United States, each series is given a name, usually from some city, village, river, or county, such as Fargo, Muscatine, Cecil, Mohave, and Ontario. There are more than 16,800 soil series in the United States.

The complete classification of a Mollisol, the Brookston series, is given in Figure 3.16. This figure illustrates how *Soil Taxonomy* can be used to show the relationship between *the* soil, a comprehensive term covering all soils, and *a* specific soil series. The figure deserves study.

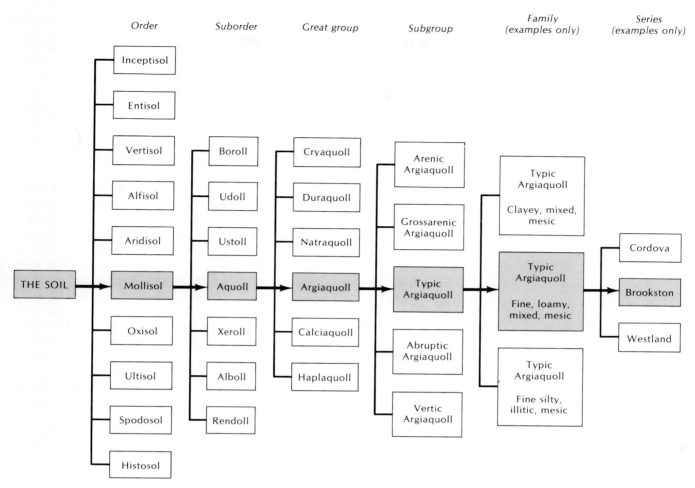

FIGURE 3.16
Diagram illustrating how one soil (Brookston) fits into the overall classification scheme. The shaded boxes show that this soil is of the Mollisols order, Aquolls suborder, Argiaquolls great group, and so on. In each category other classification units are shown.

3.10

Soil Phases, Associations, and Catenas

Soil Taxonomy is used as the basis for preparing a soil survey on which maps showing the kinds of soils present in a given area are based. Because local features and requirements will dictate the nature of these maps and, in turn, the specific soil units that are mapped, the field *mapping units* may be somewhat different from the *classification units* found in *Soil Taxonomy*. Furthermore, the mapping units may include some further differentiation of soil series or associations of two or more series as they appear in the field. Examples of such soil units follow.

SOIL PHASES

Although technically not included as a class in *Soil Taxonomy*, a *phase* is a subdivision based on some important deviation that influences the use of the soil, such as surface texture, erosion, slope, stoniness, or soluble salt content. A phase marks a departure from the normal or key classification or mapping unit already established. Thus, a Cecil sandy loam, 3–5% slope, and a Hagerstown silt loam, stony phase, are examples of soils where distinctions are made with respect to the phase.

Please note that a phase can be used at any categoric level in the system. For example, at the order level one can have a shallow Mollisol; at the suborder level, an eroded Ustoll; at the great group level, a loamy Haplustoll, and so on.

SOIL ASSOCIATIONS, COMPLEXES, AND CATENAS

A grouping of soils of different kinds, which commonly occur together in the field, is known as an *association*. The soils may be from the same soil order or they may be from different orders (see Figure 3.17). Thus, a shallow Entisol on steep uplands may be found alongside a well-drained Alfisol or a soil developed from recent alluvium. The only requirement is that the soils occur together in the same area. Soil associations are important in a practical way, since they help determine combinations of land-use patterns that must be used to support profitable agriculture.

When the soil pattern is intricate, as when contrasting soils occur adjacent to each other, delineating each kind of soil on a soil map becomes difficult if not impossible. In such cases, a soil *complex* is indicated on a soil map and an explanation of the soils present in the complex is contained in the soil survey report.

Well-drained, imperfectly drained, and poorly drained soils, all of which have developed from the same parent materials, are sometimes found

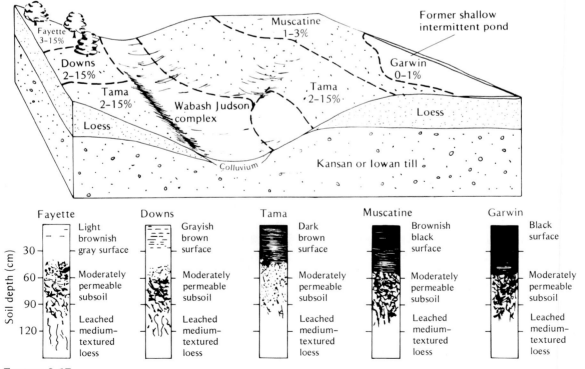

FIGURE 3.17

Association of soils in Iowa. Note the relationship of soil type to (1) parent material, (2) vegetation, (3) topography, and (4) drainage. Two Alfisols (Fayette and Downs) and three Mollisols (Tama, Muscatine, and Garwin) are shown. [From Riecken and Smith (1949)]

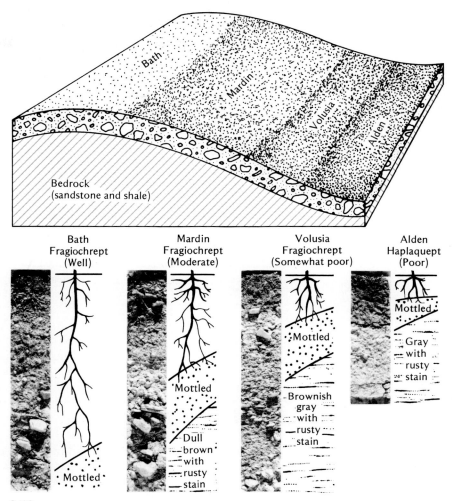

FIGURE 3.18

Monoliths showing four soils of a drainage catena (below) and a diagram showing their topographic association in the field (above). Note the decrease in the depth of the well-aerated zone (above the mottled layers) from the Bath soil (left) to the Alden (right), which remains poorly aerated throughout the growing season. These soils are all developed from the same parent material and differ only in drainage and topography. The Volusia soil as pictured was cultivated, while the others were located on virgin sites. Note that all four soils belong to the Inceptisols order.

together under field conditions. This relationship on the basis of drainage or of differences in relief is known as a *catena* and is helpful in relating the soils to the landscape in a given region. The relationship can be seen by referring to Figure 3.18, where the Bath-Mardin-Volusia-Alden catena is shown. Although all four or five members of a catena are not always found together in a given area, the diagram illustrates their relationship in respect to their drainage status.

The practical significance of the soil catena can be seen in Figure 3.19, where the tolerance of several crops to differences in soil drainage is shown. Recommendations for crop selection and management are often made on the basis of soil drainage class. Such recommendations, along with careful study of farm soil survey maps, are invaluable to farmers in making the best choices of soil and crop compatibility.

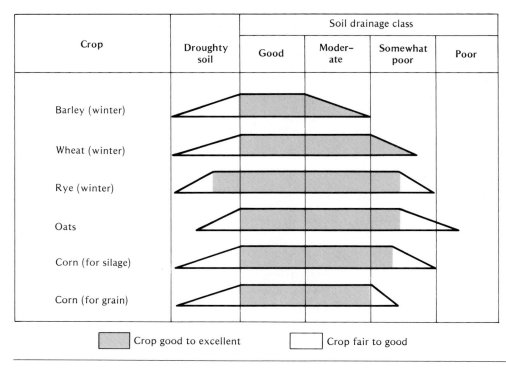

Crop	Droughty soil	Soil drainage class			
		Good	Moder-ate	Somewhat poor	Poor
Barley (winter)					
Wheat (winter)					
Rye (winter)					
Oats					
Corn (for silage)					
Corn (for grain)					

▨ Crop good to excellent   ☐ Crop fair to good

FIGURE 3.19
Relationship between soil drainage class and the growth of certain grain crops. [From *Cornell Recommends* (Anon., 1962)]

3.11

Soil Survey and Its Utilization

The classifications as outlined in the preceding pages are used in soil survey, which functions to classify, locate on a base map, and describe the nature of soils as they occur in the field. Soils in the United States are classified into *series* and *phases* on the basis of their profile characteristics. Because the work of the soil surveyor is localized, it is concerned mostly with series and phase separations, giving the broader regional distinctions only general consideration.

FIELD MAPPING

As the survey of any area progresses, the location of each soil unit is shown on a suitable base map, such as an aerial photograph (Figure 3.20). The surveyor can quickly locate a position on the map and indicate readily the soil boundaries. Using this procedure, a field map is obtained that gives not only the soil series but also the phase, which furnishes information as to slope and severity of erosion. Thus, the map becomes of even greater practical value.

The field map, after the separations have been carefully checked and correlated, is now ready for reproduction. Older maps were reproduced in color. More recent maps are reproductions of the actual aerial photograph with the appropriate soil boundaries drawn in.

When published, the map is accompanied by a bulletin containing a description of the topography, climate, agriculture, and soils of the area under consideration. Each soil series is minutely described as to profile, and suggestions for practical management for agricultural and nonagricultural uses are usually made. Such a field map is shown in Figure 3.20; by examination of the map, the relationship of topography to soil associations can be visualized.

FIGURE 3.20
Soil surveys are prepared by soil scientists, who examine soils in the field using a soil auger and other diagnostic tools (upper left). Mapping units are outlined on an aerial photograph, first in the field and then more permanently in the map room (upper right). The final map shows soil boundaries (lower) and indicates the soil name (first two letters), the slope (second capital letter), and the degree of erosion (the number). Thus, OsC2 is an Orwood silt loam (Os), with a 5–9% slope (C), and is moderately eroded. [Courtesy USDA Soil Conservation Service, Soil Survey Staff]

MODERN TECHNOLOGIES[11]

New techniques are being used to improve the quality of soil surveys and to update older surveys. These include video image analysis (VIA) by which aerial photographs are examined to distinguish up to 256 shades of gray (compared to only 32 for the human eye). These color differences are related to soil and vegetation variations. Using appropriate computer techniques, soil differences can be more clearly delineated (Figure 3.21).

Ground-penetrating radar (GPR) is also used to increase the quality and reduce the cost of soil surveys. High frequency impulses of energy are transmitted into the soil. Energy is reflected back to the surface when the

[11] Some of these techniques are discussed in Reybold and Petersen (1987).

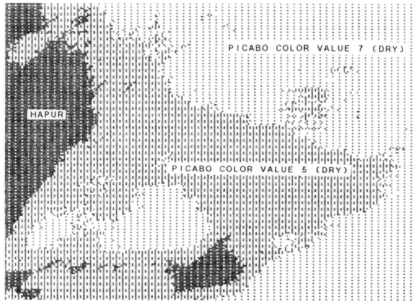

FIGURE 3.21
Illustrations of the video image analysis technique to assist soil mapping. An aerial photograph of an area in Idaho (upper) shows color shades that are related to soil differences. A video scan method was used to differentiate these soil differences (lower). The technique was checked with on-ground observations. [Photos from Harrison et al. (1987)]

impulse strikes an interface between soil particles. This reflected energy is measured and displayed on a recorder. An illustration of the field equipment used and of the resulting graphic profile is shown in Figure 3.22. This technique is not suitable for all soils since the reflectance is influenced by factors such as moisture, salt content, and type of clay. Where it is effective, however, its cost is only about one third that of the conventional soil survey.

USE OF SOIL SURVEYS

Soil survey maps and bulletins are useful in many ways. Scientists use them to facilitate research on crop production, land evaluation and zoning, and human settlements. Extension specialists and county agricultural agents often base their recommendations for land use and crop management on soil surveys and on interpretive bulletins based on them. Engineers and hydrol-

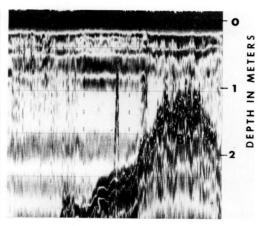

FIGURE 3.22

The ground penetration radar technique in use in the field (left), and the resulting graphic profile of an area in Florida (right) showing the depth of a Bt horizon that characterizes an Udults soil. [Photos from Doolittle (1987)]

ogists can select superior roadbeds and building sites and make estimates of water runoff and infiltration using soil survey maps and reports. Home-owners can use the reports as a basis for selecting trees and shrubs for landscaping.

The reports contain interpretive tables indicating limitations for soil use as cropland, pasture, woodland, wildlife habitat, community development, sewage disposal, recreation and engineering uses. Table 3.11 illustrates how soil surveys can be used to determine land suitability for agricultural and other purposes.

The soil survey is perhaps of greatest practical value in land classification for agricultural and other uses. This subject will now receive our attention.

3.12

Land Capability Classification

At this point, it may be well to note the difference between *soil* and *land*. Soil is the more restrictive term, referring to the natural bodies we are studying without regard to their current vegetation and use. Land is a broader term that includes not only the soil but also other physical attributes, such as water supply, existing plant cover, and location with respect to cities and means of transportation. Thus, we have *forestland, bottomland*, and *grassland*, which may include a variety of soils.

Soil survey maps and reports have become two of the bases for systems of land capability or suitability classification. These systems are based on the assumption that every hectare of land should be used in accordance with its *capability* and *limitations*. Land is classified according to the most suitable sustained use that can be made of it while providing for adequate protection from erosion or other means of deterioration.

Several land-use classification systems have been developed around the world. Some relate to agricultural use and others to forestry, recreation, wildlife use, building construction and other engineering purposes, and sewage disposal. One such land-use classification system has been developed by the USDA Soil Conservation Service. Although its primary use is for agriculture, it can be applied for other purposes.

Under the system set up by the U.S. Soil Conservation Service, eight land capability classes are recognized. These classes are numbered from I to VIII. Soils having greatest capabilities for response to management and least limitations in the ways they can be used are in Class I. Those with least capabilities and greatest limitations are found in Class VIII. This classification system is used most in soil conservation programs and will be discussed further in Section 15.16.

3.13

Soil Interpretations for Nonfarm Uses

Soil surveys also can be used to identify the potential for nonagricultural land. Examples of such soil interpretation are suitability for building sites, waste disposal, highway beds, woodlands and recreational sites, and wildlife. It is obvious that soil surveys are important tools in regional and local land-use planning. Table 3.11 illustrates the suitability of selected soil areas for both farm and nonfarm uses. Note that for some soils (e.g., Honeoye and Madrid) high quality for crop production and for nonagricultural uses are closely correlated.

TABLE 3.11

Suitability of Different Soil Areas for Production of Three Crops and for Nonfarm Uses

Based on number of 2-ha parcels per square mile with only moderate problems: >15 = good; 5–15 = fair; <5 = poor.

Soil series	Corn	Vegetables	Alfalfa	Buildings with basements	Septic tank filter fields	Summer camp sites
Honeoye	Good	Good	Good	Good	Good	Good
Madrid	Good	Good	Good	Good	Good	Good
Bath	Fair	Poor	Fair	Good	Good	Good
Volusia	Poor	Poor	Poor	Fair	Poor	Good
Chenango	Fair	Poor	Good	Good	Good	Good
Erie	Poor	Poor	Poor	Poor	Poor	Good
Ovid	Fair	Poor	Fair	Fair	Poor	Fair

[a] Modified from Cline and Marshall (1977).

Other soils, however (e.g., Bath and Chenango), rate higher for nonagricultural than for agricultural uses. This is not surprising since chemical and physical properties needed for good crop production may be quite different from those needed for waste disposal, for example. The high permeability which is often desired for waste disposal may be associated with droughtiness for crop production. These soil interpretations illustrate the usefulness of soil surveys to farmers and nonfarmers alike.

3.14

Conclusion

It is fortunate scientists have recognized that the soil which covers the earth is in fact comprised of a large number of individual soils, each with distinctive properties. Among the most important of these properties are those associated with the horizontal layers or horizons found in a soil profile. These horizons reflect the physical, chemical, and biological processes that have stimulated the genesis of individual soils and are among the most significant properties that influence how soils can and should be used.

Knowledge of the kinds and properties of soils around the world is critical to humanity's struggle for survival and well being. A soil classification system based on these properties is equally critical if we expect to use knowledge gained at one location to solve problems at other locations where similarly classed soils are found. *Soil Taxonomy*, a classification system based on known soil properties helps fill this need in some 45 countries. It is indeed worthy of study.

Study Questions

1. Assume you are traveling by car north from Georgia through Pennsylvania and New York and then on to Maine. What major changes in soil properties would you expect to encounter and to which one or more of the five major factors influencing soil formation would these properties be due? Explain. Give the same explanations if you traveled from Georgia westward to Texas and then on to Arizona.
2. The soils in glaciated areas are generally less well developed than nearby soils developed on materials not affected by the glaciation. What is the reason for this?
3. Explain how the three major processes involved in soil formation participate in the development of (a) an argillic (clay-containing) B horizon found in some soils of humid regions and (b) a Bk (carbonate concentration) horizon common in soils of semiarid and arid regions.
4. Differentiate between *a* soil and *the* soil.
5. What are the two main features of *Soil Taxonomy*? How do these assure that this system can be used throughout the world?
6. Of what importance are the diagnostic horizons in *Soil Taxonomy*?
7. Which of the epipedons are most likely to characterize soils with the highest native soil productivity? Explain.
8. Identify the soil order and suborder of soils in each of the following great groups.
(a) Natrustolls, (b) Hapludalfs, (c) Calciorthids, (d) Psammaquents, (e) Tropaquepts, (f) Paleudults, (g) Fragiaquods, (h) Haplorthox.
9. What does the name of the family—loamy, mixed, mesic Typic Ustorthents—tell you about the general properties of the soil?
10. Which of the orders of *Soil Taxonomy* would you expect to be most prominent in the soils of the following locations and why?
(a) eastern Colorado, (b) eastern Canada, (c) North Carolina piedmont, (d) western Iowa, (e) Pennsylvania, (f) wetlands of Michigan, (g) central Brazil, (h) central Oregon, (i) southern Arizona, (j) crack-forming, sticky clay areas of India.
11. Soils of which of the soil orders would you be *least* likely to select for the following?
(a) roadbed for a highway; (b) production of a crop such as alfalfa, which requires adequate water and a deep soil with a pH near neutral; (c) a shallow soil; (d) a soil low in iron oxides; (e) a soil low in organic matter in the A and B horizons; (f) a soil whose profile shows well-developed horizons; (g) a soil with adequate moisture for crop production.
12. If the land-use capability map made for a farm showed that land Class I predominated, what constraints should you place on the use of that land and why? What would be your answer if land Class VIII were predominant? Does this suggest that the *economic* value of land Class VIII is lower than that of land Class I example? Explain.
13. The Volusia soil (Table 3.11) is rated good as a base for a summer campsite (no buildings) but only fair as a site for a building foundation and poor for a septic tank filter field. What does this tell you about the properties of that soil?
14. Which section of this chapter (aside from the questions) was most useful to you? (Note this question is added in case your instructor is superstitious about the number 13).

References

ANONYMOUS. 1962. *Cornell Recommends*, an annual publication of the New York State College of Agriculture, Cornell University.

CLINE, M. G., and R. L. MARSHALL. 1977. *Soils of New York Landscapes*, Inf. Bull. 119, Physical Sciences, Agronomy 6, New York State College of Agriculture and Life Sciences, Cornell University.

DOOLITTLE, J. A. 1987. "Using ground-penetrating radar to increase the quality and efficiency of soil surveys," in W. U. Reybold and G. W. Peterson (Eds.), *Soil Survey Techniques*, SSSA Spec. Publ. No. 20 (Madison, WI: Soil Sci. Soc. Amer.).

HARPSTEAD, M. I., F. D. HOLE, and W. F. BENNETT. 1988. *Soil Science Simplified* (Ames, IA: Iowa State Univ. Press).

HARRISON, W. D., M. E. JOHNSON, and P. F. BIGGAM. 1987. "Video image analysis of large-scale vertical aerial photography to facilitate soil mapping," in W. U. Reybold and G. W. Peterson (Eds.), *Soil Survey Techniques*, SSSA Spec. Publ. No. 20 (Madison, WI: Soil Sci. Soc. Amer.).

McCRACKEN, R. J., et al. 1985. "An appraisal of soil resources in the USA," in R. F. Follett and B. A. Stewart (Eds.), *Soil Erosion and Crop Productivity* (Madison, WI: Amer. Soc. Agron.).

NSF. 1975. "All that unplowed land," *Mosaic*, **6**:17–21 (Washington, DC: National Science Foundation).

REYBOLD, W. U., and G. W. PETERSEN (Eds.). 1987. *Soil Survey Techniques*, SSSA Spec. Publ. No. 20 (Madison, WI: Soil Sci. Soc. Amer.).

RIECKEN, F. F., and G. D. SMITH. 1949. "Principal upland soils of Iowa, their occurrence and important properties," *Agron.* **49** revised, Iowa Agr. Exp. Sta.

SSSA. 1984. *Soil Taxonomy, Achievements and Challenges*, SSSA Spec. Publ. No. 14 (Madison, WI: Soil Sci. Soc. Amer.).

Soil Survey Staff. 1975. *Soil Taxonomy: A Basic System of Soil Classification for Making and Interpreting Soil Surveys* (Washington, DC: USDA Soil Conservation Service).

USDA. 1981. *Soil, Water and Related Resources in the United States*, 1980 Appraisal Part II, Soil and Water Resources Conservation Act (Washington, DC: U.S. Department of Agriculture).

Physical Properties of Mineral Soils

The physical properties of soils are extremely important in determining how soils can and should be used. They range from properties that determine a soil's suitability for a foundation for a building or a roadbed to its suitability for the production of different crop plants. Physical properties concern not only soil solids but soil water and air as well. In fact, through their influence on water movement through and off soils, physical properties also exert considerable control over the destruction of the soil itself by erosion.

A mineral soil is a physical mixture of inorganic particles, decaying organic matter, air, and water. The larger mineral fragments usually are embedded in and coated with clay and other colloidal materials. Where the larger mineral particles predominate, the soil is gravelly or sandy; where the mineral colloids are dominant, the soil is clay-like. All gradations between these extremes are found in nature. Organic matter acts as a binding agent between individual particles, thereby encouraging the formation of clumps of soil or aggregates.

Among the important physical properties of soils to be considered in this chapter are *soil texture* and *soil structure*. Soil texture involves the *size* of individual mineral particles and specifically refers to the relative proportions of various-sized particles in a given soil.

No less important is soil struture, which is the *arrangement* of soil particles into groups or aggregates. Together, these properties help determine not only the nutrient-supplying ability of soil solids but also the supply of water and air necessary for plant root activity.

4.1

Soil Texture (Size Distribution of Soil Particles)

The size of particles in mineral soil is not readily subject to change. Thus, a sandy soil remains sandy, and a clay soil remains clayey. Since the proportion of each size group in a given soil (the texture) cannot be easily altered, it is considered a basic property of a soil.

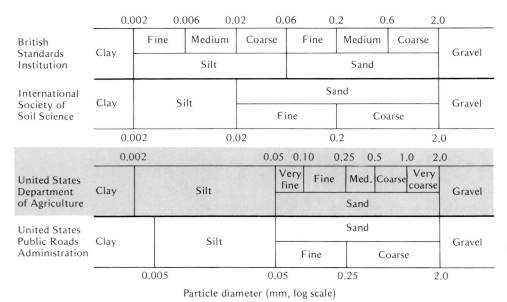

FIGURE 4.1

Classification of soil particles according to size by four systems. The U.S. Department of Agriculture system is used in this book.

To study the mineral particles of a soil, scientists separate them into groups according to size. The groups are referred to as *separates*. . The analytical procedure by which the particles are separated is called *particle-size analysis*, the determination of the particle-size distribution.

A number of different classifications have been devised. The size ranges for four of these systems are shown in Figure 4.1. The classification established by the U.S. Department of Agriculture is used in this text.

PARTICLE-SIZE ANALYSIS

A particle-size analysis is done by using sieves to mechanically separate out the very fine sand and larger separates from the finer particles. Then the weight of each separate is measured. The silt and clay contents are then determined by measuring the rate of settling of these two separates from suspension in water.

The principle involved is simple. When soil particles are suspended in water, they tend to sink. Because there is little variation in the density of most soil particles, their velocity (v) of settling is proportional to the square of the radius (r) of each particle. Thus, $v = kr^2$, where k is a constant. This equation is referred to as Stokes's law.

With knowledge of the velocity of settling, Stokes's law can be used to calculate the radius of the particles as they settle and the percentage of each size fraction in the sample. These percentages are used to identify the soil textural class, such as sand, silt, or loam.

Although stone and gravel are considered in the practical examination and evaluation of a field soil, they do not enter into the analysis of the finer particles. Their amounts are rated separately. The organic matter, comparatively small in quantity, usually is removed by oxidation before the mechanical separation.

A particle-size analysis gives a general picture of the physical properties of a soil. The analysis also is the basis for assigning each soil to a textural class. This phase is considered in more detail in Section 4.5.

4.2
Physical Nature of Soil Separates

COARSE FRAGMENTS

Fragments that range from 2 to 75 millimeters (mm) (up to 3 in.) along their greatest diameter are termed gravel or pebbles; those ranging from 75 to 250 mm (3 to 10 in.) are called cobbles (if round) or flags (if flat); and those more than 250 mm across are called stones or boulders.

SAND AND GRAVEL

Sand grains or gravel may be rounded or irregular depending on the amount of abrasion they have undergone (Figure 4.2.). Unless coated with clay and silt, such particles are not sticky even when wet. They cannot be molded as can clay and therefore are not plastic. The water-holding capacity of sand grains is low, and because of the large spaces between the separate particles, water and air pass through rapidly. Hence, soils dominated by sand or gravel possess good drainage and aeration but may be drought prone.

SILT

Silt particles are intermediate in size and properties between sand and clay particles. They are irregularly fragmental, diverse in shape, and seldom smooth or flat. Silt is essentially microsand particles, with quartz generally the dominant mineral. The silt separate, because it usually has an adhering film of clay, possesses some plasticity, cohesion (stickiness), and adsorptive capacity, but much less than the clay separate. Silt may cause the soil surface to be compact and crusty unless it is supplemented by adequate amounts of sand, clay, and organic matter.

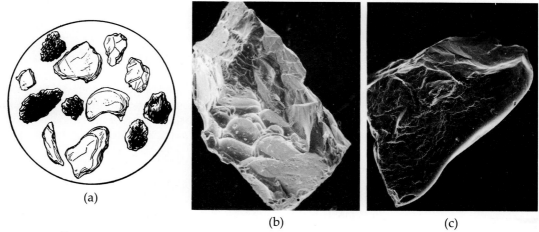

(a)   (b)   (c)

FIGURE 4.2

(a) Sand grains from soil. Note that the particles are irregular in size and shape. Quartz usually predominates, but other minerals may occur. Silt particles have about the same shape and composition, differing only in size. Scanning electron micrographs of sand grains show quartz sand (b) and a feldspar grain (c). The grains have been magnified about 40 times. [Photos courtesy J. Reed Glasmann, Union Oil Research]

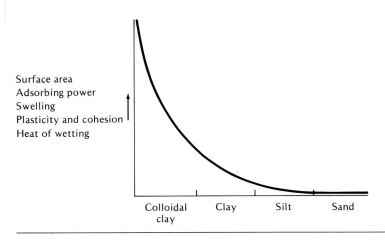

Surface area
Adsorbing power
Swelling
Plasticity and cohesion
Heat of wetting

Colloidal   Clay   Silt   Sand
clay

FIGURE 4.3
The finer the texture of a soil, the greater is the effective surface exposed by its particles. Note that adsorption, swelling, and the other physical properties cited follow the same general trend and that their intensities go up rapidly as the colloidal size is approached.

CLAY

The surface area per unit mass of clay is very high because of the small size of the individual particles. Fine colloidal clay has about 10,000 times as much surface area as the same weight of medium-sized sand. The specific surface (area per unit weight) of colloidal clay ranges from about 10 to 1000 square meters per gram (m^2/g) compared to 1 and 0.1 m^2/g for the smallest silt particle and fine sand, respectively. Since the adsorption of water, nutrients, and gas and the attraction of particles for each other are all surface phenomena, the very high specific surface of clay is significant in determining soil properties. This relationship is shown graphically in Figure 4.3.

Clay particles vary in shape from plate-like to rounded. When clay is wet, it tends to be sticky and plastic or easily molded. The presence of clay in a soil gives it a *fine texture* and slows water and air movement.

While clayey soil becomes sticky when wet, it can also be hard and cloddy when dry unless properly handled. Clay expands and contracts greatly on wetting and drying, and the water-holding capacity of soils high in clay generally is high.

4.3

Mineralogical and Chemical Compositions of Soil Separates

Although the text at this point largely emphasizes the physical properties of soil particles, the mineralogical and chemical composition of these particles is important and will be briefly surveyed.

MINERALOGICAL CHARACTERISTICS

The coarsest sand particles often are fragments of rocks as well as minerals. Quartz (SiO_2) commonly dominates sand, especially the fine sands, as well as the silt separate (Figure 4.4). In addition, differing quantities of other primary minerals, such as the various feldspars (aluminosilicates) and micas (iron and aluminum silicates), usually occur. Gibbsite (hydrous oxide of aluminum), hematite, and goethite (hydrous iron oxides) also are found, usually as coatings on the sand grains.

Some clay-size particles, especially those in the coarser clay fractions, are composed of minerals such as quartz and the hydrous oxides of iron and aluminum. The latter are particularly important in the tropics and other warmer climates. Of greater importance in temperate regions, however, are

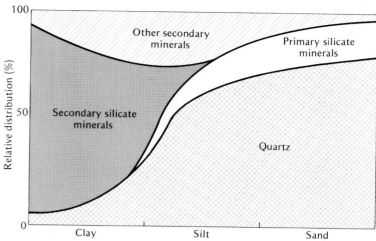

FIGURE 4.4

General relationship between particle size and kinds of minerals present. Quartz dominates the sand and coarse silt fractions. Primary silicates such as the feldspars, hornblende, and micas are present in the sands and, in decreasing amounts, in the silt fraction. Secondary silicates dominate the fine clay. Other secondary minerals, such as the oxides of iron and aluminum, are prominent in the fine silt and coarse clay fractions.

the silicate clays, which occur as plate-like particles made up of thin layers much like the pages of a book. When some of these silicate clays are wetted, water can move between the layers within a particle as well as between particles, resulting in extensive swelling of the clay. Upon drying, the process is reversed and the clay shrinks. Cracks seen in clay soils when they dry after a rain are evidence of this swelling and shrinking phenomenon. Silicate clays vary considerably in their plasticity, cohesion, and swelling/shrinking capacity.

Because of these characteristics it is important to know which clay type dominates or codominates any particular soil. The clays will be considered in greater detail in Chapter 7.

CHEMICAL MAKEUP

Since sand and silt are dominantly quartz (SiO_2) and other primary minerals known for their resistance to weathering, these two fractions have relatively low chemical activity. The primary minerals that may contain nutrient elements in their chemical makeup are generally so insoluble as to make their nutrient-supplying ability insignificant.

Chemically, the silicate clays vary widely. Some are relatively simple aluminosilicates. Others contain in their crystal structures varying quantities of iron, magnesium, potassium, and other elements. As pointed out in Section 1.14, the surfaces of all the silicate clays hold small but significant quantities of cations such as Ca^{2+}, Mg^{2+}, K^+, H^+, Na^+, NH_4^+ and Al^{3+}. The cations are exchangeable and can be released for absorption by plants.

In highly weathered soils, such as those found in the hot, humid tropics, oxides of iron and aluminum are prominent if not dominant, even in the clay size fraction. Thus, weathering can have a profound effect on the chemical and mineralogical composition of soil separates.

Since various sized soil separates differ so markedly in crystal form and chemical composition, it is not surprising that they also vary in their content of essential mineral nutrients. Logically, the sands, being mostly quartz,

TABLE 4.1

Total Phosphorus, Potassium, and Calcium Contents of Sand, Silt, and Clay Separates Commonly Found in Temperate Humid Region Soils

Separate	P (%)	K (%)	Ca (%)
Sand	0.05	1.4	2.5
Silt	0.10	2.0	3.4
Clay	0.30	2.5	3.4

would be expected to be lowest and the clay separate to be highest in nutrients. This inference is substantiated by the data in Table 4.1. The general relationships shown by these data hold true for most soils, although some exceptions may occur.

4.4

Soil Textural Classes

To convey an idea of the textural makeup of soils and to give an indication of their physical properties, *soil textural class* names are used. Three broad groups of these classes are recognized—*sands, loams,* and *clays.* Within each group specific textural class names have been devised (see Table 4.2).

SANDS

The sand group includes all soils in which the sand separates make up at least 70% and the clay separate 15% or less of the material by weight. The properties of such soils are therefore characteristically those of sand in contrast to the stickier nature of clays.

TABLE 4.2

General Terms Used to Describe Soil Texture in Relation to the Basic Soil Textural Class Names

U.S. Department of Agriculture classification system

Common names	General terms — Texture	Basic soil textural class names
Sandy soils	Coarse	⎰Sands ⎱Loamy sands
Loamy soils	Moderately coarse	⎰Sandy loam ⎱Fine sandy loam[a]
	Medium	⎰Very fine sandy loam[a] ⎱Loam, Silt loam, Silt
	Moderately fine	⎰Sandy clay loam, Silty clay loam, Clay loam
Clayey soils	Fine	⎰Sandy clay, Silty clay, Clay

[a] Although not included as class names in Figure 4.6, these soils are usually treated separately because of their fine sand content.

Two specific textural classes are recognized—*sand* and *loamy sand*—although, in practice, two subclasses are also used: *loamy fine sand* and *loamy very fine sand*.

SILTS

The silt group includes soils with at least 80% silt and 12% or less clay. Naturally the properties of this group are dominated by those of silt. Only one textural class—*silt*—is included.

CLAYS

To be designated a clay, a soil must contain at least 35% of the clay separate and, in most cases, not less than 40%. In such soils the characteristics of the clay separate are distinctly dominant, and the class names are *clay, sandy clay*, and *silty clay*. Sandy clays may contain more sand than clay. Likewise, the silt content of silty clays usually exceeds that of the clay fraction itself.

LOAMS

The loam group, which contains many subdivisions, is a more complicated soil textural class. An ideal loam may be defined as a mixture of sand, silt, and clay particles that exhibits the properties of those separates in about equal proportions.

Most soils of agricultural importance are a type of loam. They may possess the ideal makeup of equal proportions described above and be classed simply as *loam*. Usually, however, the varying quantities of sand, silt, and clay in the soil require a modified textural class name. Thus, a loam in which sand is dominant is classified as a *sandy loam*; in the same way, there may occur *silt loams, silty clay loams, sandy clay loams*, and *clay loams*.

VARIATIONS IN THE FIELD

The textural class names—*sand, loamy sand, sandy loam, loam, silt loam, sandy clay loam, silty clay loam, clay loam, sandy clay, silty clay, silt*, and *clay*—form a graduated sequence from soils that are coarse in texture and easy to handle to the clays, which are very fine and difficult to manage (Table 4.2).

While these textural class names are determined by particle-size distribution, they markedly affect other physical properties such as soil aeration and ease of tillage. For some soils, qualifying factors such as stone, gravel, and the various grades of sand become part of the textural class name.

4.5

Determination of Soil Class

"FEEL" METHOD

The common field method of determining the textural class of a soil is by its *feel*. This is ascertained by rubbing a sample of the soil, usually in a moist to wet condition, between the thumb and fingers (Figure 4.5). The way a wet soil "slicks out"—that is, develops a continuous ribbon when pressed between the thumb and fingers—indicates the amount of clay present. The longer and smoother the ribbon formed, the higher the clay content. Sand particles are gritty; silt feels like flour or talcum powder when dry and is only

(a) (b) (c)

FIGURE 4.5

The "feel" method was used to distinguish between a sand (a), a silt loam (b), and a clay soil (c). Moistened samples were rubbed between the thumb and forefinger. Note the shiny ribbon for the clay, the lack of cohesion in the sand, and the intermediate status of the silt loam.

slightly plastic and sticky when wet. Persistent cloddiness of dry soils generally is characteristic of silt and clay. (Table 4.3 gives some general criteria for field determination of soil texture.)

The "feel" method is used in soil survey and land classification. Accuracy in such a determination is of great practical value and depends largely on experience. Facility in class determination is one of the first skills a field researcher should develop.

LABORATORY METHOD

The U.S. Department of Agriculture has developed a method for naming soils based on laboratory-determined particle-size analysis, using procedures discussed briefly in Section 4.1.

TABLE 4.3

Criteria Used with the Field Method of Determining Soil Texture Classes

Criterion	Sand	Sandy loam	Loam	Silt loam	Clay loam	Clay
1. Individual grains visible to eye	Yes	Yes	Some	Few	No	No
2. Stability of dry clods	Do not form	Do not form	Easily broken	Moderately easily broken	Hard and stable	Very hard and stable
3. Stability of wet clods	Unstable	Slightly stable	Moderately stable	Stable	Very stable	Very stable
4. Stability of "ribbon" when wet soil rubbed between thumb and fingers	Does not form	Does not form	Does not form	Broken appearance	Thin, will break	Very long, flexible

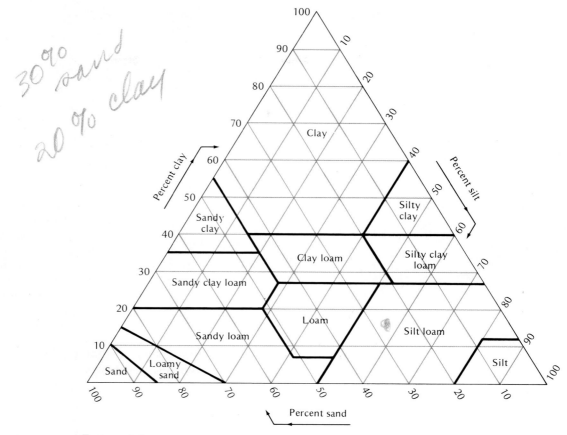

30% sand
20% clay

FIGURE 4.6

Percentages of sand, silt, and clay in the major soil textural classes. To use the diagram, locate the percentage of clay first and project inward as shown by the arrow. Do likewise for the percent silt (or sand). The point at which the two projections cross will identify the class name.

The relationship between such analyses and class names is shown diagrammatically in Figure 4.6. The diagram reemphasizes that a soil is a mixture of particles of different sizes. It illustrates how particle-size analyses of field soils can be used to check the accuracy of the soil surveyor's class designations as determined by feel. A working knowledge of how the textural triangle is used to name soils is essential. The legend of Figure 4.6 explains how to use the triangle.

The summation curves in Figure 4.7 illustrate the particle-size distribution in soils representative of three textural classes. Note the gradual change in percentage composition in relation to particle size. This figure emphasizes that there is no sharp line of demarcation in the distribution of sand, silt, and clay fractions and suggests a gradual change of properties with change in particle size.

CHANGES IN SOIL TEXTURE

The textural classes of soils are not subject to easy modification in the field. Texture of a given soil can be changed only by mixing it with another soil of a different textural class. For example, the incorporation of large quantities of sand to improve the physical properties of a clay for greenhouse or spe-

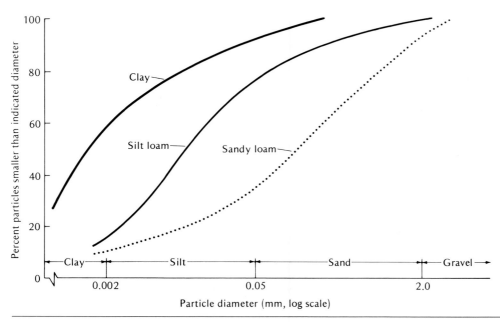

FIGURE 4.7
Particle-size distribution in three soils varying widely in their textures. Note that there is a gradual transition in the particle-size distribution in each of these soils.

cialty crops would bring about such a change. For most field crop and forest production areas, however, soil texture is not changed by cultural management.

4.6

Structure of Mineral Soils

The term *structure* relates to the grouping or arrangement of soil particles. It describes the gross, overall combination or arrangement of the primary soil separates into secondary groupings called *aggregates* or *peds*.

A profile may be dominated by a single type of structure. More often, however, several types are encountered in the different horizons.

Soil conditions and characteristics such as water movement, heat transfer, aeration, and porosity are much influenced by structure. In fact, the important physical changes imposed by the farmer in plowing, cultivating, draining, liming, and manuring his land are structural rather than textural.

TYPES OF SOIL STRUCTURE

The dominant shape of peds or aggregates in a horizon determines its structural type. The four principal types of soil structure are *spheroidal, platy, prism-like* and *block-like*. A brief description of each of these structural types (and appropriate subtypes) with schematic drawings is given in Figure 4.8. A more detailed description of each follows.

1. **Spheroidal** (*granular* and *crumb* subtypes). Rounded peds or aggregates are placed in this category. They usually lie loosely and are separated from each other. Relatively nonporous aggregates are called granules and the pattern *granular*. However, when the granules are especially porous, the term *crumb* is applied.

   Granular and crumb structures are characteristic of many surface soils, particularly those high in organic matter, and are especially prominent in grassland soils. They are the only types of aggregation that are commonly influenced by practical methods of soil management.

2. **Plate-like** (*platy*). In this structural type the aggregates (peds) are arranged in relatively thin horizontal plates, leaflets, or lenses. Platy structure is found in the surface layers of some virgin soils but may characterize the lower horizons as well.

Although most structural features are usually a product of soil-forming forces, the platy type is often inherited from the parent materials, especially those laid down by water or ice.

3. **Prism-like** (*columnar* and *prismatic* subtypes). These subtypes are characterized by vertically oriented aggregates or pillars that vary in height with different soils and may reach a diameter of 15 cm or more. Prism-like structures usually occur in subsurface horizons (especially natric horizons) in arid and semiarid regions and, when well developed, are a very striking feature of the profile. They also occur in some poorly drained soils of humid regions.

When the tops of the prisms are rounded, the term *columnar* is used. When the tops of the prisms are plane, level, and clean cut, the structural pattern is designated *prismatic*.

4. **Block-like** (*blocky* and *subangular blocky* subtypes). In this case the aggregates have been reduced to blocks, irregularly six-faced, with their three dimensions more or less equal. These fragments range from about 1 to 10 cm in thickness. In general, the design is so individualistic that identification is easy.

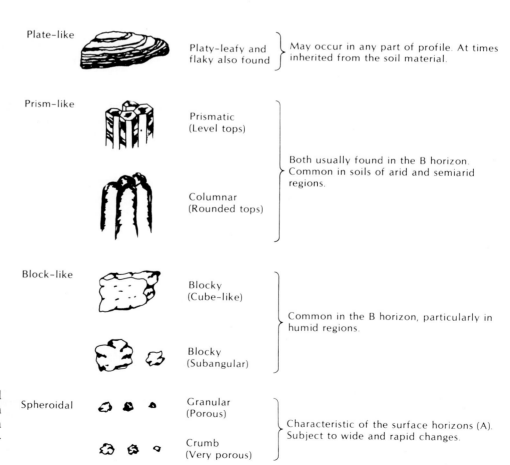

FIGURE 4.8
Various structural types found in mineral soils. Their location in the profile is suggested. In arable topsoils, a stable granular structure is prized.

When the edges of the cubes are sharp and the rectangular faces distinct, the subtype is designated *blocky*. When subrounding has occurred, the aggregates are referred to as *subangular blocky*. Block-like structures usually are confined to the subsoil, and their stage of development and other characteristics have much to do with soil drainage, aeration, and root penetration.

As already emphasized, two or more of the structural types listed usually occur in the same soil profile. In humid temperate regions, a granular aggregation in the surface horizon with a blocky, subangular blocky, or platy type in the lower horizons is usual, although granular subsurface horizons sometimes occur.

GENESIS OF SOIL STRUCTURE

The mechanics of soil structure formation are not well known. However, the permeation of the soil by plant roots tends to compress soil particles into small aggregates and to break up larger ones. Similar compression and contraction results from swelling and shrinking as soils are wetted and dried.

Plant roots (and particularly root hairs) excrete sticky organic chemicals that help bind the soil particles together into aggregates. Microbial decomposition of plant residues produces other organic materials that interact with clays and cement the aggregates together. Thus, organic materials, stimulated by microbial processes, enhance not only the formation of soil aggregates but help assure their stability. These processes are most notable in surface soils where root and animal activities, and organic matter accumulation are most noted.

The mechanism for the creation and stabilization of structural types common in lower horizons is less certain. The downward migration of silicate clays, oxides of iron and aluminum, soluble salts, and calcium carbonate is thought to encourage aggregate formation under different climatic and soil conditions. But the exact mechanism for specific aggregate formation remains obscure.

4.7

Particle Density of Mineral Soils

One means of expressing soil mass (weight) is in terms of the density of the solid particles making up the soil. Density is usually defined as the mass of a unit volume of soil *solids* and is called *particle density* (D_p). In the metric system, particle density can be expressed in terms of megagrams per cubic meter (Mg/m^3). Thus, if 1 m^3 of soil solids weighs 2.6 Mg, the particle density is 2.6 Mg/m^3.[1]

The size of the particles of a given mineral and the arrangement of the soil solids have nothing to do with the particle density. Particle density depends on the chemical composition and crystal structure of the mineral particles and is not affected by pore space.

Although there is considerable range in the density of the individual soil minerals, the figures for most mineral soils usually vary between 2.60 and

[1] Since 1 Mg = 1 million grams and 1 m^3 = 1 million cubic centimeters, this particle density can also be expressed as 2.6 g/cm^3.

$2.75 \ Mg/m^3$. This narrow range occurs because quartz, feldspar, micas, and the colloidal silicates with densities within this range usually make up the major portion of mineral soils. When unusual amounts of minerals with high particle density, such as magnetite, garnet, epidote, zircon, tourmaline, or hornblende, are present, the particle density may exceed $2.75 \ Mg/m^3$.

Organic matter weighs much less than an equal volume of mineral solids, having a particle density of $1.1–1.4 \ Mg/m^3$. Consequently, mineral surface soils, which almost always have higher organic matter content than the subsoils, usually have lower particle densities than do subsoils. Some mineral topsoils high in organic matter (15–20%) may have particle densities as low as $2.4 \ Mg/m^3$ or even below. Clearly the amount of organic matter in a soil markedly affects the particle density. Nevertheless, for general calculations, the average arable mineral surface soil (3–5% organic matter) may be considered to have a particle density of about $2.65 \ Mg/m^3$

4.8

Bulk Density of Mineral Soils

A second important mass (weight) measurement of soils is *bulk density* (D_b). It is defined as the mass (weight) of a unit volume of dry soil. This volume includes *both solids and pores*. The comparative calculations of bulk density and particle density are shown in Figure 4.9. A careful study of this figure should make clear the distinction between these two methods of expressing soil mass.

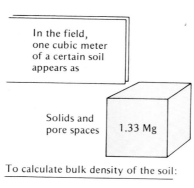

In the field, one cubic meter of a certain soil appears as

Solids and pore spaces — 1.33 Mg

To calculate bulk density of the soil:

Volume = 1 m³   Weight = 1.33 Mg
(solids + pores)   (solids only)

$$Bulk \ density = \frac{weight \ of \ oven \ dry \ soil}{volume \ of \ soil}$$
(solids + pores)

Therefore

$$Bulk \ density, D_b = \frac{1.33}{1} = 1.33 \ Mg/m^3$$

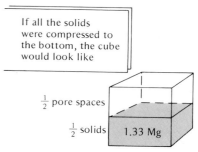

If all the solids were compressed to the bottom, the cube would look like

$\frac{1}{2}$ pore spaces
$\frac{1}{2}$ solids — 1.33 Mg

To calculate solid particle density:

Volume = 0.5 m³   Weight = 1.33 Mg
(solids only)   (solids only)

$$Solid \ particle \ density = \frac{weight \ of \ solids}{volume \ of \ soilds}$$

Therefore

$$Solid \ particle \ density, D_p = \frac{1.33}{0.5} = 2.66 \ Mg/m^3$$

FIGURE 4.9

Bulk density, D_b, and particle density, D_p, of soil. Bulk density is the weight of the solid particles in a standard volume of field soil (solids plus pore space occupied by air and water). Particle density is the weight of solid particles in a standard volume of those solid particles. Follow the calculations through carefully and the terminology should be clear. In this particular case the bulk density is one half the particle density, and the percent pore space is 50.

FACTORS AFFECTING BULK DENSITY

Since bulk density relates to the combined volumes of the solids and pore spaces, soils with a high proportion of pore space to solids have lower bulk densities than those that are more compact and have less pore space. Consequently, any factor that influences soil pore space will affect bulk density.

Fine-textured surface soils such as silt loams, clays, and clay loams generally have lower bulk densities than sandy soils. The solid particles of the fine-textured soils tend to be organized in porous grains or granules, especially if adequate organic matter is present. This condition assures high total pore space and a low bulk density. In sandy soil, however, organic matter contents generally are low, the solid particles lie close together, and the bulk densities are commonly higher than in the finer textured soils.

The bulk densities of clay, clay loam, and silt loam surface soils normally range from 1.00 Mg/m^3 to as high as 1.60 Mg/m^3, depending on their condition. A variation from $1.20–1.80 \text{ Mg/m}^3$ may be found in sands and sandy loams. Very compact subsoils may have bulk densities of 2.0 Mg/m^3 or even greater. In such compact layers there are essentially no macropores and root growth is greatly impaired, the constraint becoming most noticeable at bulk densities of 1.6 or above.

The relationship among texture, compactness, and bulk density is illustrated in Figure 4.10.

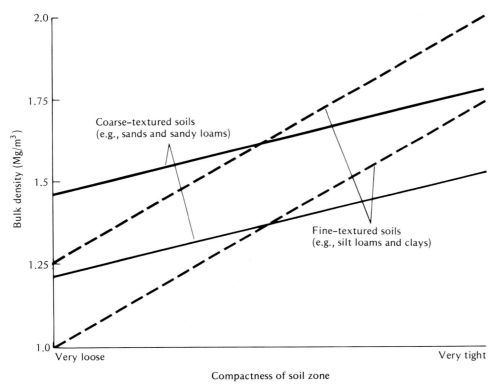

FIGURE 4.10

Generalized relationship between compactness and the range of bulk densities common in sandy soils and in those of finer texture. Sandy soils generally are less variable in their degree of compactness than are the finer textured soils. In contrast, fine-textured soils have very low bulk densities when not compacted but very high bulk densities when compacted. For all soils the surface layers are more likely to be medium to loose in compactness than are the subsoils.

TABLE 4.4

Bulk Density (D_b) Data for Several Soil Profiles

Note that sandy soils generally have higher bulk densities than those high in silt and clay. The Erie soil has a very dense subsoil, whereas the Oxisol from Brazil is loose and open.

Horizon	Grandfield fine sandy loam (Oklahoma)	Miami silt loam (Wisconsin)	Erie channery silt loam (New York)	Houston clay (Texas)	Oxisol clay (Brazil)
Plow layer	1.72	1.28	1.33	1.24	0.95
Upper subsoil	1.74	1.41	1.46	1.36	—
Lower subsoil	1.80	1.43	2.02	1.51	1.00
Parent material	1.85	1.49	—	1.61	—

Data for Grandfield soil from Dawud and Gray (1979), Miami from Nelson and Muckenhirn (1941), Houston from Yule and Ritchie (1980). Erie calculated from Fritton and Olson (1972), Oxisol from Larson et al. (1980).

Even in soils of the same surface textural class, significant differences in bulk density are found, as is shown by data on soils from several locations in Table 4.4. Moreover, the bulk density generally is higher in lower profile layers. This apparently results from a lower content of organic matter, less aggregation and root penetration, and a compaction caused by the weight of the overlying layers.

The system of crop and soil management employed on a given soil also influences its bulk density. Figure 4.11 illustrates the effect of cutting forests on bulk density of the soil, especially in the surface layers. Removal of the trees greatly increased the surface soil bulk density. The addition of crop residues or farm manure in large amounts tends to lower bulk densities of surface soils, as does a bluegrass sod. Intensive cultivation operates in the

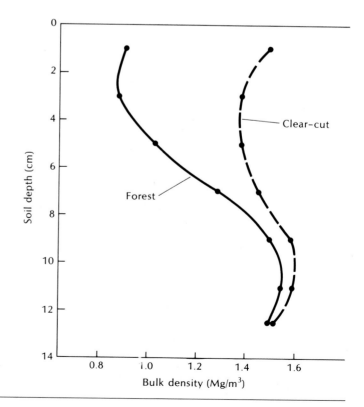

FIGURE 4.11

Effect of cutting trees (clear-cut) on the bulk density of the soil at different depths. [From McIntyre et al. (1987)]

TABLE 4.5

Bulk Density and Pore Space of Certain Cultivated Topsoils and of Nearby Uncropped Areas (One Subsoil Included)

The bulk density was increased by cropping in every case, and the pore space decreased proportionately.

Soil	Years cropped	Bulk density (Mg/m³)		Pore space (%)	
		Cropped soil	Uncropped soil	Cropped soil	Uncropped soil
Udalf (Hagerstown loam—PA)	58	1.25	1.07	50.0	57.2
Udoll (Marshall silt loam—Iowa)	50+	1.13	0.93	56.2	62.7
Aqualf (Nappanese silt loam—Ohio)	40	1.31	1.05	50.5	60.3
Ave. 19 Georgia soils	45–150	1.45	1.14	45.1	57.1
Boroll (Blaine lake silt loam—Canada)	90	1.30	1.04	50.9	60.8
Orthid (Sutherland clay—Canada)	70	1.28	0.98	51.7	63.0
Orthid (Sutherland clay, subsoil)	70	1.38	1.21	47.9	54.3

Data for Canadian soils from Tiessen et al. (1982), others from Lyon et al. (1952). Pore space percentages for the Canadian soils were calculated from bulk density data in Tiessen et al. (1982).

opposite direction, as shown in Table 4.5. These data are from longtime experiments in different locations, the soils having been under cultivation for from 40 to 150 years. Cropping increased the bulk density of the topsoils in all cases.

OTHER MASS (WEIGHT) FIGURES

Densities expressed in metric units can be converted to the English system. The dry weight, in pounds per cubit foot (lb/ft³), of clayey and silty surface soils may vary from 65 to 100 lb/ft³; sands and sandy loams may show a range of 75–110 lb/ft³. The greater the organic content, the less is this weight. Very compact subsoils, regardless of texture, may weigh as much as 125 lb/ft³.

Another figure of interest is the mass of soil in one hectare to a depth of normal plowing (15 cm). Such a *hectare–furrow slice* of a typical surface soil weighs about 2.2 million kilograms (kg). A comparable figure in the English system is 2 million pounds per *acre–furrow slice*.

4.9

Pore Space of Mineral Soils

The pore space of a soil is that portion of the soil volume occupied by air and water. The amount of pore space is determined largely by the arrangement of the solid particles.

If the particles lie close together, as in sands or compact subsoils, the total porosity is low. If they are arranged in porous aggregates, as is often the case in medium-textured soils high in organic matter, the pore space per unit volume will be high.

The validity of the above generalizations is substantiated by using a very simple formula involving particle density and bulk density figures. The derivation of the formula used to calculate the percentage of total pore space in soil follows.

Let D_b = bulk density V_s = volume of solids

D_p = particle density V_p = volume of pores

W_s = weight of soil (solids) $V_s + V_p$ = total soil volume

By definition,

$$\frac{W_s}{V_s} = D_p \quad \text{and} \quad \frac{W_s}{V_s + V_p} = D_b$$

Solving for W_s gives

$$W_s = D_p \times V_s \quad \text{and} \quad W_s = D_b(V_s + V_p)$$

Therefore

$$D_p \times V_s = D_b(V_s + V_p)$$

and

$$\frac{V_s}{V_s + V_p} = \frac{D_b}{D_p}$$

Since

$$\frac{V_s}{V_s + V_p} \times 100 = \% \text{ solid space}$$

then

$$\% \text{ solid space} = \frac{D_b}{D_p} \times 100$$

Since % pore space + % solid space = 100, and % pore space = 100 − % solid space, then

$$\% \text{ pore space} = 100 - \left(\frac{D_b}{D_p} \times 100\right)$$

Using this formula, a sandy soil having a bulk density of 1.50 and a particle density of 2.65 will be found to have 43.4% pore space. A silt loam for which the corresponding values are 1.30 and 2.65 possesses 50.9% air and water space. The latter value is close to the pore space of a normally granulated silt loam surface soil.

FACTORS INFLUENCING TOTAL PORE SPACE

The total pore space is quite variable among soils. Sandy surface soils show a range of 35–50%, whereas medium- to fine-textured soils vary from 40 to 60% and even more in cases of high organic matter and marked granulation (Figure 4.12). Pore space also varies with depth; some compact subsoils have as little as 25–30%. This accounts in part for the inadequate aeration and resistance to root penetration of such horizons.

As was the case for bulk density, past cropping exerts a decided influence on pore space of soils (Table 4.5). Data from four states and Canada show that cropping tends to lower the total pore space compared to that of uncropped soils. This reduction usually is associated with a decrease in organic matter content and a consequent lowering of granulation. Pore space in the subsoil also has been found to decrease with cropping, although to a lesser degree.

SIZE OF PORES

Two types of individual pore space—*macro* and *micro*—occur in soils. Although there is no clear-cut demarcation, pores less than about 0.06 mm in diameter are considered micropores and those larger, macropores.

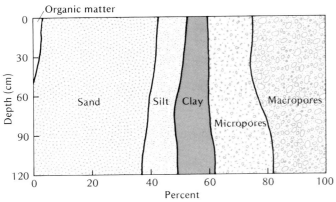

(a) Sandy loam

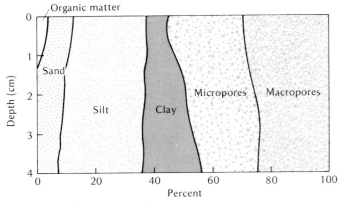

(b) Silt loam (good structure)

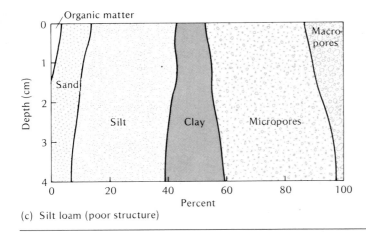

(c) Silt loam (poor structure)

FIGURE 4.12
Volume distribution of organic matter, sand, silt, clay, and pores of macro and micro sizes in a representative sandy loam (a) and in two representative silt loams, one with good soils structure (b) and the other with poor structure (c). Both silt loam soils have more total pore space than the sandy loam, but the silt loam with poor structure has a smaller volume of larger (marco) pores than either of the other two soils.

The macropores characteristically allow the ready movement of air and percolating water. In contrast, the micropores are mostly filled with water in a moist field soil and do not permit much air movement into or out of the soil. The water movement also is slow. Thus, even though a sandy soil has relatively low total porosity, the movement of air and water through such a soil is surprisingly rapid because of the dominance of the macropores.

Fine-textured soils, especially those without a stable granular structure, allow relatively slow gas and water movement despite the unusually large volume of total pore space. Here the dominating micropores often stay full of water. Aeration, especially in the subsoil, often is inadequate for satisfactory root development and desirable microbial activity. Obviously, the size of the individual pore spaces rather than their combined volume is the important consideration. The loosening and granulating of fine-textured soils promotes aeration not so much by increasing the total pore space as by raising the proportion of macrospaces.

It has already been indicated (Section 1.8 and Figure 1.6) that in a well-granulated silt loam surface soil at optimum moisture for plant growth, the total pore space will be near 50% and is likely to be shared equally by air and water. Soil aeration under such a condition is satisfactory, especially if a similar ratio of air to water extends well into the subsoil.

Cropping and Pore Size

Continuous cropping, particularly of soils originally high in organic matter, often results in a reduction of large or macropore spaces. Data from a fine-textured soil in Texas (Figure 4.13) clearly illustrate this effect. Cropping significantly reduced soil organic matter content and total pore space. But most striking is the effect of cropping on the size of the soil pores: The amount of macropore space necessary for ready air movement was reduced about one half. This severe reduction in pore size also extended into the 15–30 cm layer. In fact, samples taken as deep as 107 cm showed the same trend.

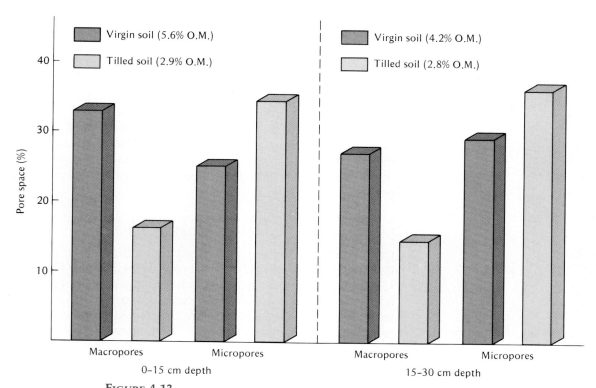

Figure 4.13
Effect of continuous cropping for at least 40–50 years (tilled soil) on the macropore and micropore spaces of an Ustert (Houston black clay in Texas). The virgin soil is not only higher in total pore space but also has much more macropore space to permit good aeration. Note that the cultivated (tilled) soil also is much lower in organic matter than the virgin untilled soil. [Data from Laws and Evans (1949)]

In recent years, *conservation tillage* practices, which minimize plowing and associated soil manipulations, have been widely adopted in the United States (Sections 4.14 and 15.12). Such practices keep crop residues on the soil surface and are remarkably effective in reducing soil erosion losses. Unfortunately, conservation tillage does not seem to increase soil porosity. In fact, some trials indicate that pore space is less with conservation tillage than with conventional tillage, a matter of some concern in soils with poor internal drainage.

4.10

Aggregation and Its Promotion in Arable Soils

In a practical sense, two sets of factors are of concern in dealing with soil aggregation: (a) those responsible for aggregate formation and (b) those that give the aggregates stability once they are formed. Since both sets of factors are operating simultaneously, it is sometimes difficult to distinguish their relative effects on stable aggregate development in soils.

GENESIS OF GRANULES AND CRUMBS

Several specific factors influence the genesis of granules and crumbs, which are so important in surface soils. The factors include (a) physical processes that encourage particle contact, (b) the influence of decaying organic matter, (c) the modifying effects of adsorbed cations, and (d) soil tillage.

PHYSICAL PROCESSES

Any action that will shift the particles back and forth and force contacts between particles will encourage aggregation. Alternate wetting and drying, freezing and thawing, the physical effects of root extension, and the mixing action of soil organisms and of tillage implements encourage such contacts and, therefore, stimulate aggregate formation.

The benefits of fall plowing of fine-textured soils, which results in the breaking of clods into smaller aggregates under the influence of a light rain, have long been used in seedbed preparation. And the aggregating influences of plant roots, earthworms, and other soil organisms must not be overlooked.

INFLUENCE OF ORGANIC MATTER

Organic matter is the major agent that stimulates the formation and stabilization of granular- and crumb-type aggregates (see Figure 4.14). As organic residues decompose, gels and other viscous microbial products along with associated bacteria and fungi encourage crumb formation. Organic exudates from plant roots also participate in this aggregating action. Organic compounds, such as polysaccharides, then chemically interact with particles of silicate clays and iron and aluminum oxides. The organic compounds orient the clays in a common plane and then form bridges between individual soil particles, thereby binding them together in water-stable aggregates.

The overall aggregating influence of organic materials has long been known. Only in recent years, however, have scientists been able to use ultrathin sectioning and photography to show direct evidence of the organo-mineral bridges that bind the particles (see Figure 4.15). The complexity of organic materials found in humus makes possible this binding process.

FIGURE 4.14
Puddled soil (left) and well-granulated soil (right). Plant roots and especially humus play the major role in soil granulation. Thus a sod tends to encourage development of a granular structure in the surface horizon of cultivated land. [Courtesy USDA Soil Conservation Service]

EFFECT OF ADSORBED CATIONS

Aggregate formation is definitely influenced by the nature of the cations adsorbed by soil colloids (see Section 7.15). For instance, when Na^+ is a prominent adsorbed ion, as in some soils of arid and semiarid areas, the particles are dispersed and an undesirable soil structure results.

By contrast, the adsorption of ions such as Ca^{2+}, Mg^{2+}, or Al^{3+} may encourage aggregate formation, starting with a process called *flocculation*.

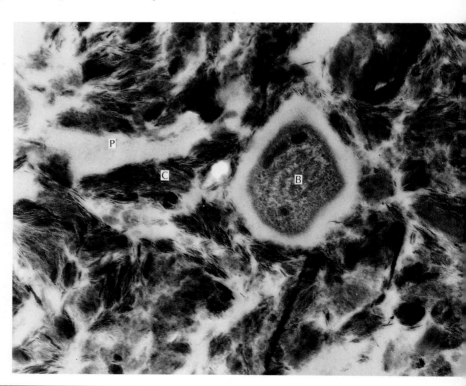

FIGURE 4.15
An ultrathin section illustrating the interaction among organic materials and silicate clays in a water-stable aggregate. The dark-colored materials (C) are groups of clay particles that are interacting with organic polysaccharides (P). A bacterial cell (B) is also surrounded by polysaccharides. Note the generally horizontal orientation of the clay particles, an orientation encouraged by the organic materials. [From Emerson et al. (1986); photograph provided by R.C. Foster, CSIRO, Glen Osmond, Australia]

These ions encourage the individual colloidal particles to come together in small aggregates called floccules. Flocculation itself, however, is only the first step because it alone does not provide for the *stabilization* of the aggregates.

INFLUENCE OF TILLAGE AND COMPACTION

Tillage has both favorable and unfavorable effects on aggregation. If the soil moisture level is favorable, the short-time effect of tillage is generally favorable because the implements break up the clods, incorporate the organic matter into the soil, kill weeds, and create a more favorable seedbed. Tillage is thus considered necessary in the normal management of some soils.

Over longer periods, however, tillage operations have detrimental effects on surface soil granules. In the first place, by mixing and stirring the soil, tillage generally hastens the oxidation of organic matter in soils. Secondly, tillage operations, especially those involving heavy equipment, tend to break down the stable soil aggregates. Compaction occurs from repeatedly driving heavy farm equipment over fields. An indication of the effect of such traffic on bulk density of soil is given in Figure 4.16. These data explain the increased interest in farming systems that drastically reduce the number of necessary tillage operations (Sections 4.14 and 15.12).

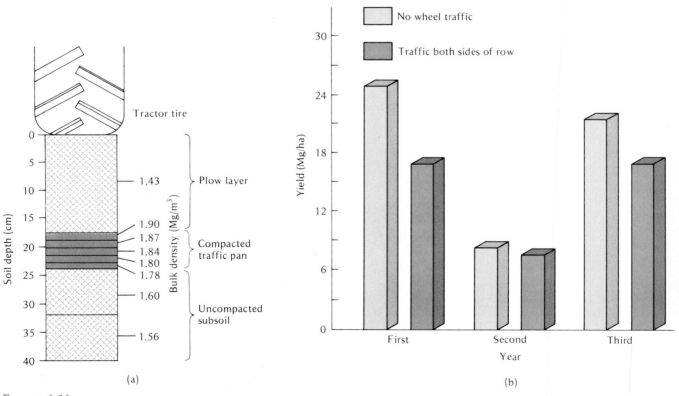

(a)   (b)

FIGURE 4.16
Heavy equipment used for tillage and other purposes compacts the soil, increases bulk density, and reduces crop yields. (a) Such equipment compacted the zone just below the plow layer on an Udult (Norfolk) soil and increased bulk density to more than 1.8 Mg/m³, the limit of penetration for cotton roots. (b) The yield of potatoes was reduced markedly for two out of three years in this test on a clay loam soil in Minnesota. No yield decrease in the second year, which was very dry. [Data from Camp and Lund (1964) and Voorhees (1984)]

Before wetting   After wetting

High O.M.   Low O.M.   High O.M.   Low O.M.

FIGURE 4.17
The aggregates of soils high in organic matter are much more stable than are those low in this constituent. The low organic matter soil aggregates fall apart when they are wetted; those high in organic matter maintain their stability.

4.11

Aggregate Stability

The stability of aggregates is of great practical importance. Some aggregates readily succumb to the beating of rain and the rough and tumble of plowing and tilling the land. Others resist disintegration, thus making the maintenance of a suitable soil structure comparatively easy (Figures 4.17 and 4.18). Three major factors appear to influence aggregate stability.

1. The temporary mechanical binding action of microorganisms, especially the thread-like filaments (mycelia) of fungi. These effects are pronounced when fresh organic matter is added to soils and are at a maximum a few weeks or months after the application.

 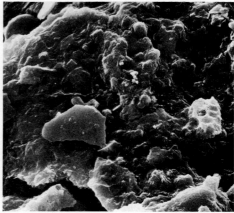

FIGURE 4.18
Scanning electron micrographs of aggregates from two topsoils, one (left) with low aggregate stability, the other (right) with high aggregate stability. Note that individual particles in the low stability aggregate are not well associated, whereas those in the stable aggregate are found together. [Photos from Craig Ross, New Zealand Soil Bureau. Used with permission.]

2. The cementing action of the intermediate products of microbial synthesis and decay, such as microbially produced gums and certain polysaccharides. Polysaccharides are shown intermixed with clay in Figure 4.15.

3. The cementing action of the more resistant stable humus components aided by similar action of certain inorganic compounds, such as iron oxides. These materials provide most of the long-term aggregate stability.

It should be emphasized that aggregate stability is not entirely an organic phenomenon. There is continual interaction between organic and inorganic components. Polyvalent inorganic cations that cause flocculation (e.g., Ca^{2+}, Mg^{2+}, Fe^{2+}, and Al^{3+}) also are thought to provide mutual attraction between the organic matter and soil clays, encouraging the development of clay–organic matter complexes. In addition, films of clay called "clay skins" often surround the soil peds and help provide stability. The noted stability of aggregates in red and yellow soils of tropical and semitropical areas is due to the hydrated oxides of iron they contain.

FIGURE 4.19
Scanning electron micrographs of soil aggregated in the upper 1 mm of a soil with stable aggregation (left) compared to one with unstable aggregates (right). Note the particles in the immediate surface have been destroyed and a surface crust has formed. [From Onofiok and Singer (1984); used with permission of Soil Science Society of America]

As a general rule, the largest of the aggregates present in any particular surface soil tend to be somewhat unstable. This is why it is difficult to build up soil aggregation beyond a certain size of granule or crumb in cultivated land.

SOIL CRUSTING

The surface soil aggregates are most vulnerable to destruction, especially by heavy rains. The dispersed surface tends to seal over and prevent water infiltration. Then when the soil dries, a hard crust is formed that impedes the emergence of seedlings. The seedlings are able to emerge only through cracks in the crust. Such crusts often spell disaster to an emerging crop. A crust-forming and a normal soil are compared in Figure 4.19.

Crust formation can be minimized by keeping some cover on the land to minimize the impact of raindrops. Also a light tillage while the soil is still moist will break up the crust before it hardens.

SOIL CONDITIONERS

Certain synthetic chemicals (polymers) are known to have the same ability to stabilize soil structure as that possessed by natural organic polymers such as polysaccharides and polyuronides. These synthetic polymers were introduced commercially in the early 1950s with great enthusiasm since they had remarkable ability to stabilize aggregates. Unfortunately, their high cost in relation to their benefits made them uneconomical, and their use was essentially abandoned.

Recently, however, new organic synthetics have been developed that are effective in helping to stabilize aggregates at concentrations as low as 0.00025%. These low concentration rates substantially reduce materials costs, which may well ensure continued interest in artificial soil conditioners. In Figure 4.20 the stabilizing effect of one of these synthetic polyacrylamides is shown.

FIGURE 4.20
The remarkable stabilizing effect of a synthetic polyacrylamide is seen in the row on the right compared to the left, untreated, row. Irrigation water broke down much of the structure of the untreated soil but had no effect on the treated row. [From Mitchell (1986)]

4.12

Structural Management of Soils

COARSE-TEXTURED SOILS

There is no practical way to greatly improve the structure of coarse-textured soils, although organic matter additions can be helpful. Organic material will not only act as a binding agent for the particles, but also will increase the water-holding capacity. Farm manures and crop residues (especially of sod crops) are good sources for the organic matter.

FINE-TEXTURED SOILS

The structural management of a fine-textured soil such as a clay or sandy clay is not as simple as that of a sandy soil. Because of its high plasticity and cohesion, if a clay soil is tilled when wet, the aggregates are broken down, the individual particles tend to act independently, the macropores disappear, and the soil is said to be *puddled*. In this condition it is nearly impervious to air and water. When the soil dries, it usually becomes hard and dense. On the other hand, if a clay or sandy clay soil is plowed too dry, large hard clods, which are difficult to work into a good seedbed, are turned up. Obviously, the tillage of clay soils must be timed carefully with respect to their moisture content.

Some clay soils of tropical regions are much more easily managed than those just described. The clay fraction of these soils is dominated by hydrous oxides of iron and aluminum, which are not as sticky, plastic, and difficult to work. When properly fertilized and managed, these oxide-type clays can be productive. For example, they are among the world's best soils for pineapple production.

As was the case for sandy soils, organic matter maintenance plays a key role in the management not only of clay soils, but of much larger areas of medium-textured soils such as silt loams and loams. The return of crop residues and inclusion of sod crops in the rotation helps maintain soil organic matter and, in turn, good soil structure.

The data in Table 4.6 illustrate the importance of growing other than row crops if water-stable aggregates are to be maintained. The detrimental effect of continuous corn is obvious. Where animals are included in the farming systems, manures should be returned to the land along with those crop residues that are not consumed.

TABLE 4.6
Water-Stable Aggregation of a Udoll (Marshall Silt Loam) near Clarinda, Iowa, under Different Cropping Systems

| | Water-stable aggregates (%) | |
Crop	Large (1 mm and above)	Small (less than 1 mm)
Corn continuously	8.8	91.2
Corn in rotation	23.3	76.7
Meadow in rotation	42.2	57.8
Bluegrass continuously	57.0	43.0

From Wilson et al. (1947).

4.13

Soil Consistence

Soil *consistence* is a term used to describe the resistance of a soil at various moisture contents to mechanical stresses or manipulations. It is a composite expression of those forces of mutual attraction among soil particles that determine the ease with which a soil can be reshaped or ruptured. It is commonly measured by feeling and manipulating the soil by hand or by pulling a tillage instrument through it.

The consistence of soils is generally described at three soil moisture levels: wet, moist, and dry. Terms used to describe soil consistence at these three moisture levels are shown in Table 4.7.

TABLE 4.7
Terms Used to Describe the Consistence of Soils

Conditions of least coherence are represented by terms at the top of each column. Greater coherence characterizes terms as you move down the column.

	Wet soils		Moist soils	Dry soils
	Stickiness	*Plasticity*		
Increasing coherence →	Nonsticky	Nonplastic	Loose	Loose
	Slightly sticky	Slightly plastic	Very friable	Soft
	Sticky	Plastic	Friable	Slightly hard
	Very sticky	Very plastic	Firm	Hard
			Very firm	Very hard
			Extremely firm	Extremely hard

FIELD EXAMPLES

Wet clay soils are commonly termed *slightly sticky*, *sticky*, or *very sticky*, and their plasticity would be indicated as *slightly plastic*, *plastic*, or *very plastic*.

A wet sand, on the other hand, would be *nonsticky* and *nonplastic*. Gradations in stickiness and plasticity would be found with wet medium-textured soils such as clay loams, silt loams, and some loams.

The consistence of moist soils varies from *loose* to *extremely firm* (see Table 4.7). Sandy soils (and some high in silt) are mostly loose, whereas soils high in sticky clays are usually extremely firm. Moist soils that are easily crushed and yet show some coherence when pressed together are termed *friable* or *very friable*, conditions that are prized for ease of manipulation and management.

Dry soils with little coherence also will be *loose*. Those with high coherence are *extremely hard* and are so resistant to pressure that clods cannot be broken by hand. The degree of cementation of soil materials, independent of soil moisture, also is considered in identifying soil consistence. Terms such as *weakly cemented*, *strongly cemented*, and *indurated* are used to define categories of cementation.

Consistence has importance for the practical use of soils. The terms used to describe this soil property are relevant to concerns such as soil tillage and compaction by farm machinery. These subjects are covered in the next section.

4.14

Tilth and Tillage

Although frequent mention has been made of plowing and cultivation in relation to soil structure, attention must be given to seedbed preparation and to the maintenance of stable soil structure throughout the crop-growing season. The term "tilth" is central to such a discussion.

TILTH

Simply defined, *tilth* refers to "the physical condition of the soil in relation to plant growth" and, hence, includes all soil physical conditions that influence crop development.

Tilth depends not only on granule formation and stability, but also on such factors as bulk density, soil moisture content, degree of aeration, rate of water infiltration, drainage, and capillary-water capacity. As might be expected, tilth often changes rapidly and markedly. For instance, the working properties of fine-textured soils may be altered abruptly by a change in consistence caused by a slight change in moisture.

CONVENTIONAL TILLAGE AND CROP PRODUCTION

Tillage involves the mechanical manipulation of the soil with the objective of promoting good tilth and, in turn, higher crop production.

For centuries farmers have tilled the soil for three primary reasons: (1) to control weeds, (2) to present a suitable seedbed for crop plants, and (3) to incorporate organic residues into the soil.

Since the Middle Ages, the moldboard plow has been the *primary tillage* implement most used in the Western world. Its purpose is to lift, twist, and turn the soil while incorporating crop residues and animal wastes into the plow layer.

In more recent times, the moldboard plow has been supplemented by the disk plow, which is used to cut up the residues and partially incorporate them into the soil. In conventional practice such primary tillage has been followed by a number of *secondary tillage* operations, such as harrowing to kill weeds and break up clods and thereby prepare a suitable seedbed.

After the crop is planted, the soil may receive further secondary tillage to control weeds and break up crusting of the immediate soil surface. In modern agriculture, all conventional tillage operations are performed with tractors and other heavy equipment that often pass over the land several times before the crop is finally harvested.

SHORT-TERM VERSUS LONG-TERM EFFECTS ON SOIL TILTH

The immediate effects of most conventional tillage operations are beneficial. Crop residues are broken down more quickly if they are cut up and incorporated into the soil by tillage implements. Immediately after plowing, the soil is loosened and the total pore space is higher than before plowing. Tillage can provide excellent seedbeds and is a good means of weed control. However, conventional tillage also leaves the soil surface bare and subject to wind and water erosion.

The long-term effects of conventional tillage, especially plowing, are generally undesirable. Rapid breakdown of organic residues can hasten the reduction of soil organic matter content and, in turn, of aggregate stability (see Figure 4.17). By passing over the field frequently, tractors and other

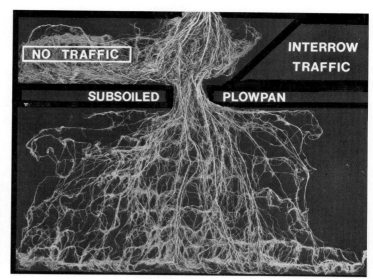

FIGURE 4.21
Root distribution of a cotton plant. On the right, interrow tractor traffic and plowing have caused a plowpan that restricts root growth. Roots are more prolific on the left where there had been no recent tractor traffic and the plowpan had been broken up by a subsoiling implement. [Courtesy USDA National Tillage Machinery Laboratory]

heavy equipment compact the soil and encourage the development of a dense zone (plow pan) immediately below the plowed layer (Figure 4.21).

In recent years, land management systems have been developed that minimize the need for soil tillage. Since these systems also leave considerable plant residues on or near the soil surface, they protect the soil from erosion. For this reason the tillage practices followed are called *conservation tillage*.

Figure 4.22 illustrates an extreme "no-till" operation where one crop is planted in the residue of another with no primary tillage. Other "minimum" tillage systems permit some stirring of the soil but still leave a high proportion of the crop residues on the surface. These organic residues protect the

FIGURE 4.22
An illustration of one conservation tillage system. Wheat is being harvested and soybeans are planted with no intervening tillage operation. This no-tillage system permits double cropping, saves fuel costs and time, and helps conserve the soil. [Courtesy Allis-Chalmers Corporation]

soil from the beating action of raindrops and the abrasive action of the wind, thereby reducing water and wind erosion. Because of this association with erosion control, conservation tillage will be considered further in Section 15.12.

4.15

Engineering Implications

The influence of the physical condition of soils on plant growth and associated agricultural practices has been emphasized, but the physical properties have equal effects on the use of soils and soil materials for nonagricultural purposes. For example, physical properties largely determine the usefulness of soil materials as a highway bed, as a foundation for a house or other building, as a material for an earthen reservoir dam, or for other engineering purposes.

Perhaps the most important property of a soil for engineering uses is *soil strength*. This is a measure of the capacity of a soil to withstand stresses without giving way to those stresses by collapsing or becoming deformed. The failure of a soil to withstand stress can be seen where building foundations are not well supported or where earthen dams give way under pressure from the impounded reservoir waters.

Soil strength is determined by a number of related soil characteristics, among which are soil *compressibility*, soil *compactability*, and *bearing resistance*. Some soil particles, such as those of some colloidal silicate clays and of micas of all sizes, can be compressed when a load is placed upon them. When that load is removed, however, these particles tend to regain their original shape, in effect reversing their compression. As a result, these particles are not conducive to a relatively permanent compaction state, a property desirable to provide the soil strength needed for most engineering purposes. For example, a roadbed or foundation base that will not remain compacted is not very stable.

Soil compressibility and compaction also influence the bearing resistance or the ability of a soil to resist the penetration of an object (such as a building foundation) into that soil. Bearing resistance is essential to prevent the penetration of the surface materials of a highway into the underlying roadbed. This is a property of great practical importance, and apparently it is dependent on some of the more fundamental physical properties of the individual soil particles.

Soil strength is determined by a number of factors, such as soil moisture content, particle-size distribution (soil texture), and the mineralogy of the different soil particles. The influence of soil moisture on a soil's response to stress was covered in the discussion of soil consistence (Section 4.13). A "hard" dry soil has greater resistance to deformation than a wet "plastic" soil, especially if the silicate clay content is high. The concept of soil consistence is as relevant to engineering uses of soils as it is to soil tillage.

In general, coarser textured materials have higher soil strength than those with small particle size. For example, quartz sand grains are subject to little compressibility, whereas the silicate clays are more easily compressed. As a consequence, most materials high in clay are not considered suitable for highway beds, dams, or building foundations. The effects of freezing and thawing on materials high in clay (and moisture) further complicate the use of these materials for engineering purposes. This subject will receive more attention in Chapter 6.

The mineralogical makeup of soil particles also affects engineering purposes. The equidimensional grains of minerals such as quartz and feldspars, of which the sand fraction is composed, are conducive to high soil strength. In contrast, ease of compressibility, with subsequent expansion when load is removed, is characteristic of the slate-like silicate clays. This in turn gives low soil strength.

The adverse effect of some minerals on soil strength is not limited to the clay size fractions, however. For example, mica particles of all sizes are associated with low soil strength, poor compactibility, and high compressibility. Soil materials high in mica do not make good roadbeds.

Note that the various factors influencing soil strength do not act independently. The example just given of the effects of mineralogy of some soil particles on soil strength without regard to particle size could be cited. Likewise, the interaction between soil moisture and particle size is important. Thus, a moist sand is fairly stable for vehicle traffic, whereas a vehicle's tires may sink into a dry sand or a sand completely saturated with water. Apparently, a small amount of moisture results in sufficient cohesion between particles to give some soil strength. With no moisture or with the particles completely saturated with water, the sand particles tend to act independently and do not resist the traffic stress.

The complexities of engineering aspects of the physical properties of soils are beyond the purview of this text. At the same time, engineering implications must be recognized along with those relating to agriculture.

4.16

Conclusion

The physical properties of soils control to a marked extent their use for both agricultural and nonagricultural purposes. The nature and properties of the individual particles, their size distribution, and their arrangement in soils have profound effects on plant growth and on all kinds of soil manipulations and use. These effects in turn influence total nonsolid pore space as well as pore size, thereby imparting water and air relationships.

The properties of individual particles and their proportionate distribution (soil texture) are subject to little human control in field soils. Some control is possible, however, on the arrangement of these particles into aggregates (soil structure) and on the stability of these aggregates. Proper crop selection, rotation, and management help assure this control. To these management practices have been added, in recent years, conservation tillage, which minimizes soil manipulations while it decreases soil erosion and water runoff.

Study Questions

1. Compare the physical, chemical and mineralogical characteristics of sand and clay particles.
2. A gardening "expert" in his radio program advised a listener to add barnyard manure to improve the texture of his soil. Do you agree with his recommendation? Why or why not?
3. Which soil structural type(s) are most common in the plow layer (upper 15 cm) and in the lower (subsoil) layers of soils in humid temperate regions? Which are most common in the lower horizons of soil in arid regions?
4. Differentiate between particle density and bulk density. Which is most subject to human control?
5. One cubic meter of a clay soil weighs 1.4 Mg. Calculate the bulk density of this soil.
6. Calculate the percentage pore space in a soil with a bulk density of 1.2 Mg/m^3 and a particle density of 2.6 Mg/m^3.
7. Assume you have management responsibilities for three soils characterized in the following manners: a sandy loam, loam (well granulated), and clay (poorly granulated). How do you think these soils would

compare with respect to

    a. The lowest clay content each soil could have
    b. Probable particle densities and bulk densities
    c. Probable total pore space (1) immediately after a rain and (2) two weeks after a rain
    d. Probable volume of macropores and micropores
    e. Probable stability of soil aggregates

8. How are soil aggregates formed, and which constituents are responsible for their stability?

9. How do you explain the fact that clay soils, which commonly have higher total pore space than sandy soils, are often more poorly aerated than the sandy soils?

10. Under most circumstances the addition of crop residues and farm manures are sound management practices to encourage aggregation in soils and to maintain its stability. Explain how these organic materials perform these functions.

11. What is conservation tillage, and what are its advantages and disadvantages?

12. A farmer has used conservation tillage on a well-drained field successfully but has been disappointed with its use on a nearby poorly drained soil. To what would you attribute the difference?

13. How do the physical properties of soil influence the use of soils for engineering purposes?

References

Camp, C. R., and J. F. Lund. 1964. "Effects of soil compaction on cotton roots," *Crops and Soils*, **17**:13–14.

Dawud, A. Y., and F. Gray. 1979. "Establishment of the lower boundary of the sola of weakly developed soils that occur in Oklahoma," *Soil Sci. Soc. Amer. J.*, **43**:1201–07.

D'Itri, F. M. (Ed.). 1985. *A Systematic Approach to Conservation Tillage* (Chelsea, MI: Lew Publishers, Inc).

Emerson, W. W., R. C. Foster, and J. M. Oades. 1986. Organomineral complexes in relation to soil aggregation and structure," in P. M. Huang and M. Schnitzer (Eds.), *Interaction of Soil Minerals With Natural Organics and Microbes*, SSSA Special publication No. 17 (Madison, WI: Soil Sci. Soc. Amer.).

Faulkner, E. H. 1943. *Plowman's Folly* (Norman, OK: Univ. of Oklahoma Press).

Fritton, D. D., and G. W. Olson. 1972. "Bulk density of a fragipan soil in natural and disturbed profiles," *Soil Sci. Soc. Amer. Proc.*, **36**:686–89.

Griffith, D. R., J. V. Mannering, and C. B. Richey. 1976. "Energy requirements and areas of adaptation for eight tillage-planting systems for corn," *Proc. Conf. on Energy and Agr.* (St. Louis, MO: Center for the Biol. Natural Systems, Washington Univ.).

Larson, W. E., S. C. Gupta, and R. A. Useche. 1980. "Compression of agricultural soils from eight soil orders," *Soil Sci. Soc. Amer. J.*, **44**:450–57.

Laws, W. D., and D. D. Evans. 1949. "The effects of long-time cultivation on some physical and chemical properties of two rendzina soil," *Soil Sci. Soc. Amer. Proc.*, **14**:15–19.

Lyon, T. L., and H. O. Buckman. 1952. *The Nature and Properties of Soils*, 5th ed. (New York: Macmillan), p. 60.

McIntyre, S. C., et al. 1987. "Using cesium-137 to estimate soil erosion on a clearcut hillside," *J. Soil Water Conserv.*, **42**:117–20.

Mitchell, A. R., 1986. "Polyacrylamide application in irrigation water to increase infiltration," *Soil Sci.*, **141**:353–58.

Nelson, L. B., and R. J. Muckenhirn. 1941. "Field percolation rates of four Wisconsin soils having different drainage characteristics," *J. Amer. Soc. Agron.*, **33**: 1028–36.

Onofiok, O., and M. J. Singer. 1984. "Scanning electron microscope studies of surface crusts formed by simulated rainfall," *Soil Sci. Soc. Amer. J.*, **48**:1137–43.

Phillips, R. E., and S. H. Phillips. 1984. *No Tillage Agriculture: Principles and Practices* (New York: Van Nostrand Reinhold).

Soane, B. D., and J. C. Pigeon. 1975. "Tillage requirement in relation to soil physical properties," *Soil Sci.*, **119**:376–84.

Tiessen, H., J. W. B. Stewart, and J. R. Bettany. 1982. "Cultivation effects on the amounts and concentration of carbon, nitrogen, and phosphorus in grassland soils," *Agron. J.*, **74**:831–35.

Unger, P. W., and T. M. McCalla. 1980. "Conservation tillage systems," *Adv. Agron.*, **33**:1–58.

Van Doren, D. M., Jr., G. B. Triplett, and J. E. Henry. 1976. "Influence of long-term tillage, crop rotation and soil type combinations on corn yield," *Soil Sci. Soc. Amer. J.*, **40**:100–105.

Voorhees, W. B. 1984. Soil compaction, a curse or a cure?" *Solutions*, **28**:42–47 (Peoria, IL: Solutions Magazine Inc.).

Wilson, H. A., R. Gish, and G. M. Browning. 1947. "Cropping systems and season as factors affecting aggregate stability," *Soil Sci. Soc. Amer. Proc.*, **12**: 36–43.

Yule, D. F., and J. T. Ritchie. 1980. "Soil shrinkage relationships of Texas Vertisols: I. Small cores," *Soil Sci. Soc. Amer. J.*, **44**:1285–91.

Soil Water: Characteristics and Behavior

We are interested in soil-water relationships for several reasons. First, large quantities of water must be supplied to satisfy the requirements of growing plants because water is continually lost by evaporation from leaf surfaces. Thus water must be available when the plants need it, and most of it must come from the soil. Second, water is the solvent that together with the dissolved nutrients makes up the soil solution from which plants absorb essential elements. Third, soil moisture helps control two other important factors essential to normal plant growth—soil air and soil temperature. And last but not least, the control of the disposition of water as it strikes the soil determines to a large extent the incidence of soil erosion—that devastating menace that lifts soil particles from the soil surface and carries them into streams, lakes, and the oceans.

5.1

Structure and Related Properties of Water

Water participates directly in dozens of soil and plant reactions and indirectly affects many others. Its ability to do so is determined primarily by its structure. Water is a simple compound, its individual molecules containing one oxygen atom and two much smaller hydrogen atoms. The elements are bonded together covalently, each hydrogen atom sharing its single electron with the oxygen.

The resulting molecule is not symmetrical, however. Instead of the atoms being arranged linearly (H—O—H), the hydrogen atoms are attached to the

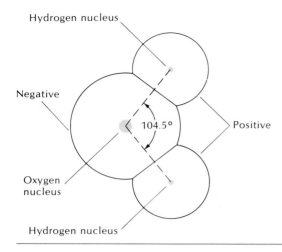

Hydrogen nucleus

Negative

Oxygen
nucleus

Hydrogen nucleus

104.5°

Positive

FIGURE 5.1
Two-dimensional representation of a water molecule showing a large oxygen atom and two much smaller hydrogen atoms. The HOH angle of 104.5° results in an asymmetrical arrangement. One side of the water molecule (that with the two hydrogens) is electropositive; the other is electronegative. This accounts for the polarity of water.

oxygen in sort of a V-shaped arrangement at an angle of only 104.5°. As shown in Figure 5.1, this results in an asymmetrical molecule with the shared electrons closer to the oxygen than to the hydrogen. Consequently, the side of the water molecule on which the hydrogen atoms are located tends to be electropositive and the opposite side electronegative. Molecules whose positive and negative charge centers do not coincide are termed *polar* molecules. The polarity of water accounts for many reactions so important in soil and plant science.

POLARITY

The property of polarity helps explain how water molecules interact with each other. Each water molecule does not act completely independently but rather is coupled with other neighboring molecules. The hydrogen (positive) end of one molecule attracts the oxygen (negative) end of another, resulting in a chain-like (polymer) grouping.

Polarity also accounts for a number of other important properties of water. For example, it explains why water molecules are attracted to electrostatically charged ions. Cations such as H^+, Na^+, K^+, and Ca^{2+} become hydrated through their attraction to the oxygen (negative) end of water molecules. Likewise, negatively charged clay surfaces attract water, this time through the hydrogen (positive) end of the molecule. Polarity of water molecules also encourages the dissolution of salts in water since the ionic components have greater attraction for water molecules than for each other.

When water molecules become attracted to electrostatically charged ions or clay surfaces, they are more closely packed than in pure water. In this packed state their energy level is lower than in pure water. Thus, when ions or clay particles become hydrated, energy must be released. That released energy is evidenced as *heat of solution* when ions hydrate or as *heat of wetting* in the case of hydrating clay particles.

HYDROGEN BONDING

The phenomenon by which hydrogen atoms act as links between water molecules is called *hydrogen bonding*. This is a relatively low-energy coupling in which a hydrogen atom is shared between two electronegative atoms such as O and N. Because of its high electronegativity, an O atom in one water molecule exerts some attraction for the H atom in a neighboring water molecule. This type of bonding accounts for the polymerization of water.

Hydrogen bonding also accounts for the relatively high boiling point, specific heat, and viscosity of water compared to the same properties of other hydrogen-containing compounds, such as H_2S, which has a greater molecular weight but no hydrogen bonding. It is also responsible for the structural rigidity of some clay crystals and for the structure of some organic compounds, such as proteins.

COHESION VERSUS ADHESION

Hydrogen bonding accounts for two basic forces responsible for water retention and movement in soils: the attraction of water molecules for each other (cohesion) and the attraction of water molecules for solid surfaces (adhesion). By *adhesion* (also called adsorption), some water molecules are held rigidly at the soil solid surfaces. In turn, these tightly bound water molecules hold by cohesion other water molecules further removed from the solid surfaces. Together, the forces of adhesion and cohesion make it possible for soil solids to retain water and control its movement and use. Adhesion and cohesion also make possible the property of plasticity possessed by clays (see Section 4.2).

SURFACE TENSION

One other important property of water that markedly influences its behavior in soils is that of surface tension. This phenomenon is commonly evidenced at liquid–air interfaces and results from the greater attraction of water molecules for each other (cohesion) than for the air above (Figure 5.2). The net effect is an inward force at the surface that causes water to behave as if its surface were covered with a stretched elastic membrane. Because of the relatively high attraction of water molecules for each other, water has a high surface tension compared to that of most other liquids. As we shall see, surface tension is an important property, especially as a factor in the phenomenon of capillarity, which determines how water moves and is retained in soil.

5.2

Capillary Fundamentals and Soil Water

The phenomenon of capillarity is a common one, the classic example being the movement of water up a wick when the lower end is immersed in water.

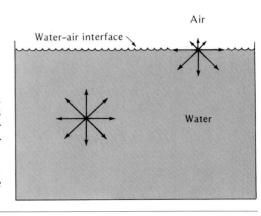

FIGURE 5.2

Comparative forces acting on water molecules at the surface and beneath the surface. Forces acting below the surface are equal in all directions since each water molecule is attracted equally by neighboring water molecules. At the surface, however, the attraction of the air for the water molecules is much less than that of water molecules for each other. Consequently, there is a net downward force on the surface molecules, and the result is something like a compressed film or membrane at the surface. This phenomenon is called *surface tension*.

Capillarity is due to two forces: (a) the attractive force of water for the solids on the walls of channels through which it moves (adhesion or adsorption) and (b) the surface tension of water, which is due largely to the attraction of water molecules for each other (cohesion).

CAPILLARY MECHANISM

Capillarity can be demonstrated by placing one end of a fine glass tube in water. The water rises in the tube, and the smaller the tube bore, the higher the water rises (Figure 5.3). The water molecules are attracted to the sides of the tube and start moving up the tube in response to this attraction. The cohesive force between individual water molecules assures that water not directly in contact with the side walls is also pulled up the tube. This continues until the weight of water in the tube counterbalances the cohesive and adhesive forces.

The height of rise in a capillary tube is inversely proportional to the tube diameter and directly proportional to the surface tension, which, in turn, is determined largely by cohesion between water molecules. The capillary rise can be approximated as

$$h = \frac{2T}{rdg}$$

where h is the height of capillary rise in the tube, T is the surface tension, r is the radius of the tube, d, is the density of the liquid, and g is the force of gravity. For water, this equation reduces to the simple expression

$$h = \frac{0.15}{r}$$

This equation emphasizes the inverse relation between height of rise and size of the tube through which the water rises.

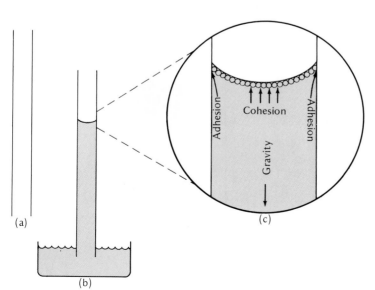

FIGURE 5.3

Diagrams illustrating the phenomenon of capillarity. (a) The situation just before lowering a fine glass tube to the water surface. (b) When the tube is inserted in the liquid, water moves up the tube owing to (c) the attractive forces between the water molecules and the wall of the tube (adhesion) and to the mutual attraction of the water molecules for each other (cohesion). Water will move up the tube until the downward pull of gravity equals the attractive forces of cohesion and adhesion.

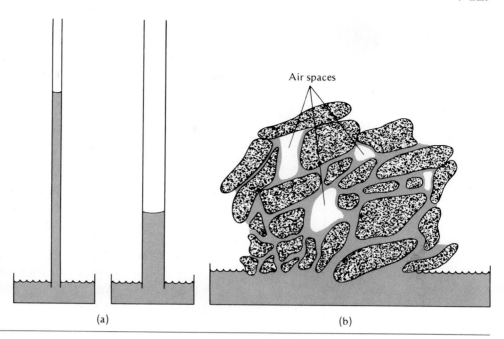

Figure 5.4
Upward movement by capillarity (a) in glass tubes of different sizes and (b) in soils. Although the mechanism is the same in the tubes and in the soil, adjustments are extremely irregular in soil because of the tortuous shape and variability in size of the soil pores and because of entrapped air.

Height of Rise in Soils

Capillary forces are at work in all moist soils. However, the rate of movement and the rise in height are less than one would expect on the basis of soil pore size. One reason is that soil pores are not straight, uniform openings like glass tubes. Furthermore, some soil pores are filled with air, which may be entrapped, slowing down or preventing the movement of water by capillarity (Figure 5.4).

The upward movement due to capillarity in soils is illustrated in Figure 5.5. Usually the height of rise resulting from capillarity is greater with fine-textured soils if sufficient time is allowed and the pores are not too small. This is readily explained on the basis of the capillary size and the continuity of the pores. With sandy soils the adjustment is rapid, but so many of the pores are noncapillary that the height of rise cannot be great.

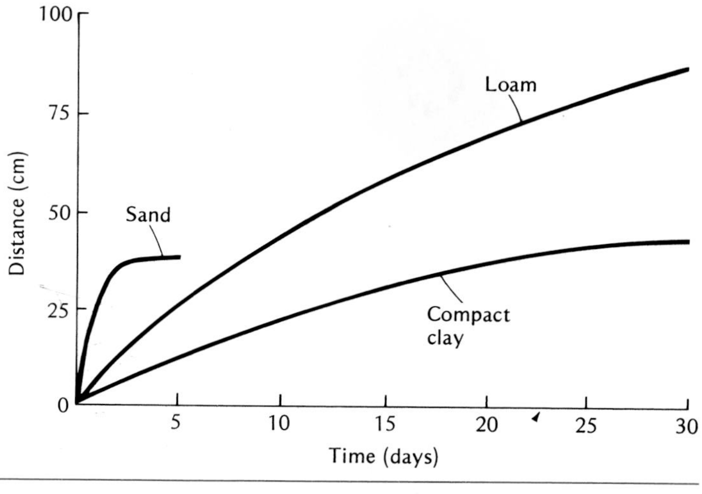

Figure 5.5
Upward movement of moisture from a water table through soils of different textures and structures. Note the very rapid rise in sand but the moderate height attained. Apparently, the pores of loam are more favorable for movement than those in compact clay.

Capillarity is traditionally illustrated as an upward adjustment. But movement in any direction takes place since the attractions between soil pores and water are as effective with horizontal pores as with vertical ones. The significance of capillarity in controlling water movement in small pores will become evident as we turn to soil water energy concepts.

5.3

Soil Water Energy Concepts

The retention and movement of water in soils, its uptake and translocation in plants, and its loss to the atmosphere are all energy-related phenomena. Different kinds of energy are involved, including potential, kinetic, and electrical. In the discussions that follow, however, *free energy* is the term we shall use to characterize the energy status of water. This is appropriate since free energy is sort of a summation of all other forms of energy available to do work. Also, the free energy level in a substance is a general measure of the tendency of that substance to change.

As we consider energy, we should keep in mind that all substances, including water, have a tendency to move or change from a state of higher to one of lower free energy. For example, all other conditions being equal, water movement in soils will generally be from a zone where the free energy of the water is high (wet soil) to one where its free energy is low (dry soil). Therefore, if we know the pertinent energy levels at various points in a soil, we can predict the direction of water movement. It is the *differences* in energy levels from one contiguous site to another that influence this water movement.

FORCES AFFECTING FREE ENERGY

The discussion of the structure and properties of water in the previous section suggests three important forces affecting the free energy of soil water. Adhesion, or the attraction of the soil solids (matrix) for water, provides a *matric* force (responsible for adsorption and capillarity) that markedly reduces the free energy of the adsorbed water molecules and even some of those held by cohesion. Likewise, the attraction of ions and other solutes for water, resulting in *osmotic* forces, tends to reduce the free energy of water in the soil solution. Osmotic movement of pure water across a semipermeable membrane into a solution is evidence of the lower free energy state of the solution.

The third major force acting on soil water is *gravity*, which tends to pull the water downward. The free energy of soil water at a given elevation in the profile is thus higher than that of pure water at some lower elevation. Such a difference in free energy level causes water to flow.

SOIL WATER POTENTIAL

While free energy levels are important, the *difference* in free energy from one water condition to another is of greater practical significance. The difference between the free energy of soil water and that of pure water in a standard reference state is termed *soil water potential* (ψ). The components of soil water potential due to differences in free energy resulting from gravitational, matric, and osmotic forces are termed *gravitational potential* (ψ_g), *matric potential* (ψ_m), and *osmotic potential* (ψ_o), respectively. Each potential results from

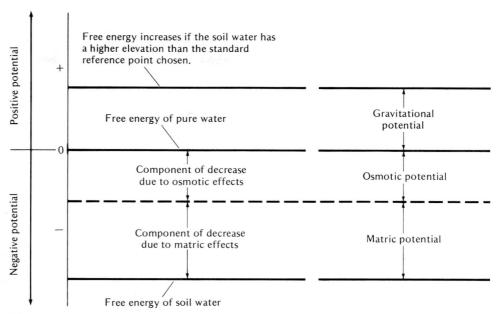

Figure 5.6
Relationship between the free energies of pure water and of soil water and the effect of elevation on free energy to illustrate the gravitational potential. Note that osmotic effects and the effects of attraction of the soil solids (matrix) for water both decrease the free energy of soil water. The component of the total decrease due to osmotic effects is the osmotic potential, whereas the component due to matric effects gives a measure of the matric potential. The effect of gravity is to increase the free energy if the standard reference point assigned to free water is at a lower elevation than the soil water in the profile. As shown, both osmotic and matric potentials are negative; this explains why they are sometimes referred to as *suction* or *tension*. The gravitational potential is generally positive. The behavior of soil water at any one time will be affected by each of these three potentials.

a different force and they are technically not additive. They act simultaneously, however, to influence water behavior in soils. The general relationship of soil water potential to free energy is shown in Figure 5.6.

GRAVITATIONAL POTENTIAL

The force of gravity acts on soil water the same as it does on any other body, the attraction being toward the Earth's center. The gravitational potential (ψ_g) of soil water may be expressed mathematically as

$$\psi_g = gh$$

where g is the acceleration due to gravity and h is the height of the soil water above a reference elevation. The reference elevation is usually chosen within the soil profile or at its lower boundary, to assure that the gravitational potential of soil water above the reference point will always be positive.

Gravity plays an important role in removing excess water from the upper rooting zones following heavy precipitation or irrigation. It will be given further attention when the movement of soil water is discussed (see Section 5.7).

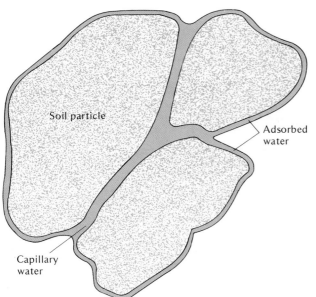

FIGURE 5.7

Two "forms" of water that together give rise to matric potential (adsorbed and capillary). The soil solids tightly adsorb water, whereas capillary forces are responsible for water's being held in the capillary pores.

MATRIC AND OSMOTIC POTENTIALS

Matric potential (ψ_m) is the result of two phenomena, adhesion (or adsorption) and capillarity (Figure 5.7). The attraction of soil solids and their exchangeable ions for water was mentioned in a previous section, as was the loss of energy (heat of wetting) when the water is adsorbed. This attraction, along with the surface tension of water, also accounts for the capillary force (see Section 5.2). The net effect of these phenomena is to reduce the free energy of soil water as compared to that of unadsorbed pure water. Consequently, matric potentials (ψ_m) *are always negative.*

The matric potential (ψ_m) exerts its effect not only on soil moisture retention but on soil water movement as well. Differences between the ψ_m of two adjoining zones of a soil encourage the movement of water. Water moves, for example, from a moist zone (high free energy) to a dry zone (low free energy). Although this movement may be slow, it is extremely important, especially in supplying water to plant roots.

The osmotic potential (ψ_o) is attributable to the presence of solutes in the soil—in other words, to the soil solution. The solutes may be inorganic salts or organic compounds. They reduce the free energy of water, primarily because the solute ions or molecules attract the water molecules. The process of osmosis is illustrated in Figure 5.8. This figure should be studied carefully.

Unlike the matric potential (ψ_m), the osmotic potential (ψ_o) has little effect on the mass movement of water in soils. Its major effect is on the uptake of water by plant roots. In soils high in soluble salts, ψ_o may be greater in the soil solution than in the plant root cells. This leads to constraints in the uptake of water by the plants. Since water vapor pressure is lowered by the presence of solutes, ψ_o also affects the movement of water vapor. The relationship between the matric and osmotic components of total soil water potential is shown in Figure 5.9. This figure should be studied carefully to be certain the meaning of these two potentials is clear.

METHODS OF EXPRESSING ENERGY LEVELS

Several units have been used to express differences in energy levels of soil water. A common means of expressing potential is in terms of the height in

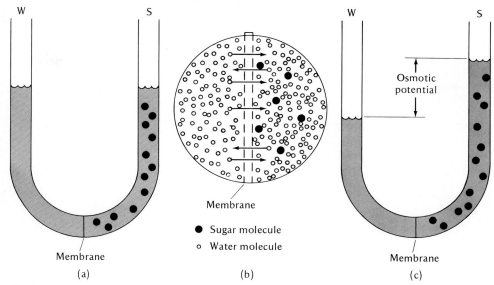

FIGURE 5.8

Illustration of the process of osmosis and of osmotic pressure. (a) A U tube containing water (W) in the left arm and a solution (S) of sugar in water in the right arm. These liquids are separated by a membrane that is permeable to water molecules but not to the dissolved sugar. (b) Enlarged portion of the membrane with water molecules moving freely from the water side to the solution side and vice versa. The sugar molecules, in contrast, are unable to penetrate the membrane. Since the effect of sugar is to decrease the free energy of the water on the solution side, more water passes from left to right than from right to left. (c) At equilibrium sufficient water has passed through the membrane to bring about a significant difference in the heights of liquid in the two arms. The difference in the levels in the W and S arms represents the osmotic potential. [Modified from Keeton (1972)]

centimeters of a unit water column whose weight just equals the potential under consideration. The greater the centimeter height, the greater is the potential measured.

Soil water potential (ψ) also is commonly expressed in terms of standard atmospheric pressure at sea level, which is 14.7 lb/in.^{2}, 760 mm Hg, or 1020 cm of water. The unit termed *bar* approximates that of a standard atmosphere. A *millibar* is 1/1000 bar. Ten bars is equivalent to the SI unit *megapascal* (MPa), which, along with the bar, will be used in this text. The relationship among these three means of expressing soil water potential is seen in Table 5.1.

TABLE 5.1

Approximately Equivalents of Common Means of Expressing Differences in Energy Levels of Soil Water

Height of unit column of water (cm)	Soil water potential (bars)	Soil water potential (MPa)[a]
0	0	0
10.2	−0.01	−0.001
102	−0.1	−0.01
306	−0.3	−0.03
1,020	−1.0	−0.1
15,300	−15	−1.5
31,700	−31	−3.1
102,000	−100	−10.0

[a] The SI unit megapascal (MPa) is equivalent to 10 bars.

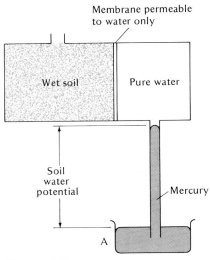
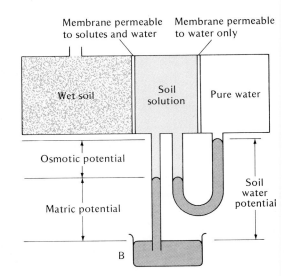

FIGURE 5.9

Relationship among osmotic, matric, and combined soil water potentials. Assume a container of soil separated from pure water by a membrane permeable only to the water; the pure water is connected with a vessel of mercury through a tube as shown (left). Water will move into the soil in response to the attractive forces associated with soil solids (matric) and with solutes (osmotic). At equilibrium the height of rise in the mercury in the tube above vessel A is a measure of this combined soil water potential (matric plus osmotic). If a second container were to be placed between the pure water and the soil and if it were separated from the soil by a membrane permeable to *both* water and solutes (right), ions would move from the soil into this container, eventually giving a concentration not too different from the soil solution. The difference between the free energies of the pure water and of the soil solution gives a measure of the *osmotic potential*. The *matric potential* is the difference between the combined and osmotic potentials and is measured by the height of rise of the mercury in vessel B. The gravity potential is not shown in this diagram. [Modified from Richards (1965)]

5.4

Soil Moisture Content and Soil Water Potential

The previous discussions suggest an inverse relation between the water content of soils and the tenacity with which the water is held. Water is more apt to flow out of a wet soil than from one low in moisture. As we might expect, many factors affect the relationship between soil water potential (ψ) and moisture content (θ). A few examples will illustrate this point.

SOIL MOISTURE VERSUS ENERGY CURVES

The relationship between soil water potential (ψ) and moisture content (θ) of three soils of different textures is shown in Figure 5.10. The absence of sharp breaks in the curves indicates a gradual change in potential with increased soil water and vice versa. The clay soil holds much more water at a given potential level than does loam or sand. Likewise, at a given moisture content, the water is held much more tenaciously in the clay than in the other two soils. As we shall see, however, as much as half or even more of the water held by clay soils is held so tightly that it cannot be removed by growing plants. In any case, the influence of texture on soil moisture retention is obvious.

The structure of a soil also influences its soil moisture–energy relationships. A well-granulated soil has more total pore space than one with poor granulation or one that has been compacted. The reduced pore space may be reflected in a lower water-holding capacity. The compacted soil also may have a higher proportion of small and medium-sized pores, which tend to hold the water with greater tenacity than do the larger pores.

The soil moisture versus potential curves in Figure 5.10 have marked practical significance. They illustrate retention–energy relationships that influence various field processes, such as the movement of water in soils and the uptake and use of water by plants. The curves should be referred to frequently as the applied aspects of soil water behavior are considered in the following sections.

5.5

Measuring Soil Moisture Content and Tension

Two general types of measurements are applied to soil water. The moisture content may be measured, directly or indirectly, or the soil moisture potential (tension) may be determined.

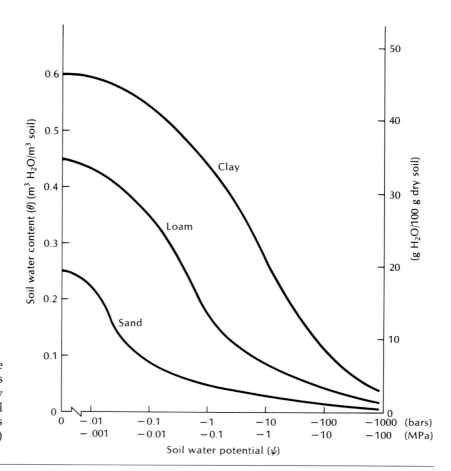

FIGURE 5.10
Soil moisture/potential curves for three representative mineral soils. The curves show the relationship obtained by slowly drying completely saturated soils. The soil moisture potential ψ (which is negative) is expressed in terms of bars (upper scale) and megapascals (MPa) (lower scale).

MOISTURE CONTENT

A number of methods are used to measure the amount of moisture in a soil. *Gravimetric* procedures permit the direct measurement of the amount of water associated with a given mass (and, indirectly, volume) of dry soil solids. A sample of moist soil, usually taken in cores in the field, is weighed, then dried in an oven at a temperature of 100–110 °C, and finally weighed again. The water lost by the soil represents the soil moisture in the moist sample.

The most common means of expressing soil water content is the mass or volume of water associated with a given mass or volume of soil solids. The *volumetric water content* (θ) is defined as the volume of water associated with a given volume (usually a cubic meter) of dry soil solids (see Figure 5.10). A comparable expression is the *mass water content*, or the mass of water associated with a given mass (usually 1 kg) of soil. Either of these two means of expression is acceptable; however, we shall use the volumetric water content (θ) in this text.

Another practical unit for measuring water added to soil (especially by irrigation) is that of the *acre foot*. This is the amount of water needed to cover

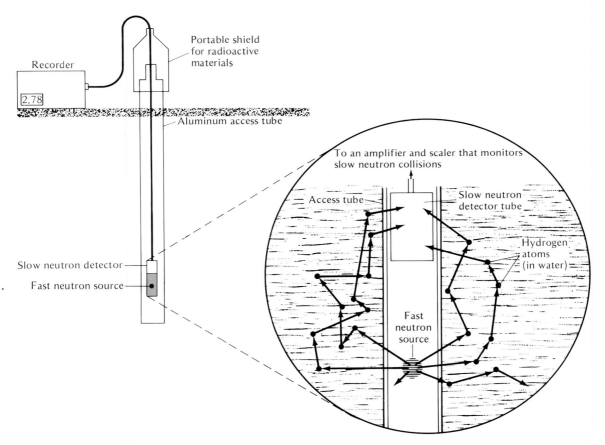

FIGURE 5.11

How a neutron moisture meter operates. The probe, containing a source of fast neutrons and a slow neutron detector, is lowered into the soil through an access tube. Neutrons are emitted by the source (e.g., radium or americium–beryllium) at a very high speed (fast neutrons). When these neutrons collide with a small atom such as the hydrogen contained in soil water, their direction of movement is changed and they lose part of their energy. These "slowed" neutrons are measured by a detector tube and a scalar. The reading is related to the soil moisture content.

an acre of land to a depth of 1 ft. Similarly, an acre inch represents the water needed to cover an acre to a depth of 1 in. Such terminology is often used in ascertaining the amount of irrigation water needed or used.

Two indirect methods of measuring soil water content will be mentioned. First is the *resistance* method, which takes advantage of the fact that the electrical resistance of certain porous materials such as gypsum, nylon, and fiberglass is related to their water content. When blocks of these materials, suitably imbedded with electrodes, are surrounded by soil, they absorb water in proportion to the soil's moisture content. In turn, the electrical resistance of the blocks is related to the amount of water they absorb. After being calibrated in the laboratory, blocks are placed in moist soil in the field, from which they absorb water. When equilibrium is reached, the electrical resistance is read. This resistance gives a measure of the water absorbed and, in turn, of the water content of the surrounding soil. Using these blocks, changes in soil moisture content in the field can be monitored without disturbing the soil. These blocks give reasonably accurate moisture readings in the range of −1 to −15 bars potential.

Another indirect method of determining soil moisture in the field involves *neutron scattering*. The principle of the neutron moisture meter is based on the ability of hydrogen atoms to reduce drastically the speed of fast-moving neutrons and to scatter them (Figure 5.11). These meters are versatile and give accurate results in mineral soils where water is the primary source of combined hydrogen. In organic soils, however, the method has some constraints since much of the hydrogen in these soils is combined in organic substances rather than in water.

MOISTURE TENSION

The tension or suction with which water is held in soils is an expression of soil water potential (ψ), except it is expressed in positive rather than negative terms. *Field tensiometers*, such as the one shown in Figure 5.12, measure this

FIGURE 5.12
Tensiometer method of determining moisture stress in the field. (a) Cross section showing the essential components of a tensiometer. Water moves through the porous end of the instrument in response to the pull of the soil. (b) A tensiometer in place in the field showing a portable meter to measure the tension and in turn the potential in millibars (mbar). [Photo courtesy A. M. Wierenga, Soil Measurement Systems, Las Cruces, NM]

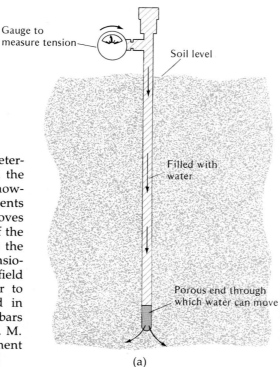

Gauge to measure tension

Soil level

Filled with water

Porous end through which water can move

(a)

(b)

tension. The tensiometer is filled with water and then placed in the soil. Its effectiveness is based on the principle that water in the tensiometer is drawn through a fine porous cup into the adjacent soil until equilibrium is reached, at which time the potential in the soil is the same as that in the tensiometer. Tensiometers are used successfully in determining the need for irrigation when the soil is to be kept well-supplied with water. Their range of usefulness is between 0 and −0.8 bar potential, a range comparable with that attained using laboratory tensiometers called *tension plates*.

A *pressure membrane* apparatus (Figure 5.13) is used to measure matric potential–moisture content relations at potential values as low as −100 bars. This important laboratory tool makes possible simultaneous accurate measurement of energy–soil moisture relations of a number of soil samples over a wide energy range in a relatively short time.

5.6

Types of Soil Water Movement

Three types of movement within the soil are recognized—*saturated flow, unsaturated flow,* and *vapor movement.* Both saturated and unsaturated flow involve liquid water in contrast to vapor flow. We shall consider liquid water flow first.

The flow of liquid water is due to a *gradient* in matric potential (ψ_m) from one soil zone to another. The direction of flow is from a zone of higher ψ_m to one of lower ψ_m moisture potential. Saturated flow takes place when the soil pores are completely filled (or saturated) with water. Unsaturated flow occurs when the pores in even the wettest soil zones are only partially filled with water. In each case, moisture flow is due to energy–soil relationships. This will be evident as we consider the three types of movement.

5.7

Saturated Flow Through Soils

In most soils, at least some soil pores contain air as well as water; that is, they are unsaturated. Under some conditions, however, at least part of a soil profile may be completely saturated; that is, all pores, large and small, are filled with water. The lower horizons of poorly drained soils are often saturated as are portions of well-drained soils above stratified layers of clay. During and immediately following a heavy rain or irrigation application, pores in the upper soil zones are often filled entirely with water.

The flow of water under saturated conditions is determined by two major factors: the *hydraulic force* driving the water through the soil (commonly gravity) and the *hydraulic conductivity*, or the ease with which the soil pores permit water movement.

Hydraulic conductivity can be expressed mathematically as

$$V = Kf$$

where V is the total volume of water moved per unit time, f is the water moving force and K is the hydraulic conductivity of the soil. The hydraulic conductivity of a uniform saturated soil is essentially constant and is dependent on the size and configuration of the soil pores. This is in contrast to the values in an unsaturated soil, where hydraulic conductivity decreases with the moisture content.

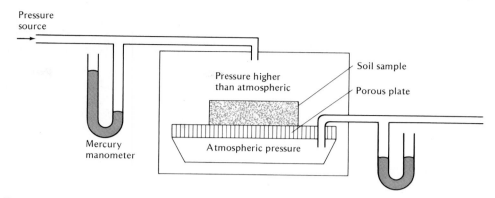

FIGURE 5.13
Pressure membrane apparatus used to determine moisture content–matric potential relations in soils. An outside source of gas creates a pressure inside the cell. Water is forced out of the soil through a porous plate into a cell at atmospheric pressure. The applied pressure when the downward flow ceases gives a measure of the moisture potential in the soil. This apparatus will measure much lower soil moisture potential values (drier soils) than will tensiometers or tension plates.

An illustration of vertical saturated flow is shown in Figure 5.14. The driving force, known as the *hydraulic gradient*, is the difference in height of water above and below the soil column. The volume of water moving down the column will depend on this force as well as on the hydraulic conductivity of the soil.

It should not be inferred from Figure 5.14 that saturated flow occurs only down the profile. The hydraulic force also will cause horizontal and upward flow. The rate of such flow is usually not quite as rapid, however, since the

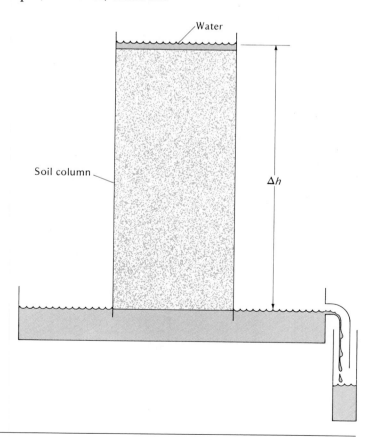

FIGURE 5.14
Saturated (percolation) flow in a column of soil. All soil pores are filled with water. The force drawing the water through the soil is Δh, the difference in the heights of water above and below the soil layer. This same force could be applied horizontally. The water is shown running off into a side container to illustrate that water is actually moving down the profile.

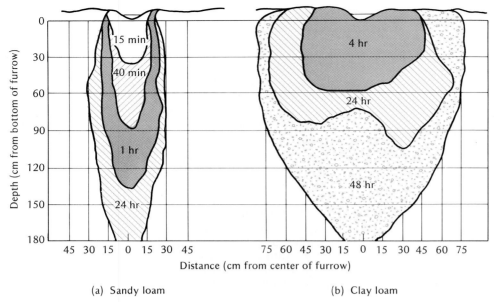

FIGURE 5.15
Comparative rates of irrigation water movement into a sandy loam and a clay loam. Note the much more rapid rate of movement in the sandy loam, especially in a downward direction. [Redrawn from Cooney and Peterson (1955)]

force of gravity does not assist horizontal flow and hinders upward flow. Downward and horizontal flow is illustrated in Figure 5.15, which records the flow of irrigation water into two soils, a sandy loam and a clay loam. Some of the water movement was likely by saturated flow. The water moved down much more rapidly in the sandy loam than in the clay loam. On the other hand, horizontal movement was much more evident in the clay loam.

FACTORS INFLUENCING THE HYDRAULIC CONDUCTIVITY OF SATURATED SOILS

Any factor affecting the size and configuration of soil pores will influence hydraulic conductivity. The total flow rate in soil pores is proportional to the fourth power of the radius. Thus, flow through a pore 1 mm in radius is equivalent to that in 10,000 pores with a radius of 0.1 mm, even though it takes only 100 pores of radius 0.1 mm to give the same cross-sectional area as a 1-mm pore. Obviously, the macropore spaces will account for most of the saturated water movement.

The texture and structure of soils are the properties to which hydraulic conductivity is most directly related. Sandy soils generally have higher saturated conductivities than finer textured soils. Likewise, soils with stable granular structure conduct water much more rapidly than do those with unstable structural units, which break down upon being wetted. Fine clay and silt can clog the small connecting channels of even the larger pores. Fine-textured soils that crack during dry weather at first allow rapid water movement; later, the cracks swell shut, thereby drastically reducing water movement.

5.8

Unsaturated Flow in Soils

In saturated soils, the relatively rapid water movement is through the large and continuous pores. But in unsaturated soils, these macropores are filled with air, leaving only the finer pores to accommodate water movement, which, as would be expected, is slow. This fact is illustrated in Figure 5.16, which shows the general relationship between matric potential (ψ_m) (and, in turn, moisture level) and hydraulic conductivity. Note that at or near zero potential (which characterizes the saturated flow region), the hydraulic conductivity is thousands of times greater than at potentials that characterize typical unsaturated flow (−0.1 bar and below).

At high potential levels (high moisture contents), hydraulic conductivity is higher in the sand than in the clay. The opposite is true at low potential values (low moisture contents). This relationship is to be expected since the dominance of large pores in the coarse-textured soil encourages saturated flow, whereas the prominence of finer (capillary) pores in the clay soil encourages more unsaturated flow than in the sand.

From the preceding discussion we see that unsaturated flow is governed by the same general principles affecting saturated flow—that is, by the hydraulic conductivity and the driving force. In this case, the force is primarily the *matric potential gradient* or the difference in the matric potential of the moist soil areas and the drier areas into which the water is moving. Movement will be from a zone of thick moisture films (higher matric potential, e.g., −0.001 MPa) to one of thin films (lower matric potential, e.g., −0.1 MPa).

The influence of potential gradient on water movement is illustrated by the moisture curves shown in Figure 5.17. The researchers measured in the laboratory the rate of water movement from three moist soil samples to adjacent dry soil samples. Water moved more rapidly from the sample with highest moisture content. The higher the water content in the moist soil, the greater the matric potential gradient between the moist and dry soil and, in

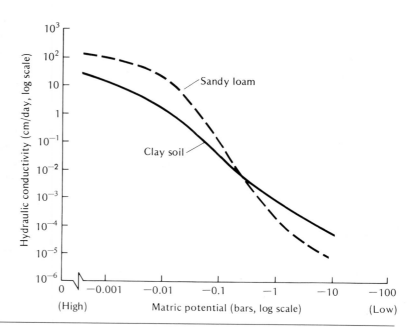

FIGURE 5.16
Generalized relationship between matric potential and hydraulic conductivity for a sandy soil and a clay soil (note log scales). Saturation flow takes place at or near zero potential, while most of the unsaturated flow occurs at a potential of −0.1 bar or below.

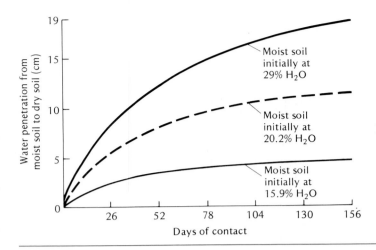

FIGURE 5.17
Rate of water movement from a moist soil at three moisture levels to a drier one. The higher the water content of the moist soil, the greater is the potential gradient and the more rapid the delivery. Water adjustment between two slightly moist soils at about the same water content will be exceedingly slow. [After Gardner and Widtsoe (1921)]

turn, the more rapid the flow. In this case, the rate of movement obviously is a function of the matric potential gradient.

5.9

Water Movement in Stratified Soils

The discussion up to now has dealt almost entirely with soils that are assumed to be uniform in texture and structure. In the field, however, layers differing in physical makeup from the overlying horizons are common. These layers have a profound influence on water movement and deserve specific attention.

Various kinds of stratification are found in many soils. Impervious silt or clay pans are common, as are sand and gravel lenses or other subsurface layers. In all cases, the effect on water movement is similar—that is, the downward movement is impeded. The influence of layering is represented in Figure 5.18, where a layer of sand is seen to impede downward movement of water in an otherwise fine-textured soil. The macropores of the sand offer less attraction for the water than does the finer textured material. Only when the downward-moving water builds up at the interlayer interface will the water be so loosely held by the fine-textured material that the sand below can attract the water and permit it to move downward. Study Figure 5.18 to learn why this sandy layer can reduce downward movement of water.

The significance of stratification is obvious. For example, because it retards downward movement, stratification definitely influences the amount of water the upper part of the soil holds in the field. The stratified layer acts as a moisture barrier until a relatively high moisture level is built up. This gives a much higher field moisture level than that normally encountered in freely drained soils.

5.10

Water Vapor Movement in Soils

Two types of water vapor movement occur in soils, *internal* and *external*. Internal movement takes place within the soil, that is, in the soil pores.

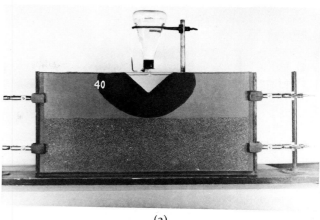

(a)

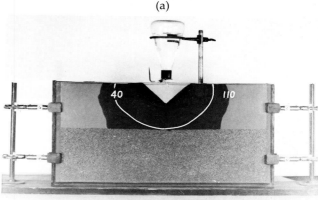

(b)

FIGURE 5.18
Downward water movement in soils with a stratified layer
of coarse material. (a) Water is applied to the surface of a
medium-textured topsoil. Note that after 40 min downward movement is no greater than movement to the sides,
indicating that in this case the gravitational force is insignificant compared to the matric potential gradient between
dry and wet soil. (b) The *downward* movement stops when
a coarse-textured layer is encountered. After 110 min no
movement into the sandy layer has occurred. The macropores of the sand provide less attraction for water than the
finer texture soil above. Only when the moisture (and in
turn the matric potential gradient) is raised sufficiently will
the moisture move into the sand. (c) After 400 min the moisture content of the overlying layer becomes sufficiently
high to give a moisture potential of about −0.5 atm or
more and downward movement into the coarse material
takes place. Thus sandy layers, as well as compact silt and
clay, influence downward moisture movement in soils.
[Courtesy W.H. Gardner, Washington State University]

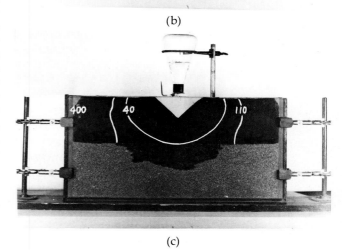

(c)

External movement occurs at the land surface, and water vapor is lost by
surface evaporation (see Section 14.4).

Water vapor moves from one point to another within the soil in response
to differences in vapor pressure. Thus, water vapor will move from a moist
soil where the soil air is nearly 100% saturated with water vapor (high vapor
pressure) to a drier soil where the vapor pressure is somewhat lower.
Likewise, if the temperature of one part of a uniformly moist soil is lowered,
the vapor pressure will decrease and water vapor will tend to move toward
this cooler part. Heating will have the opposite effect. Figure 5.19 illustrates
the relationships.

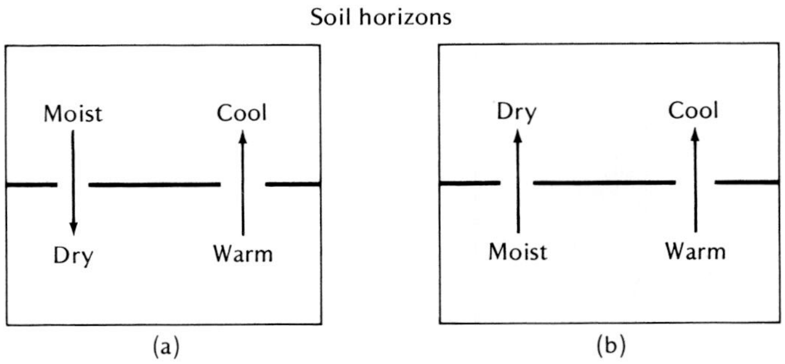

Soil horizons

FIGURE 5.19
Vapor movement tendencies that may be expected between soil horizons differing in temperature and moisture. In (a) the tendencies more or less negate each other, but in (b) they are coordinated and considerable vapor transfer might be possible if the liquid water in the soil capillaries does not interfere.

The actual amount of water vapor in a soil at optimum moisture for plant growth is surprisingly small, being at one time perhaps no more than 10 kg in the upper 15 cm of a hectare of soil. This compares with 375,000 kg of liquid water in the same soil volume.

Because the amount of water vapor is small, its movement in soils is of limited practical importance if the soil moisture is kept near optimum for plant growth. In dry soils, however, vapor moisture movement may be of considerable significance, especially in supplying moisture to drought-resistant desert plants, many of which can exist at extremely low moisture levels.

5.11

Retention of Soil Moisture in the Field

Keeping in mind the energy–soil moisture relations covered in previous sections, we now turn to some more practical considerations. We shall start by following the moisture and energy relations of a soil during and after a heavy rain or the application of irrigation water.

MAXIMUM RETENTIVE CAPACITY

During a heavy rain or while being irrigated, a soil may become saturated with water and ready downward drainage will occur. At this point, the soil is said to be *saturated* with respect to water (Figure 5.20) and at its *maximum retentive capacity*. The matric potential is high, being nearly the same as that of pure water.

FIELD CAPACITY

Following the rain or irrigation, there will be a continued relatively rapid downward movement of some of the water in response to the hydraulic gradient, mostly gravity. After two or three days, this rapid downward movement will become negligible. The soil then is said to be at its *field capacity*. At this time, water has moved out of the *macropores*, and its place has been taken by air. The *micropores* or *capillary pores* are still filled with water and will supply the plants with needed moisture. The matric potential will vary slightly from soil to soil but generally ranges from -0.1 to -0.3 bar, assuming drainage into a less moist zone of similar porosity. Moisture movement will continue to take place, but the rate of movement (unsaturated flow) is slow because it now is due primarily to capillary forces, which are effective only in micropores (Figure 5.20).

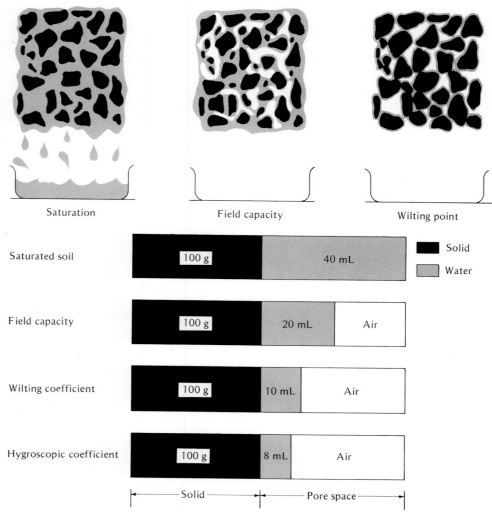

FIGURE 5.20
Volumes of water and air associated with 100 g of a well-granulated silt loam at different moisture levels. The top bar shows the situation when a representative soil is completely saturated with moisture. This situation will usually occur for short periods of time during a rain or when the soil is being irrigated. Water will soon drain out of the larger pores (*macropores*). The soil is then said to be at the *field capacity*. Plants will remove moisture from the soil quite rapidly until they begin to wilt. When permanent wilting of the plants occurs, the soil moisture is said to be at the *wilting coefficient*. There is still considerable moisture in the soil, but it is held too tightly to permit its absorption by plant roots. A further reduction in moisture content to the *hygroscopic coefficient* is illustrated in the bottom bar. At this point the water is held very tightly, mostly by the soil colloids. [Top drawings modified from *Irrigation on Western Farms* published by the U.S. Departments of Agriculture and Interior]

PERMANENT WILTING PERCENTAGE OR WILTING COEFFICIENT

As plants absorb water from a soil, they lose most of it through evaporation at the leaf surfaces (transpiration). Some water also is lost by evaporation directly from the soil surface. These two losses occur simultaneously, and the combined loss is termed evapotranspiration.

As the soil dries, plants begin to wilt to conserve moisture during the daytime. At first the plants will regain their vigor at night, but ultimately they will remain wilted night and day. Although not dead, the plants are now in a permanently wilted condition and will die if water is not provided. Under this condition, a measure of soil water potential (ψ) shows a value of about -15 bars for most crop plants. Some xerophytes can continue to remove water at this and even more negative potentials.

The soil moisture content of the soil at this stage is called the *wilting coefficient* or *permanent wilting percentage*. The water remaining in the soil is found in the smallest of the micropores and around individual soil particles (Figure 5.20). Obviously, a considerable amount of the water in soils is not available to higher plants.

HYGROSCOPIC COEFFICIENT

As soil moisture is lowered below the wilting point, the water molecules that remain are very tightly held, mostly being adsorbed by colloidal soil surface. This state is approximated when the atmosphere above a soil sample is essentially saturated (98% relative humidity) with water vapor and equilibrium is established. The water is held so tightly (-31 bars) that much of it is considered nonliquid and can move only in the vapor phase. The moisture content of the soil at this point is termed the *hygroscopic coefficient*. Soils high in colloidal materials will hold more water under these conditions than will sandy soils and those low in clay and humus (Table 5.2).

POTENTIAL AND MOISTURE CONTENT

As soil moisture is reduced from the field capacity to the hygroscopic coefficient, the potential–moisture curves described in Section 5.4 are pertinent. This relationship is illustrated in Figure 5.21, which shows the moisture content–matric potential relationship for a loam soil and identifies the ranges in potential for each of the field soil conditions just described. The diagram at the right of this figure also suggests physical and biological classification schemes for soil water. However, this diagram is coupled intentionally with the moisture–potential curve to emphasize the fact that there are no clearly identifiable "forms" of water in soil. There is only a gradual change in potential with moisture content. This should be kept in mind in the next section as we discuss some commonly used soil moisture classification schemes.

TABLE 5.2

Volumetric Water Content (θ) at Field Capacity and Hygroscopic Coefficient for Three Representative Soils and the Calculated Capillary Water

Note that the clay soil retains most water at the field capacity, but much of that water is held tightly in the soil at -31 bars potential by soil colloids (hygroscopic coefficient).

	Volume % (θ)		
Soil	Field capacity (-0.3 bar)	Hygroscopic coefficient (-31 bars)	Capillary water (col. 1 − col. 2)
Sandy loam	12	3	9
Silt loam	30	10	20
Clay	35	18	17

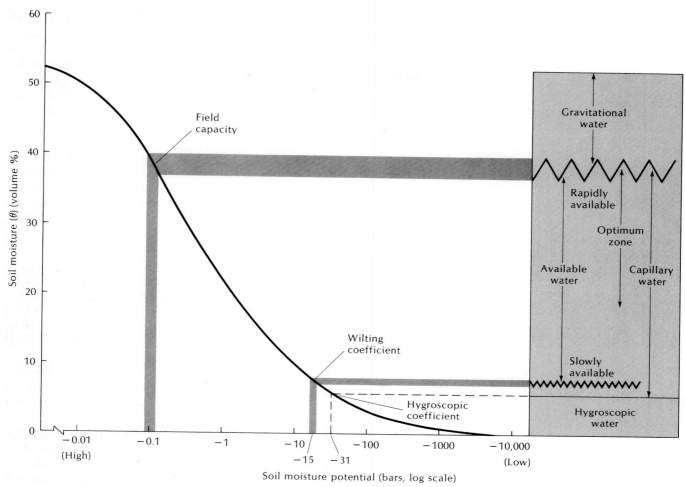

FIGURE 5.21
Potential moisture curve of a loam soil as related to different terms used to describe water in soils. The wavy lines in the diagram to the right suggest that measurements such as field capacity are not very quantitative. The gradual change in potential with soil moisture change discourages the concept of different "forms" of water in soils. At the same time, such terms as *gravitational* and *available* assist in the qualitative description of moisture utilization in soils.

5.12

Conventional Soil Moisture Classification Schemes

On the basis of observations of soil-water-plant relations, two types of soil water classification schemes have been developed: physical and biological. These schemes are useful in a practical way even though they lack the scientific basis that characterizes the preceding moisture and energy discussions.

PHYSICAL CLASSIFICATION

From a physical point of view, the terms *gravitational, capillary,* and *hygroscopic* waters are identified in Figure 5.21. Water in excess of the field capacity (-0.1 to -0.3 bar and upward) is termed gravitational. Even though energy of retention is low, gravitational water is of limited use to plants because it is present in the soil for only a short time, and, while in the soil, it occupies the larger pores, thereby reducing soil aeration. The removal of gravitational water from the soil by drainage is generally a requisite for optimum plant growth.

As the name suggests, capillary water is held in the pores of capillary size and behaves according to laws governing capillarity. Such water includes most of the water taken up by growing plants and exerts potentials between -0.1 and -0.31 bar.

Hygroscopic water is that bound tightly by the soil solids at potential values lower than -31 bars. It is essentially nonliquid and moves primarily in vapor form. Higher plants cannot absorb hygroscopic water, but some microbial activity has been observed in soils containing only hygroscopic water.

BIOLOGICAL CLASSIFICATION

There is a definite relationship between moisture retention and its use by plants. Gravitational water is of limited use and may be harmful. In contrast, moisture retained in the soil between the field capacity (-0.1 to -0.3 bar) and the wilting coefficient (-15 bars) is said to be usable by plants and as such is *available water*. Water held at a potential lower than -15 bars is called *unavailable* to most plants (Figure 5.21).

In most soils, optimum growth of plants takes place when the soil moisture content is kept near the field capacity with a moisture potential of -1 bar or higher. Thus, the moisture zone for optimum plant growth does not extend over the complete range of moisture availability.

The various terms employed to describe soil water physically and biologically are useful in a practical way, but at best they are only semiquantitative. For example, measurement of field capacity tends to be somewhat arbitrary because the value obtained is affected by such factors as the initial soil moisture in the profile before wetting, the removal of water by plants, and surface evaporation during the period of downward flow. Also, the determination as to when the downward movement of water due to gravity has "essentially ceased" is rather arbitrary. These facts stress once again that there is no clear line of demarcation between different "forms" of soil water.

5.13

Factors Affecting Amount of Plant-Available Soil Moisture

The amount of soil water available for plant uptake is determined by a number of factors, including moisture potential relations (matric and osmotic), soil depth, and soil stratification or layering. Each will be discussed briefly.

MATRIC POTENTIAL

Matric potential (ψ_m) influences the amount of soil moisture plants can take up because it affects the amounts of water at the field capacity and at the wilting coefficient. These two characteristics, which determine the quantity of water a given soil can supply to growing plants, are influenced by the texture, structure, and organic matter content of the soil.

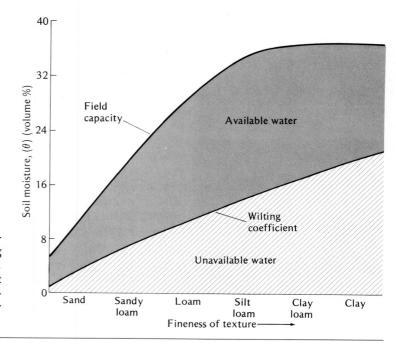

FIGURE 5.22
General relationship between soil moisture characteristics and soil texture. Note that the wilting coefficient increases as the texture becomes finer. The field capacity increases until we reach the silt loams, then levels off. Remember these are representative curves; individual soils would probably have values different from those shown.

The general influence of texture is shown in Figure 5.22. Note that as fineness of texture increases, there is a general increase in available moisture storage from sands to loams and silt loams. However, clay soils frequently provide less available water than do well-granulated silt loams. The comparative available water-holding capacities are also shown by this graph.

The influence of organic matter deserves special attention. The available moisture content of a well-drained mineral soil containing 5% organic matter will probably be higher than that of a comparable soil with 3% organic matter. One might erroneously assume that this favorable effect is a result of the higher available water content of the organic matter. Such is not the case. Rather, most of the benefit of organic matter is attributable to its favorable influence on soil structure and, in turn, on the volume of soil pores. Although humus has high moisture content at field capacity, its wilting coefficient is also proportionately high. Thus, the contribution of humus toward available moisture is primarily indirect through its effect on soil structure.

OSMOTIC POTENTIAL

The presence of salts in soils, either from applied fertilizers or as naturally occurring compounds, can influence soil water uptake. Osmotic potential (ψ_o) effects in the soil solution tend to reduce the range of available moisture in such soils by increasing the amount of water left in the soil at the time the plants wilt permanently (wilting coefficient). For soils high in salts, the total moisture stress will include the matric potential as well as the osmotic potential of the soil solution. In most humid region soils these osmotic potential effects are insignificant. In contrast, they become of considerable importance in some soils of arid and semiarid regions, especially those with significant salt contents.

SOIL DEPTH AND LAYERING

All other factors being equal, deep soils will have greater total available moisture-holding capacities than shallow ones. For deep-rooted plants, this is of practical significance, especially in those subhumid and semiarid regions where supplemental irrigation is not possible.

Soil stratification or layering will influence markedly the available water and its movement in the soil. Impervious layers slow down drastically the rate of water movement and also restrict the penetration of plant roots, thereby reducing the soil depth from which moisture can be drawn. Sandy layers also act as barriers to soil moisture movement from the finer textured layers above, as explained in Section 5.9 and Figure 5.18.

The development of wide cracks that extend downward into the profile of dry soils high in clay (especially Vertisols) markedly affects water infiltration into these soils. Initially, the infiltration of rain or irrigation water is very rapid, as the water literally pours downward into the cracks. Such movement permits a considerable amount of water to enter the soil and to move down deep into the profile. Very soon, however, the clays are wetted and swell, forcing the cracks to close. Further infiltration is very slow. Fortunately, however, considerable soil moisture is already in the soil for subsequent use by plants.

The capacity of soils to store available moisture determines to a great extent their usefulness in practical agriculture. This capacity is often the buffer between an adverse climate and crop production. It becomes more significant as the use of water for all purposes—industrial and domestic as well as agricultural—begins to tax the supply of this all-important natural resource.

5.14

How Plants are Supplied with Water—Capillarity and Root Extension

At any one time, only a small proportion of the soil water is adjacent to the absorptive plant root surfaces. How then do the plant roots get access to the immense amount of water (see Section 14.5) needed to offset transpiration by vigorously growing crops? Two phenomena seem to account for this access: the capillary movement of the soil water to plant roots and the growth of the roots into moist soil.

RATE OF CAPILLARY MOVEMENT

When plant rootlets absorb water, they reduce the moisture content, thus reducing the potential in the soil immediately surrounding them. In response to this lower potential, water tends to move toward the plant roots. The rate of movement depends on the magnitude of the potential gradients developed and the conductivity of the soil pores. With some sandy soils, the adjustment may be comparatively rapid and the flow appreciable. In fine-textured and poorly granulated clays, the movement will be sluggish and only a meager amount of water will be delivered.

The total distance that water flows by capillarity on a day-to-day basis may be only a few centimeters. This would lead one to believe that capillary movement is not a significant means of enhancing moisture uptake by plants. However, if the roots have penetrated much of the soil volume, movement over greater distances may not be necessary. Even during periods of hot, dry weather when evaporative demand is high, capillary

TABLE 5.3
Length of Soybean Roots at Different Soil Depths in a Captina Silt Loam (Typic Fragiudult) in Arkansas

Soil depth (cm)	Root length (km/m^3)	
	Nonirrigated	Irrigated
0–16	76	89
16–32	30	37
32–48	21	27
48–64	14	16

Calculated from Brown et al. (1985).

movement can be an important means of providing water to plants. It is of special significance during periods of low moisture content when plant root extension is minimized.

RATE OF ROOT EXTENSION

Capillary movement of water is complemented by rapid rates of root extension, which assure that new root–soil contacts are constantly being established. Such root penetration may be rapid enough to take care of most of the water needs of a plant growing in a soil at optimum moisture. The mat of roots, rootlets, and root hairs in a meadow is an example of the viable root systems of plants. Table 5.3 provides data on the length of roots of soybeans in one experiment. To the figures given must be added the length of thousands of root hairs which penetrate the soil.

The primary limitation of root extension is the small proportion of the soil with which roots are in contact at any one time. Even though the root surface is considerable, as shown in Table 5.3, root–soil contacts commonly account for less than 1% of the total soil surface area. This suggests that most of the water must move from the soil to the root even though the distance of movement may be no more than a few millimeters. It also suggests the complementarity of capillarity and root extension as means of providing soil water for plants.

ROOT DISTRIBUTION

The distribution of roots in the soil profile determines to a considerable degree the plant's ability to absorb soil water. Some plants, such as corn and soybeans, have most of their roots in the upper 25–30 cm of the profile (Table 5.4). In contrast, perennial crops such as alfalfa and fruit trees have

TABLE 5.4
Percentage of Root Mass of Three Crop Plants Found in the Upper 30 cm Compared with Deeper Depths (30–180 cm)

Crop	Percentage of roots	
	Upper 30 cm	30–180 cm
Soybeans	71	29
Corn	64	36
Sorghum	86	14

From Mayaki et al. (1976).

deep root systems and are able to absorb a considerable proportion of their moisture from subsoil layers. Even in these cases, however, it is likely that much of the root absorption is from the upper layers of the soil, provided these layers are well supplied with water. On the other hand, if the upper soil layers are moisture-deficient, even crops such as corn and soybeans will absorb much of their water from the lower horizons.

ROOT–SOIL CONTACT

As roots grow into the soil, they move into pores of sufficient size to accommodate them. Contact between the outer cells of the root and the soil permits ready movement of water from the soil into the plant in response to differences in energy levels (Figure 5.23). When the plant is under moisture stress, however, the roots tend to shrink in size in response to this stress. Such conditions exist during a hot, dry spell and are most severe during the daytime when transpiration from plant leaves is at a maximum. The diameter of roots under these conditions may shrink by 30–50%. The shrinkage reduces considerably the direct root–soil contact as well as the movement of liquid water and nutrients into the plants. Although water vapor can still be absorbed by the plant, its rate of absorption is too low to do more than keep the most drought-tolerant plants alive.

5.15

Conclusion

The water molecule has a polar structure, one end of the molecule having a partial positive charge and the other a partial negative charge. This polar property stimulates the electrostatic attraction of water to both soluble cations and soil solids. These attractive forces tend to reduce the free energy level of soil water below that of pure water. The extent of this reduction, called soil water potential (ψ), has a profound influence on a number of soil properties but especially on the movement of soil water and its uptake by plants.

The water potential due to the attraction of soil solids (matrix) for water (matric potential, ψ_m) combines with the force of gravity to largely control water movement. This movement is relatively rapid in soils high in moisture and with an abundance of macropores. In drier soils, however, the adsorption of water is so strong that its movement in the soil and its uptake by plants are minimized. As a consequence, plants may die for a lack of water when there are still significant quantities of water in the soil because that water is unavailable to plants.

Water is supplied to plants by capillary movement toward the root surfaces and by growth of the roots into moist soil areas. Both processes are important. Vapor movement takes place in soils but is of significance only in supplying water for drought-resistant desert plants.

The osmotic potential (ψ_o) becomes significant in soils with high soluble salt levels that can constrain the uptake of water from the soil. Such conditions occur most often in soils with restricted drainage occurring in areas of low rainfall.

The characteristics and behavior of soil water are very complex. As we have gained more knowledge, however, it has become apparent that soil water is governed by relatively simple basic physical principles. Further-

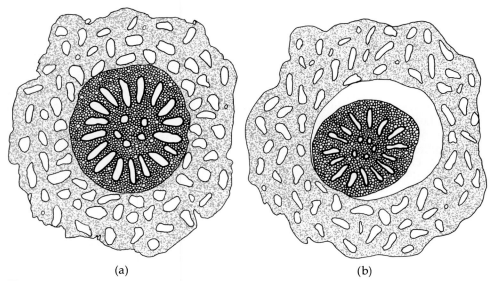

(a)   (b)

FIGURE 5.23

Cross section of a corn root surrounded by soil. (a) During periods of low soil moisture stress on the plant, the root completely fills the soil pore. (b) When the plant is under severe moisture stress, such as during a hot dry period, the root shrinks, significantly reducing root–soil contact. Such shrinkage of roots may occur on a hot summer day, even when the soil moisture content is high. [From Huck et al. (1970)]

more, researchers are discovering the similarity between these principles and those governing the uptake and use of soil moisture by plants, which are the subject of Chapter 14.

Study Questions

1. Explain how the structure of the water molecule is responsible for its attraction to soil solids and to inorganic ions.
2. What is the difference between adhesion and cohesion as they relate to soil water, and how does each influence water movement in soils?
3. How high will water move up a capillary tube whose radius is uniformly 0.001 mm?
4. What are the forces responsible for (a) matric potential (ψ_m), (b) osmotic potential (ψ_o), and (c) gravitational potential (ψ_g)?
5. Two soil samples with similar texture are placed in intimate contact; one has a matric potential (ψ_m) of -0.1 bar and the second -0.001 bar. Which likely has the higher moisture content (θ), in which direction will water move, and why?
6. What is the principle of the neutron moisture meter? Of a field tensiometer?
7. Why is the rate of water movement in saturated soils faster than that in unsaturated soils?

8. Why does the presence of a stratified layer of sand in the lower horizons of a loam soil impede the downward (and upward from below the sand layer) movement of water?
9. The moisture percentages of two field soils are 20% and 30%, respectively. How do they compare in terms of their "available" moisture contents and their matric potential levels? Explain.
10. What two phenomena are primarily responsible for the supply of water to plant roots?
11. Why is the absorption of soil water by plants constrained by the presence of inorganic salts in the soil water?
12. A clay soil at a field capacity of 35 kg water/100 kg soil may provide less water to plants than a loam soil of equal depth at a field capacity of 25 kg water/100 kg soil. What is the likely explanation for this difference?

References

BROWN, E. A., C. E. CAVINESS, and D. A. BROWN. 1985. "Response of selected soybean cultivars to soil moisture deficit," *Agron. J.*, **77**:274–78.

COONEY, J. J., and J. E. PETERSON. 1955. *Avocado Irrigation*, Leaflet 50, California Agr. Ext. Serv.

GARDNER, W., and J. A. WIDTSOE. 1921. "The movement of soil moisture," *Soil Sci.*, **11**:230.

HUCK, M. G., B. KLEPPER, and H. M. TAYLOR. 1970. "Diurnal variations in root diameter," *Plant Physiol.*, **45**:529.

KEETON, W. T. 1972. *Biological Science*, 2nd ed. (New York: Norton).

MAYAKI, W. C., L. R. STONE, and I. D. TEARE. 1976. "Irrigated and nonirrigated soybean, corn and grain sorghum root systems," *Agron. J.*, **68**:532–34.

RICHARDS, L. A. 1965. "Physical condition of water in soil," in *Agron. 9: Methods of Soil Analysis*, Part I (Madison, WI: Amer. Soc. Agron.).

Soil Air and Soil Temperature

As indicated in Chapters 1 and 4, approximately half of the total volume of a representative mineral surface soil is occupied by water and gases. Chapter 5 placed major emphasis on the soil moisture phase but emphasized throughout the interrelationship of soil air and soil water; one cannot be affected without changing the other. We now focus on soil aeration and on soil temperature, two important physical characteristics significantly affected by soil water.

6.1

Soil Aeration Characterized

Soil aeration is a vital process because it largely controls the soil levels of two life-sustaining gases, oxygen and carbon dioxide. These gases take part in the *respiration* of the roots of plants as well as soil microorganisms. This respiration involves the oxidation of organic compounds as follows (using sugar as an example of organic compounds).

$$C_6H_{12}O_6 + 6 O_2 \rightarrow 6 CO_2 + 6 H_2O$$
Sugar

Through *photosynthesis* this reaction is reversed. Carbon dioxide and water are combined by green plants to form sugars, and oxygen is released to be consumed by humans and other animals.

Soil aeration is a critical component of this overall system. For respiration to continue in the soil, oxygen must be supplied and carbon dioxide removed. Through aeration, there is an exchange of these two gases between the soil and the atmosphere. In a well-aerated soil, this exchange is sufficiently rapid to prevent the deficiency of oxygen or the toxicity of excess carbon dioxide. For most land plants, the supply of oxygen in the soil air must be kept above 10%. In turn, carbon dioxide concentration and that of other potentially toxic gases such as methane must not be allowed to build up excessively.

6.2

Soil Aeration Problems in the Field

Under field conditions, poor soil aeration occurs under two conditions: (a) when the moisture content is so high that there is little or no room for gases and/or (b) when the exchange of gases with the atmosphere is so slow that desirable levels of soil gases cannot be maintained. The latter may occur even when sufficient *total air space* is available.

EXCESS MOISTURE

The first case is characterized in the extreme by a waterlogged condition in the soil. A low spot in a field and a flat area where water tends to stand for a short while are good examples of this condition. In well-drained soils water-logging may also occur during a heavy rainstorm, when excess irrigation water is applied, or if the soil has been compacted by plowing or by heavy machinery.

Such complete saturation of the soil with water can be disastrous for certain plants in only a short period of time, a matter of a few hours being critical in some cases. This emphasizes the importance of artificial drainage of heavy soils and of other means of maintaining good soil aeration.

GASEOUS INTERCHANGE

The second case relates to gas exchange. The more rapid the usage of oxygen and the corresponding release of carbon dioxide, the greater is the need for the exchange of gases between the soil and the atmosphere above. This exchange is facilitated by two mechanisms: *mass flow* and *diffusion*. Mass flow of air, which is due to pressure differences between the atmosphere and the soil air, is less important than diffusion in determining the total exchange that occurs. It is enhanced, however, by fluctuations in soil moisture content. As water moves into the soil during a rain or from irrigation, air must be forced out. Likewise, when soil water is lost by evaporation from the soil surface or is taken up by plants, air is drawn into the soil. Mass flow also is modified slightly by other factors such as temperature, barometric pressure, and wind movements.

Most of the gaseous interchange in soils occurs by diffusion. Through this process, each gas moves in a direction determined by its own partial pressure (Figure 6.1). The partial pressure of a gas in a mixture is simply the pressure this gas would exert if it alone were present in the volume occupied by the mixture. Thus, if the pressure of air is 1 atmosphere (~1 bar), the partial pressure of oxygen, which makes up about 21% of the air by volume, is approximately 0.21 bar.

Diffusion allows extensive gas movement from one area to another even though there is no overall pressure gradient for the total mixture of gases. There is, however, a concentration gradient for each individual gas, which may be expressed as a *partial pressure gradient*. As a consequence, even though the total soil–air pressure and that of the atmosphere may be the same, a higher concentration of oxygen in the atmosphere will result in a net movement of this particular gas into the soil. Carbon dioxide and water

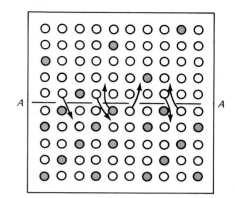

FIGURE 6.1

How the process of diffusion takes place. The total gas pressure is the same on both sides of boundary A–A. The partial pressure of oxygen is greater, however, in the top portion of the container. Therefore oxygen tends to diffuse into the lower portion, where fewer oxygen molecules are found. The carbon dioxide molecules, on the other hand, move in the opposite direction owing to the higher partial pressure of this gas in the lower half. Eventually equilibrium will be established when the partial pressures of oxygen and carbon dioxide are the same on both sides of the boundary.

vapor normally move in the opposite direction since the partial pressures of these two gases are generally higher in the soil air than in the atmosphere. A representation of the principles involved in diffusion is given in Figure 6.1.

6.3

Means of Characterizing Soil Aeration

The aeration status of a soil can be characterized conveniently in three ways: (a) the content of oxygen and other gases in the soil atmosphere; (b) the oxygen diffusion rate (ODR); and (c) the oxidation–reduction (redox) potential. Each will receive brief attention.

GASEOUS OXYGEN OF THE SOIL

The atmosphere above the soil contains nearly 21% O_2, 0.035% CO_2 and more than 78% N_2. In comparison, soil air has about the same level of N_2 but is consistently lower in O_2 and higher in CO_2. The O_2 content may be only slightly below 20% in the upper layers of a soil with a stable structure and an ample quantity of macropores. It may drop to less than 5% or even to near zero in the lower horizons of a poorly drained soil with few macropores.

Low O_2 contents also are found in low-lying areas where water has accumulated. Even in well-drained soils, marked reductions in the O_2 content of soil air may follow a heavy rain, especially if plants are growing rapidly or if large quantities of manure or other decomposing organic residues are present (Figure 6.2).

It is fortunate that the water in many soils contains small but significant quantities of dissolved O_2. As a consequence, even when all the soil pores are filled with water, soil microorganisms can extract some of the O_2 dissolved in the water for metabolic purposes. This small amount of dissolved O_2 soon is used up, however, and if the excess water is not removed, even microbial activities are jeopardized.

Since the N_2 content of soil air is relatively constant, there is a general inverse relationship between the contents of the other two major components of soil air—O_2 and CO_2—O_2 decreasing as CO_2 increases. Although the actual differences in CO_2 amounts may not be impressive, they are significant, comparatively speaking. Thus, when the soil air contains only

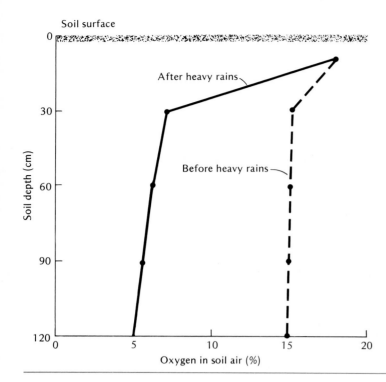

Soil surface

FIGURE 6.2
Oxygen content of soil air before and after heavy rains in a soil on which cotton was being grown. The rainwater replaced most of the soil air. The small amount of oxygen remaining was consumed in the respiration of crop roots and soil organisms. The carbon dioxide content probably increased accordingly. [Redrawn from Patrick (1977); used with permission of the Soil Science Society of America]

0.30% CO_2, this gas is more than eight times as concentrated as it is in the atmosphere. In cases where the CO_2 content becomes as high as 10%, there is nearly 300 times as much present as is found in the air above. At such levels, carbon dioxide may be toxic to some plant processes.

OXYGEN DIFFUSION RATES

Perhaps the best measurement of the aeration status of a soil is the *oxygen diffusion rate* (ODR). The ODR largely determines the rate at which oxygen can be replenished when it is used by respiring plant roots or by microorganisms or when it is forced out of the soil pores by water.

Figure 6.3 is a graph showing how ODR decreases with soil depth. Even when the oxygen level at the surface was 20%, the ODR rate at 97 cm (38 in.) was less than half that at 11 cm (4.3 in.). When a lower oxygen concentration at the soil surface was used, the ODR decreased even more rapidly with depth.

Researchers have found that the ODR is of critical importance to growing plants. For example, the growth of roots of most plants ceases when the ODR drops to about 20×10^{-8} g/cm^2 per minute (see Figure 6.3). Top growth is normally satisfactory as long as the ODR remains above $30–40 \times 10^{-8}$ g/cm^2 per minute. In Table 6.1 are found some field measurements of ODR, along with comments about the condition of the plants. Note the general tendency for difficulty if the ODR gets below the critical level.

OXIDATION–REDUCTION (REDOX) POTENTIAL

One important chemical characteristic of soils related to soil aeration is the reduction and oxidation states of the chemical elements in these soils. If a soil is well-aerated, oxidized states such as those of ferric iron (Fe^{3+}), manganic manganese (Mn^{4+}), nitrate (NO_3^-), and sulfate (SO_4^{2-}) dominate. In poorly drained and poorly aerated soils, the reduced forms of such elements

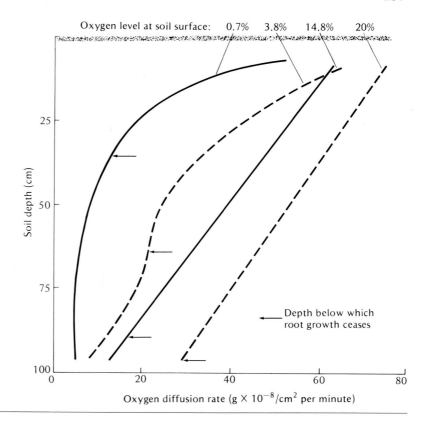

FIGURE 6.3

Effect of soil depth and oxygen concentration at the surface on the oxygen diffusion rate (ODR). Arrows indicate snapdragon root penetration depth. Even with a 20% O_2 level at the surface, the diffusion rate at ~95 cm is less than half that at the surface. Note that when the ODR drops to about 20×10^{-8} g/cm^2 per minute, root growth ceases. [Redrawn from Stolzy et al. (1961)]

are found, for example, ferrous iron (Fe^{2+}), manganous manganese (Mn^{2+}), ammonium (NH_4^+), and sulfides (S^{2-}). The presence of these reduced forms is an indication of restricted drainage and poor aeration.

An indication of the oxidation and reduction states of systems (including those in soils) is given by the *oxidation–reduction potential* or *redox potential* (E_h). It provides a measure of the tendency of a system to reduce or oxidize chemicals and is usually measured in volts or millivolts. If E_h is positive and high, strong oxidizing conditions exist. If it is low and even negative, elements are found in reduced forms.

TABLE 6.1

Relationship Between Oxygen Diffusion Rate (ODR) and the Condition of Different Plants

When the ODR drops below about 40×10^{-8} g/cm^2 per minute, the plants appear to suffer. Sugar beets require high ODR even at 30 cm depth. ODR values in 10^{-8} g/cm^2 per minute.

Plant	Soil texture	ODR at three soil depths			Remarks
		10 cm	20 cm	30 cm	
Broccoli	Loam	53	31	38	Very good growth
Lettuce	Silt loam	49	26	32	Good growth
Beans	Loam	27	27	25	Chlorotic plants
Sugar beets	Loam	58	60	16	Stunted tap root
Strawberries	Sandy loam	36	32	34	Chlorotic plants
Cotton	Clay loam	7	9	—	Chlorotic plants
Citrus	Sandy loam	64	45	39	Rapid root growth

From Stolzy and Letey (1964).

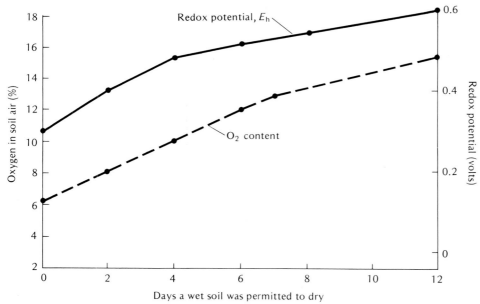

FIGURE 6.4

Relationship between oxygen content of soil air and the redox potential (E_h). Measurements were taken at a 28 cm depth in a soil that had been irrigated continuously for 14 days prior to the drying. Note the general relationship between these two parameters. [Data selected from Meek and Grass (1975); used with permission of Soil Science Society of America]

The positive correlation in one soil between O_2 content of soil air and E_h (redox potential) is shown in Figure 6.4. In a well-drained soil, the E_h is in the 0.4–0.7 volt (V) range. As aeration is reduced, the E_h declines to a level of about 0.3–0.35 V when gaseous oxygen is depleted.

The E_h value at which oxidation–reduction reactions occur varies with the specific chemical to be oxidized or reduced. In Table 6.2 are listed oxidized and reduced forms of several elements of importance in soils along with the approximate redox potentials at which the oxidation–reduction reactions occur. Note that gaseous oxygen is reduced to water at E_h levels of 0.38–0.32 V. At lower E_h values, microorganisms use combined oxygen for their

TABLE 6.2

Oxidized and Reduced Forms of Certain Elements in Soils and the Redox Potentials (E_h) at Which Change in Forms Commonly Occurs

Note that gaseous oxygen is depleted at E_h levels of 0.38–0.32 V. At lower E_h levels microorganisms utilize combined oxygen for their metabolism and thereby reduce the elements.

Oxidized form	Reduced form	E_h at which change of form occurs
O_2	H_2O	0.38 to 0.32
NO_3^-	N_2	0.28 to 0.22
Mn^{4+}	Mn^{2+}	0.28 to 0.22
Fe^{3+}	Fe^{2+}	0.18 to 0.15
SO_4^{2-}	S^{2-}	−0.12 to −0.18
CO_2	CH_4	−0.2 to −0.28

From Patrick and Reddy (1978).

metabolism and thereby reduce the elements. These data emphasize that soil aeration helps determine the specific chemical species present, and in turn essential nutrient availability as well as chemical toxicities, in soils.

OTHER GASES

Soil air usually is much higher in water vapor than the atmosphere, being essentially saturated except at or very near the surface of the soil. This fact already has been stressed in connection with the movement of water. Also, under waterlogged conditions, the concentration of gases, such as methane and hydrogen sulfide, which are formed as organic matter decomposes, is notably higher in soil air.

6.4

Factors Affecting Soil Aeration

The drainage of excess water from a soil is a primary factor in determining its aeration status. Air can enter the soil pores only when at least some of the water is removed. The oxygen level in soil pores just above a water table is generally low.

The most important factors influencing the aeration of well-drained soils are those that determine the volume of the soil's macropores. Macropore content definitely affects the total air space as well as gaseous exchange and biochemical reactions. Soil texture, bulk density, aggregate stability, and organic matter content are among the soil properties that help determine macropore content and, in turn, soil aeration.

The concentrations of both oxygen and carbon dioxide are definitely affected by microbial decomposition of organic residues. Incorporation of large quantities of manure, crop residues, or sewage sludge may alter the soil air composition appreciably (Figure 6.5). Respiration by higher plant roots and by soil organisms around the roots is also a significant process.

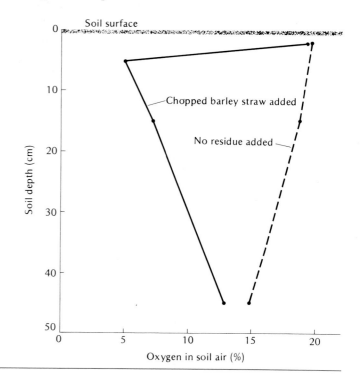

FIGURE 6.5

The effect of soil depth and straw residue application on the oxygen content of a Xerorthent (Yolo loam). Ten tons of chopped barley straw was mixed in the upper 10 cm of soil 2 months before the measurements were made. Note that the oxygen content at the 5 and 15 cm depths of soil was greatly reduced, the gas having been used in the decomposition of the straw. [From Rolston et al. (1982); used with permission of Soil Science Society of America]

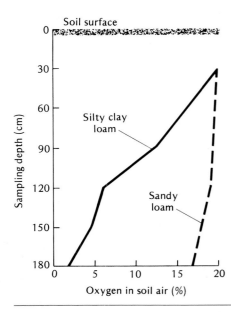

FIGURE 6.6

Average oxygen content of two orchard soils during the months of May and June. Normal root growth and functions would occur at much deeper levels on the sandy soil. [From Boynton (1938)]

SUBSOIL VERSUS TOPSOIL

Subsoils are usually more deficient in oxygen than are top soils. The total pore space as well as the macropore space is generally much lower in the deeper horizons. Average oxygen contents at different soil depths at two orchard sites in the months of May and June, as shown in Figure 6.6, illustrate this point. The soil moisture potential, which is likely higher in the clay loam soil, is associated with low oxygen and high carbon dioxide contents. Carbon dioxide percentages as high as 14.6 were recorded for the lower horizons of the finer textured soil.

SOIL HETEROGENEITY

The aeration status of different components varies greatly in a given soil. Thus, poorly aerated zones or pockets may be found in an otherwise well-drained and well-aerated soil. The poorly aerated component may be a heavy-textured or compacted soil layer, or it may be located inside a soil structural unit where tiny pores may limit ready air exchange (Figure 6.7). For these reasons, oxidation reactions may be occurring within a few centimeters of another location where reducing conditions exist. This heterogeneity of soil aeration should be kept in mind.

SEASONAL DIFFERENCES

There is marked seasonal variation in the composition of soil air. In the springtime in temperate humid regions, soils are apt to be wet and cold, and opportunities for ready gas exchange are poor. In the summer months when the soils are normally drier, opportunity for gaseous exchange is increased. This commonly results in relatively high oxygen and low carbon dioxide levels. Some exceptions to this rule may be found, however. Since high summer temperatures also encourage rapid microbial release of carbon dioxide, a given soil containing easily decomposable organic matter may have higher carbon dioxide levels in the summer than in the winter (Figure 6.8). The dependence of soil air composition on soil moisture and soil temperature is of vital significance.

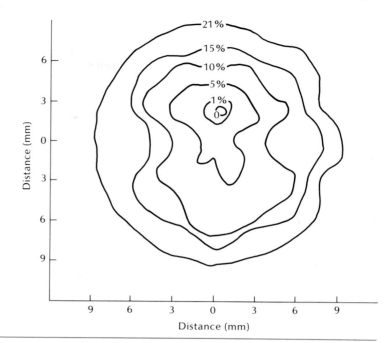

FIGURE 6.7

A "map" showing the oxygen content of soil air in a wet aggregate from an Aquic Hapludoll (Muscatine silty clay loam) from Iowa. The measurements were made with a unique microelectrode. Note that the oxygen content near the aggregate center was zero, while that near the edge of the aggregate was 21%. Thus pockets of oxygen deficiency can be found in a soil whose overall oxygen content may not be low. [From Sextone et al. (1985)]

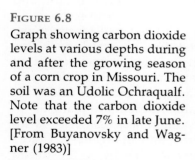

FIGURE 6.8

Graph showing carbon dioxide levels at various depths during and after the growing season of a corn crop in Missouri. The soil was an Udolic Ochraqualf. Note that the carbon dioxide level exceeded 7% in late June. [From Buyanovsky and Wagner (1983)]

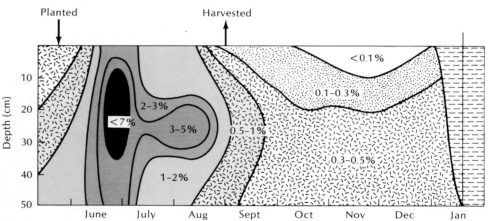

6.5

Effects of Soil Aeration

EFFECTS ON SOIL REACTIONS AND PROPERTIES

Soil aeration influences many soil reactions and, in turn, soil properties. The most obvious of these reactions are associated with the microbial breakdown of organic residues. Poor aeration slows down the rate of decay as evidenced by the relatively high organic matter level of poorly drained soils, especially in swampy areas.

The nature as well as the rate of organic decay is determined by the oxygen gas content of the soil. Where gaseous oxygen is present, *aerobic* organisms are active, and oxidation reactions such as the following occur. Sugar is used as an example of organic compounds.

$$C_6H_{12}O_6 + 6\,O_2 \rightarrow 6\,CO_2 + 6\,H_2O$$
$$\text{Sugar}$$

In the absence of gaseous oxygen, *anaerobic* organisms take over. Much slower breakdown occurs through reactions such as the following.

$$C_6H_{12}O_6 \rightarrow 3\,CO_2 + 3\,CH_4$$
$$\text{sugar} \qquad\qquad \text{Methane}$$

Poorly aerated soils are significant sources of methane gas, a pollutant of the upper atmosphere that is in part responsible for long-term global warming.

Less complete decomposition than that described in the two preceding equations may yield certain organic acids and ethylene gas (C_2H_4), which can be toxic to higher plants. Obviously, the presence or absence of oxygen gas completely modifies the nature of the decay process and its effect on plant growth.

OXIDATION–REDUCTION OF INORGANIC ELEMENTS

The level of soil oxygen largely determines the forms of several inorganic elements, as shown in Table 6.3. The oxidized states of nitrogen and sulfur are readily utilizable by higher plants. In general, the oxidized forms of other elements are also more desirable for most common crops grown on acid soils of humid regions. This is because the reduced forms of iron and manganese are sometimes present at toxic levels in moist, acid soils.

In drier areas, the opposite is generally true, and reduced forms of elements such as iron and manganese are preferred. In the neutral to alkaline soils of these drier areas, oxidized forms of iron and manganese are tied up in highly insoluble compounds, resulting in deficiencies of these elements. Such differences illustrate the interaction of aeration and soil acidity or alkalinity (indicated by pH) in supplying available nutrients to plants.

In addition to the chemical aspect of these differences, *soil color* is influenced markedly by the oxidation state of iron and manganese. Colors such as red, yellow, and reddish brown are encouraged by well-oxidized conditions. More subdued shades such as grays and blues predominate if insufficient oxygen is present. Soil color can be used in field methods for determining the need for drainage. Imperfectly drained soils are characterized by alternate streaks of oxidized and reduced materials. The *mottled* condition indicates a zone of alternate good and poor aeration, a condition not conducive to proper plant growth.

TABLE 6.3
Oxidized and Reduced Forms of Several Important Elements

Element	Normal form in well-oxidized soils	Reduced form found in waterlogged soils
Carbon	CO_2	CH_4
Nitrogen	NO_3^-	N_2, NH_4^+
Sulfur	SO_4^{2-}	H_2S, S^{2-}
Iron	Fe^{3+} (ferric oxides)	Fe^{2+} (ferrous oxides)
Manganese	Mn^{4+} (manganic oxides)	Mn^{2+} (manganous oxides)

TABLE 6.4

Comparative Tolerance by Different Crop Plants of a High Water Table and Accompanying Restricted Aeration

Ladino clover grows well with a water table near the soil surface, but wheat and other cereals grow best when the water table is at least 100 cm below the surface.

Plants tolerant to constant table at each depth			
15–30 cm	*40–60 cm*	*75–90 cm*	*100+ cm*
Ladino clover	Alfalfa	Corn	Wheat
Orchard grass	Potatoes	Peas	Barley
Fescue	Sorghum	Tomato	Oats
	Mustard	Millet	Peas
		Cabbage	Beans
		Snap beans	Horse beans
			Sugar beets
			Colza

From several sources reviewed by Williamson and Kriz (1970).

EFFECTS ON ACTIVITIES OF HIGHER PLANTS

Higher plants are adversely affected in at least three ways by conditions of poor aeration: (a) the growth of the plant, particularly the roots, is curtailed; (b) the absorption of nutrients and water is decreased; and (c) as discussed in the previous section, the formation of certain inorganic compounds toxic to plant growth is favored.

Plant Growth. Different plant species vary in their ability to tolerate poor aeration (Table 6.4). Wheat and barley, for example, require high air porosities for best growth. In contrast, ladino clover can grow with very low air porosity, and rice can grow with its roots submerged in water. Furthermore, the tolerance of a given plant to low porosity may be different for seedlings than for rapidly growing plants. A case in point is the tolerance of red pine to restricted drainage during its early development and its poor growth or even death on the same site at later stages.

In spite of these wide variations in soil air porosity limitation, soil physicists generally consider that if the soil volume occupied by air is reduced below 10–12%, most plants are likely to suffer.

Nutrient and Water. Oxygen deficiency is known to curtail the absorption by plants of both nutrients and water. This is because low oxygen levels constrain root respiration, a process that provides the energy for nutrient and water absorption. It is ironic that an oversupply of water in the soil can reduce the amount of water absorbed by plants. The effect of aeration on nutrient absorption by cotton is illustrated by the data in Table 6.5. It is no wonder that plants growing on poorly drained soils may show nutrient-deficiency symptoms even though the soils are fairly well supplied with available nutrient elements.

SOIL COMPACTION AND AERATION

The negative effects of soil compaction are not all owing to poor aeration. Soil layers can become so dense as to impede the growth of roots even if an adequate oxygen supply is available. For example, some compacted soil layers adversely affect cotton more by preventing root penetration than by lowering available oxygen content.

TABLE 6.5

Effect of Water Table Level on Oxygen Content of Soil Air in a Typic Torrifluvent (Hoytville Silty Clay), Yield of Cotton, and Plant Uptake of N, P, and K

Oxygen content at a depth of 23 cm in this heavy soil was sharply reduced as the water table was raised from 90 to 30 cm depth, and cotton yields and nutrient uptake were decreased accordingly.

Depth of water table (cm)	Oxygen content at 23 cm[a] (%)	Yield of cotton (g)	Nutrient uptake by five plants (mg)		
			N	P	K
30	1.6	57	724	85	1091
60	8.3	108	1414	120	2069
90	13.2	157	2292	156	3174

From Meek et al. (1980).
[a] On June 23, 1976.

6.6

Aeration in Relation to Soil and Crop Management

Both surface and subsurface drainage are essential if an aerobic soil environment is to be maintained. The removal of excess quantities of water must take place if sufficient oxygen is to be supplied. The importance of surface runoff and tile drainage will be discussed later (see Sections 14.15 and 14.16).

The maintenance of a stable soil structure is an important means of augmenting good aeration. Pores of macrosize, which are encouraged by large stable aggregates, are soon drained of water following a rain, thereby allowing gases to move into the soil from the atmosphere. Maintenance of organic matter by addition of farm manure and crop residues and by growth of legumes is perhaps the most practical means of encouraging aggregate stability, which, in turn, encourages good drainage and better aeration.

In poorly drained, heavy-textured soils, however, it is often impossible to maintain optimum aeration without resorting to some cultivation of the soil. Thus, in addition to controlling weeds, cultivation of heavy-textured soils has a second very important function—that of aiding soil aeration. Consequently, no tillage or minimum tillage practices, which are quite satisfactory on well-drained soils, have limitations on poorly drained soils. Also yields of row crops, especially those with large tap roots such as sugar beets and rutabagas, often are increased by frequent light cultivations that do not injure the fibrous roots. Part of this increase, undoubtedly, is due to aeration.

CROP–SOIL ADAPTATION

In addition to the direct methods of controlling soil aeration, the selection of crops tolerant of low oxygen is important. Alfalfa, fruit and forest trees, and other deep-rooted plants require deep, well-aerated soils; such plants are sensitive to a deficiency of oxygen, even in the lower soil horizons. In contrast, shallow-rooted plants, such as grasses and alsike and ladino clovers, do well on soils that tend to be poorly aerated, especially in the subsoil. The rice plant flourishes even when the soil is submerged in water. These facts are significant in deciding what crops should be grown and how they are to be managed in areas where aeration problems are acute.

The dominating influence of moisture on the aeration of soils is universally apparent, and control of the moisture means at least a partial control of the aeration. Now let us consider a second important physical property at times influenced by soil water—that of soil temperature.

6.7

Soil Temperature

The temperature of a soil greatly affects the physical, biological, and chemical processes occurring in that soil. In cold soils, chemical and biological rates are slow. Biological decomposition can come to a near standstill, thereby limiting the rate at which nutrients such as nitrogen, phosphorus, sulfur, and calcium are made available. Also, absorption and transport of water and nutrient ions by higher plants are adversely affected by low temperature.

PLANT PROCESSES

Plants vary widely in the soil temperature at which they grow best, and there is variability in the optimum temperature for different plant processes. For example, corn germination requires a soil temperature of at least 7–10 °C (45–50 °F), and optimum root growth occurs at about 25 °C (Figure 6.9). Potato tubers develop best when the soil temperature is 16–21 °C (60–70 °F). Wheat grows best at about 24 °C. Optimum vegetative growth of apples and peaches is obtained when the soil temperature is about 18 °C. A comparable figure for citrus is 25 °C.

FIGURE 6.9
The influence of soil temperature on the early growth of corn tops and roots when the air temperature was kept optimum for plant growth. Obviously, corn is quite sensitive to soil temperature differences.

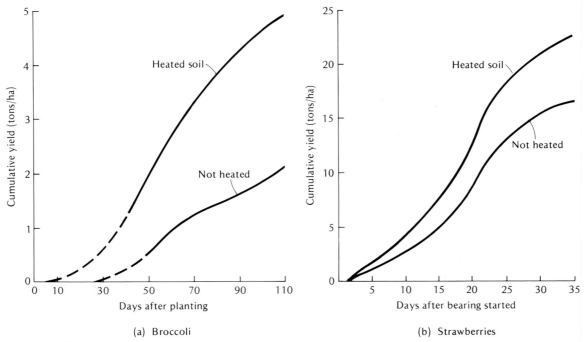

FIGURE 6.10

The influence of heating the soil in an experiment in Oregon on the yield of broccoli and strawberries. Heating cables were buried about 92 cm deep. They increased the average temperature of the 0–100 cm layer by about 10 °C, although the upper 10 cm of soil was warmed by only about 3 °C. [From Rykbost et al. (1975); used with permission of American Society of Agronomy]

In cool temperate regions, the yields of some vegetable and small fruit crops are increased by warming the soil (Figure 6.10). The life cycle of some flowers and ornamentals also is influenced greatly by soil temperature. For example, the tulip bulb requires "chilling" to develop flower buds in early winter, although flower development is suppressed until the soil warms up the following spring. In warm regions, root growth near the surface may be encouraged by shading the soil, either with vegetation or through the use of a mulch.

MICROBIAL PROCESSES

Some microbiological processes are influenced markedly by soil temperature changes. For example, the microbial oxidation of ammonium ions to nitrate ions, which occurs most readily at temperatures of 27–32 °C (80–90 °F), is neglible when the soil temperature is lowered to about 10 °C (50 °F). Advantage is taken of this fact in nitrogen fertilizer management. Thus, anhydrous ammonia fertilizers can be injected into cold soils in the spring, and the ammonium ions thus added will not readily be oxidized to nitrate ions until the soil temperature rises. The practice is valuable because it conserves nitrate ions, which (in contrast to ammonium ions) are subject to ready leaching from soils. When the soil warms up later in the year, nitrates will being to appear but they will be readily absorbed by rapidly growing plants.

Advantage can be taken of the sensitivity of some microbes to high soil temperatures to control certain plant diseases. In environments with hot summers [maximum daily air temperatures >35 °C (>95 °F)], covering the ground with transparent plastic sheeting can raise the temperature of the upper few centimeters of soil to 50 °C (122 °F) and even higher. Such temperatures markedly reduce certain wilt-causing fungal diseases of vegetables and fruits and adversely affect some weed seeds and insects. This heating process, called *soil solarization*, is used to control pests and diseases in some high-value crops.

Freezing and Thawing

In addition to the direct influence of temperature on plant and animal life, the effect of freezing and thawing must be considered. Alternate freezing and thawing subject the soil aggregates to pressures as ice crystals form and expand and alter the physical structure of the soil. Freezing and thawing of the upper layers of soil can also result in *heaving* of perennial forage crops such as alfalfa (Figure 6.11). This action, which is most severe on bare, imperfectly drained soils, can kill alfalfa plants, as well as those of some clovers, and trefoil. Changes in soil temperature have the same effect on shallow house foundations and roads with fine material as a base. Such structures show the effects of heaving in the spring.

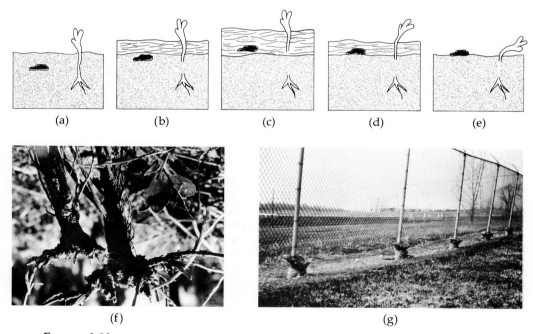

(a)   (b)   (c)   (d)   (e)

(f)   (g)

FIGURE 6.11
The diagrams illustrate how frost action in soils can move (heave) plants and stones upward. (a) Position of stone and plant before the soil freezes. (b) Frost action has started: thin lenses of ice have formed in the upper few centimeters of soil. Since water expands when it freezes, the lenses of ice expand and move the soil as well as the rock and plant upward, breaking the plant roots. (c) Maximum soil freezing with rock and upper plant parts moved farther up. (d) Partial thawing. (e) Final thawing, which leaves the stone on or near the soil surface and the plant upper roots exposed. (f) Alfalfa plants pushed out of the ground by frost action. (g) Fence posts encased in concrete "jacked" out of the ground by frost action. [Photos courtesy J. C. Henning, University of Missouri (f), and R. L. Berg, Corps of Engineers, Cold Regions Research and Engineering Laboratory, Hanover, NH (g)]

The temperature of soils in the field is dependent directly or indirectly on at least three factors: (a) the net amount of heat the soil absorbs; (b) the heat energy required to bring about a given change in the temperature of a soil; and (c) the energy required for changes such as evaporation, which are constantly occurring at or near the surface of soils. These factors will now be considered.

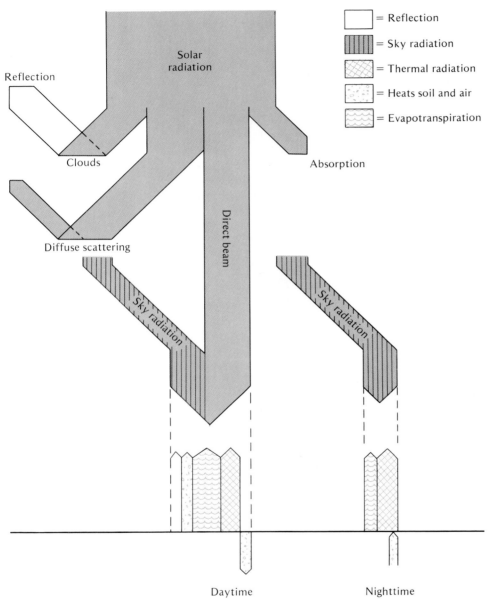

FIGURE 6.12
Schematic representation of the radiation balance in daytime and nighttime in the spring or early summer in a temperate region. About half the solar radiation reaches the Earth, either directly or indirectly, from sky radiation. Most radiation that strikes the Earth in the daytime is used as energy for evapotranspiration or is radiated back to the atmosphere. Only a small portion, perhaps 10%, actually heats the soil. At night the soil loses some heat, and some evaporation and thermal radiation occur.

6.8

Absorption and Loss of Solar Energy

Solar radiation is the primary source of energy to heat soils. But clouds and dust particles intercept the sun's rays and absorb, scatter, or reflect most of their energy (Figure 6.12). Only about 35–40% of the solar radiation actually reaches the Earth in cloudy humid regions and 75% in cloud-free arid areas. The global average is about 50%.

Little of the solar energy reaching the earth actually results in soil warming. The energy is used primarily to evaporate water from the soil or from leaf surfaces, or it is radiated or reflected back to the sky. Only about 10% is absorbed by the soil and can be used to warm it. Even so, this energy is of critical importance to soil processes and to higher plants growing on the soil.

In addition to solar radiation, other factors influence the amount of energy absorbed by soils, including soil color, the slope, and vegetative cover. Dark-colored soils absorb more energy than lighter colored ones. This does not necessarily imply, however, that dark soils are always warmer. In fact, the opposite is often true. Dark soils usually are high in organic matter and hold large amounts of water, which requires comparatively more energy to be warmed than soil and also cools the soil when it is evaporated.

The angle at which the sun's rays strike the soil also influences soil temperature. If the incoming path of the rays is perpendicular to the soil surface, energy absorption (and soil temperature increase) is greatest (Figure 6.13). As an example, three soils, one on a southerly slope of 20°, one on a nearby level field, and one on a northerly slope of 20°, will receive energy from the sun's rays on June 21 (at the 42nd Parallel North) in the proportion of 106:100:81.

Whether the soil is bare or is covered with vegetation or a mulch is another factor that markedly influences the amount of solar radiation reaching the soil. The effect of a dense forest is universally recognized. Even an ordinary field crop such as bluegrass has a very noticeable effect, especially on temperature fluctuations. Bare soils warm up more quickly and cool off more rapidly than those covered with vegetation or with plastic mulches. Frost penetration during the winter is considerably greater in bare noninsulated land.

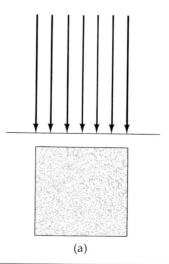
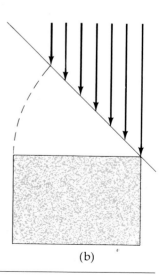

FIGURE 6.13
Effect of the angle at which the sun's rays strike the soil on the area of soil that is warmed. (a) If a given amount of radiation from the sum strikes the soil at right angles, the radiation is concentrated in a relatively small area, and the soil warms quite rapidly. (b) If the same amount of radiation strikes the soil at a 45° angle, the area affected is larger by about 40%, the radiation is not so concentrated, and the soil warms up more slowly. This is one of the reasons why north slopes tend to have cooler soils than south slopes. It also accounts for the colder soils in winter than in summer.

(a)

(b)

6.9

Specific Heat of Soils

A dry soil is more easily heated than a wet one. This is because the amount of energy required to raise the temperature of water by 1 °C (its *heat capacity*) is much greater than that required to warm soil solids by 1 °C. When heat capacity is expressed per unit mass—for example, in calories per gram (cal/g)—it is called *specific heat*. The specific heat of pure water is 1.00 cal/g or 1000 cal/kg [4.18 joules per gram (J/g)]; that of dry soil is about 0.2 cal/g (0.8 J/g).

The specific heat is an important soil property. It largely controls the degree to which soils warm up in the spring. For example, consider two soils with comparable characteristics, one a dry soil with only 10 kg water/100 kg soil solids, the other a wet soil with 30 kg water/100 kg soil solids. The drier soil has a specific heat of 273 cal/kg, whereas the wet soil has a specific heat of 385 cal/kg.[1] Obviously, the wet soil will warm up much more slowly than the drier one. Furthermore, if water cannot be drained freely from the wet soil, the water must be evaporated, a process that is very energy-consuming, as the next section will show.

6.10

Heat of Vaporization

The evaporation of water from soil surfaces requires a large amount of energy, 540 kilocalories (kcal) or 2.257 joules for every kilogram of water vaporized. This energy must be provided by solar radiation or it must come from the surrounding soil. In either case, evaporation has the potential of cooling the soil. For example, if the amount of water associated with 100 g of dry soil was reduced by evaporation from 25 to 24 g and if all the thermal energy needed to evaporate the water came from the moist soil, the soil would be cooled by about 12 °C. Such a figure is hypothetical because only a part of the heat of vaporization comes from the soil itself. Nevertheless, it indicates the tremendous cooling effect of evaporation.

The low temperature of a wet soil is due partially to evaporation and partially to high specific heat. The temperature of the upper few centimeters of wet soil is commonly 3–6 °C (6–12 °F) lower than that of a moist or dry soil. This is a significant factor in the spring in a temperate zone when a few degrees will make the difference between the germination or lack of germination of crop seeds.

[1] These figures can be easily verified. For example, consider the soil with 30 kg water/100 kg dry soil, or 0.3 kg water/kg dry soil. The number of calories required to raise the temperature of 0.3 kg of water by 1 °C is

$$0.3 \text{ kg} \times 1000 \text{ cal/kg} = 300 \text{ cal}$$

The corresponding figure for the kilogram of soil solids is

$$1 \text{ kg} \times 200 \text{ cal/kg} = 200 \text{ cal}$$

Thus, a total of 500 cal is required to raise the temperature of 1.3 kg of the moist soil by 1 °C. Since the *specific heat* is the number of calories required to raise the temperature of 1 kg of wet soil by 1 °C, in this case, it is

$$500/1.3 = 385 \text{ cal/kg}$$

6.11

Movement of Heat in Soils

As mentioned in Section 6.8, some of the solar radiation that reaches Earth slowly penetrates the profile, largely by *conduction*; this is the same process by which heat moves along an iron pipe, one end of which is placed in a fire. The rate of conduction in soil is influenced by a number of factors, the most important of which is the moisture content of the soil layers. Heat passes from soil to water about 150 times more easily than from soil to air. As the water content increases in a soil, the air content decreases, and the transfer resistance is lowered decidedly. When sufficient water is present to form a bridge between most of the soil particles, further additions will have little effect on heat conduction. Here again the major role of soil moisture is obvious.

The significance of conduction with respect to field temperatures is not difficult to comprehend. It provides a means of temperature adjustments, but, because it is slow, changes in subsoil temperatures lag behind those of the surface layers. Moreover, temperature changes are always less in the subsoil. In temperate regions, surface soils in general are expected to be warmer in summer and cooler in winter than the subsoil, especially the lower horizons of the subsoil.

Mention also should be made of the effect of rain or irrigation water on soil temperature. For example, in temperate zones, spring rains definitely warm the surface soil as the water moves into it. Conversely, in the summer the rainfall cools the soil because it is often cooler than the soil it penetrates. In practice, however, the spring rains, by increasing the amount of water to be removed by evaporation, often accentuate low temperature problems.

6.12

Soil Temperature Data

The temperature of the soil at any time depends on the ratio of the energy absorbed to that being lost. The constant change in this relationship is reflected in the *seasonal, monthly*, and *daily* temperatures. The accompanying data (Figures 6.14 and 6.15) from College Station, Texas, and Lincoln, Nebraska, are representative of average seasonal and monthly temperatures in relation to soil depth in humid temperature regions.

It is apparent from these figures that considerable seasonal and monthly variations of soil temperature occur, even at the lower depths. The surface layer temperatures vary more or less according to the temperature of the air, although these layers are generally warmer than the air throughout the year.

In the subsoil the seasonal temperature increases and decreases lag behind changes registered in the surface soil and in the air. Accordingly, the temperature data for March at College Station suggest that the surface soil temperatures have already begun to respond to the warming of the spring while temperatures of the deep subsoil seem still to be responding to the cold winter weather.

The subsoil temperatures were less variable than the air and surface soil temperatures, although there was some temperature variation even at the 300 cm depth. The subsoil was generally warmer in the late fall and winter and cooler in the spring and summer than the surface soil layers and the air. This is to be expected since the subsoils are not subject to the direct effects of solar radiation.

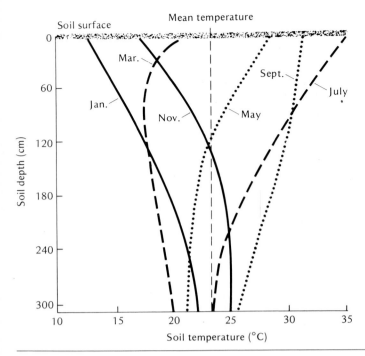

FIGURE 6.14
Average monthly soil temperatures for six of the twelve months of the year at different soil depths at College Station, TX (1951–1955). Note the lag in soil temperature change at the lower depths. [From Fluker (1958)]

The low variation in subsoil temperatures is an advantage not only from the standpoint of agriculture but for other sectors as well. Thus, the deep subsoils are less apt to freeze in the winter months, thereby protecting the roots of perennial crop plants and trees. Likewise, frost action on houses (Figure 6.11) or roads is not as apt to occur if the house or road foundation is based in the subsoil.

The temperature lag of subsoils is also of practical importance in temperature regulation in houses that use a so-called heat pump. The pumps circulate water in pipes imbedded in a subsoil field surrounding a house and take advantage of the fact that subsoils are cooler than the atmosphere in summer, and warmer in winter, to help reduce temperature extremes in the house.

DAILY VARIATIONS

With a clear sky, the air temperature in temperate regions rises from lowest in the morning to a maximum at about 2:00 P.M. The surface soil, however, does not reach its maximum until later in the afternoon because of the usual lag. This retardation is greater and the temperature change is less as the depth increases. The lower subsoil shows little daily or weekly fluctuation; the variation in the subsoil, as already emphasized, is a slow monthly or seasonal change.

6.13

Soil Temperature Control

The temperature of field soils is not subject to radical human regulation. However, two kinds of management practice have significant effects on soil temperature: those that keep some type of cover or mulch on the soil and

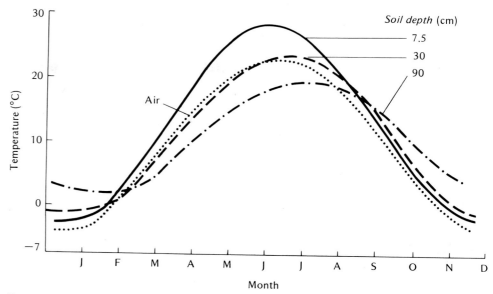

FIGURE 6.15
Average monthly air and soil temperatures at Lincoln, NE (12 years). Note that the 7.5 cm soil layer is consistently warmer than the air above and that the 90 cm soil horizon is cooler in spring and summer, but warmer in the fall and winter, than surface soil.

those that reduce excess soil moisture. These effects have meaningful biological implications.

MULCH AND TILLAGE

Soil temperatures are influenced by soil cover and especially by organic residues or other types of mulch placed on the soil surface. Figure 6.16 shows that mulches are definitely soil temperature modifiers. In periods of hot weather, they keep the surface soil cooler than where no cover is used; in contrast, during cold spells in the winter, they moderate rapid temperature declines. Mulches tend to buffer extremes in soil temperatures.

Until fairly recently, the significant use of mulches in modifying soil temperature extremes was limited mostly to home gardens and flower beds. Although these uses are still important, the effects of mulches have been extended to field-crop culture in areas that have adopted "conservation" tillage practices. Conservation tillage leaves much if not all the crop residues at or near the soil surface.

The influence of surface residues on soil temperature is illustrated by data shown in Figure 6.17. Soil temperatures were consistently lower during July and August when the "no tillage" practice was followed. Crop residues left on the surface undoubtedly account for this effect.

The soil temperature depressing effect of some mulch practices has a serious negative impact on corn production in the northern Great Plains states. The lower temperatures in May and early June resulting from these practices inhibit seed germination, seedling performance, and often the yields of corn. Ridging the soil, permitting water to drain out of the ridge, and then planting on top of the ridge is an innovative way to alleviate this problem. In warm regions, mulching may provide the double benefit of increased rooting in the more fertile topsoil plus decreased evaporation of water from the soil surface.

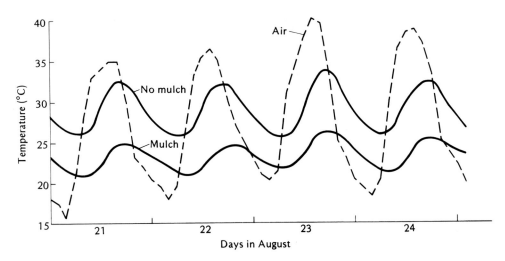

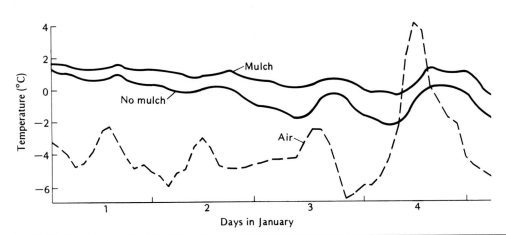

FIGURE 6.16
(a) Influence of straw mulch (8 tons/ha) on air temperature at a depth of 10 cm during an August hot spell in Bushland, TX. Notice the soil temperatures in the mulched area are consistently lower than where no mulch was applied. (b) During a cold period in January the soil temperature was higher in the mulched than in the unmulched area. [Redrawn from Unger (1978); used with permission of American Society of Agronomy]

MOISTURE CONTROL

A second means of exercising some control over soil temperature is through management practices that influence soil moisture. Poorly drained soils in temperate regions that are wet in the spring have temperatures 3–6 °C lower than comparable well-drained soils. Only by removing this water can temperature depression be alleviated. Water removal can be attempted by the installation of appropriate drainage systems, using ditches and underground tile (Sections 14.15 and 14.16). Where this is not feasible, the innovative ridging systems of tillage just referred to must be used. The water simply must be removed from the soil zone in which plant seeds and roots are located.

As was the case with soil air, the controlling influence of soil water on soil temperature is apparent everywhere. Whether a problem concerns acquisition of insolation, loss of energy to the atmosphere, or the movement of heat back and forth within the soil, the amount of water present is always important. Water regulation seems to be the key to what little practical temperature control it is possible to exert on field soils.

Covering the soil with transparent plastic sheeting can increase soil temperatures by the process of soil solarization (see Section 6.7). Soil

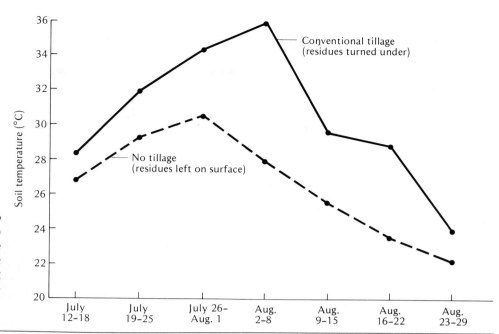

FIGURE 6.17
Effect of plowing under crop residues (conventional tillage) or leaving them on the surface (no tillage) on the temperature of an Ultic Hapludalf (Wheeling fine sandy loam) in West Virginia. [Data from Bennett et al. (1976)]

temperature increases of 8–10 °C (13–18 °F) can be achieved using this method, and the influence is noted to depths of more than 30 cm. The method is used for disease and moisture control in the production of some high-value vegetable and fruit crops in areas with low rainfall such as California and Israel.

6.14

Conclusion

Soil aeration and soil temperature are two physical properties that critically affect the quality of soils for agricultural purposes. Plants have definite requirements for soil oxygen along with limited tolerance for carbon dioxide, methane, and other such gases found in poorly aerated soils. Some microbes such as the nitrifiers and general purpose decay organisms are also constrained by low levels of soil oxygen.

Plants as well as microbes are quite sensitive to differences in soil temperature, particularly in temperate climates where low temperatures can limit essential biological processes. Soil temperature also impacts on the use of soils for engineering purposes, again primarily in the cooler climates. Frost action, which can move perennial plants such as alfalfa out of the ground, can do likewise for building foundations, fence posts, sidewalks, and highways.

Soil water can provide considerable control over both soil air and soil temperature. It competes with soil air for the occupancy of soil pores. Soil water also resists changes in soil temperature by virtue of its high specific heat and high energy requirement for evaporation. Moisture control in soils does more to influence soil aeration and soil temperature than any other soil management tool. Soil drainage and innovative tillage practices are our best means of bringing about temperature control.

Study Questions

1. A heavy application (40 Mg/ha) of farmyard manure was made in July to a loam soil and worked into the topsoil. What would you expect the short term and longer term effect to be on soil aeration?

2. Describe the two major processes by which gases are exchanged in soils. Which is more significant on a day to day basis?

3. Describe three means of characterizing soil aeration.

4. Assume you examine a soil profile and find a mixture of brown and gray colored materials in the B horizon. What does this tell you about the aeration status of this horizon? To what chemicals are the colors probably due?

5. If the redox potential of a soil is 0.22 V, what forms of iron, manganese, and sulfur would you expect to find in it? Explain.

6. What are the products of aerobic and of anaerobic decomposition of organic matter? Which of these systems suggest conditions that are best for most crop plants?

7. Give some examples of negative effects of freezing and thawing of soils and explain why these effects occur.

8. About what percent of the total solar radiation is absorbed by soils in cloudy humid areas? In arid areas?

9. Explain how the slope of land and soil color affect soil temperature.

10. What is the specific heat of a moist soil with 25 kg water/100 kg dry soil?

11. If moisture loss by evaporation from a soil reduced the water content of the furrow slice from 18 to 15 kg water/100 kg dry soil and if one fourth of the energy for the vaporization came from the soil in the furrow slice, what would be the approximate temperature reduction?

12. Conservation tillage has many advantages but has been found to reduce corn yields on poorly drained soils in the northern parts of the United States. Explain how soil temperature differences may account for this reduction.

13. What soil temperature principles are used in heat-pump systems for heating and cooling homes?

References

BENNETT, O. L., E. L. MATHIAS, and C. B. SPEROW. 1976. "Double cropping for hay and no-tillage corn production as affected by sod species with rates of atrazine and nitrogen," *Agron J.*, **68**:250–54.

BOYNTON, D., et al. 1938. "Are there different critical oxygen concentrations for the different phases of root activity?" *Science*, **88**:569–70.

BUYANOVSKY, G. A., and G. H. WAGNER. 1983. "Annual cycles of carbon dioxide level in soil air," *Soil Sci. Soc. Amer. J.*, **47**:1139–45.

FLUKER, B. J. 1958. "Soil temperature," *Soil Sci.*, **86**:35–46.

MEEK, B. D., and L. B. GRASS. 1975. "Redox potential in irrigated desert soils as an indicator of aeration status," *Soil Sci. Soc. Amer. J.*, **39**:870–75.

MEEK, B. D., et al. 1980. "Cotton yield and nutrient uptake in relation to water table depth," *Soil Sci. Soc. Amer. J.*, **44**:301–305

PATRICK, W. H., Jr. 1977. "Oxygen content of soil air by a field method," *Soil Sci. Soc. Amer. J.*, **41**:651–52.

PATRICK, W. H., Jr., and C. N. REDDY. 1978. "Chemical changes in rice soils," *Soils and Rice* (Los Banos, Laguna, Philippines: The Int. Rice Res. Inst.), pp. 361–79.

ROLSTON, D. E., et al. 1982. "Field measurement of denitrification: III. Rates during irrigation cycles," *Soil Sci. Soc. Amer. J.*, **46**:289–96.

RYKBOST, K. A., et al. 1975. "Yield response to soil warming: vegetable crops," *Agron. J.*, **67**:738–43.

SEXSTONE, A. J., et al. 1985. "Direct measurement of oxygen profiles and denitrification rates in soil aggregates," *Soil Sci. Soc. Amer. J.*, **49**:645–51.

STOLZY, L. H., et al. 1961. "Root growth and diffusion rates as functions of oxygen concentration," *Soil Sci. Soc. Amer. Proc.*, **2**:463–67.

STOLZY, L. H., and J. LETEY. 1964. "Characterizing soil oxygen conditions with a platinum microelectrode," *Adv. Agron.*, **16**:249–79.

UNGER, P. W. 1978. "Straw mulch effects on soil temperatures and sorghum germination and growth," *Agron. J.*, **70**:858–64.

WILLIAMSON, R. E., and G. J. KRIZ. 1970. "Response of agricultural crops to flooding, depth of water table and soil gaseous composition," *Amer. Soc. Agr. Eng. Trans.*, **13**:216–20.

Soil Colloids: Their Nature and Practical Significance

Next to photosynthesis and respiration, probably no process in nature is as vital to plant and animal life as the exchange of ions between soil particles and growing plant roots. These *cation* and *anion exchanges* take place mostly on the surfaces of the finer or colloidal fractions of both the inorganic and organic matter—clays and humus.

Such colloidal particles serve very much as does a modern bank. They are the sites within the soil where ions of essential mineral elements such as calcium, potassium, and sulfur are held and protected from excessive loss by percolating rain or irrigation water. Subsequently, the essential ions can be "withdrawn" from the colloidal "bank" sites and taken up by plant roots. In turn, these elements can be "deposited" or returned to the colloids through the addition of commercial fertilizers, lime, manures, and plant residues.

The colloidal particles (clays and humus) are the seat of most chemical, physical, and biological properties of soils. Consequently, no understanding of soils is complete without at least a general knowledge of the nature and constitution of these tiny particles. Their common properties will be covered first and then their individual characteristics will receive attention.

7.1

General Properties of Soil Colloids

SIZE

The most important common property of inorganic and organic colloids is their extremely small size. They are too small to be seen with an ordinary light microscope. Only with an electron microscope can they be photo-graphed. Most are smaller than 2 micrometers (μm) in diameter.[1]

SURFACE AREA

Because of their small size, all soil colloids expose a large *external* surface per unit mass. The external surface area of 1 g of colloidal clay is at least 1000 times that of 1 g of coarse sand. Some colloids, especially certain silicate clays, have extensive *internal* surfaces as well. These internal surfaces occur between plate-like crystal units that make up each particle and often greatly exceed the external surface area. The total surface area of soil colloids ranges from 10 m^2/g for clays with only external surfaces to more than 800 m^2/g for clays with extensive internal surfaces. The colloid surface area in the upper 15 cm of a hectare of a clay soil could be as high as 700,000 km^2 (270,000 mi^2).

SURFACE CHARGES

Soil colloidal surfaces, both external and internal, characteristically carry negative and/or positive charges. For most soil colloids, electronegative charges predominate, although some mineral colloids in very acid soils have a net electropositive charge. The presence and intensity of the particle charges influence the attraction and repulsion of the particles toward each other, thereby influencing both physical and chemical properties. The source of these charges will be considered later.

ADSORPTION OF CATIONS AND WATER

An important consequence of the charges on soil colloids is the attraction of ions of an opposite charge to the colloidal surfaces. Such attraction is of particular significance for negatively charged colloids. The colloidal parti-cles, referred to as *micelles* (microcells), attract hundreds of thousands of positively charged ions, or *cations*,[2] such as H$^+$, Al^{3+}, Ca^{2+}, and Mg^{2+}. This gives rise to an *ionic double layer* (Figure 7.1). The colloidal particle constitutes the *inner* ionic layer, being essentially a huge anion, with both external and internal layers that are negative in charge. The *outer* layer is made up of a swarm of rather loosely held (adsorbed) cations attracted to the negatively charged surfaces. Thus, a colloidal particle is accompanied by a swarm of cations that are *adsorbed* or held on the particle surfaces

In addition to the adsorbed cations, a large number of water molecules are associated with soil colloidal particles. Some are attracted to the adsorbed cations, each of which is hydrated. Others are held in the internal surfaces of the colloidal particles. These water molecules play a critical role in determin-ing both the physical and chemical properties of soil.

---

[1] Since the upper limit in diameter for the colloidal state is generally considered to be about 1 μm, some clay particles (the upper limit for which is 2 μm) are technically not colloids. Even so they have colloid-like properties. For a recent review of colloids in soils, see Dixon and Weed (1989).

[2] In water solutions the H$^+$ ion is always hydrated, giving rise to the *hydronium* ion H$_3$O$^+$ [sometimes written H$^+$(aq)]. For simplicity however, the H$^+$ ion designation will be used in this text.

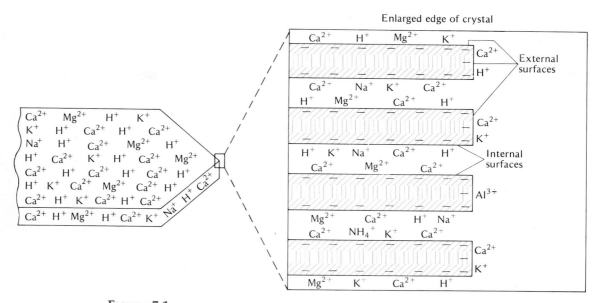

FIGURE 7.1
Diagrammatic representation of a silicate clay crystal (micelle) with its sheet-like structure, its innumerable negative charges, and its swarm of adsorbed cations. The enlarged schematic view of the edge of the crystal illustrates the negatively charged internal surfaces of this particular particle, to which cations and water are attracted. Note that each crystal unit has definite mineralogical structure. (See Figure 7.4 for a more detailed drawing of structure.)

7.2

Types of Soil Colloids

There are four major types of colloids present in soils: layer silicate clays; iron and aluminum oxide clays; allophane and associated amorphous clays; and humus. Although each group possesses the general colloidal characteristics described in the preceding section, each also has specific characteristics that make it distinctive and useful.

LAYER SILICATE CLAYS

The most important property of these clays is their crystalline, layer-like structures. Each particle is made up of a series of layers much like the pages of a book (Figure 7.2). The layers comprise primarily horizontally oriented sheets of silicon, aluminum, magnesium, and/or iron atoms surrounded and held together by oxygen and hydroxy groups. The formula of one of these clays, kaolinite ($Si_2Al_2O_5(OH)_4$), illustrates their general chemical makeup.

The exact chemical composition and the internal arrangement of the atoms in a given colloidal crystal account for its surface charge and for other associated properties. These include the capacity of each silicate colloid to hold and exchange ions as well as its physical properties. For example, some colloidal clays are highly sticky and plastic while others are only mildly so.

Silicate clay minerals are the most prominent inorganic colloids in soils of

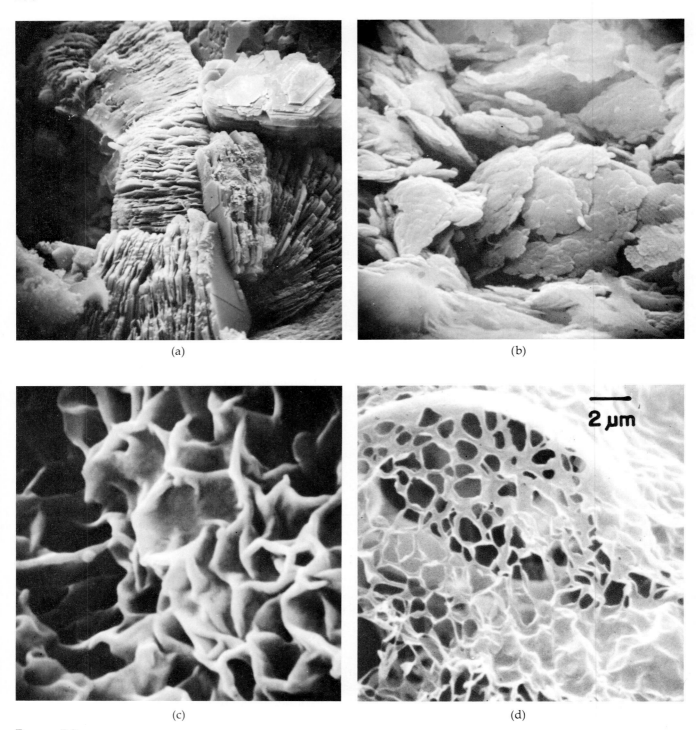

FIGURE 7.2
Crystals of three silicate clay minerals and a photomicrograph of humic acid found in soils. (a) Kaolinite from Illinois magnified about 1900 times (note hexagonal crystal at upper right). (b) A fine-grained mica from Wisconsin magnified about 17,600 times. (c) Montmorillonite (a smectite group mineral) from Wyoming magnified about 21,000 times. (d) Fulvic acid (a humic acid) from Georgia magnified about 23,000 times. [(a)–(c) Courtesy Dr. Bruce F. Bohor, Illinois State Geological Survey; (d) from Dr. Kim H. Tan, University of Georgia; used with permission of Soil Science Society of America]

temperate areas and are prominent in soils of the tropics as well. Each of the types of silicate clays will be given more detailed consideration later.

HYDROUS OXIDES OF IRON AND ALUMINUM

These clays are commonly dominant in the highly weathered soils of the tropics and semitropics and are also present in significant quantities in the soils of some temperate regions. Properties of the red and yellow soils so common in the southeastern United States are determined to a large degree by these clays.

Examples of iron and aluminum oxides common in soils are gibbsite $(Al_2O_3 \cdot 3H_2O)$ and goethite $(Fe_2O_3 \cdot H_2O)$. The formulas also may be written in the hydroxide form, that is, gibbsite as $Al(OH)_3$ and geothite as $FeOOH$. For simplicity, they will be referred to as the Fe, Al oxide clays.

Less is known about these clays than about the layer silicates. Some have definite crystalline structures, but others are amorphous. The Fe, Al clays are not as sticky and plastic as the layer silicate clays. At high pH values, the micelles carry a small negative charge and are surrounded by a corresponding equivalent of cations. In very acid soils, however, some Fe, Al oxides are positively charged and attract negatively charged anions instead of cations (anion exchange is covered in Section 7.14).

ALLOPHANE AND OTHER AMORPHOUS MINERALS

In many soils there are significant quantities of colloidal matter that is either amorphous or has a crystalline structure not sufficiently ordered to be detected by X rays. Because they lack ordered three-dimensional crystalline structure, they are sometimes referred to as *short-range order minerals*. They are more difficult to study than well-crystallized minerals and, consequently, less is known about them.

Perhaps the most significant amorphous silicate colloid is allophane. This mineral is a somewhat poorly defined aluminum silicate with a general composition approximating $Al_2O_3 \cdot 2SiO_2H_2O$. It is most prevalent in soils developed from volcanic ash such as some found in the northwestern part of the United States. Allophane has a high capacity to adsorb cations but also adsorbs anions (see Section 7.14).

ORGANIC SOIL COLLOIDS

The colloidal organization of humus has some similarities to that of clay. A highly charged micelle is surrounded by a swarm of cations. The humus colloids are not crystalline, however. They are composed basically of carbon, hydrogen, and oxygen rather than of silicon, aluminum, iron, oxygen, and hydroxy groups. The organic colloidal particles vary in size, but they may be at least as small as the silicate clay particles.

The negative charges of humus are associated with partially dissociated

enolic (—OH), carboxyl (—COOH), and phenolic (—⬡—OH) groups; these

groups in turn, are associated with central units of varying size and complexity. The relationship is illustrated in Figure 7.3.

As is the case for the Fe, Al oxides, the negative charge associated with humus is dependent on the soil pH. Under very acid conditions, the negative charge is not very high, lower than that of some of the silicate clays. With a rise in pH, however, the hydrogen ions dissociate from first the carboxyl groups and then the enolic and phenolic groups. This leaves a

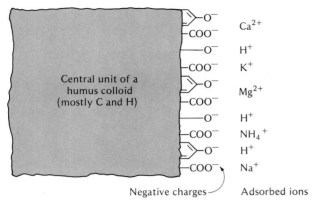

FIGURE 7.3

Adsorption of cations by humus colloids. The phenolic hydroxy groups are attached to aromatic rings,

⬡—OH

Other —OH and the carboxyl (—COOH) groups are bonded to carbon atoms in the central unit. Note the general similarity to the adsorption situation in silicate clays.

greatly increased negative charge on the colloid. Under neutral to alkaline conditions, the electronegativity of humus per unit weight greatly exceeds that of the silicate clays. In these high pH soils, the adsorbed hydrogen is replaced by calcium, magnesium, and other cations (Figure 7.3).

7.3

Adsorbed Cations

In humid regions, the cations of calcium, aluminum, and hydrogen are by far the most numerous, whereas in an arid region soil, calcium, magnesium, potassium, and sodium predominate (see Table 7.1). A colloidal complex may be represented in the following simple and convenient way for each region.

a Ca^{2+}
b Al^{3+}
c H^+ | Micelle |
d M

(humid region)

e Ca^{2+}
f Mg^{2+}
g K^+ | Micelle |
h M

(arid region)

TABLE 7.1

Typical Proportions of Major Adsorbed Cations in the Surface Layers of Different Soil Orders

The percentage figures in each case are based on the sum of the cation equivalents taken as 100.

Soil order[a]	Typical location	H^+ and Al^{3+} (%)[b]	Ca^{2+} (%)	Mg^{2+} (%)	K^+ (%)	Na^+ (%)
Oxisols	Hawaii	85	10	3	2	tr[c]
Spodosols	New England	80	15	3	2	tr
Ultisols	Southeast U.S.	65	25	6	3	1
Alfisols	Pennsylvania to Wisconsin	45	35	13	5	2
Vertisols	Alabama to Texas	40	38	15	5	2
Mollisols	Midwest U.S.	30	43	18	6	3
Aridisols	Southwest U.S.	—	65	20	10	5

[a] See Chapter 3 for soil descriptions.
[b] Al^{3+} adsorption includes that of complex aluminum hydroxy ions.
[c] tr = trace.

The M stands for the small amounts of other "base-forming" cations adsorbed (e.g., Na^+, NH_4^+) by the colloids. The a through h indicate that the numbers of cations are variable.

These examples illustrate that soil colloids and their associated exchangeable ions can be considered, although perhaps in an oversimplified way, as *complex salts* with a large anion (micelle) surrounded by numerous cations.

CATION PROMINENCE

Two major factors will determine the relative proportion of the different cations adsorbed by clays. First, these ions are not all held with equal tightness by the soil colloids. The order of strength of adsorption,[3] when the ions are present in equivalent quantities, is $Al^{3+} > Ca^{2+} > Mg^{2+} > K^+ = NH_4^+ > Na^+$.

Second, the relative concentration of the cations in the soil solution will help determine the degree to which adsorption occurs. Thus, in the soil solution of very acid soils, the concentrations of both H^+ and Al^{3+} are high, and these ions dominate the adsorbed cations. At neutral pH and above, however, the concentrations in the soil solution of both H^+ and Al^{3+} are very low, and, consequently, the adsorption of these ions is minimal. In neutral to moderately alkaline soils, Ca^{2+} and Mg^{2+} dominate. In some poorly drained soils of arid regions, salts high in sodium accumulate, and the adsorption of Na^+ becomes much more prominent. The typical proportions of cations in different soils are shown in Table 7.1.

CATION EXCHANGE

As pointed out in Chapter 1, the surface-held adsorbed cations are subject to exchange with other cations held in the soil solution. For example, a calcium ion held on the colloidal surface is subject to exchange with two H^+ ions in the soil solution.

$$\boxed{\text{Micelle}}\,Ca^{2+} \quad + \quad 2\,H^+ \quad \rightleftharpoons \quad \boxed{\text{Micelle}}\begin{matrix} H^+ \\ H^+ \end{matrix} \quad + \quad Ca^{2+}$$

(colloid)   (soil solution)   (colloid)   (soil solution)

Thus, soil colloids are focal points for cation-exchange reactions, which have profound effects on soil–plant relations. These will be discussed in greater detail after consideration is given to the specific types and characteristics of soil colloids (see Section 7.10).

7.4

Fundamentals of Layer Silicate Clay Structure

Now that the general characteristics of soil colloids and their associated cations are understood, we turn to a more detailed consideration of silicate clays. The use of X-ray electron microscopy and other techniques has demonstrated that silicate clay particles are crystalline and that each particle comprises individual layers or sheets (Figure 7.1). The mineralogical organization of these layers varies from one type of clay to another and markedly

[3] The strength of adsorption of the H^+ ion is difficult to determine since hydrogen-dominated mineral colloids break down to form aluminum-saturated colloids.

affects the properties of the mineral. For this reason, some attention will be given to the fundamentals of silicate clay structure before consideration of specific silicate clay minerals.

SILICA TETRAHEDRAL AND ALUMINA-MAGNESIA OCTAHEDRAL SHEETS

The most important silicate clays are known as *phyllosilicates* (Gr. *phyllon*, leaf) because of their leaf-like or platelet structure. As shown in Figure 7.4, they are comprised of two kinds of horizontal *sheets*, one dominated by silicon, the other by aluminum and/or magnesium.

The basic building block for the silica-dominated sheet is a unit composed of one silicon atom surrounded by four oxygen atoms. It is called the silica *tetrahedron* because of its four-sided configuration (Figure 7.4). An interlocking array of a series of these silica tetrahedra tied together horizontally by shared oxygen anions gives a *tetrahedral sheet*.

Aluminum and/or magnesium ions are the key cations in the second type of sheet. An aluminum (or magnesium) ion surrounded by six oxygen atoms or hydroxy groups gives an eight-sided building block termed *octahedron* (Figure 7.4). Numerous octahedra linked together horizontally comprise the *octahedral* sheet. An aluminum-dominated sheet is known as a *dioctahedral* sheet, whereas one dominated by magnesium is called a *trioctahedral* sheet. The distinction is due to the fact that *two* aluminum ions in a *di*octahedral sheet satisfy the same negative charge from surrounding oxygens and hydroxys as *three* magnesium ions in a *tri*octahedral sheet. As will be seen later, numerous intergrades occur where both cations are present.

The tetrahedral and octahedral sheets are the fundamental structural units of silicate clays. They, in turn, are bound together within the crystals by shared oxygen atoms into different *layers*. The specific nature and combination of sheets in these layers vary from one type of clay to another and largely control the physical and chemical properties of each clay. The relationship between sheets and layers shown in Figure 7.4 is important and should be well understood.

ISOMORPHOUS SUBSTITUTION

The structural arrangements just described suggest a very simple relationship among the elements making up silicate clays. In nature, however, more complex formulas result. The weathering of a wide variety of rocks and minerals permits cations of comparable size to substitute for silicon, aluminum, and magnesium ions in the respective tetrahedral and octahedral sheets.

The ionic radii of a number of ions common in clays are listed in Table 7.2 to illustrate this point. Note that aluminum is only slightly larger than silicon. Consequently, aluminum can fit into the center of the tetrahedron in the place of the silicon without changing the basic structure of the crystal. The process, called *isomorphous substitution*, is common and accounts for the wide variability in the nature of silicate clays.

Isomorphous substitution also occurs in the octahedral sheet. Note from Table 7.2 that ions such as iron and zinc are not too different in size from aluminum and magnesium ions. As a result, these ions can fit into the position of either the aluminum or magnesium as the central ion in the octahedral sheet. It should be emphasized that some layer silicates are characterized by isomorphous substitution in either or both of the tetrahedral or octahedral sheets.

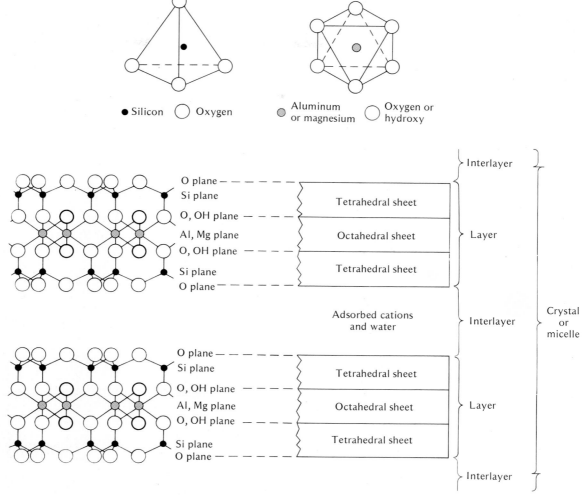

FIGURE 7.4

The basic molecular and structural components of silicate clays. (a) A single *tetrahedron*, a four-sided building block comprised of a silicon ion surrounded by four oxygen atoms, and a single eight-sided *octahedron*, in which an aluminum (or magnesium) ion is surrounded by six hydroxy groups or oxygen atoms. (b) In clay crystals thousands of these tetrahedral and octahedral building blocks are connected to give horizontal planes of silicon and aluminum (or magnesium) ions. These planes alternate with planes of oxygen atoms and hydroxy groups (heavy circles). The silicon plane and associated oxygen/hydroxy planes make up a *tetrahedral sheet*. Similarly, the aluminum/magnesium plane and associated oxygen/hydroxy planes comprise the *octahedral sheet*. Different combinations of tetrahedral and octahedral sheets are termed *layers*. In some silicate clays these layers are separated by *interlayers* in which water and adsorbed cations are found. Many layers are found in each *crystal* or *micelle* (microcell).

SOURCE OF CHARGES

Isomorphous substitution is of vital importance because it is the primary source of both negative and positive charges of silicate clays. For example, the substitution of one Al^{3+} for a Si^{4+} in the tetrahedral sheet leaves one unsatisfied negative charge. Likewise, the substitution of one Al^{3+} for one

TABLE 7.2

Ionic Radii of Elements Common in Silicate Clays and an Indication of Which Are Found in the Tetrahedral and Octahedral Sheets

Note that Al, Fe, O, and OH can fit in either.

Ion	Radius (nm)[a]	Found in
Si^{4+}	0.042	Silica tetrahedra
Al^{3+}	0.051	Silica tetrahedra
Fe^{3+}	0.064	Silica tetrahedra
Mg^{2+}	0.066	Alumina octahedra
Zn^{2+}	0.074	Alumina octahedra
Fe^{2+}	0.070	Exchange sites
Na^+	0.097	Exchange sites
Ca^{2+}	0.099	Exchange sites
K^+	0.133	Exchange sites
O^{2-}	0.140	Both sheets.
OH^-	0.155	Both sheets.

[a] 1 nm = 10^{-9}m.

Mg^{2+} in a trioctahedral sheet results in one excess positive charge. The net charge associated with a clay micelle is the balance between the positive and negative charges. In most clays, the negative charges predominate. This subject will receive more attention later (see Section 7.8).

7.5

Mineralogical Organization of Silicate Clays

On the basis of the number and arrangement of tetrahedral (silica) and octahedral (alumina–magnesia) sheets contained in the crystal units or layers, silicate clays are classified into three different groups: (a) 1:1-type minerals—one tetrahedral (Si) to one octahedral (Al) sheet; (b) 2:1-type minerals—two tetrahedral (Si) to one octahedral (Al) sheet; and (c) 2:1:1-type minerals. For illustrative purposes, each of these is discussed briefly in terms of the main member of the group.

1:1-TYPE MINERALS

The *layers* of the 1:1-type minerals are made up of one tetrahedral (silica) sheet combined with one octahedral (alumina) sheet—hence the terminology 1:1-type crystal (Figure 7.5). In soils, *kaolinite* is the most prominent member of this group, which includes *halloysite, nacrite,* and *dickite*.

The tetrahedral and octahedral sheets in a layer of a kaolinite crystal are held together tightly by oxygen atoms, which are mutually shared by the silicon and aluminum cations in their respective sheets. These layers, in turn, are held together by *hydrogen bonding* (see Section 5.1). Consequently, the structure is *fixed*, and no expansion ordinarily occurs between layers when the clay is wetted. Cations and water do not enter between the structural layers of a 1:1-type mineral particle. The effective surface of kaolinite is thus restricted to its outer faces or to its external surface area. Also, there is little isomorphous substitution in this 1:1-type mineral. Along with the relatively low surface area of kaolinite, this accounts for its low capacity to adsorb cations.[4]

---

[4] Since the adsorbed cations may be freely exchanged with other cations, the capacity to adsorb cations is usually referred to as *cation exchange capacity* (see Section 7.11).

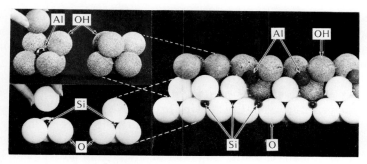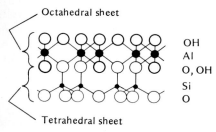

FIGURE 7.5

Models of ions that constitute a layer of the 1:1-type clay kaolinite. Note that each layer consists of alternating octahedral (alumina) and tetrahedral (silica) sheets—hence the designation 1:1. Aluminum ions surrounded by six hydroxy groups and oxygen atoms make up the octahedral sheet (upper left). Smaller silicon ions associated with four oxygen atoms constitute the tetrahedral sheet (lower left). The octahedral and tetrahedral sheets are bound together (center) by mutually shared oxygen atoms. The result is a layer with hydroxys on one surface and oxygens on the other. A schematic drawing of the ionic arrangement (right) shows a cross-sectional view of a crystal layer. The kaolinite mineral is comprised of a series of these flat layers tightly held together with no interlayer spaces.

Kaolinite crystals usually are hexagonal in shape (Figure 7.2). In comparison with other clay particles, they are large in size, ranging from 0.10 to 5 μm across, with the majority falling within the 0.2–2 μm range. Because of the strong binding forces between their structural layers, kaolinite particles are not readily broken down into extremely thin plates.

In contrast with the other silicate groups, kaolinite exhibits very little plasticity (capability of being molded), cohesion, shrinkage, and swelling. Its restricted surface and limited adsorptive capacity for cations and water molecules suggest that kaolinite does not exhibit colloidal properties of a high order of intensity (Table 7.3).

2:1-TYPE MINERALS

The crystal units (layers) of these minerals are characterized by an octahedral sheet sandwiched between two tetrahedral sheets. Three general groups have this basic crystal structure. Two of them, *smectite* and *vermiculite*, include expanding-type minerals, while the third, *fine-grained micas* (*illite*), is relatively nonexpanding.

TABLE 7.3

Major Properties of Selected Silicate Clay Minerals and of Humus

Property	Smectite	Vermiculite[a] (Dioctahedral)	Fine mica	Chlorite	Kaolinite	Humus
Size (μm)	0.01–1.0	0.1–5.0	0.2–2.0	0.1–2.0	0.5–5.0	0.1–1.0
Shape	Flakes	Plates; flakes	Flakes	Variable	Hexagonal crystals	Variable
External surface (m^2/g)	70–120	50–100	70–100	70–100	10–30	500–800
Internal surface (m^2/g)	550–650	500–600	—	—	—	
Intralayer spacing[b](nm)	1.0–2.0	1.0–1.5	1.0	1.4	0.7	—
Net negative charge (cmol/kg)[c]	80–120	100–180	15–40	15–40	2–5	200–750

[a] Dioctahedral vermiculite is more common in soils than trioctahedral vermiculite.
[b] From the top of one layer to the top of the next similar layer; 1 nm = 10^{-9} m.
[c] Centimoles per kilogram (1 cmol = 0.01 mol), a measure of the cation exchange capacity (see Section 7.11).

Expanding Minerals. The *smectite* group is noted for *interlayer expansion*, which occurs by swelling when the minerals are wetted, the water entering the interlayer space and forcing the layers apart. *Montmorillonite* is the most prominent member of this group in soils, although *beidellite, nontronite,* and *saponite* also are found.

The flake-like crystals of smectites (Figure 7.2) are composed of 2:1-type layers, as shown in Figure 7.6. In turn, these layers are loosely held together by very weak oxygen-to-oxygen and cation-to-oxygen linkages. Exchangeable cations and associated water molecules are attracted between layers (the interlayer space), causing *expansion* of the crystal lattice. The *internal surface*, then exposed, by far exceeds the external surface area of these minerals. For example, the *specific surface* or total surface area per unit mass (external and internal) of one smectite mineral (montmorillonite) is 700–800 m²/g. A comparable figure for kaolinite is only 5–20 m²/g (Table 7.3). Commonly, these smectite crystals range in size from 0.01–1 μm, much smaller than the average kaolinite particle.

Isomorphous substitution of Mg^{2+} for some of the Al^{3+} ions in the dioctahedral sheet accounts for most of the negative charge for smectites, although some substitution of Al^{3+} for Si^{4+} has occurred in the tetrahedral sheet. The smectites commonly show a high cation exchange capacity, perhaps 20–40 times that of kaolinite (Table 7.3).

Smectites also are noted for their high plasticity and cohesion and their

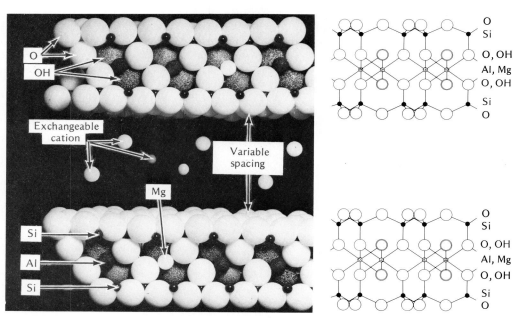

FIGURE 7.6

Model of two crystal layers and an interlayer characteristic of montmorillonite, a smectite expanding-lattice 2:1-type clay mineral. Each layer is made up of an octahedral (alumina) sheet sandwiched between two tetrahedral (silica) sheets. There is little attraction between oxygen atoms in the bottom tetrahedral sheet of one unit and those in the top tetrahedral sheet of another. This permits a ready and variable space between layers, which is occupied by water and exchangeable cations. This internal surface far exceeds the surface around the outside of the crystal. Note that magnesium has replaced aluminum in some sites of the octahedral sheet. Likewise, some silicon atoms in the tetrahedral sheet may be replaced by aluminum (not shown). These substitutions give rise to a negative charge, which accounts for the high cation exchange capacity of this clay mineral.

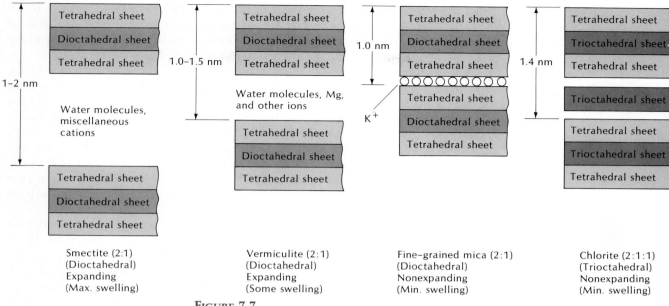

FIGURE 7.7

Schematic drawing illustrating the organization of tetrahedral and octahedral sheets in three 2:1-type minerals (smectite, vermiculite, and fine-grained mica) and one 2:1:1-type, chlorite. The aluminum-dominated dioctahedral minerals are illustrated for the 2:1 types because such minerals are most commonly found in soils. Trioctahedral chlorite in which magnesium ions dominate is the most common form of this mineral in soils. Note maximum interlayer expansion in smectite, with reduced expansion in vermiculite due to moderate binding power of numerous Mg^{2+} ions. Fine-grained mica and chlorite do not expand because K^+ ions (fine-grained mica) and an extra trioctahedral sheet (chlorite) tightly bind the 2:1 layers together. The interlayer spacings are shown in nanometers (1 nm = 10^{-9} m).

marked swelling when wet and shrinkage on drying. Wide cracks commonly form as smectite-dominated soils (e.g., Vertisols) are dried (Figure 7.17). The dry aggregates or clods are very hard, making such soils difficult to till.

Vermiculites are also 2:1-type minerals in that an octahedral sheet occurs between two tetrahedral sheets. In most soil vermiculites, the octahedral sheet is aluminum-dominated (dioctahedral), although magnesium-dominated (trioctahedral) vermiculites are also common. In the tetrahedral sheet of most vermiculites, considerable substitution of aluminum for silicon has taken place. This accounts for most of the very high net negative charge associated with these minerals.

Water molecules, along with magnesium and other ions, are strongly adsorbed in the interlayer space of vermiculites (Figure 7.7). However, they act primarily as bridges holding the units together rather than as wedges driving them apart. The degree of swelling is, therefore, considerably less for vermiculites than for smectites. For this reason, vermiculites are considered *limited-expansion* clay minerals, expanding more than kaolinite but much less than the smectites.

The cation exchange capacity of vermiculites usually exceeds that of all other silicate clays, including montmorillonite and other smectites (Table 7.3), because of the very high negative charge in the tetrahedral sheet. Vermiculite crystals are larger than those of the smectites but much smaller than those of kaolinite.

ILLITE

Nonexpanding Minerals. *Micas* are the type minerals in this group. Muscovite and biotite are examples of unweathered micas often found in the sand and silt separates. Weathered minerals similar in structure to these micas are found in the clay fraction of soils. They are called *fine-grained micas.*[5]

Like smectites, fine-grained micas have a 2:1-type crystal. However, the particles are much larger than those of the smectites. Also, the major source of charge is in the tetrahedral sheet where about 20% of the silicon sites are occupied by aluminum atoms. This results in a high net negative charge in the tetrahedral sheet, even higher than that found in vermiculites. To satisfy this charge, potassium ions are strongly attracted in the interlayer space and are just the right size to fit snugly into certain spaces in the adjoining tetrahedral sheets (Figures 7.7 and 7.8). The potassium thereby acts as a binding agent, preventing expansion of the crystal. Hence, fine-grained micas are quite *nonexpansive.*

Such properties as hydration, cation adsorption, swelling, shrinkage, and plasticity are much less intense in fine-grained micas than in smectites (Table 7.3). The fine-grained micas exceed kaolinite with respect to these characteristics, but this may be due in part to the presence of interstratified layers of smectite or vermiculite. In size, too, fine-grained mica crystals are intermediate between the smectites and kaolinites (Table 7.3). Their specific surface area varies from 70 to 100 m^2/g, about one eighth that for the smectites.

2:1:1-TYPE MINERALS

This silicate group is represented by *chlorites*, which are common in a variety of soils. Chlorites are basically iron-magnesium silicates with some aluminum present. In a typical chlorite clay crystal, 2:1 layers, such as are found in vermiculites, alternate with a magnesium-dominated trioctahedral sheet, giving a 2:1:1 ratio (Figure 7.7). Magnesium also dominates the trioctahedral sheet in the 2:1 layer of chlorites. Thus, the crystal unit contains two silica tetrahedral sheets and two magnesium-dominated trioctahedral

[5] In the past the term *illite* has been used to identify these clay minerals.

FIGURE 7.8

Model of a 2:1-type nonexpanding lattice mineral of the fine-grained mica group. The general constitution of the layers is similar to that in the smectites, one octahedral (alumina) sheet between two tetrahedral (silica) sheets. However, potassium ions are tightly held between layers, giving the mineral a more or less rigid type of structure that prevents the movement of water and cations into the space between layers. The internal surface and exchange capacity of fine-grained micas are thus far below those of the smectites.

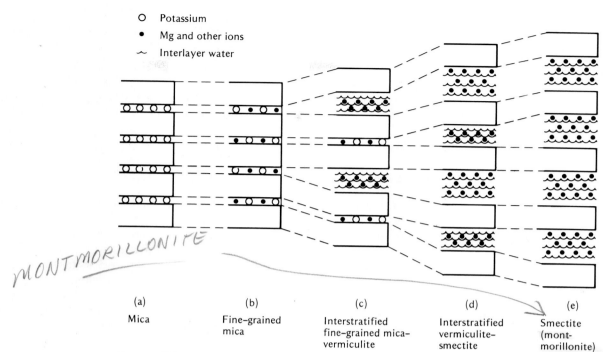

○ Potassium
● Mg and other ions
～ Interlayer water

MONTMORILLONITE

(a)	(b)	(c)	(d)	(e)
Mica	Fine-grained mica	Interstratified fine-grained mica-vermiculite	Interstratified vermiculite-smectite	Smectite (mont-morillonite)

FIGURE 7.9

Structural differences among silicate minerals and their mixtures. Potassium-containing micas (a) with rigid crystals lose part of their potassium and weather to fine-grained mica, which is less rigid and attracts exchangeable cations to the interlayer space (b). At a more advanced stage of weathering (c), the potassium is leached from between some of the 2:1 layers; water and magnesium ions bind the layers together and an interstratified fine-grained mica–vermiculite is present. Further weathering removes more potassium (d), water and exchangeable ions push in between the 2:1 layers, and an interstratified vermiculite–smectite is formed. More weathering produces smectite (e), the highly expanded mineral. Smectite in turn is subject to weathering to kaolinite and iron and aluminum oxides. Although most smectites actually are formed by other processes, the sequence here illustrates the structural relationships of smectites to the other minerals.

sheets, giving rise to the term 2:1:1 or 2:2-type structure.

The negative charge of chlorites is about the same as that of fine-grained micas and considerably less than that of the smectites or vermiculites. Like fine micas, chlorites may be interstratified with vermiculites or smectites in a single crystal. Particle size and surface area for chlorites are also about the same as for fine-grained micas. There is no water adsorption between the chlorite crystal units, which accounts for the nonexpansive nature of this mineral.

MIXED AND INTERSTRATIFIED LAYERS

Specific groups of clay minerals do not occur independently of one another. In a given soil, it is common to find several clay minerals in an intimate mixture. Furthermore, some mineral colloids have properties and composition intermediate between those of any two of the well-defined minerals just described. Such minerals are termed *mixed layer* or *interstratified* because the individual layers within a given crystal may be of more than one type. Terms such as "chlorite-vermiculite" and "fine-grained mica–smectite" are used to describe mixed-layer minerals (Figure 7.9). In some soils, they are more common than single-structured minerals such as montmorillonite.

7.6

Genesis of Silicate Clays

The silicate clays are developed from the weathering of a wide variety of minerals by at least two distinct processes: (a) a slight physical and chemical *alteration* of certain primary minerals and (b) a *decomposition* of primary minerals with the subsequent *recrystallization* of certain of their products into the silicate clays. These processes will each be given brief consideration.

ALTERATION

The changes that occur as muscovite mica is altered to fine-grained mica represent a good example of alteration. Muscovite is a 2:1-type primary mineral with a nonexpanding crystal structure and a formula of $KAl_2(Si_3Al)O_{10}(OH)_2$. As weathering occurs, the mineral is broken down in size to the colloidal range, part of the potassium is lost, and some silicon is added from weathering solutions. The net result is a less rigid crystal structure and an electronegative charge. The fine mica colloid that emerges still has a 2:1-type structure, only having been *altered* in the process. Some of these changes, perhaps oversimplified, can be shown as

$$KAl_2(Al\ Si_3)O_{10}(OH)_2 + 0.2\ Si^{4+} + 0.1\ M^+ \xrightarrow{\ H_2O\ }$$

Muscovite
(rigid crystal)   (soil solution)

$$M_{0.1}^+(K_{0.7})Al_2(Al_{0.8}Si_{3.2})O_{10}(OH)_2 + 0.3\ K^+ + 0.2\ Al^{3+}$$

Fine mica
(semirigid crystal)   (soil solution)

RECRYSTALLIZATION

This process involves the complete breakdown of the crystal structure and recrystallization of clay minerals from products of this breakdown. It is the result of much more intense weathering than that required for the alteration process just described.

An example of recrystallization is the formation of kaolinite (a 1:1-type clay mineral) from solutions containing soluble aluminum and silicon that came from the breakdown of primary minerals having a 2:1-type structure. Such recrystallization makes possible the formation of more than one kind of clay from a given primary mineral. The specific clay mineral that forms depends on weathering conditions and the specific ions present in the weathering solution as crystallization occurs.

RELATIVE STAGES OF WEATHERING

The more specific conditions conducive to the formation of important clay types are shown in Figure 7.10. Note that fine-grained micas and magnesium-rich chlorites represent earlier weathering stages of the silicates, and kaolinite and ultimately iron and aluminum oxides the most advanced stages. The smectites (e.g., montmorillonite) represent intermediate stages.

GENESIS OF INDIVIDUAL CLAYS

There are a variety of processes by which individual clays are formed. For example, fine-grained micas and chlorite are commonly formed from the alteration of muscovite and biotite micas, respectively. Vermiculites also can be formed through this process, although they as well as the smectites can

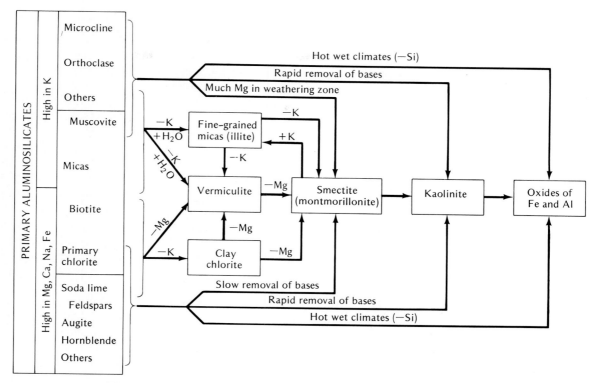

FIGURE 7.10

General conditions for the formation of the various layer silicate clays and oxides of iron and aluminum. Fine-grained micas, chlorite, and vermiculite are formed through rather mild weathering of primary aluminosilicate minerals, whereas kaolinite and oxides of iron and aluminum are products of much more intense weathering. Conditions of intermediate weathering intensity encourage the formation of smectite. In each case silicate clay genesis is accompanied by the removal of soluble elements such as K, Na, Ca, and Mg.

be products of weathering of the fine-grained micas and chlorites. Smectites may also result from the process of recrystallization under neutral to alkaline weathering conditions. Kaolinite is formed by recrystallization under reasonably intense acid weathering conditions, which remove most of the metallic cations. Under the hot humid conditions of the tropics, the intense weathering usually produces the oxides or iron and aluminum (Figure 7.10).

7.7

Geographic Distribution of Clays

The clay of any particular soil is generally made up of a mixture of different colloidal minerals. In a given soil, the mixture may vary from horizon to horizon. This occurs because the kind of clay that develops depends not only on climatic influences and profile conditions but also on the nature of the parent material. The situation may be further complicated by the presence in the parent material itself of clays that were formed under a preceding and perhaps an entirely different type of climatic regime. Nevertheless, some general deductions seem possible, based on the relationships shown in Figure 7.10.

ZONAL SOIL DIFFERENCES

The well-drained and well-weathered Oxisols of humid and subhumid tropics tend to be dominated by the oxides of iron and aluminum. These clays are also prominent in the Ultisols of the southeastern part of the United States. Kaolinite is commonly the dominant silicate mineral in Ultisols (Table 7.4) and is also found along with the hydrous oxide clays in more tropical areas.

The smectite, vermiculite, and fine-grained mica groups are more prominent in Alfisols, Mollisols, and Vertisols where weathering is less intense. These clays are common in the northern part of the United States, Canada, and regions with similar temperatures throughout the world. Where either the parent material or the soil solution surrounding the weathering minerals is high in potassium, fine-grained micas are apt to be formed. Parent materials high in metallic cations (particularly magnesium) or subject to restricted drainage, which discourages the leaching of metallic cations, encourage smectite formation. Thus, fine-grained micas and smectites are more likely to be prominent in Aridisols than in soils prominent in the more humid areas.

The strong influence of parent material on the geographic distribution of clays can be seen in the "black belt" Vertisols of Alabama, Mississippi, and Texas. These soils, which are dark in color, have developed from base-rich marine parent materials and are dominated by smectite clays. The surrounding soils, which have developed from different parent materials, are high in kaolinite and hydrous oxides, clays that are more representative of this warm, humid region. Similar situations exist in central India and Sudan.

Data in Table 7.4 show the dominant clay minerals in different soil orders, the descriptions of which were given in Chapter 3. These data tend to substantiate the generalization just discussed. For example, Oxisols and Ultisols are characteristic of areas of intense weathering, and Aridisols are found in desert areas. The dominant clay minerals for these areas are as expected on the basis of Figure 7.10.

Although a few broad generalizations relating to the geographic distribution of clays are possible, these examples suggest that local parent materials and weathering conditions tend to dictate the kinds of clay minerals found in soils.

TABLE 7.4

Prominent Occurrence of Clay Minerals in Different Soil Orders in the United States and Typical Locations for These soils

Soil order[a]	General weathering intensity	Typical location in U.S.	Hydrous oxides	Kaolinite	Smectite	Fine-grained mica	Vermiculite	Chlorite	Intergrades
Aridisols	Low	Dry areas			××	××	×		
Vertisols[b]	↑	Alabama, Texas			×××				×
Mollisols		Central		×	××	×	×	×	×
Alfisols		Ohio, Pennsylvania, New York		×	×	×	×	×	×
Spodosols	↓	New England	×	×					
Ultisols		Southeast	××	×××			×	×	×
Oxisols	High	Hawaii, Puerto Rico	×××	××					

[a] See Chapter 3 for soil descriptions.
[b] By definition these soils have swelling-type clays, which account for the dominance of smectites.

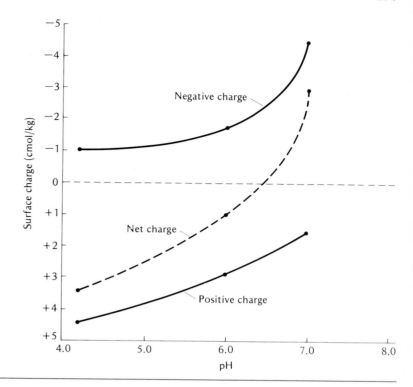

FIGURE 7.13
Surface charge on an Oxisol subsoil in relation to pH. At low pH values the positive charges predominate. Only at pH 6.5 did the net charge become negative. [From Van Raij and Peech (1972)]

in soil, it is not surprising that positive and negative charges may be exhibited at the same time. In most soils of temperate regions, the negative charges far exceed the positive (Table 7.6). However, in some acid soils of the tropics, high in iron and aluminum hydroxides, the overall net charge may be positive. The effect of soil pH on positive and negative charges on such soils is illustrated in Figure 7.13.

The charge characteristics of selected soil colloids are shown in Table 7.6. Note the high percentage of constant negative charges in some 2:1-type clays (e.g., smectites and vermiculites). Humus, kaolinite, allophane, and Fe, Al oxides have mostly variable (pH-dependent) negative charges and exhibit modest positive charges at very low pH values.

CATION AND ANION ADSORPTION

The charges associated with soil particles attract simple and complex ions of opposite charge. Thus, a given colloidal mixture may exhibit not only a maze of positive and negative surface charges but an equally complex complement of simple cations and anions such as Ca^{2+} and SO_4^{2-} that are attracted by the particle charges.

Figure 7.14 illustrates how soil colloids attract the mineral elements so important for plant growth.[6] The adsorbed anions are commonly present in smaller quantities than the cations because the negative charges generally predominate on the soil colloid, especially in soils of temperate regions. Consideration now will be given to the exchange of ions between micelles and the soil solution, starting first with cation exchange.

[6] In addition to the simple inorganic cations and anions, more complex organic compounds as well as charged organo-mineral complexes are adsorbed by the charged soil colloids.

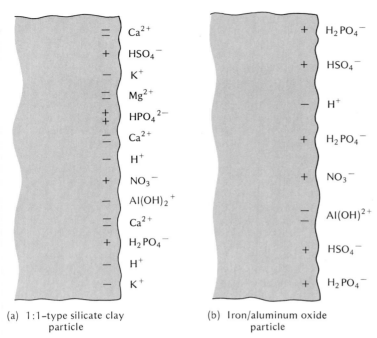

(a) 1:1–type silicate clay
      particle

(b) Iron/aluminum oxide
      particle

FIGURE 7.14
Illustration of adsorbed cations and anions in an acid soil with (a) 1:1-type silicate clay and (b) iron and aluminum oxide particles. Note the dominance of the negative charges on the silicate clay and of the positive charges on the aluminum oxide particle.

7.10

Cation Exchange

In Section 7.3 the various cations adsorbed by the exchange complex[7] were shown to be subject to replacement by other cations through a process called *cation exchange*. For example, hydrogen ions generated as organic matter decomposes (see Section 3.2) can displace calcium and other metallic cations from the colloidal complex. This can be shown simply as follows, where only one adsorbed calcium ion is being replaced.

$$Ca^{2+}\boxed{Micelle} + 2\ H^+ \rightleftharpoons \begin{matrix} H^+ \\ H^+ \end{matrix}\boxed{Micelle} + Ca^{2+}$$

The reaction takes place rapidly, and the interchange of calcium and hydrogen is chemically equivalent. As shown by the double yield arrows, the reaction is reversible and will go to the left if calcium is added to the system.

CATION EXCHANGE UNDER NATURAL CONDITIONS

As it usually occurs in humid region surface soils the cation exchange reaction is somewhat more complex, but the principles illustrated in the simple equation apply. Assume, for the sake of simplicity, that the numbers of Ca^{2+}, Al^{3+}, H^+, and all other metallic cations such as Mg^{2+} and K^+ (represented as M^+) are in the ratio 40, 20, 20, and 20 per micelle, respec-

[7] This term includes all soil colloids, inorganic and organic, capable of holding and exchanging cations.

tively. (The metallic cations represented as M^+ will be considered monovalent in this case.) Hydrogen from carbonic acid (H_2CO_3) will react with the micelles as follows.

$$
\begin{array}{c}
Ca_{40} \\
Al_{20} \\
H_{20} \\
M_{20}
\end{array}
\boxed{\text{Micelle}} + 5\,H_2CO_3 \rightleftharpoons
\begin{array}{c}
Ca_{38} \\
Al_{20} \\
H_{25} \\
M_{19}
\end{array}
\boxed{\text{Micelle}} + 2\,Ca(HCO_3)_2 + M(HCO_3)
$$

(soil solids)   (in soil solution)   (soil solids)   (in soil solution)

Where sufficient rainfall is available to leach calcium and other metallic cations, the reaction tends to go toward the right and the soils tend to become more acid. In regions of low rainfall, however, calcium and other salts are more plentiful since they are not easily leached from the soil. The metallic ions drive the reaction to the left, keeping the soil neutral or at an even higher pH. In addition, some of the calcium will be precipitated as $CaCO_3$, especially in the subsoil at the lower limit of rainfall penetration. The interaction of climate, biological processes, and cation exchange thus helps determine the chemical properties of soils.

Important

INFLUENCE OF LIME AND FERTILIZER

Cation exchange reactions are reversible; therefore if some basic calcium compound such as limestone is applied to an acid soil, the preceding reaction will be driven to the left. Calcium ions will replace hydrogen and other cations, the H^+ ions will be neutralized by OH^- or CO_3^{2-} ions, and the soil pH will be raised. If, on the other hand, acid-forming chemicals, such as sulfur, are added to an alkaline soil in a dryland area, the H^+ ions would replace the metal cations on the soil colloids, and the soil pH would decrease.

One more illustration of cation exchange is the reaction that may occur when a fertilizer containing potassium chloride is added to soil.

$$
\begin{array}{c}
Ca_{40} \\
Al_{20} \\
H_{40} \\
M_{20}
\end{array}
\boxed{\text{Micelle}} + 7\,KCl \rightleftharpoons
\begin{array}{c}
K_7 \\
Ca_{38} \\
Al_{20} \\
H_{39} \\
M_{18}
\end{array}
\boxed{\text{Micelle}} + 2\,CaCl_2 + HCl + 2\,MCl
$$

(soil solids)   (in soil solution)   (soil solids)   (in soil solution)

The added potassium is adsorbed on the colloid and replaces an equivalent quantity of calcium, hydrogen, and other cations that appear in the soil solution. The adsorbed potassium remains largely in an available condition but is less subject to leaching than are most fertilizer salts. Hence, cation exchange is an important consideration, not only for nutrients already present in soils but also for those applied in commercial fertilizers and in other ways.

7.11

Cation Exchange Capacity

Previous sections have dealt qualitatively with cation exchange. We now consider cation exchange quantitatively by describing the *cation exchange capacity*. This property, which is defined simply as "the sum total of the exchangeable cations that a soil can adsorb," is easily determined. By stan-

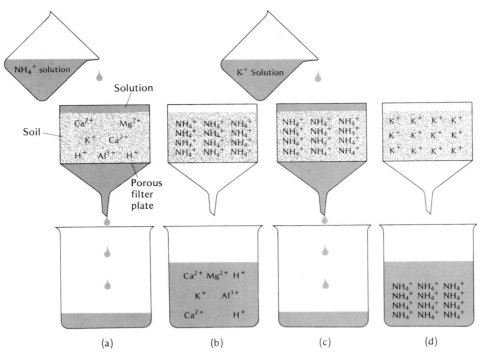

FIGURE 7.15
Illustration of a method for determining the cation exchange capacity of soils. (a) A given mass of soil containing a variety of exchangeable cations is leached with an ammonium (NH_4^+) salt solution. (b) The NH_4^+ ions replace the other adsorbed cations, which are leached into the container below. (c) After the excess NH_4^+ salt solution is removed with an organic solvent, such as alcohol, a K^+ salt solution is used to replace and leach the adsorbed NH_4^+ ions. (d) The amount of NH_4^+ released and washed into the lower container can be determined, thereby measuring the chemical equivalent of the cation exchange capacity (i.e., the negative charge on the soil colloids).

dard methods all the adsorbed cations in a soil are replaced by a common ion, such as barium, potassium, or ammonium; then the amount of adsorbed barium, potassium, or ammonium is determined (Figure 7.15).

MEANS OF EXPRESSION[8]

The cation exchange capacity (CEC) is expressed in terms of moles of positive charge adsorbed per unit mass. For the convenience of expressing CEC in whole numbers, we shall use *centimoles of positive charge per kilogram of soil* (cmol/kg). Thus, if a soil has a cation exchange capacity of 10 cmol/kg, 1 kg of this soil can adsorb 10 cmol of H^+ ion, for example, and can exchange it with 10 cmol of another monovalent cation, such as K^+ or Na^+, or with 5 cmol of divalent cation, such as Ca^{2+} or Mg^{2+}. In each case, the 10 cmol of negative charge associated with 1 kg of soil is attracting 10 cmol of positive charges, whether they come from H^+, K^+, Na^+, NH_4^+, Ca^{2+}, Mg^{2+}, Al^{3+}, or any other cation.

Note that the cations are adsorbed and exchanged on a *chemically equivalent* basis. One mole of charge is provided by 1 mole of H^+, K^+ or any other

[8] In the past, cation exchange capacity generally has been expressed as milliequivalents per 100 g soil. In this text, the International System of Units (SI) is being used. Fortunately, since one milliequivalent per 100 g soil is equal to 1 centimole (cmol) of positive or negative charge per kilogram of soil, it is easy to compare soil data using either of these methods of expression.

monovalent cation, by $\frac{1}{2}$ mole of Ca^{2+}, Mg^{2+}, or other divalent cation and by $\frac{1}{3}$ mole of Al^{3+} or other trivalent cation.

With this chemical equivalency in mind, it is easy to express cation exchange in practical field terms. For example, when an acid soil is limed and calcium ions replace part of the adsorbed hydrogen ions, the reaction that occurs between the unicharged H^+ ion and the di-charged Ca^{2+} ion is

$$\boxed{Micelle}\begin{matrix}H^+\\H^+\end{matrix} + Ca^{2+} \rightleftharpoons \boxed{Micelle}\,Ca^{2+} + 2\,H^+$$

Note that the 2 moles of charge associated with two H^+ ions are replaced by the equivalent charge associated with one Ca^{2+}. In other words, 1 mole of H^+ ion (1 g) would exchange with $\frac{1}{2}$ mole of Ca^{2+} ion (40/2 = 20 g). Accordingly, to replace 1 *centimole* H^+/kg would require 20/100 = 0.2 g Ca^{2+}/kg soil. The amount of Ca^{2+} required for a hectare–furrow slice (2.2 million kg) is $0.2 \times 2.2 \times 10^6$ = 440,000 g or 440 kg. This can be expressed in terms of the quantity of limestone ($CaCO_3$) to supply the Ca^{2+} by multiplying by the ratio $CaCO_3/Ca^{2+}$ = 100/40 = 2.5. Thus, 440×2.5 = 1100 kg limestone per hectare–furrow slice would exchange with 1 mole H^+/kg soil. A comparable figure in the English system is 1000 pounds per acre–furrow slice. These are practical relationships worth remembering.

CATION EXCHANGE CAPACITIES OF SOILS

The cation exchange capacity (CEC) of a given soil is determined by the relative amounts of different colloids in that soil and by the CEC of each of these colloids. Thus, sandy soils have lower CECs than clay soils because the coarse-textured soils are commonly lower in both clay and humus content. Likewise, a clay soil dominated by 1:1-type silicate clays and Fe, Al oxides will have a much lower CEC than will one with similar humus content dominated by smectite clays. See Table 7.3 for the exchange capacities of the major soil colloids.

The generalizations of the preceding paragraph can be verified by examination of Table 7.7. The CECs of a number of soils from different locations in the United States demonstrate the general relationship between soil texture (mostly the clay content) and CEC. Note that the Cecil clay from Alabama, dominated by 1:1-type clays and Fe, Al oxides, has a CEC of only 4.0 cmol/kg. The CEC of a second clay soil from the same state (Susquehanna clay), in which 2:1-type colloids are dominant, is much higher (34.2 cmol/kg).

TABLE 7.7
Cation Exchange Capacities of a Wide Variety of Surface Soils from Various Parts of the United States

Soil order (series, state)	Exchange capacity[a]	Soil order (series, state)	Exchange capacity[a]	Soil order (series, state)	Exchange capacity[a]
Sand		**Loam**		**Clay and Clay Loam**	
Udult (Sassafras, NJ)	2	Ochrept (Hoosic, NJ)	11	Udult (Cecil clay loam, AL)	4
Psamment (Plainfield, WI)	3	Ochrept (Dover, NJ)	14	Udult (Cecil clay, AL)	5
		Udult (Collington, NJ)	16	Xeroll (Gleason clay loam, CA)	32
Sandy Loam		**Silt Loam**			
Udult (Sassafras, NJ)	3	Udalf (Fayette, MN)	13	Udalf (Susquehanna clay, AL)	34
Udult (Norfolk, AL)	3	Boralf (Spencer, WI)	14		
Udult (Cecil, SC)	6	Ustoll (Dawes, NE)	18	Xeroll (Sweeney clay, CA)	58
Udult (Coltsneck, NJ)	10	Udalf (Penn, NJ)	20		
		Udalf (Miami, WI)	23		
		Udoll (Grundy, IL)	26		

Data compiled from Lyon et al. (1952).   [a]Centimoles of charge per kilogram of dry soil.

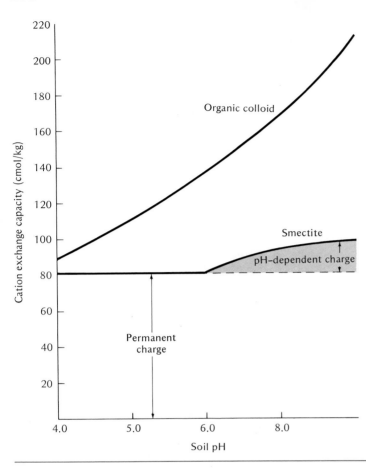

FIGURE 7.16
Influence of pH on the cation exchange capacity of smectite and humus. Below pH 6.0 the charge for the clay mineral is relatively constant. This charge is considered permanent and is due to ionic substitution in the crystal unit. Above pH 6.0 the charge on the mineral colloid increases slightly because of ionization of hydrogen from exposed hydroxy groups at crystal edges. In contrast to the clay, essentially all of the charges on the organic colloid are considered pH dependent. [Smectite data from Coleman and Mehlich (1957); organic colloid data from Helling et al. (1964)]

pH and Cation Exchange Capacity

In previous sections the cation exchange capacity of most soils was demonstrated to increase with pH (Figure 7.16). At very low pH values, the cation exchange capacity is also generally low. Under these conditions, only the "permanent" charges of the 2:1 type clays (see Section 7.8) and a small portion of the pH-dependent charges of organic colloids, allophane, and some 1:1 clays hold exchangeable ions.

As the pH is raised, the negative charges on some 1:1-type silicate clays, allophane, humus, and even Fe, Al oxides increases, thereby increasing the cation exchange capacity. To obtain a measure of the maximum retentive capacity, the CEC is usually determined at a pH of 7.0 or above. At neutral or slightly alkaline pH, the CEC reflects most of the pH-dependent charges as well as the permanent charge.

Representative Figures

From the preceding discussions, it is possible to approximate the CEC of a soil in a given area if the type and contents of the dominant clay colloids and the humus are known. For example, the CEC of the dominant clays in surface soils of the humid temperate region in the United States averages about 0.5 cmol/kg for each 1% clay in the soil. A comparable figure for humus is 2.0 cmol/kg for each 1% of humus content. By using these figures, it is possible to calculate roughly the cation exchange capacity of a repre-

TABLE 7.8
Cation Exchange Data for Representative Mineral Surface Soils in Different Areas

Characteristics	Humid region soil (Alfisol)	Semiarid region soil (Aridisol)	Arid region soil (Natrargids)[a]
Exchangeable calcium (cmol/kg)	6–9	14–17	12–14
Other exchangeable bases (cmol/kg)	2–3	5–7	8–12
Exchangeable hydrogen and/or aluminum (cmol/kg)	4–8	1–2	0
Cation exchange capacity (cmol/kg)	12–18	20–26	20–26
Base saturation (%)	66.6	90–95	100
Probable pH	5.6–5.8	~7	8–10

[a] Significant sodium saturation.

sentative humid temperate region surface soil from the percentages of clay and organic matter present. For soils dominated by kaolinite and Fe, Al oxides, comparable figures might be 0.1 for each 1% of clay and 2.0 for each 1% of organic matter. Such approximations can be made in any area where the types and contents of the dominant clays and the humus are known.

7.12
Exchangeable Cations in Field Soils

The specific exchangeable cations associated with soil colloids differ from one climatic region to another—Ca^{2+}, H^+, Al^{3+}, and complex aluminum hydroxy ions being most prominent in humid regions and Ca^{2+}, Mg^{2+}, and Na^+ dominating in low rainfall areas (Table 7.8). The cations that dominate the exchange complex have a marked influence on soil properties.

The proportion in any soil of the cation exchange capacity satisfied by a given cation is termed the *percentage saturation* for that cation. Thus, if 50% of the CEC were satisfied by Ca^{2+} ions, the exchange complex is said to have a *percentage calcium saturation* of 50.

This terminology is especially useful in identifying the relative proportions of sources of acidity and alkalinity in the soil solution. Thus, the percentage saturation with H^+ and Al^{3+} ions gives an indication of the acid conditions, while increases in the nonacid cation percentage (commonly referred to as the *percentage base saturation*[9]) indicate the tendency toward neutrality and alkalinity. The percentage base saturation is an important soil property, especially since it is inversely related to soil acidity. These relationships will be discussed further in Chapter 8.

The percentage cation saturation of essential elements such as calcium and potassium also greatly influences the uptake of these elements by growing plants. This now will receive attention along with other cation exchange–plant nutrition interactions.

[9] Technically speaking, nonacid cations such as Ca^{2+}, Mg^{2+}, K^+, Na^+, etc., are not bases. When adsorbed by soil colloids in the place of H^+ ions, however, they reduced acidity and increase the soil pH. For that reason, they are referred to as bases and the portion of the CEC that they satisfy is usually termed percentage base saturation.

7.13

Cation Exchange and Availability of Nutrients

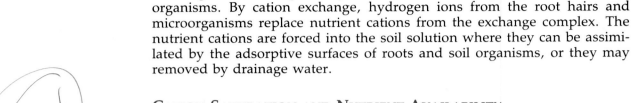

Exchangeable cations generally are available to both higher plants and microorganisms. By cation exchange, hydrogen ions from the root hairs and microorganisms replace nutrient cations from the exchange complex. The nutrient cations are forced into the soil solution where they can be assimilated by the adsorptive surfaces of roots and soil organisms, or they may removed by drainage water.

CATION SATURATION AND NUTRIENT AVAILABILITY

Several factors operate to expedite or retard the release of nutrients to plants. First, there is the percentage saturation of the exchange complex by the nutrient cation in question. For example, if the percentage calcium saturation of a soil is high, the displacement of this cation is comparatively easy and rapid. Thus, 6 cmol/kg of exchangeable calcium in a soil whose exchange capacity is 8 cmol/kg (75% calcium saturation) probably would mean ready availability, but 6 cmol/kg of exchangeable calcium when the total exchange capacity of a soil is 30 cmol/kg (20% calcium saturation) produces quite the opposite condition. This is one reason that, for a "calcium-loving" crop such as alfalfa, the calcium saturation of at least part of the soil should approach or even exceed 80%.

INFLUENCE OF COMPLEMENTARY ADSORBED CATIONS

A second factor influencing the plant uptake of a given cation is the complementary ions held on the colloids. As was discussed in Section 7.3, the strength of adsorption of different cations is in the following order.

$$Al^{3+} > Ca^{2+} > Mg^{2+} > K^+ = NH_4{}^+ > Na^+$$

Consequently, a nutrient cation such as K^+ is less tightly held by the colloids if the complementary ions are Al^{3+} and H^+ (acid soils) than if they are Mg^{2+} and Na^+ (neutral to alkaline soils). The loosely held K^+ ions are more readily available for absorption by plants or for leaching in acid soils.

There are also some nutrient "antagonisms," which in certain soils cause inhibition of uptake of some cations by plants. Thus, potassium uptake by plants is limited by high levels of calcium in some soils. Likewise, high potassium levels are known to limit the uptake of magnesium even when significant quantities of magnesium are present in the soil.

EFFECT OF COLLOID TYPE

Third, differences exist in the tenacity with which several types of colloidal micelles hold specific cations and in the ease with which they exchange cations. At a given percentage base (nonacid) saturation, smectites—which have a high charge density per unit of colloid surface—hold calcium much more strongly than does kaolinite (low charge density). As a result, smectite clays must have about 70% base saturation before calcium will exchange easily and rapidly enough to satisfy most plants. In contrast, a kaolinite clay exchanges calcium much more readily, serving as a satisfactory source of this constituent at a much lower percentage base saturation. Obviously, the need to add limestone to the two soils will be somewhat different, partly because of this factor.

7.14

Anion Exchange

Anion exchange was discussed briefly in Section 7.9. The positive charges associated with hydrous oxides of iron and aluminum, some 1:1-type clays, and amorphous materials, such as allophane, make possible the adsorption of anions (Figure 7.14). In turn, these anions are subject to replacement by other anions, just as cations replace each other. The *anion exchange* that takes place, although it may not approach cation exchange quantitatively, is important as a means of providing to higher plants readily available nutrient anions.

The basic principles of cation exchange apply as well to anion exchange, except that the charges on the colloids are positive and the exchange is among negatively charged anions. A simple example of an anion exchange reaction is

$$\boxed{\text{Micelle}}\,NO_3^- \ + \ Cl^- \ \rightleftharpoons \ \boxed{\text{Micelle}}\,Cl^- \ + \ NO_3$$

(soil solid)   (in soil solution)   (soil solid)   (in soil solution)

Just as in cation exchange, *equivalent* quantities of NO_3^- and Cl^- are exchanged, the reaction can be reversed, and plant nutrients can be released for plant adsorption.

Although simple reactions such as this are common, note that the adsorption and exchange of some anions including phosphates, molybdates, and sulfates are somewhat more complex. The complexity is owing to specific reactions between the anions and soil constituents. For example, the $H_2PO_4^-$ ion may react with the protonated hydroxy group rather than remain as an easily exchanged anion.

$$>\!Al\!-\!OH_2^+ \ + \ H_2PO_4^- \ \rightarrow \ >\!Al\!-\!H_2PO_4 \ + \ H_2O$$

(soil solid)   (in soil solution)   (soil solid)   (soil solution)

This reaction actually reduces the net positive charge on the soil colloid. Also, the H_2PO_4 is held very tightly by the soil solids and is not readily available for plant uptake.

Despite these complexities, anion exchange is an important mechanism for interactions in the soil and between the soil and plants. Together with cation exchange it largely determines the ability of soils to provide nutrients to plants promptly.

7.15

Physical Properties of Colloids

Soil colloids differ widely in their physical properties, including plasticity, cohesion, swelling, shrinkage, dispersion, and flocculation. These properties greatly influence the usefulness of soils for both agricultural and nonagricultural purposes.

PLASTICITY

Soils containing more than about 15% clay exhibit plasticity—that is, pliability and the capability of being molded. This property is probably due to the plate-like nature of the clay particles and the combined lubricating and

TABLE 7.9
Plastic and Liquid Limits of Several Soils and of Na- and Ca-Saturated Smectite

Clay soils with large amounts of smectites (Susquehanna and Bashaw) have high liquid limits as do Na-saturated clays.

Soil	Location	Plastic limit	Liquid limit
Davidson (Paleudult)	Georgia	19	27
Cecil (Hapludult)	Georgia	29	49
Putnam (Albaqualf)	Georgia	24	37
Susquehanna (Paleudalf)	Georgia	29	57
Sliprock (Haplumbrept)	Oregon	46	59
Jory (Haplohumult)	Oregon	30	45
Bashaw (Pelloxerert)	Oregon	18	71
Na-saturated smectite	—	—	950
Ca-saturated smectite	—	—	360
Na-saturated kaolinite	—	—	36

Data for Georgia soils from Hammel et al. (1983), Oregon soils from McNabb (1979), clays from Warkentin (1961).

binding influence of the adsorbed water. Thus, the particles of plastic soils easily slide over each other, much like panes of glass with films of water between them.

Plasticity is exhibited over a range moist-to-wet soil levels referred to as plasticity limits. At the lower of these levels termed *plastic limit* the soil begins to exhibit plasticity but molded pieces crumble easily when a little pressure is applied. The plastic limit is the lowest moisture content at which a soil cannot be deformed without cracking (Table 7.9). Soils should not be tilled at moisture contents above the plastic limit.

The upper plasticity limit or *liquid limit* is the moisture content at which soil ceases to be plastic, becomes semifluid (like softened butter), and tends to flow much like a liquid. While these limits have only modest use for agricultural purposes, they have special meaning in the classification of soils for engineering purposes, such as the bearing strength for a building or a highway bed. Obviously, a soil that tends to flow when wet does not make a suitable foundation for a building or for a highway.

The plastic and liquid limits of several soils and two clay samples are shown in Table 7.9. These data illustrate the levels to be expected in temperate soils. Soils with wide ranges between the plastic and liquid limits are difficult to handle in the field. The soils with the highest liquid limits (Susquehanna and Bashow) contain high levels of smectite clays. Smectites generally have high liquid limits, especially if saturated with sodium. Kaolinite clays, in contrast, have low liquid limit values.

Plasticity is of practical importance because of its influence on tillage operations. Thus, the cultivation of a fine-textured soil when it is too wet will result in a "puddled" condition detrimental to suitable aeration and drainage. With clayey soils, especially those of the smectite type, plasticity presents a significant problem. Stable granular structure is often difficult to establish and maintain in clayey soils high in smectites.

COHESION

A second characteristic, *cohesion*, indicates the tendency of clay particles to stick together. This tendency is due primarily to the attraction of the clay particles for the water molecules held between them. Hydrogen bonding between clay surfaces and water and also among water molecules is the

FIGURE 7.17
A field scene showing the cracks that result when a soil high in clay dries out. The type of clay in this case was probably a smectite. [Courtesy USDA Soil Conservation Service]

attractive force responsible for the cohesion. It accounts for the presence in some soils of hard clods that resist being broken down even with repeated tillage.

As one might expect, smectites and fine-grained micas exhibit a much more noticeable degree of cohesion than do kaolinite or hydrous oxides. Humus, by contrast, tends to reduce the attraction of individual clay particles for each other.

SWELLING AND SHRINKAGE

The third and fourth major characteristics of silicate clays are *swelling* and *shrinkage*. Some clays such as the smectites swell when wet and shrink when dry. After a prolonged dry spell, soils high in smectites (e.g., Vertisols) often are criss-crossed by wide, deep cracks that, at first, allow rain to penetrate rapidly (Figure 7.17). Later, because of swelling, such a soil is likely to close up and become much more impervious than one dominated by kaolinite, chlorite, or fine-grained micas. Vermiculite is intermediate in its swelling and shrinking characteristics.

Some swelling is owing to the penetration of water between crystal layers, resulting in intracrystal expansion. Most of the swelling, however, results from water attracted to the colloids and to ions adsorbed by them and from tiny air bubbles entrapped as the water moves into the extremely small pores of those soils (Figure 7.18).

Apparently swelling, shrinkage, cohesion, and plasticity are closely related. They depend not only on the clay mixture present in a soil and the dominant adsorbed cation but also on the nature and amount of humus that

209

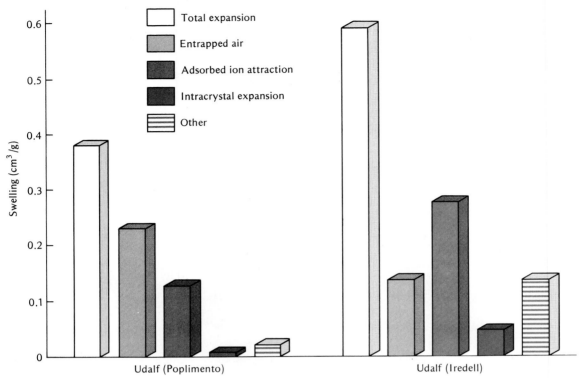

FIGURE 7.18

Total expansion (swelling) of two soils upon wetting and the approximate proportion of the swelling due to various factors. The Iredell soil with 63% smectite plus
vermiculite swells more than the Poplimento with only 15% of these two clays. Note
that the attraction of adsorbed ions for water and the air entrapped as the soils are
wetted are primarily responsible for the swelling. [Modified from Parker et al. (1982);
used with permission of Soil Science Society of America]

accompanies the inorganic colloids. These properties of soils are largely
responsible for the development and stability of soil structure, as has been
stressed previously (see Section 4.10). Swelling, shrinkage, cohesion, and
plasticity also account for the instability of highway beds and of shallow
building foundations as the soil gets wet and dries out.

DISPERSION AND FLOCCULATION

When a sample of clay soil is shaken up in a cylinder, the clay fraction will
tend to stay dispersed while the silt and sand fractions settle to the bottom.
The *dispersion* of clays is owing to the repulsion of negatively charged
particles for each other, each particle moves away from every other. Dispersion is encouraged by the large number of water molecules associated with
each micelle and with the adsorbed cations.

Highly hydrated monovalent cations, such as Na^+, that are not very
tightly held by the micelles enhance clay dispersion; more tightly held
cations, such as Ca^{2+} and Al^{3+}, inhibit it. Apparently, the loosely held Na^+
ions do not effectively reduce the electronegativity of the micelle; thus the
individual micelles continue to repel each other and stay in dispersion.

The electronegativity of a Na-saturated clay dispersion can be reduced
by replacing the Na^+ with other cations such as H^+, Ca^{2+}, Mg^{2+}, and Al^{3+}.
The addition of simple salts, which increases the concentration of cations

around the micelle, will also reduce the effective electronegative charge on the micelle (Section 7.9). Lowering the soil pH is another method of reducing electronegativity.

Each of these techniques tends to encourage the opposite of dispersion, *flocculation*. From the standpoint of agriculture, flocculation is generally beneficial because it is a first step in the formation of stable aggregates or granules. The ability of common cations to flocculate soil colloids is in the general order of $Al^{3+} > H^+ > Ca^{2+}\ Mg^{2+} > K^+ > Na^+$. This is fortunate, since the colloidal complexes of humid and subhumid region soils are dominated by aluminum, hydrogen, and calcium and those of semiarid regions are high in calcium ions. All of these cations enhance flocculation.

In areas of arid regions where sodium-containing water accumulates and is evaporated, Na^+ ions may become prominent on the exchange complex. The dominance of Na^+ ions results in a dispersed condition of the soil colloids, eliminates the macropores, and makes the soils largely impervious to water penetration. Most plants will not grow under these conditions. The Na^+ ions must be leached away and replaced, before normal growth will occur.

It is clear that the six colloidal properties we have discussed are of great importance in the practical management of arable soils. The field control of soil structure must take them into account. It would be worthwhile to review the discussion relating to structural management of cultivated lands (given in Section 4.11) with the colloidal viewpoint in mind.

6 colloidal properties

7.16

Conclusion

Colloidal materials control most of the chemical and physical properties of soils. The specific types of inorganic and organic colloids in soils vary greatly in their capacities to adsorb and exchange nutrients and in their ability to be physically manipulated for both agricultural and nonagricultural purposes. It is important that we learn something about the makeup of these colloidal materials so that we can better understand why they differ from each other and what steps must be taken to use them properly.

Study Questions

1. Describe the "soil colloidal complex," indicate its various components, and explain how it tends to serve as a "bank" for plant nutrients.
2. How do you account for the difference in surface area associated with a grain of kaolinite clay compared to that of montmorillonite, a smectite?
3. Contrast the difference in crystalline structure among kaolinite, smectites, fine-grained micas, vermiculites, and chlorites.
4. Assume that you discover that significant isomorphous substitution has taken place in both the trioctahedral and tetrahedral sheets of a theoretical 2:1-type mineral. In the trioctahedral sheet (dominated by magnesium), 0.2 equivalent mass of aluminum has substituted for a chemically equivalent quantity of magnesium. In the tetrahedral sheet, 0.2 equivalent

mass of aluminum has substituted for silicon. Would you expect the colloid to have an electronegative or electropositive charge? Would you expect to find such a mineral in soils? Give reasons for your answers.
5. Four colloids—humus, smectite, kaolinite, and Fe, Al oxides—are chemically characterized at two pH levels, 4.0 and 8.0. How would you expect them to compare at each pH level in terms of (a) extent of electronegativity, (b) extent of electropositivity, and (c) cation exchange capacity?
6. There are two basic processes by which silicate clays are formed by weathering of primary minerals. Which of these would likely be responsible for the formation of (a) fine-grained mica and (b) kaolinite from muscovite mica?

7. Which of the silicate clay minerals would most likely be found in (a) the tropics, (b) arid and semiarid temperate regions, and (c) humid temperate regions?

8. Which of the silicate clay minerals would be *most* and *least* desired if one were interested in (a) a good foundation for a building, (b) a high cation exchange capacity, (c) an adequate source of potassium, and (d) a soil on which hard clods form after plowing?

9. Which of the silicate clays would you most likely find in Vertisols? Explain.

10. Which of the following would you expect to be *most* and *least* sticky and plastic when wet: (a) a soil with significant sodium saturation in a semiarid area, (b) a soil high in exchangeable calcium in a subhumid temperate area, and (c) a well-weathered acid soil in the tropics. Explain your answer.

11. What is the cation exchange capacity of a soil with 40 cmol/kg of net negative charge? Explain.

12. Assume that 4 cmol/kg Ca^+ were replaced from the exchange complex by H^+. How much H^+, in cmol, would be adsorbed in the process? Explain.

13. A soil contains 4% humus, 10% montmorillonite, 10% vermiculite, and 10% Fe, Al oxides. What is its approximate cation exchange capacity?

References

BUSECK, P. R. 1983. "Electron microscopy of minerals," *Amer. Scientist*, **71**:175–85.

COLEMAN, N. T., and A. MEHLICH. 1957. "The chemistry of soil pH," *The Yearbook of Agriculture (Soil)* (Washington, DC: U.S. Department of Agriculture).

DIXON, J. B., and S. B. WEED. 1989. *Minerals in Soil Environments*, 2nd ed. (Madison, WI: Soil Sci. Soc. Amer.).

HAMMEL, J. E., M. E. SUMMER and J. BUREMA. 1983. "Atterberg limits as indices of external areas of soils," *Soil Sci. Soc. Amer. J.*, **47**:1054–56.

HELLING, C. S., et al. 1964. "Contribution of organic matter and clay to soil cation exchange capacity as affected by the pH of the saturated solution," *Soil Sci. Soc. Amer. Proc.*, **28**:517–20.

LYON, T. L., H. O. BUCKMAN, and N. C. BRADY. 1952. *The Nature and Properties of Soils*, 5th ed. (New York: Macmillan).

McNABB, D. H. 1979. "Correlation of soil plasticity with amorphous clay constituents," *Soil Sci. Soc. Amer. J.*, **43**:613–16.

PARKER, J. C., D. F. AMOS, and L. W. ZELANZNY. 1982. "Water adsorption and swelling of clay minerals in soil systems," *Soil Sci. Soc. Amer. J.*, **46**:450–56.

VAN RAIJ, B., and M. PEECH. 1972. "Electrochemical properties of some Oxisols and Alfisols of the tropics," *Soil Sci. Soc. Amer. Proc.*, **36**:587–93.

WARKENTIN, B. P. 1961. "Interpretation of the upper plastic limit of clays," *Nature*, **190**:287–88.

Soil Reaction: Acidity and Alkalinity

an acid soil has a pH of less than 7

Perhaps the most outstanding characteristic of the soil solution is its reaction—that is, whether it is acidic, alkaline, or neutral. Microorganisms and higher plants respond markedly to soil reaction because it tends to control so much of their chemical environment.

Soil acidity is common in all regions where precipitation is high enough to leach appreciable quantities of exchangeable base-forming cations (Ca^{2+}, Mg^{2+}, K^+, and Na^+) from the surface layers of soils. The condition is so widespread and its influence on plants is so pronounced that acidity has become one of the most discussed properties of soils.

Alkalinity occurs when there is a comparatively high degree of saturation with base-forming cations. The presence of calcium, magnesium, and sodium carbonates also can result in a preponderance of hydroxy ions over hydrogen ions in the soil solution.[1] Under such conditions, the soil is alkaline, sometimes very strongly so. If sodium carbonate is present, a pH of 9 or 10 may be reached in some soils. Alkaline soils are, of course, characteristic of arid and semiarid regions. The discussion of alkaline soils will follow that of acid soils.

8.1

Source of Hydrogen and Hydroxide Ions

Two adsorbed cations—hydrogen and aluminum—are largely responsible for soil acidity. The mechanisms by which these two cations exert their influence depends on the degree of soil acidity and on the source and nature of the soil colloids.

---

[1] When salts of strong bases and weak acids, such as Na_2CO_3, K_2CO_3, and $MgCO_3$, dissolve, they undergo hydrolysis and develop alkalinity. For Na_2CO_3, the reaction is

$$2\,Na^+ + CO_3^{2-} + HOH \rightleftharpoons 2\,Na^+ + OH^- + HCO_3^-$$

The OH^- ion greatly enhances alkalinity.

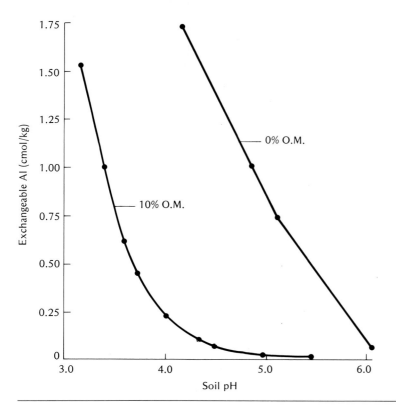

FIGURE 8.1
The effect of soil pH on exchangeable aluminum in a soil–sand mixture (0% O.M.) and a soil–peat mixture (10% O.M.). Apparently, the organic matter binds the aluminum in an unexchangeable form. This organic matter–aluminum interaction helps account for better plant growth at low pH values on soils high in organic matter. [Modified from Hargrove and Thomas (1981); used with permission of American Society of Agronomy]

STRONGLY ACID SOILS

Under very acid soil conditions (pH less than 5.0), much aluminum[2] becomes soluble (Figure 8.1) and is either tightly bound by organic matter or is present in the form of aluminum or aluminum hydroxy cations. These exchangeable ions are adsorbed in preference to other cations by the negative charges of soil colloids, which at low pH values are dominantly permanent charges associated with 2:1-type silicate clays.

The adsorbed aluminum is in equilibrium with aluminum ions in the soil solution, and aluminum ions contribute to soil acidity through their tendency to hydrolyze. Two simplified reactions illustrate how adsorbed aluminum can increase acidity in the soil solution.

$$\boxed{\text{Micelle}} \, Al^{3+} \quad \rightleftharpoons \quad Al^{3+}$$

Adsorbed	Aluminum ion
aluminum	(in soil solution)

The aluminum[3] ions in the soil solution are then hydrolyzed.

$$Al^{3+} + H_2O \rightarrow Al(OH)^{2+} + H^+$$

The H^+ ions thus released lower the pH of the soil solution and are the major source of hydrogen in most very acid soils.

Adsorbed hydrogen ions are a second but much more limited source of

---

[2] Iron (Fe^{3+}) also is solubilized under acid conditions and forms hydroxy cations just as does aluminum. However, since the acidity generated by iron is much less than that generated by aluminum, only the aluminum involvement will be considered.

[3] The Al^{3+} ion is actually highly hydrated, being present in forms such as $Al(H_2O)_6^{3+}$. For simplicity, however, it will be shown as the simple Al^{3+} ion.

H^+ ions in very acid soils. Much of the hydrogen in very acid soils, along with some iron and aluminum, is bound so tightly by covalent bonds in the organic matter and on clay crystal edges that it contributes only modestly to the soil solution acidity (see Section 7.9). Only on the strong acid groups of humus and some of the permanent charge exchange sites of the clays are H^+ ions held in an exchangeable form. These H^+ ions are in equilibrium with the soil solution. A simple equation to show the release of adsorbed hydrogen to the soil solution is

$$\boxed{\text{Micelle}}\, H^+ \quad \rightleftharpoons \quad H^+$$

Adsorbed   Hydrogen ion
hydrogen   (in soil solution)

Thus, it can be seen that the effect of both adsorbed hydrogen and aluminum ions is to increase the H^+ ion concentration in the soil solution.

MODERATELY ACID SOILS

Aluminum and hydrogen compounds also account for soil solution H^+ ions in moderately acid soils, with pH values between 5.0 and 6.5, but by different mechanisms. Moderately acid soils have somewhat higher percentage base saturations than the strongly acid soils. The aluminum can no longer exist as ions but is converted to aluminum hydroxy ions[4] by reactions such as

$$Al^{3+} + OH^- \rightarrow \quad Al(OH)^{2+}$$
$$Al(OH)^{2+} + OH^- \rightarrow \quad Al(OH)_2{}^+$$

Aluminum hydroxy
ions

Some of the aluminum hydroxy ions are adsorbed and act as exchangeable cations. As such, they may be equilibrium with the soil solution just as is the Al^{3+} ion is in very acid soils. In the soil solution they produce hydrogen ions by the following hydrolysis reactions, using again as examples the simplest formulas for aluminum hydroxy ions.

$$Al(OH)^{2+} + H_2O \rightarrow Al(OH)_2{}^+ + H^+$$
$$Al(OH)_2{}^+ + H_2O \rightarrow Al(OH)_3 + H^+$$

In some 2:1-type clays, particularly vermiculite, the aluminum hydroxy ions (as well as iron hydroxy ions) play another role. They move into the interlayer space of the crystal units and become very tightly adsorbed, preventing intracrystal expansion and blocking some of the exchange sites. Raising the soil pH results in the removal of these ions and the release of the exchange sites. Thus aluminum (and iron) hydroxy ions are partly responsible for the "pH-dependent" charge of soil colloids.

In moderately acid soils, the small amount of readily exchangeable hydrogen contributes to soil acidity in the same manner as shown for very acid soils. In addition, with the rise in pH, some hydrogen atoms that at low pH were bound tenaciously through covalent bonding by the organic matter, Fe Al oxides, and 1:1-type clays are now subject to release as H^+ ions. These hydrogen ions are associated with the pH-dependent sites previously men-

---

[4] The actual aluminum hydroxy ions are much more complex than those shown. Formulas such as $[Al_6(OH)_{12}]^{6+}$ and $[Al_{10}(OH)_{22}]^{8+}$ are examples of the more complex ions.

tioned (see Section 7.9). Their contribution to the soil solution might be illustrated as follows.

$$\boxed{\text{Micelle}\ {}^{H}_{H}}\quad +\quad Ca^{2+}\quad \rightarrow\quad \boxed{\text{Micelle}\ }Ca^{2+}\quad +\quad 2\ H^{+}$$

Bound hydrogen Calcium ion   Calcium ion   Hydrogen ion
(not dissociated) (in soil solution)   (exchangeable) (in soil solution)

Again, the colloidal control of soil solution pH has been demonstrated, as has the dominant role of the aluminum and base-forming cations.

NEUTRAL TO ALKALINE SOILS (pH 7 AND ABOVE)

Soils that are neutral to alkaline are not dominated by either hydrogen or aluminum ions. The permanent charge exchange sites are now occupied mostly by exchangeable Ca^{2+}, Mg^{2+} and other base-forming cations. Both the hydrogen and aluminum hydroxy ions have been largely replaced. Most of the aluminum hydroxy ions have been converted to gibbsite by reactions such as

$$Al(OH)_2{}^+ + OH^- \rightarrow Al(OH)_3$$
Gibbsite
(insoluble)

More of the pH-dependent charges have become available for cation exchange, and the H^+ ions released therefrom move into the soil solution and react with OH^- ions to form water. The place of hydrogen on the exchange complex is taken by Ca^{2+}, Mg^{2+} and other base-forming cations.

SOIL pH AND CATION ASSOCIATIONS

Figure 8.2 summarizes the distribution of ions in a hypothetical soil as affected by pH. It illustrates the change in cation dominance around soil

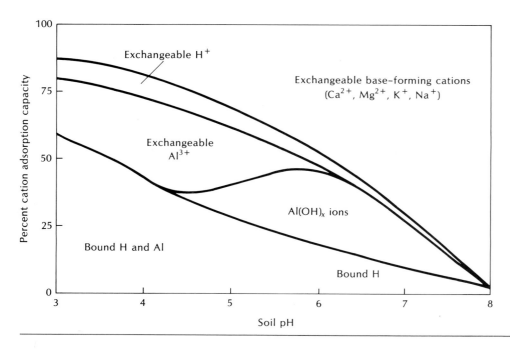

FIGURE 8.2
General relationship between soil pH and the cations held by soil colloids. Under very acid conditions, exchangeable aluminum and hydrogen ions and bound hydrogen and aluminum dominate. At higher pH values, the exhangeable bases predominate, while at intermediate values, aluminum hydroxy ions such as $Al(OH)^{2+}$ and $Al(OH)_2{}^+$ are prominent. This diagram is for average conditions; any particular soil would likely give a modified distribution.

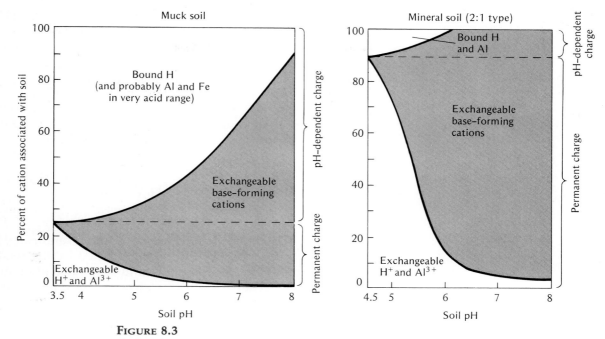

FIGURE 8.3

Relationship between pH and the association of hydrogen, aluminum, and base-forming cations with an organic (muck) soil and a mineral soil dominated by 2:1-type silicate clays. Note the dominance of the permanent charge in the mineral soil and of the pH-dependent charge in the organic soil. The CEC of the muck soil (the sum of all exchangeable ions) declined rapidly as the pH was reduced. [Redrawn from Mehlich (1964)]

colloids as the pH is changed. Study it carefully, keeping in mind that for any particular soil the distribution of ions might be quite different.

The effect of pH on the distribution of Ca^{2+}, Mg^{2+} and other base-forming cations and of H^+ and Al^{3+} ions in a muck and in a soil dominated by 2:1 clays is shown in Figure 8.3. Note that permanent charges dominate the exchange complex of the mineral soil, whereas pH-dependent charges account for most of the adsorption in the muck soil. Consequently, the exchange capacity of the muck declines rapidly as the pH is lowered, whereas there is little such decline in the case of the 2:1-type clay. The effect of pH on the cation exchange capacity of kaolinite and related clays is similar to that shown by the organic soil.

Note that in Figures 8.2 and 8.3 two forms of hydrogen and aluminum are shown. That tightly held by the pH-dependent sites (covalent bonding) is termed *bound* hydrogen and aluminum. In contrast, the hydrogen and aluminum ions associated with permanent negative charges on the colloids are *exchangeable*. Although only the exchangeable ions have an immediate effect on soil pH, both forms are very much involved in determining how much lime is needed to change soil pH (see Section 8.2).

While the factors responsible for soil acidity are far from simple, two dominant groups of elements are in control. The H^+ ions and different aluminum ions (Al^{3+}, $Al(H_2O)^{3+}$, and so on) generate acidity, and most of the other cations combat it. This simple statement is worth remembering.

SOURCE OF HYDROXIDE IONS

If a liming material such as $Ca(OH)_2$ is added to an acid soil, hydrogen and aluminum ions are replaced on the exchange complex by Ca^{2+} ions and the

H^+ and Al^{3+} ion concentrations in the soil solution will decrease. The concentration of OH^- ions will simultaneously increase, since there is an inverse relationship between the H^+ and OH^- ions in water-based solutions. Thus, the base-forming cations can become indirect sources of OH^- ions as they are adsorbed on soil colloids.

Cations such as Ca^{2+}, Mg^{2+} and Na^+, which are dominant in soils of low rainfall regions, also have a more direct effect on the OH^- ion concentration of the soil solution. A definite alkaline reaction results from the hydrolysis of colloids saturated with these cations. For example,

$$Ca^{2+}\ \boxed{Micelle}\ +\ 2\,H_2O \rightleftharpoons\ {}^{H^+}_{H^+}\boxed{Micelle}\ +\ Ca^{2+}\ +\ 2\,OH^-$$

(soil solid)   (in soil solution)   (soil solid)   (in soil solution)

In a soil that is highly saturated with base-forming cations, OH^- ion levels are high. The ultimate soil pH will be determined by the relative proportions of base-forming cations and of H^+ and associated aluminum ions. As will be discussed in Section 8.3, these relative proportions are best expressed as the percentage base saturation.

8.2

Classification of Soil Acidity

Research suggests three kinds of acidity: (a) *active acidity* due to the H^+ ion in the soil solution; (b) *salt-replaceable acidity*, represented by the hydrogen and aluminum that are *easily exchangeable* by other cations in a simple unbuffered salt solution such as KCl; and (c) *residual acidity*, which can be neutralized by limestone or other alkaline materials but cannot be detected by the salt-replaceable technique. Obviously, these types of acidity all add up to the *total acidity* of a soil.

ACTIVE ACIDITY

The active acidity is a measure of the H^+ ion activity in the soil solution at any given time. However, the quantity of H^+ ions owing to active acidity is very small compared to the quantity in the exchange and residual acidity forms. For example, only about 2 kg of calcium carbonate would be required to neutralize the active acidity in a hectare–furrow slice of an average mineral soil at pH 4 and 20% moisture. Even though the concentration of hydrogen ions owing to active acidity is extremely small, it is important because this is the environment to which plants and microbes are exposed.

SALT-REPLACEABLE (EXCHANGEABLE) ACIDITY

This type of acidity is primarily associated with the exchangeable aluminum and hydrogen ions that are present in largest quantities in very acid soils. (Figure 8.2). These ions can be released into the soil solution by an unbuffered salt such as KCl.

$$\boxed{Micelle}{}^{Al^{3+}}_{H^+}\ +\ 4\,KCl \rightleftharpoons\ \boxed{Micelle}{}^{K^+\,K^+}_{K^+\,K^+}\ +\ AlCl_3\ +\ HCl$$

(soil solid)   (soil solution)   (soil solid)   (soil solution)

In moderately acid soils, the quantity of easily exchangeable aluminum and hydrogen is quite limited (Figure 8.2). Even in these soils, however, the limestone needed to neutralize this type of acidity is commonly more than 100 times that needed for the soil solution (active acidity).

At a given pH value, exchangeable acidity is generally highest for smectites, intermediate for vermiculites, and lowest for kaolinite. In any case, however, it accounts for only a small portion of the total soil acidity as the next section will verify.

RESIDUAL ACIDITY

Residual acidity is that which remains in the soil after active and exchange acidity has been neutralized. Residual acidity is generally associated with aluminum hydroxy ions and with hydrogen and aluminum atoms that are bound in nonexchangeable forms by organic matter and silicate clays (see Figure 8.2). If lime is added to a soil, the pH increases and the aluminum hydroxy ions are changed to uncharged gibbsite as follows.

$$Al(OH)^{2+} \xrightarrow{\ OH^-\ } Al(OH)_2{}^+ \xrightarrow{\ OH^-\ } Al(OH)_3$$

In addition, as the pH increases bound hydrogen and aluminum can be released by calcium and magnesium in the lime materials [$Ca(OH)_2$ is used as an example of the reactive calcium liming material].

$$\boxed{\text{Micelle}\genfrac{}{}{0pt}{}{Al}{H}} \; + \; 2\,Ca(OH)_2 \;\rightarrow\; \boxed{\text{Micelle}\genfrac{}{}{0pt}{}{Ca^{2+}}{Ca^{2+}}} \; + \; Al(OH)_3 + H_2O$$

Bound H and Al (not exchangeable)   Exchangeable Ca^{2+}

The residual acidity is commonly far greater than either the active or salt-replaceable acidity. Conservative estimates suggest the residual acidity may be 1000 times greater than the soil solution or active acidity in a sandy soil and 50,000 or even 100,000 times greater in a clayey soil high in organic matter. The amount of ground limestone recommended to at least partly neutralize residual acidity is commonly 4–8 metric tons (Mg) per hectare–furrow slice (1.8–3.6 tons/AFS). It is obvious that the pH of the soil solution is only "the tip of the iceberg" in determining how much lime is needed.

8.3

Colloidal Control of Soil Reaction

The previous sections clearly illustrate the colloidal control of soil pH. This control is exerted through (a) the proportion of the cation exchange capacity (CEC) that is satisfied by base-forming cations, (b) the nature of the micelles, (c) the kind of adsorbed bases, and (d) the level of soluble salts in the soil solution.

PERCENTAGE BASE SATURATION

The percentage of the CEC that is satisfied by the base-forming cations is termed *percentage base saturation*.

$$\% \text{ base saturation} = \frac{\text{exchangeable base-forming cations (cmol/kg)}}{\text{CEC (cmol/kg)}}$$

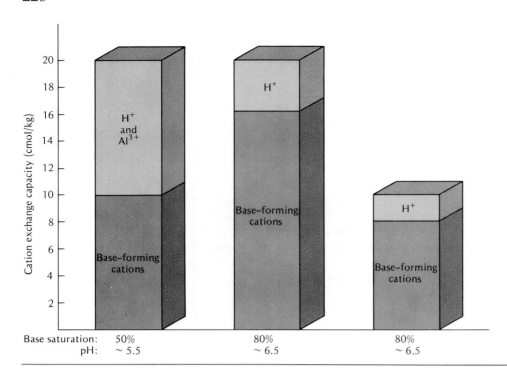

FIGURE 8.4
Three soils with percentage base saturations of 50, 80, and 80, respectively. The first is a clay loam; the second, the same soil satisfactorily limed; and the third, a sandy loam with a CEC of only 10 cmol/kg. Note especially that soil pH is correlated more or less closely with percentage base saturation. Also note that the sandy loam (right) has a higher pH than the acid clay loam (left), even though clay contains more exchangeable bases.

Figure 8.4 should be helpful in illustrating this relationship. A low percentage base saturation means acidity, whereas a percentage base saturation of 50–80 will result in neutrality or alkalinity. In general, humid region soils dominated by *silicate clays* and *humus* are acid if their percentage base saturation is much below 80. When such soils have a percentage base saturation of 80 or above, they usually are neutral or alkaline.

NATURE OF THE MICELLE

There is considerable variation in the pH of different types of colloids at the same percentage base saturation. This is due to differences in the ability of the colloids to furnish hydrogen ions to the soil solution. For example, the dissociation of adsorbed H^+ ions from the smectites is much higher than that from the hydrous Fe, Al oxides clays. Consequently, the pH of soils dominated by smectites is appreciably lower than that of the oxides at the same percentage base saturation. The dissociation of adsorbed hydrogen from 1:1-type silicate clays and organic matter is intermediate between that from smectites and from the hydrous oxides.

In acid soils the interaction among the soil colloids, especially those involving iron and aluminum compounds, markedly affects the pH–base saturation relations. This relationship will receive attention when buffering is considered (see Section 8.4).

KIND OF ADSORBED BASE-FORMING CATIONS

The comparative quantities of each of the base-forming cations present in the colloidal complex is another factor influencing soil pH. Soils with high sodium saturation have much higher pH values than those dominated by calcium and magnesium. Thus, at a percentage base saturation of 90, the presence of calcium, magnesium, potassium, and sodium ions in the ratio 4:1:1:9 would certainly result in a higher pH than if the ratio were 10:3:1:1. In the first instance, a sodium-calcium complex is dominated by the sodium; in the second, calcium is dominant.

NEUTRAL SALTS IN SOLUTION

The presence in the soil solution of neutral salts, such as the sulfates and chlorides of sodium, potassium, calcium, and magnesium, has a tendency to increase the activity of the H^+ ion in solution and, consequently, decrease the soil pH. For example, if calcium chloride ($CaCl_2$) were added to a slightly acid soil, the following reaction would occur.

$$\begin{array}{c} H^+ \\ H^+ \end{array} \boxed{\text{Micelle}} + Ca^{2+} + 2\,Cl^- \;\rightleftharpoons\; Ca^{2+} \boxed{\text{Micelle}} + 2\,H^+ + 2\,Cl^-$$

(soil solid) (soil solution) (soil solid) (soil solution)

Obviously, the quantity of H^+ ion in the soil solution has increased.

The addition of neutral salts to an alkaline soil also will reduce pH, but by a different mechanism. In alkaline soil the salts depress the hydrolysis of colloids saturated with base-forming cations. Thus, the presence of sodium chloride (NaCl) in a soil with high sodium saturation will tend to reverse the reaction shown.

$$Na^+ \boxed{\text{Micelle}} + H_2O \;\rightleftharpoons\; H^+ \boxed{\text{Micelle}} + Na^+ + OH^-$$

(soil solid) (soil solution) (soil solid) (soil solution)

The Na^+ ion in the added salt will drive the reaction to the left, thereby reducing the concentration of OH− ion in the soil solution and increasing that of the H^+ ion. This reaction is of considerable practical importance in certain alkaline soils of arid regions because it keeps the pH from rising to toxic levels. At the same time osmotic potential of the soil solution is decreased by the addition of salt, and water may become less available to the plant.

Since the reaction of the soil solution is influenced by the four distinct and uncoordinated factors just discussed, a close correlation would not always be expected between percentage base saturation and pH when comparing soils at random. Yet with soils of similar origin, texture, and organic content, a rough correlation does exist.

8.4

Buffering of Soils

As previously indicated, there is a distinct resistance to a change in pH of the soil solution. This resistance is called *buffering*. It can be explained very simply in terms of the equilibrium that exists among the active, salt-replaceable, and residual acidities (see Figure 8.5). This general relationship can be illustrated as follows.

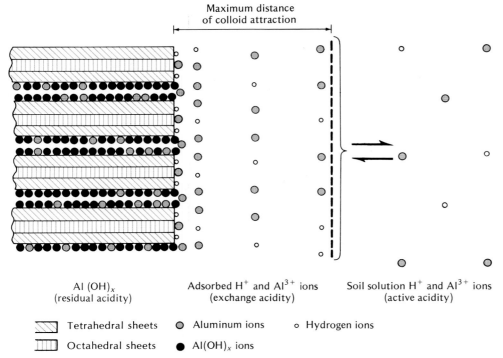

Maximum distance
of colloid attraction

Al (OH)$_x$
(residual acidity)

Adsorbed H$^+$ and Al^{3+} ions
(exchange acidity)

Soil solution H$^+$ and Al^{3+} ions
(active acidity)

▨	Tetrahedral sheets	◉ Aluminum ions	○ Hydrogen ions
▥	Octahedral sheets	● Al(OH)$_x$ ions	

FIGURE 8.5

Equilibrium relationship among residual salt-replaceable (exchangeable) and soil solution (active) acidity on a 2:1 colloid. Note that the adsorbed and residual ions are much more numerous than those in the soil solution even when only a small portion of the clay crystal is shown. The Al(OH)$_x$ ions are held tightly in internal spaces and are not exchangeable. Remember that the aluminum ions, by hydrolysis, also supply hydrogen ions to the soil solution. It is obvious that neutralizing only the hydrogen and aluminum ions in the soil solution will be of little consequence. They will be quickly replaced by ions associated with the colloid. This means high buffering capacity.

If just enough lime (a base) is applied to neutralize the hydrogen ions in the soil solution, they are replenished as the reactions move to the right. The result is a resistance to change in the soil solution pH. In other words, the buffering effect is being exercised. The rise in soil pH with the addition of base is negligibly small until enough lime is added to deplete appreciably the hydrogen and aluminum of the exchangeable and residual acidities (see Figure 8.6).

Buffering is equally important in preventing a rapid lowering of the pH of soils. Hydrogen ions are generated as organic acids are formed during organic decay, and a temporary increase in the hydrogen ion concentration in the soil solution occurs. In this case, the equilibrium reaction shown would immediately shift to the left so that more hydrogen ions would become adsorbed or bound on the micelle. Again, the resultant pH change, this time a lowering, in the soil solution would be very small because of buffering.

These two examples show clearly the principles involved in buffering and show that the basis of buffer capacity lies in the adsorbed and bound cations of the complex. Hydrogen and aluminum ions, together with the adsorbed base-forming cations, largely control buffering just as they indirectly control the pH of the soil solution.

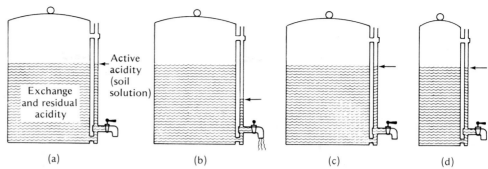

FIGURE 8.6

The buffering action of a soil can be likened to that of a coffee dispenser. (a) The active acidity, which is represented by the coffee in the indicator tube on the outside of the urn, is small in quantity. (b) When hydrogen ions are removed, this active acidity falls rapidly. (c) The active acidity is quickly restored to near the original level by movement from the exchange and residual acidity. By this process, there is considerable resistance to the change of active acidity. A second soil with the same active acidity (pH) level but a much smaller exchange and residual acidity (d) would have a lower buffering capacity.

8.5

Buffer Capacity of Soils

The higher the exchange capacity of a soil, the greater its buffer capacity, other factors being equal. This relationship exists because more reserve acidity must be neutralized to effect a given increase in the percentage base saturation. Thus, the higher the clay and organic matter contents, the more lime is required for a given change in soil pH.

BUFFER CURVES

Buffer capacity of soils is normally not the same throughout the percentage base saturation range. Titration curves such as the one shown in Figure 8.7, which is based on a large number of Florida soils, illustrate this point. Note that for these soils the degree of buffering is highest between pH 4.5 and 6.0 and is relatively uniform in this range. (The buffer capacity drops off below pH 4.5 and above pH 6.0.) This means that the same amount of lime would be required to raise the pH of these soils from 5.0 to 5.5 as from 5.5 to 6.0.

The curve in Figure 8.7 is a composite for many soils and does not represent the buffer capacity for any particular soil. However, it does illustrate the practical importance of the buffering curves in estimating the lime needed to bring about change in pH in a soil.

VARIATION IN TITRATION CURVES

The relationship of pH to percent base saturation varies from one colloid to another (see Figure 8.8). At a given percent base saturation, the pH is normally highest in hydrous oxide clay materials, intermediate in kaolinite and humus, and lowest in 2:1-type minerals. Obviously, the type of clay definitely affects the pH–base saturation percentage relationship.

Aluminum and iron compounds have an effect on the buffering of soils. For example, at low pH values, Al^{3+} and hydroxy aluminum ions react with

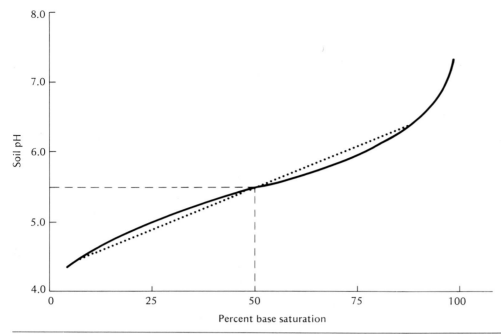

FIGURE 8.7
Theoretical titration curve based on a large number of Florida soils. The dotted line indicates the zone of greatest buffering. The maximum buffering should occur at approximately 50% base saturation. [From Peech (1941)]

and bind or block exchange sites in silicate clays and in humus, thereby reducing the CEC of the colloids. If the soil is then limed, these ions are removed, the CEC increases, and more calcium, magnesium, and other base-forming cations can be adsorbed. The net result is an increase in the quantity of lime needed to increase the soil pH. Thus, the presence of aluminum compounds tends to enhance the buffer capacity of a soil.

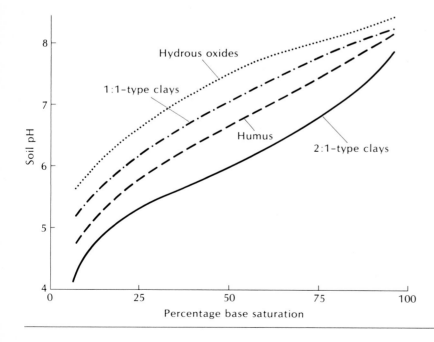

FIGURE 8.8
Theoretical titration curves of representative soil colloids. The hydrous oxides exhibit the highest pH values at a given percent base saturation; the 2:1-type clays, the lowest; and the 1:1 type and organic matter in minerals soils, in between. The reaction of Al^{3+} and Fe^{3+} ions and of Fe, Al oxides with silicate clays and organic matter markedly influence the titration curves of these colloids.

IMPORTANCE OF SOIL BUFFERING

Soil buffering is important for two primary reasons. First, it tends to assure reasonable stability in the soil pH, preventing drastic fluctuations that might be detrimental to higher plants and to soil microorganisms. Second, it influences the amount of chemicals, such as lime or sulfur, that are commonly added to change the soil pH. Buffering is indeed a significant soil property.

8.6

Changes in Soil pH

ACID-FORMING FACTORS

As organic matter decomposes, both organic and inorganic acids are formed. The simplest and perhaps the most widely found is carbonic acid (H_2CO_3), a product of the reaction of carbon dioxide and water. The slow but persistent solvent action of carbonic acid (H_2CO_3) on the mineral constituents of the soil is responsible for the removal of large quantities of base-forming cations (e.g., Ca^{2+} and Mg^{2+}) by dissolution and leaching. Much stronger organic acids, from the very simple to more complex, are products of microbial decay and are exudates from plant roots.

Inorganic acids such as sulfuric acid (H_2SO_4) and nitric acid (HNO_3) are potent suppliers of hydrogen ions in the soil. In fact, these acids, along with strong organic acids, encourage the development of moderately and strongly acidic conditions (Figure 8.9). Sulfuric and nitric acids are formed not only by the organic decay processes, but also from the microbial action on certain inorganic sulfur- and nitrogen-containing materials such as elemental sulfur, ammonium nitrate, and ammonium sulfate.

Large quantities of sulfuric and nitric acids are formed in the atmosphere from oxides of nitrogen and sulfur emitted from the combustion of coal, gasoline, and other fossil fuels at sites near large cities or around large industrial complexes. The precipitation of these substances is called "acid rain" since it has a pH value of 4.0–4.5 and, in extreme cases, may be as low as 2.0 (see Section 18.11). This contrasts with "normal" rainfall, with a pH

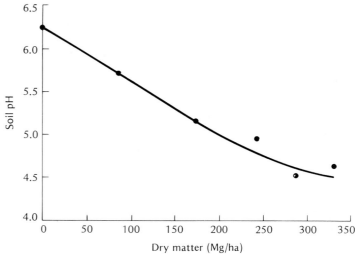

FIGURE 8.9
Effect of large applications of sewage sludge over a period of 6 years on the pH of a Paleudult (Orangeburg fine sandy loam). The reduction was due to the organic and inorganic acids, such as HNO_3, formed during decomposition and oxidation of the organic matter. [Data from Robertson et al. (1982)]

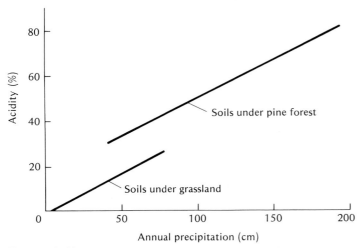

FIGURE 8.10

Effect of annual precipitation on the percent acidity of untilled California soils under grassland and pine forests. Note that the degree of acidity goes up as the precipitation increases. Also note that the forest produced a higher degree of soil acidity than did the grassland. [From Jenny et al. (1968)]

commonly 5.0–5.6 at equilibrium with atmospheric carbon dioxide. Although the hydrogen ions added to the soil in acid rain are not sufficient to bring about significant pH changes at once, over a long period of time their addition may have a significant acidifying effect, especially for soils with low buffering capacity.

Leaching also encourages acidity because it removes the base-forming cations that might compete with hydrogen and aluminum on the exchange complex. The effect of leaching on the acidity of soils developed under grassland and forests is shown in Figure 8.10.

BASE-FORMING FACTORS

Any process that will encourage high levels of the exchangeable base-forming cations, such as Ca^{2+}, Mg^{2+}, K^+, and Na^+, will contribute toward a reduction in acidity and an increase in alkalinity. Weathering processes release these exchangeable cations from minerals and make them available for adsorption. The addition of limestone to acid soils augments nature's supply of base-forming cations. Irrigation waters also contain salts of these cations, which when adsorbed by soil colloids, can increase soil alkalinity, sometimes excessively when Na^+ low is dominant.

Conditions that permit the exchangeable base-forming cations to remain in the soil will encourage high pH values. For example, the recycling of these cations by deep-rooted plants brings them to the surface, and soil organisms incorporate them into the topsoil. This recycling accounts for the relatively high pH of soils of the semiarid and arid regions, and prevents some soils in the humid tropics from becoming excessively acid.

In arid regions, leaching waters do not remove most of the base-forming cations as they are weathered from soil minerals. Consequently, the percentage base saturation of these soils remains high. In general, this situation is favorable for crop production. When the pH is too high, however, deficiencies of iron, manganese, and other micronutrients occur. Likewise when sodium is the dominant cation, plant growth is negatively affected, not only because of high pH but because the soil particles are dispersed and few micropores are present (see Section 8.18).

MINOR FLUCTUATIONS IN SOIL pH

There are several causes of minor fluctuations in soil pH. For instance, the movement of salts into and out of soil zones as soil moisture moves up and down the profile will influence pH. Similarly, the pH of mineral soils declines during the crop-growing seasons as a result of the acids produced by microorganisms and by the roots of higher plants. When soil temperatures are low, an increase in pH often is noted because biotic activities during these times are considerably slower.

HYDROGEN ION VARIABILITY

There is considerable variation in the pH of the soil solution in different parts of the soil. Differences in pH are noted from one site in the soil to one only a few centimeters away. Such variation may result from local microbila action due to the uneven distribution of organic residues in the soil.

The variability of the soil solution is important in many respects. For example, it affords microorganisms and plant roots a great variety of solution environments. Organisms unfavorably influenced by a given hydrogen ion concentration may find, at an infinitesimally short distance away, a different environment that is more satisfactory. The variety of environments may account in part for the great diversity in microbial species present in normal soils.

8.7

Soil Reaction—Correlations

Soil pH significantly influences other soil chemical properties as well as biological organisms. Figure 8.11 shows one example: the effect of soil acidity (and aluminum toxicity) on the growth of soybean roots in a mineral soil. Only a few other examples will be considered here to illustrate this point further.

CHEMICAL PROPERTIES

The soil pH significantly affects the availability of most of the chemical elements of importance to plants and microbes. The general relationship between pH and the levels of base-forming cations (Ca^{2+}, Mg^{2+}, K^+, and Na^+) as well as Al^{3+} ion has already been discussed. Likewise, the tendency for toxic levels of elements such as iron, manganese, and aluminum to be established at low pH (Figure 8.12)—and deficiencies of iron and manganese at high pH—have been mentioned. The availabilities of nitrogen, sulfur, and molybdenum are somewhat restricted at low pH values, whereas that of phosphorus is best at intermediate pH levels. More specific information on availabilities of nutrients will be given in Chapters 12 and 13.

The tolerance of different soil microbes to soil acidity and alkalinity will be covered in Chapter 9. Suffice it to say that most general purpose bacteria and actinomycetes function best at intermediate and high pH values. Fungi seem to be particularly versatile, flourishing satisfactorily over a wide pH range. In most soils, therefore, the activities of fungi tend to predominate in acid soils, whereas at intermediate and higher pH they meet stiff competition from actionmycetes and bacteria.

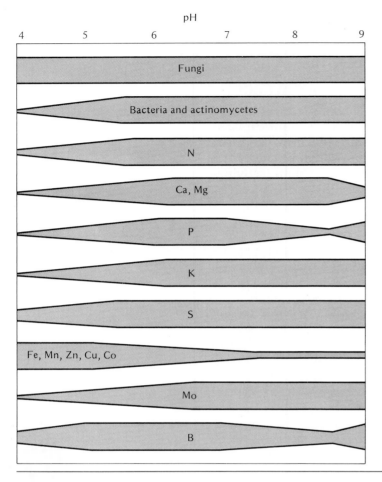

FIGURE 8.11

Relationships existing in mineral soils between pH on the one hand and the activity of microorganisms and the availability of plant nutrients on the other. The wide portions of the bands indicate the zones of greatest microbial activity and the most ready availability of nutrients. When the correlations as a whole are taken into consideration, a pH range of about 6–7 seems to promote best the availability of plant nutrients. In short, if soil pH is suitably adjusted for phosphorus, the other plant nutrients, if present in adequate amounts, will be satisfactorily available in most cases.

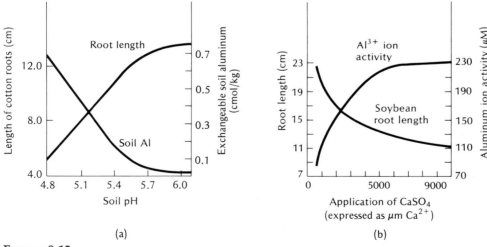

FIGURE 8.12

High concentrations of exchangeable or soil solution aluminum in acid soils are toxic to plant roots, a toxicity that can be reduced somewhat by adding $CaSO_4$. (a) Cotton root length is increased as soil pH is increased and exchangeable aluminum is decreased. (b) Adding $CaSO_4$ to a nutrient solution (pH = 2) reduces the Al^{3+} activity and permits an increase in soybean root length. [(a) From Adams and Lund (1966); (b) drawn from data in Nobel et al. (1988)]

All in all, a soil in the intermediate pH range presents the most satisfactory biological environment. Nutrient conditions are favorable without being extreme, and phosphorus availability is at a maximum.

HIGHER PLANTS AND pH

Plants vary considerably in their tolerance of acid and/or alkaline conditions (see Figure 8.13). For example, legume crops such as alfalfa and sweet clover grow best in near neutral or alkaline soils, and most humid region mineral soils must be limed to grow these crops satisfactorily.

Rhododendrons and azaleas are at the other end of the scale. They apparently require a considerable amount of iron, which is abundantly

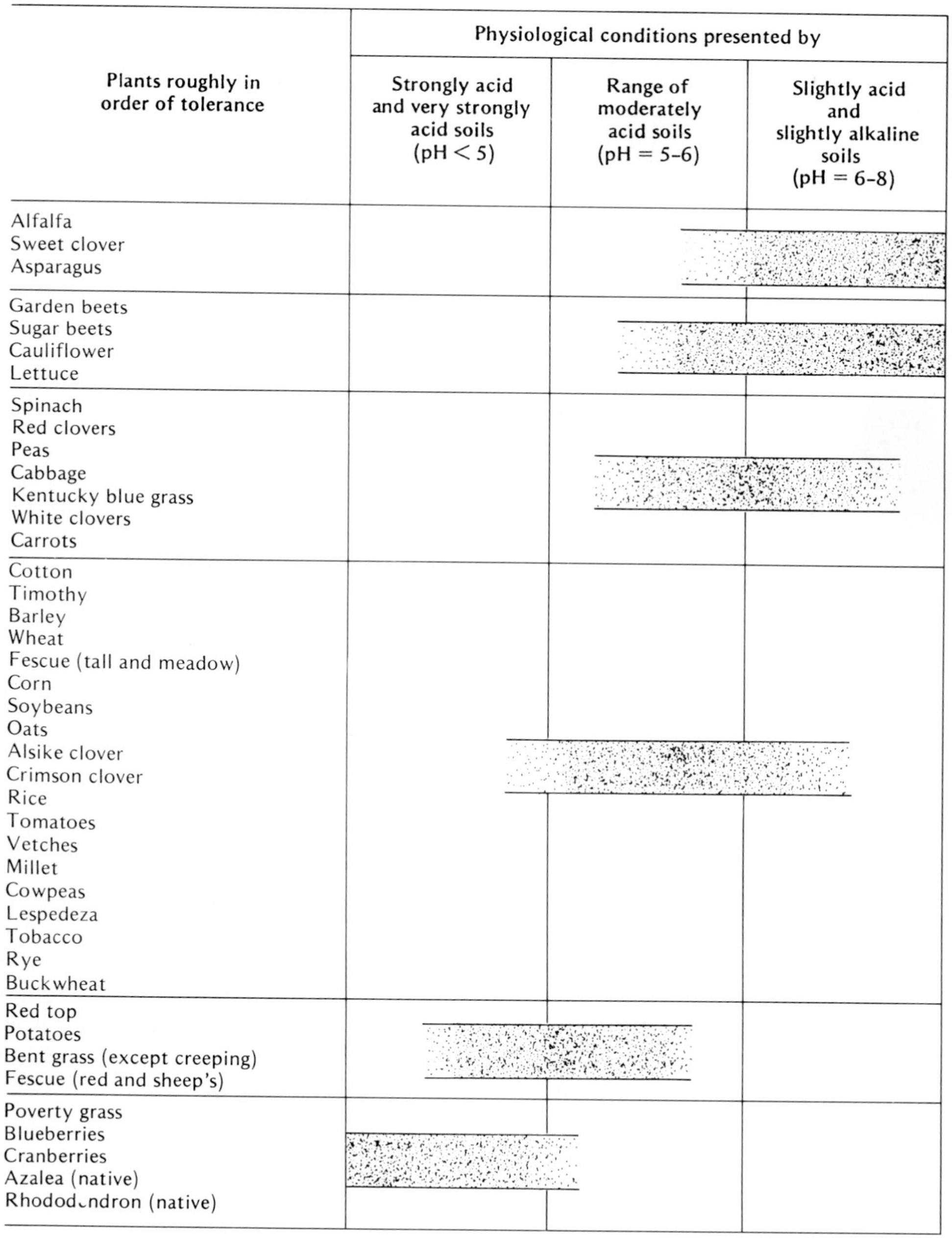

FIGURE 8.13

Relation of higher plants to the physiological conditions presented by mineral soils of different reactions. Note that the correlations are very broad and are based on pH ranges. The fertility level will have much to do with the actual relationship in any specific case. Such a chart is of great value in deciding whether or not to apply chemicals such as lime or sulfur to change the soil pH.

available only at low pH values. If the pH and the percentage base saturation are not low enough, these plants will show chlorosis (yellowing of the leaves) and other symptoms indicative of an unsatisfactory nutritive condition such as a deficiency of iron.

Most forest trees seem to grow well over a wide range of soil pH levels, which indicates they have at least some tolerance of acid soil. Many species, particularly the conifers, tend to intensify soil acidity. Forests exist as natural vegetation in regions of acid soils.

It is fortunate that most productive arable soils have intermediate pH values, being not too acidic and not too alkaline. Most cultivated crop plants grow well on these soils. Since pasture grasses, many legumes, small grains, intertilled field crops, and a large number of vegetables are included in this broadly tolerant group, mild soil acidity or alkalinity is not a deterrent to their growth. In terms of pH, a range of perhaps 6.0–7.2 is most suitable for most crop plants.

8.8

Determination of Soil pH

Soil pH is tested routinely, and the test is easy and rapid to make. Soil samples are collected in the field. The pH is measured directly, or the samples are brought to the laboratory for more accurate pH determination.

ELECTROMETRIC METHOD

The most accurate method of determining soil pH is with a pH meter. In this method a sensing glass electrode (referenced with a standard calomel electrode) is inserted into a soil–water mixture that simulates the soil solution. The difference between the H^+ ion activities in the wet soil and in the glass electrode gives rise to an electrometric potential difference that is related to the soil solution pH. The instrument gives very consistent results, and its operation is simple.

Traditionally, the pH measurement has been made in a suspension of soil in water (usually a ratio of 1:1 or 1:2). It also can be made in a similar suspension of soil in a dilute, unbuffered KCl solution. The former measurement is for the active acidity; the latter gives the salt-replaceable or exchangeable acidity.

DYE METHODS

Dye methods take advantage of the fact that certain organic compounds change color as the pH is increased or decreased. Mixtures of such dyes provide significant color changes over a wide pH range (3–8). A few drops of the dye solutions are placed in contact with the soil, usually on a white porcelain plate (see Figure 8.14). After standing a few minutes, the color of the dye is compared to a color chart that indicates the approximate pH.

In other dye methods, porous strips of paper are impregnated with the dye or dyes. When brought in contact with a mixture of water and soil, the paper absorbs the water and by color change indicates the pH. Such dye methods are accurate within about 0.2 pH unit.

pH OF SALT-AFFECTED SOILS

The procedures just described for measuring the soil pH are less satisfactory for soils containing significant soluble salts. At soil-to-water ratios of 1:1 or

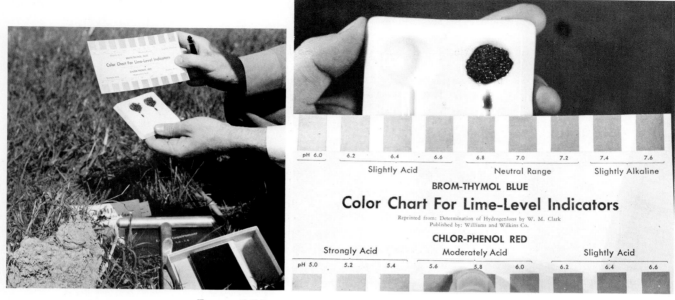

FIGURE 8.14
The indicator method for determining pH is widely used in the field. It is simple and is accurate enough for most purposes. [Courtesy New York State College of Agriculture and Life Sciences, Cornell University]

1:2, the pH measured is significantly higher than is found in the field. Consequently, the pH of salt-affected soils is commonly determined on a saturated paste of the soil. The moisture content of a paste is sufficiently near that of the soil in the field to make the measured pH more meaningful.

LIMITATIONS OF pH VALUES

Laboratory pH measurements can be made rather precisely. However, the interpretation of this measurement must be applied in the field with some caution. First, there is considerable variation in the pH from one spot in a given field to another. Localized effects of fertilizers may give sizable pH variations within the space of a few inches. Also, interpretations with respect to the management of a whole field, involving several million kilograms of soil, are made on the basis of the pH of a mere 10 g sample of soil. Obviously, care must be taken to minimize sampling errors.

Because of these limitations, the heavy reliance placed on the pH of soils might seem peculiar. However, soil pH is correlated with so many chemical and biological factors that influence both soils and plants. Thus, more may be inferred regarding the physiological condition of a soil from its pH value than from any other single measurement.

8.9

Methods of Intensifying Soil Acidity in Humid Regions

Although the usual desire of farmers and gardeners in humid regions is to increase soil pH, some plants such as rhododendrons and azaleas grow best on soils at pH of 5.0 and below. To accommodate these plants and to control some soil-borne diseases, it is sometimes necessary to increase the acidity of even acid soils. This is done by adding acid-forming organic and inorganic materials.

ACID ORGANIC MATTER

As organic matter decomposes, organic and inorganic acids are formed that can reduce the soil pH, especially if the organic matter is low in base-forming cations. Leaf mold, pine needles, tan bark, sawdust, and acid moss peat are quite satisfactory organic materials to add around ornamental plants. Farm manures may be alkaline, however, and should be used with caution.

USE OF CHEMICALS

When the addition of acid organic matter is not feasible, chemicals may be used. For rhododendrons, azaleas, and other plants that require considerable iron, such as blueberries and cranberries, ferrous sulfate sometimes is recommended. Upon hydrolysis, the ferrous ion enhances acidity by reactions such as

$$Fe^{2+} + H_2O \rightleftharpoons Fe(OH)_2 + 2\,H^+$$

$$4\,Fe^{2+} + 6\,H_2O + O_2 \rightleftharpoons 4\,Fe(OH)_2{}^+ + 4\,H^+$$

The H^+ ions released drastically lower the pH and liberate some of the iron already present in the soil. At the same time, soluble and available iron (Fe^{2+}) is being added in the ferrous sulfate. Such a chemical thus serves a double purpose in effecting a change in the physiological condition of a soil.

Other materials often used to increase soil acidity are elemental sulfur and, in some irrigation systems, sulfuric acid. Sulfur usually undergoes rapid microbial oxidation in the soil (see Section 11.25), and sulfuric acid is produced.

$$2\,S + 3\,O_2 + 2\,H_2O \rightarrow 2\,H_2SO_4$$

Under favorable conditions, sulfur is four to five times more effective kilogram-for-kilogram in developing acidity than is ferrous sulfate. Although ferrous sulfate brings about more rapid plant response, sulfur is less expensive, easy to obtain, and is often used on the farm for other purposes.

The quantities of ferrous sulfate or sulfur that should be applied will depend upon the buffering capacity of the soil and its original pH level.

Control of Potato Scab. Sulfur also is effective in the control of potato scab because the actinomycetes responsible are discouraged by acidity. Ordinarily, when the pH is lowered to 5.3, the virulence of actinomycetes is much reduced. However, using sulfur for this purpose or to increase soil acidity affects the management of the land, especially crop rotation. Consideration must be given to choosing succeeding crops that will not be adversely affected.

8.10

Decreasing Soil Acidity—Liming Materials

Soil acidity is commonly decreased by adding carbonates, oxides, or hydroxides of calcium and magnesium, compounds that are referred to as *agricultural limes*.[5]

---

[5] For a recent discussion of lime and its use, see Adams (1984).

FIGURE 8.15
Alfalfa was seeded in this field trial. In the foreground the soil was unlimed (pH = 5.2). In the background 9 Mg/ha (4 tons/acre) of limestone was applied before seeding.

CARBONATE FORMS

There are a number of sources of carbonate of lime including marl, oyster shells, basic slag, and precipitated carbonates, but ground limestone is the most common and is by far the most widely used of all liming materials. The two important minerals included in limestones are *calcite*, which is mostly calcium carbonate ($CaCO_3$), and *dolomite*, which is primarily calcium-magnesium carbonate [$CaMg(CO_3)_2$]. These minerals occur in varying proportions in limestones. When little or no dolomite is present, the limestone is referred to as *calcitic*. As the magnesium increases, this grades into a *dolomitic limestone*. Finally, if the stone is almost entirely composed of calcium-magnesium carbonate and impurities, the term *dolomite* is used. Most of the crushed limestone on the market today is calcitic and/or dolomitic.

Ground limestone is effective in increasing crop yields (see Figure 8.15). It varies in amount of calcium and magnesium carbonates from approximately 75 to 99%. The average total carbonate level of the representative crushed limestone is about 94%.

OXIDE FORMS

Commercial oxide of lime is normally referred to as *burned lime, quicklime,* or often simply as the *oxide*. It is produced by heating limestone in large commercial kilns in which reactions such as the following take place.

$$CaCO_3 + \text{heat} \rightarrow CaO + CO_2 \uparrow$$
Calcite

$$CaMg(CO_3)_2 + \text{heat} \rightarrow CaO + MgO + 2\ CO_2 \uparrow$$
Dolomite

Oxide of lime is considerably more costly than limestone. It is also considerably more caustic than limestone and consequently is difficult to handle, but it reacts much more rapidly with the soil than does limestone.

HYDROXIDE FORM

The hydroxide form of lime is commonly referred to as *hydrated lime*, since it is produced by adding water to burned lime. The reaction is

$$CaO + MgO + 2\ H_2O \rightarrow Ca(OH)_2 + Mg(OH)_2$$

Hydroxide of lime appears on the market as a white powder and is more caustic than burned lime. Like the oxide, it requires bagging, preferably in waterproof bags. It is used where a rapid rate of reaction is desired and/or where a high soil pH is necessary. Like burned lime hydrated lime is quite expensive compared to limestone, and its use is confined largely to home gardens and specialty crops. Representative samples of hydrated lime are generally about 95% calcium and magnesium hydroxide.

8.11

Reactions of Lime in the Soil

When liming materials are added to a soil, the calcium and magnesium compounds react with carbon dioxide and with the acid colloidal complex.

REACTION WITH CARBON DIOXIDE

When applied to an acid soil, all liming materials—whether the oxide, hydroxide, or carbonate—react with carbon dioxide and water to yield the bicarbonate form. The carbon dioxide partial pressure in the soil, usually several hundred times greater than that in atmospheric air, is generally high enough to drive such reactions. For the purely calcium limes, the reactions are as follows.

$$CaO + H_2O + 2\,CO_2 \rightarrow Ca(HCO_3)_2$$
$$Ca(OH)_2 + 2\,CO_2 \rightarrow Ca(HCO_3)_2$$
$$CaCO_3 + H_2O + CO_2 \rightarrow Ca(HCO_3)_2$$

REACTION WITH SOIL COLLOIDS

All liming materials will react with acid soils, the calcium and magnesium replacing hydrogen and aluminum on the colloidal complex. The adsorption with respect to calcium may be indicated as follows, assuming hydrogen ions are replaced.

$$\begin{array}{l} H^+ \\ \;\; \boxed{Micelle} \;+\; Ca(OH)_2 \;\rightleftharpoons\; Ca^{2+}\boxed{Micelle} + 2\,H_2O \\ H^+ \end{array}$$

$$\begin{array}{l} H^+ \\ \;\; \boxed{Micelle} \;+\; Ca(HCO_3)_2 \;\rightleftharpoons\; Ca^{2+}\boxed{Micelle} + 2\,H_2O + 2\,CO_2 \\ H^+ \quad\quad\quad\;\; \text{(in solution)} \end{array}$$

$$\begin{array}{l} H^+ \\ \;\; \boxed{Micelle} \;+\; CaCO_3 \;\rightleftharpoons\; Ca^{2+}\boxed{Micelle} + H_2O + CO_2 \\ H^+ \quad\quad\;\; \text{(solid phase)} \end{array}$$

As these reactions proceed, carbon dioxide is freely evolved. In addition, the adsorption of the calcium and magnesium ions raises the percentage base saturation of the colloidal complex, and the pH of the soil solution increases correspondingly.

COMPOUNDS OF CALCIUM AND MAGNESIUM IN LIMED SOIL

Soon after limestone is added to a soil, three forms of calcium and magnesium are found: (a) undissolved solid calcium and calcium-magnesium carbonates; (b) exchangeable bases adsorbed by the colloidal matter; and

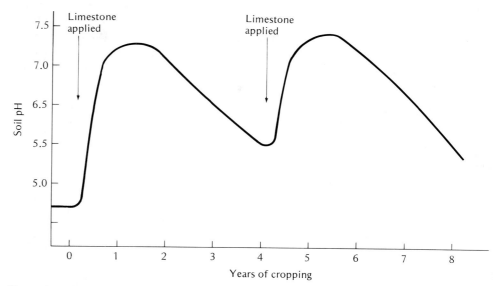

FIGURE 8.16
Influence of limestone applications on pH of a cropped soil. The initial rate of limestone application was assumed to be 8–10 Mg/ha (3.5–4.5 tons/acre). Note that it takes about 1 year for most of the limestone to react. Leaching and crop removal deplete the calcium and magnesium supplied in the limestone, and in time the pH decreases until limestone is added again.

(c) dissociated cations in the soil solution, mostly in association with bicarbonate ions. Later, when the calcium and calcium-magnesium carbonates have completely dissolved, the system becomes somewhat simpler, involving only the exchangeable cations and those in soil solution.

DEPLETION OF CALCIUM AND MAGNESIUM

As the soluble calcium and magnesium compounds are removed from the soil by the growing crop or by leaching, the percentage base saturation and pH are gradually reduced; eventually another application of lime is necessary. This type of cyclic activity is typical of much of the calcium and magnesium added to arable soils in humid regions (see Figure 8.16).

The need for repeated applications of limestone in humid regions suggest significant losses of calcium and magnesium from the soil. Table 8.1 illustrates losses of these elements by leaching compared with those from crop removal and soil erosion. Note that the total loss from all three causes, expressed in the form of carbonates, approaches 1 Mg/ha per year. While

TABLE 8.1

Calcium and Magnesium Losses from a Soil by Erosion, Crop Removal, and Leaching in a Humid-Temperate Region

Values are in kilograms per hectare per year.

	Calcium expressed as			Magnesium expressed as		
Manner of removal	*Ca*	*CaO*	*CaCO$_3$*	*Mg*	*MgO*	*MgCO$_3$*
By erosion, Missouri experiments, 4% slope	95	133	238	33	55	115
By the average crop of a standard rotation	50	70	125	25	42	88
By leaching from a representative silt loam	115	161	288	25	42	88
Total			651			291

there is considerable variation in these losses among different farming systems and in different locations, it is well to recognize that the liming of soils is not a one-time venture but must be repeated with some degree of regularity. The data in Table 8.1 suggest that magnesium as well as calcium must be provided by liming, or by other means, if appropriate nutrient balance is to be maintained.

8.12

Lime Requirements—Quantities Needed

The amount of liming material required to bring about a desired pH change is determined by several factors, including (a) the change in pH required, (b) the buffer capacity of the soil, (c) the chemical composition of the liming materials used, and (d) the fineness of the liming materials.

The range of soil pH optimum for major crops discussed in Section 8.7, suggests the increase in soil pH that may be desired. Clearly a higher pH is needed for alfalfa than for corn. In Section 8.4 differences in buffer capacity of soils was covered. The relationship of buffering to limestone requirements of soils with different textures is shown in Figure 8.17. Clearly, the lime requirements for an acid fine-textured clay are much higher than those for a sand or a loam with the same pH value.

8.13

Influence of Chemical Composition of Liming Materials

The chemical composition of liming materials affects the rate of reaction of these compounds with soils. For example, burned and hydrated limes react quickly with soils, bringing about significant changes in soil pH in only a few

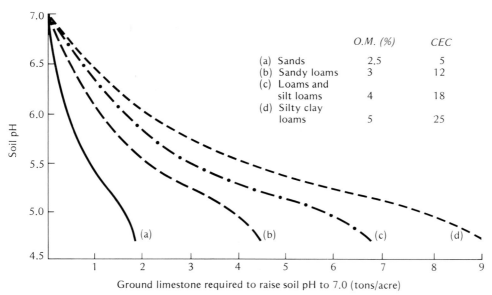

	O.M. (%)	CEC
(a) Sands	2.5	5
(b) Sandy loams	3	12
(c) Loams and silt loams	4	18
(d) Silty clay loams	5	25

FIGURE 8.17

Relationship between soil texture and the amount of limestone required to raise the pH of New York soils to 7.0. Representative organic matter (O.M.) and cation exchange capacity (CEC) levels are shown. [From Peech (1961)]

weeks. In contrast, limestones, and particularly those high in dolomite, react much more slowly, taking a year and even longer to react fully with soil colloids.

Chemical Guarantees

In addition to affecting reaction rates, the chemical composition of limestones determines their long-term effects on soil pH. The effects of liming are so important that they have been recognized in laws governing the sale of liming materials. These laws require *guarantees* as to the chemical composition of limes, the composition usually being listed in terms of one or more of the following.

1. Content of elemental calcium and magnesium.
2. Conventional oxide content (percentages of CaO and MgO).
3. Calcium oxide equivalent (neutralizing ability of all compounds expressed in terms of calcium oxide).
4. Total carbonates (used for limestones, the sum of calcite and dolomite).
5. Calcium carbonate equivalent or total neutralizing power (neutralizing ability of all compounds expressed in terms of calcium carbonate).

The first two designations of composition have the advantage of indicating the magnesium content, which is important because liming materials are significant sources of this element. The last two designations are used for limestones. Although the calcium carbonate equivalent is technically a more accurate way of stating total neutralizing power, the difference between this figure and total carbonates is not very great for most limestones of equal purity.

Table 8.2 compares the compositions of four liming materials: a burned lime (oxide) of 95% purity, a 98% pure hydrated lime (hydroxide), and two 95% pure limestones, one dolomitic, one calcitic. Note the much greater acid neutralizing ability of the burned and hydrated limes (refer to the CaO or $CaCO_3$ equivalent). Also note that dolomitic limestone has a slightly higher neutralizing power ($CaCO_3$ equivalent) than calcitic limestone with the same degree of purity. These figures illustrate the importance of chemical compositions in determining which lime to use and how much is to be applied.

TABLE 8.2
A Comparison of Different Means of Expressing the Calcium and Magnesium Contents in Typical Burned Lime (Oxide), Hydrated Lime (Hydroxide), and Two Limestones (One Calcitic and One Dolomitic)

Material	Actual chemicals (%)	Element (%) Ca	Element (%) Mg	Conventional oxide (%) CaO	Conventional oxide (%) MgO	CaO equivalent (%)	Total carbonates (%)	$CaCO_3$[a] equiv. (%)
Burned lime	77 CaO	55.0	—	77.0	—	} 102.2	—	182.5
	18 MgO	—	10.9	—	18.0			
Hydrated lime	75 Ca(OH)$_2$	40.5	—	58.6	—	} 78.9	—	140.8
	23 Mg(OH)$_2$	—	9.6	—	15.9			
Calcitic limestone	95 CaCO$_3$	38.0	—	53.2	—	53.2	95	95
Dolomitic limestone	35 CaCO$_3$ 60 CaMg(CO$_3$)$_2$ }	(14.0) (13.0) 27.0	— (7.9) 7.9	(19.6) (18.2) 37.8	— (13.1) 13.1	(19.6) (36.4) 56.0	95	(35) (65) 100

[a] Sometimes referred to as total neutralizing power.

CHEMICAL EQUIVALENT OF LIMING MATERIALS

The data in Table 8.2 suggest also that there is chemical equivalency among the different liming materials. Thus, one molecule of $CaCO_3$ will ultimately neutralize the same acidity as one molecule of any of the other liming compounds, such as CaO, MgO, $Ca(OH)_2$, and so on. To calculate the chemical equivalency of these materials the ratio of their molecular masses must be used. For example, to calculate the $CaCO_3$ equivalent of a pure burned lime (CaO) simply multiply by the molecular ratio of $CaCO_3$ to CaO.

$$\frac{CaCO_3}{CaO} = \frac{100}{56} = 1.786$$

Then 100 kg of pure CaO has a $CaCO_3$ equivalent of $100 \times 1.786 = 178.6$. If the burned lime is only 95% pure, 100 kg would supply only 95 kg of CaO so that 95% burned lime would have a $CaCO_3$ equivalency of $95 \times 1.786 = 169.6$. The use of other appropriate molecular ratios such as those following makes possible comparisons of the acid neutralizing ability of all liming materials.

$$\text{Ca eq of } CaCO_3 \quad \frac{Ca}{CaCO_3} = \frac{40}{100} = 0.40$$

$$\text{CaO eq of } CaCO_3 \quad \frac{CaO}{CaCO_3} = \frac{56}{100} = 0.56$$

$$\text{CaO eq of } MgCO_3 \quad \frac{CaO}{MgCO_3} = \frac{56}{84} = 0.67$$

$$\text{CaCO}_3 \text{ eq of } MgCO_3 \quad \frac{CaCO_3}{MgCO_3} = \frac{100}{84} = 1.19$$

$$\text{Mg eq of } MgCO_3 \quad \frac{Mg}{MgCO_3} = \frac{24}{84} = 0.29$$

$$\text{Mg eq of } MgO \quad \frac{Mg}{MgO} = \frac{24}{40} = 0.60$$

As an example, let us use these ratios to calculate how much of a limestone ($CaCO_3$ equivalent = 90) would be required to neutralize the same acidity as 1 metric ton (Mg) of a burned lime with a CaO equivalent of 98.

1. The burned lime has the neutralizing ability of

$$1000 \times 0.98 = 980 \text{ kg of pure } CaO$$

2. This 980 kg of CaO is equivalent to

$$980 \times 1.79 \ (CaCO_3/CaO) = 1750 \text{ kg of pure } CaCO_3$$

3. The limestone has a $CaCO_3$ equivalent of only 90, so the amount of limestone needed is

$$1750/0.90 = 1944 \text{ kg}$$

8.14

Limestone Fineness and Reactivity

The finer a liming material, the more rapidly it will react with the soil. The oxide and hydroxide of lime usually appear on the market as powders, so their fineness is always satisfactory, but different limestones may vary considerably in particle size as well as hardness. The inefficiency of coarser and harder limestones has necessitated legal requirements for a *fineness guarantee*.

MEASURING FINENESS

The fineness of limestones is measured by passing a sample through a series of standard sieves (screens) with openings of designated size. A number 10 mesh sieve has 10 wires per inch and opening sizes of 2 mm. Similarly a 60 mesh screen opening is 0.25 mm and a 200 mesh screen opening 0.075 mm. Remember that the size opening of a given screen represents the *maximum* diameter of particles that can pass through; thus all smaller particles will also pass through the screen.

SOIL REACTION AND CROP YIELD

The effect of the limestone size on the rate of reaction with soil is shown in Figure 8.18. At the end of 3 months nearly 90% of the calcite in the finest fraction (100 mesh) had reacted with the soil, whereas less than 20% of the 20 mesh size particles had reacted. Note that at all size fractions dolomite was less reactive than calcite. The data emphasize the interaction between chemical composition and fineness in ascertaining liming rates.

Crop response to liming is also influenced markedly by limestone fineness. Figure 8.19 presents a summary of data from many locations to illus-

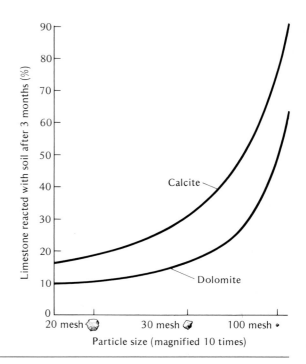

FIGURE 8.18
Relationship between particle size of calcite and dolomite and the rate of reaction of these minerals with the soil. Note that calcite particles of a given size react more rapidly than do corresponding dolomite particles. The coarse fractions of both minerals neutralize the soil acidity very slowly. [Data calculated from Schollenberger and Salter (1943)]

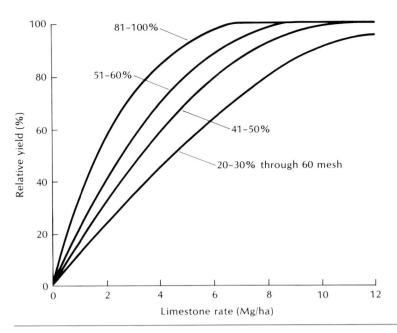

FIGURE 8.19

The effect of limestone fineness on the response of crops to increased rates of limestone application. These are average data from a number of field experiments. [Redrawn from Barber (1984)]

trate this point. To obtain at least 80% of the highest yields, 8 Mg/ha of limestone would be required if only 20–30% of the particles passed through a 60 mesh sieve. Less than 4 Mg/ha would be needed if 81–100% of the limestone particles could pass through a 60 mesh sieve. These data and others obtained from field experiments throughout the humid regions of the United States suggest that *limestones with at least 50% passing through a 60 mesh screen are quite satisfactory for most agricultural purposes.* Although finer materials may give slightly higher yields, the additional cost of grinding to produce the finer limestone may negate the benefits achieved.

8.15

Practical Considerations

The soil and lime characteristics along with cost factors largely determine the quantity and nature of lime to be applied. However, in ordinary practice it is seldom wise to apply more than about 7–9 Mg/ha (3–4 tons/acre) of finely

TABLE 8.3

Suggested Total Amounts of Finely Ground Limestone That Should Be Applied per Hectare–Furrow Slice of Mineral Soil for Alfalfa in Rotation[a]

	Limestone (kg/ha)	
Need for lime	*Sandy loam*	*Silt loam*
Moderate	2000–3000	3000–6000
High	3000–5000	6000–8000

[a] In calculating these rates, the assumption is made that the amounts suggested will be used as *initial* applications. After the pH of the soil has been raised to the desired level, smaller *maintenance* rates may be satisfactory.

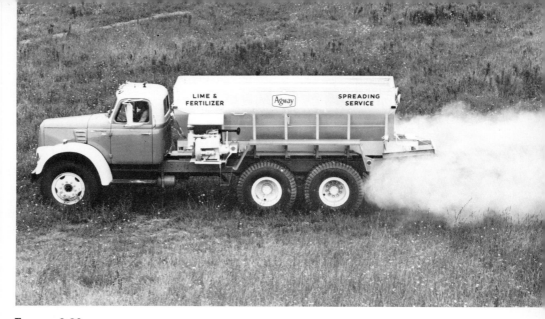

FIGURE 8.20
Bulk application of limestone by specially equipped trucks is the most widespread method of applying liming materials. Because of the weight of such machinery, much limestone is applied to sod land and plowed under. In many cases this same method is used to spread commercial fertilizers. [Courtesy Harold Sweet, Agway Inc., Syracuse, NY]

ground limestone to a mineral soil at any one time—and even lower rate limits may be appropriate on sandy soils. Data given in Table 8.3 suggest reasonable rates that may be applied on coarse- and fine-textured soils.

In the choice of liming materials attention should be given to the need for magnesium. Everything else being equal, a magnesium-containing limestone should be favored to help maintain appropriate nutritional balance.

The lime should be spread with or ahead of the crop that gives the most satisfactory response. Thus, in a rotation with corn, wheat, and 2 years of alfalfa, the lime may be applied after the wheat harvest to favorably influence the alfalfa, which follows. However, since most lime is bulk spread using heavy trucks (Figure 8.20), applications commonly take place on the sod or hay crop. This prevents adverse soil compaction by the truck tires, which might occur if the lime were applied on recently tilled land.

Another practical consideration is the danger of *overliming* so that the resultant pH values will be too high for optimal crop growth. Overliming is not very common on fine-textured soils with high buffer capacities, but it can occur easily on coarse-textured soils that are low in organic matter. The detrimental results of excess lime include deficiencies of iron, manganese, copper, and zinc; reduced availability of phosphate; and restraints on the uptake of boron. Overliming should be avoided.

8.16

Lime and Soil Fertility Management

The maintenance of satisfactory soil fertility levels in humid regions depends considerably on the judicious use of lime (see Figure 8.21). Liming not only maintains the levels of exchangeable calcium and magnesium but, in so doing, also provides a chemical and physical environment that encourages the growth of most common crop plants. Furthermore, lime counteracts the

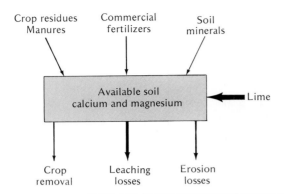

FIGURE 8.21
Important ways by which *available* calcium and magnesium are supplied to and removed from soils. The major losses are through leaching and erosion. These losses are largely replaced by lime and fertilizer applications. Fertilizer additions are much higher than is generally realized, for some phosphate fertilizers contain large quantities of calcium.

acid-forming tendency of nitrogen-containing fertilizers whose use has increased markedly in the past quarter century. Lime is truly a foundation for much of modern temperature zone agriculture.

8.17

Normal Neutral and Alkaline Soils of Dry Areas

Primary focus up to now has been on soil acidity and its control. But the problems and opportunities relating to neutral and alkaline soils of arid areas are also important.

In soils of arid and semiarid regions, lack of extensive leaching leaves the level of base-forming cations quite high. As a result, the pH is commonly 7 or above, positively charged aluminum and aluminum hydroxy ions are absent, and hydrogen ions are at extremely low levels. Adsorbed Ca^{2+} and Mg^{2+} ions dominate the exchange complex, although concentrations of Na^+ and K^+ ions are generally higher than in acid soils.

The 2:1-type silicate clays are prominent, although 1:1-type clays, hydrous oxides, and allophane are of importance locally. Organic matter is lower than in soils of humid regions. For a given texture class, the cation exchange capacities of soils of dry areas are commonly as high or higher than those of comparable acid soils. This is because the pH-dependent negative charges on humus, 1:1-type minerals, allophane, and Fe, Al oxides are effective at the higher pH levels. However, in some alkaline soils iron and aluminum are present in highly insoluble forms that participate only sparingly in reactions with silicate clays, organic matter, and higher plants. As a result, iron deficiencies are common in plants grown on these soils.

A well-developed profile in low rainfall areas usually carries at some point (usually in the C horizon) a calcium carbonate accumulation greater than that of its parent material. The lower the rainfall, the nearer the surface this layer will be (Figure 3.3). As a result, these soils may have alkaline subsoils and alkaline or neutral surface layers. When enough leaching has occurred to free the solum of calcium carbonate, a mild acidity may develop in the surface horizons. The genetic classification of soils characterized by low moisture with a pertinent description of each group is presented in Section 3.7.

Mention has been made of the bad affects of overliming soils of humid areas. In arid areas such conditions of high soil pH and consequent nutrient imbalance occur naturally. Some soils with pH values of 7.5 and above are deficient in micronutrients such as iron and zinc. Supplemental micronutrients applied through fertilizers are quickly tied up in insoluble forms unless selected protective organic complexes called *chelates* are used (see Section 13.6) or high rates of fertilizer are applied and incorporated into the root zone. In some cases foliar applications of these micronutrients are used, especially for fruit trees.

Some of the most important agricultural soils in low rainfall areas are being irrigated. As a result, their moisture relations during periods of crop production are not greatly different than those of soils in humid regions. This human-induced application of water and its subsequent evaporation can contribute to the accumulation of salts, an activity that natural processes have brought about in some arid and semiarid areas. Salt-affected soils will be covered next.

8.18

Characteristics of Saline and Sodic Soils[6]

Salts accumulate in some surface soils of arid and semiarid regions because there is insufficient rainfall to flush them from the upper soil layers. About one third of the soils in these regions in the United States have some degree of salinity accumulations. The salts are primarily chlorides and sulfates of calcium, magnesium, sodium, and potassium. The source of these salts is the weathering of rocks and minerals, rainfall, ground waters, and irrigation. Once deposited or released in the soil, the salts are brought to or near the surface by upward-moving water, which then evaporates, leaving the salts behind. Unfortunately, high levels of these salts cannot be tolerated by most crop plants, a fact that severely limits the use of some salt-affected soils.

MEASURING SALINITY AND ALKALINITY

Research has shown that the detrimental effects on plants stem not only from the high salt contents but also from the level of sodium in the soil, especially in relation to levels of calcium and magnesium. High exchangeable sodium levels are detrimental both physically and chemically. This situation has led to the development of techniques to measure three primary soil properties that, along with soil pH, can be used to characterize salt-affected soils. Each technique will be discussed briefly.

Salinity First, the salt concentration of the soil is estimated by methods based on the ability of the salt in the soil solution to conduct electricity. Laboratory measurements of the *electrical conductivity (EC)* of the soil solution extracted from a saturated sample of soil give an indication of the salt levels. Recently, two field techniques based on conductivity have been developed to measure soil salinity (see Rhodes and Corwin, 1984). These techniques make field mapping of salinity a possibility. For convenience, the salinity is commonly expressed in terms of the electrical conductivity (EC), which is measured in decisiemens per meter[7] (dS/m).

---

[6] See Bresler et al. (1982) for a discussion of these soils.
[7] Formerly expressed as millimhos per centimeter (mmho/cm). Since 1 S = 1 mho, 1 dS/m = 1 mmho/cm.

Sodium Status Two means are used to characterize the sodium status of these soils. The *exchangeable sodium percentage (ESP)* identifies the degree to which the exchange complex is saturated with sodium.

$$ESP = \frac{\text{exchangeable sodium (cmol/kg)}}{\text{cation exchange capacity (cmol/kg)}}$$

ESP levels of 15 yield pH values of 8.5 and above. Higher levels may bring the pH to at least 10.

The ESP is complemented by a second more easily measured characteristic, the sodium adsorption ratio (SAR), which gives information on the comparative concentrations of Na^+, Ca^{2+}, and Mg^{2+} in soil solutions. It is calculated as follows.

$$SAR = \frac{[Na^+]}{\sqrt{\frac{1}{2}([Ca^{2+}] + [Mg^{2+}])}}$$

where $[Na^+]$, $[Ca^{2+}]$, and $[Mg^{2+}]$ are the concentrations (in cmol/kg) of the sodium, calcium, and magnesium ions in the soil solution. SAR of a soil extract takes into consideration that the adverse effect of sodium is moderated by the presence of calcium and magnesium ions. The SAR also is used to characterize the irrigation water added to these soils. This property is becoming more widely used in characterizing salt-affected soils.

Using EC, EPS, and SAR characteristics and soil pH, salt-affected soils are classified as *saline, saline-sodic*, and *sodic* (Table 8.4). These will be considered in order.

SALINE SOILS[8]

Saline soils contain a concentration of neutral soluble salts sufficient to interfere seriously with the growth of most plants. The electrical conductivity (EC) of a saturated extract of the soil solution is more than 4 decisiemens per meter (dS/m).[9] The exchangeable sodium percentage (ESP) is less than about 15, and the pH usually is less than 8.5 because the salts are neutral, the chlorides and sulfates of the base-forming cations dominating. Saline soils have been called *white alkali* soils because a surface incrustation, if present, is white in color (Figure 8.22).

---

[8] Salinization is a term used in reference to the natural processes that result in the accumulation of neutral soluble salts such as NaCl in soils.

[9] Some fruit and vegetable crops are adversely affected when the EC is lower than 4 dS/m, leading some scientists to recommend that 2 dS/m be the level above which a soil could be classed as saline.

TABLE 8.4
Properties of Normal Soils Compared to Acid, Saline, Sodic, and Saline-Sodic Soils

Soil	Common pH	Electrical conductivity (EC) (dS/m)	Sodium adsorption ratio (SAR)[a]
Normal	6.5–7.2	<4	<13–15
Acid	<6.5	<4	<13–15
Saline	<8.5	>4	<13–15
Saline-sodic	<8.5	>4	>13–15
Sodic	>8.5	<4	>13–15

[a] Exchangeable sodium percentage (ESP) and sodium adsorption ratio (SAR) are closely related in most soils.

FIGURE 8.22
(a) White "alkali" spot in a field of alfalfa under irrigation. Because of upward capillarity and evaporation, salts have been brought to the surface where they have accumulated. (b) White salt crust on a saline soil from Colorado. The white salts are in contrast with the darker colored soil underneath. (Scale is shown in inches and centimeters.) [(a) Courtesy USDA Soil Conservation Service]

Although the Na^+ ion concentration may be somewhat higher than those of Ca^{2+} and Mg^{2+}, the sodium adsorption ratio (SAR) is less than 13 on saline soils. These soils can be reclaimed by merely leaching out the salts, an important consideration in their management. Of course, the leaching water must have a low SAR to prevent an increase in sodium adsorption, and, in turn, in soil pH.

SALINE-SODIC SOILS

The saline-sodic soils contain appreciable quantities of neutral soluble salts and enough sodium ions to seriously affect most plants. The exchangeable sodium percentage (ESP) is greater than about 15, and the electrical conductivity (EC) of a saturated extract is more than 4 dS/m. The pH is commonly 8.5 or less because of the presence of neutral salts. The sodium adsorption ratio (SAR) is at least 13 in these soils.

245

Leaching of saline-sodic soils, unlike the result with saline soils, will markedly raise the pH, unless calcium or magnesium salt concentrations are high in the soil or in the irrigation water. Increase of the pH occurs because once the neutral soluble salts are removed, the exchangeable sodium readily hydrolyzes and thereby sharply increases the OH^- ion concentration of the soil solution, as follows.

$$Na^+ \boxed{Micelle} + H_2O \rightleftharpoons H^+ \boxed{Micelle} + Na^+ + OH^-$$

(soil solid)   (soil solution)   (soil solid)   (soil solution)

In the presence of Na-containing neutral salts, this reaction is driven back to the left, OH^- ion concentration in the soil solution is reduced, and the pH is prevented from rising above about 8. If the neutral salts are removed, the pH rises above 8.5 and the mineral colloids become dispersed, resulting in a tight, impervious soil structure. At the same time, sodium toxicity to plants is increased.

SODIC SOILS

Sodic soils do not contain any great amount of soluble salts. The detrimental effects of these soils on plants is due not only to the toxicity of Na^+, HCO_3^-, and OH^- ions but also to reduced water infiltration and aeration. The high pH is largely due to the hydrolysis of sodium carbonate.

$$2\,Na^+ + CO_3^{2+} + H_2O \rightleftharpoons 2\,Na^+ + HCO_3^- + OH^-$$

The sodium complex also undergoes hydrolysis.

$$Na^+ \boxed{Micelle} + H_2O \rightleftharpoons H^+ \boxed{Micelle} + Na^+ + OH^-$$

The exchangeable sodium percentage (ESP) of sodic soils is decidedly more than 15 and the SAR is more than 13. The adsorbed sodium is free to hydrolyze because the concentration of neutral soluble salts is rather low. An increase in hydroxide ion and pH results. The pH is always above 8.5, often rising to 10.0 or higher. The electrical conductivity of a saturated extract is less than 4 dS/m.

Owing to the deflocculating influence of the sodium, sodic soils usually are in an unsatisfactory physical condition. As already stated, the leaching of a saline-sodic soil can readily produce a similar physical condition unless the leaching waters are high in calcium salts.

Because of the extreme alkalinity resulting from the high sodium content, the surface of sodic soils is usually discolored by the dispersed humus carried upward by the capillary water and deposited when it evaporates. Hence, the name *black alkali* has been used to describe these soils. Sometimes located in small areas called *slick spots*, sodic soils may be surrounded by soils that are relatively productive.

8.19

Growth of Plants on Saline and Sodic Soils

Saline and saline-sodic soils with their relatively low pH (usually less than 8.5) detrimentally influence plants largely because of their high soluble salt concentration (Figure 8.22). When the soil solution containing a relatively large amount of dissolved salts is brought into contact with a plant cell, water will pass by osmosis from the cell into the more concentrated soil solution. The cell then collapses. The kind of salt, the plant species, and the rate of salinization are factors that determine the concentration at which the cell succumbs. The adverse physical condition of the soils, especially the saline-sodic soils, also may be a factor.

Sodic soils, dominated by active sodium, exert a detrimental effect on plants in four ways: (a) caustic influence of the high pH induced by the sodium carbonate and bicarbonate; (b) toxicity of the bicarbonate and other anions; (c) the adverse effects of the active sodium ions on plant metabolism and nutrition and the low micronutrient availablity due to high pH; and (d) oxygen deficiency due to breakdown of soil structure in sodium-dominated soils.

8.20

Tolerance of Higher Plants to Saline and Sodic Soils

The capacity of higher plants to grow satisfactorily on salty soils depends on a number of interrelated factors, including the physiological constitution of the plant, its stage of growth, and its rooting habits. It is interesting to note that old alfalfa is more tolerant of salt-affected soils than young alfalfa, and that deep-rooted legumes show a greater resistance to such soils than the shallow-rooted ones.

In terms of the soil, the nature of the various salts, their proportionate amounts, their total concentration, and their distribution in the solum must be considered. The structure of the soil and its drainage and aeration are also important.

While it is difficult to forecast accurately the tolerance of crops, the comparative data presented in Table 8.5 may be helpful.

TABLE 8.5
Relative Tolerance of Certain Plants to Salty Soils

Tolerant	Moderately tolerant	Moderately sensitive	Sensitive
Barley, grain	Barley, forage	Alfalfa	Apple
Bermuda grass	Beet, garden	Broad bean	Apricot
Bougainvillea	Broccoli	Cauliflower	Bean
Cotton	Brome grass	Cabbage	Blackberry
Date	Clover, berseem	Clover, alsike, Ladino,	Carrot
Natal plum	Fig	red, strawberry	Celery
Mutall alkali grass	Orchard grass	Corn	Grapefruit
Rescue grass	Oats	Cowpea	Lemon
Rosemary	Rye, hay	Cucumber	Onion
Sugar beet	Rye grass, perennial	Lettuce	Orange
Salt grass	Sorghum	Pea	Peach
Wheat grass, crested	Sudan grass	Peanut	Pear
	Trefoil, birdsfoot	Rice, paddy	Pineapple; Guava
Wheat grass, fairway	Wheat	Soybean	Potato
	Wheat grass, western	Sweet clover	Raspberry
Wheat grass, tall		Timothy	Rose
Wild rye, altai			Strawberry
Wild rye, Russian			Tomato

Modified from Carter (1981).

8.21

Management of Saline and Sodic Soils

Since saline and sodic soils are found mostly in arid regions, their use for agricultural purposes commonly requires irrigation water. The quality of that water, especially in relation to its salt content and to its SAR, is extremely important in the management of the soils. Water high in sodium salts can bring about harmful effects unless these salts are counterbalanced by soluble calcium and magnesium salts. Knowledge of the quality of irrigation water is a requisite for good management of saline and sodic soils.

Three kinds of general management practices have been used to maintain or improve the productivity of saline and sodic soils. The first is *eradication* or removal of the salts; the second is a *conversion* of some of the salts to less injurious forms; and the third is designated *control*. In the first two methods, an attempt is made to eliminate by various means some of the salts or to render them less toxic. In the third, the salts are kept so well distributed throughout the soil solum that there is no toxic concentration within the root zone.

ERADICATION (SALT REMOVAL)

The most common methods used to free the soil of excess salts are installation of drainage systems and leaching or flushing. A combination of the two, flooding after field drainage ditches have been installed, is the most thorough and satisfactory. When this method is used in irrigated regions, heavy and repeated applications of water can be made. The salts that dissolve are leached from the solum and drained away. The irrigation water used must not be high in soluble salts, especially those containing sodium.

The leaching method works especially well with pervious saline soils, whose soluble salts are largely neutral and high in calcium and magnesium. Of course, little exchangeable sodium should be present. Leaching saline-sodic soils (and even sodic soils if the water will percolate) with waters very high in salts but low in sodium may be effective. Conversely, treating sodic and saline-sodic soils with water low in salts may intensify the alkalinity of the soil. The removal of the neutral soluble salts allows an increase in the percent sodium saturation, thereby increasing the concentration of hydroxide ions in the soil solution.

CONVERSION

The use of gypsum ($CaSO_4 \cdot 2H_2O$) on sodic soils is commonly recommended for the purpose of exchanging Ca^{2+} for Na^+ on the micelle and removing bicarbonates from the soil solution.

$$2\,NaHCO_3 + CaSO_4 \rightarrow CaCO_3 + Na_2SO_4 + CO_2$$

$$Na_2CO_3 + CaSO_4 \rightleftharpoons CaCO_3 + Na_2SO_4$$
$$\text{(leachable)}$$

$$\begin{matrix} Na^+ \\ Na^+ \end{matrix} \boxed{\text{Micelle}} + CaSO_4 \rightleftharpoons Ca^{2+} \boxed{\text{Micelle}} + Na_2SO_4$$
$$\text{(leachable)}$$

Several tons of gypsum per hectare are usually necessary. The soil must be kept moist to hasten the reaction, and the gypsum should be throughly mixed into the surface by cultivation, not simply plowed under. The treatment must be supplemented later by a thorough leaching of the soil with irrigation water to leach out some of the sodium sulfate.

Elemental sulfur and sulfuric acid can be used to advantage on salty lands, especially where sodium carbonate abounds. The sulfur, upon biological oxidation, yields sulfuric acid, which not only changes the sodium carbonate to the less harmful sodium sulfate but also decreases the alkalinity. The reactions of sulfuric acid with the compounds containing sodium may be shown as follows.

$$Na_2CO_3 + H_2SO_4 \rightleftharpoons CO_2 \uparrow + H_2O + \underset{\text{(leachable)}}{Na_2SO_4}$$

$$\begin{matrix} Na^+ \\ Na^+ \end{matrix} \boxed{\text{Micelle}} + H_2SO_4 \rightleftharpoons \begin{matrix} H^+ \\ H^+ \end{matrix} \boxed{\text{Micelle}} + \underset{\text{(leachable)}}{Na_2SO_4}$$

Not only is the sodium carbonate changed to sodium sulfate, a mild neutral salt, but the carbonate radical is removed from the system. When gypsum is used, however, a portion of the carbonate may remain as a calcium compound ($CaCO_3$).

CONTROL

The retardation of evaporation is an important feature in the control of salty soils. Slowing evaporation will not only save moisture but also will retard the translocation upward of soluble salts into the root zone.

Where irrigation is practiced, an excess of water should be avoided unless it is needed to free the soil of soluble salts. Frequent light irrigations are often necessary, however, to keep the salts sufficiently dilute to allow normal plant growth.

The timing of irrigation is extremely important on salty soils, particularly during the spring planting season. Since young seedlings are especially sensitive to salts, irrigation often precedes or follows planting to move the salts downward. After the plants are well established, their salt tolerance is somewhat greater.

The practice in recent years of adding nitrogen as anhydrous ammonia to irrigation water as it is applied to a field has created some soil problems (see Gardner and Roth, 1984). The high pH brought about by the ammonia causes some of the calcium in the water to precipitate and, thus, raises the sodium adsorption ratio (SAR) and the hazard of increased exchangeable sodium percentage (ESP). To counteract these difficulties, sulfuric acid is sometimes applied to the irrigation water to reduce its pH as well as that of the soil. This practice may well spread where there are economical sources of sulfuric acid and where the personnel applying it have been trained and alerted to serious hazards of using this strong acid.

The use of salt-resistant crops is another important feature of the successful management of saline and alkali lands. Sugar beets, cotton, sorghum, barley, rye, sweet clover, and alfalfa are particularly tolerant (Table 8.4). Moreover, a temporary alleviation of alkali will allow less resistant crops to be established. Farm manure is very useful in such an attempt. A crop such as alfalfa, once it is growing vigorously, may maintain itself in spite of the salt concentrations that may develop later. The root action of tolerant plants is exceptionally helpful in improving sodic soils in a poor physical condition. Aggregation is improved, and root channels are left through which water and oxygen can penetrate the soil.

8.22

Conclusion

No other single chemical soil characteristic is more important in determining the chemical environment of higher plants and soil microbes than the pH. There are few reactions involving any component of the soil or of its biological inhabitants that are not sensitive to soil pH. This sensitivity must be recognized in any soil managment system.

Soil pH is largely controlled by soil colloids and their associated exchangeable cations. Aluminum and hydrogen enhance soil acidity, whereas calcium and other base-forming cations (especially sodium) encourage soil alkalinity. The colloids are also the mechanism for soil buffering, which resists rapid and violent changes in soil reaction, giving stability to most plant–soil systems. Knowing how pH is controlled, how it influences the supply and availability of essential plant nutrients as well as toxic elements, and how it affects higher plants and human beings is truly significant, and is a goal worth seeking.

Study Questions

1. Describe the role of aluminum in enhancing soil acidity, identifying the ionic species involved, and the effect of these species on the CEC of soils.

2. If you could somehow extract the soil solution from a hectare–furrow slice of an acid soil (pH = 5), only a few kilograms of limestone would be needed to neutralize the soil solution. Yet under field conditions up to 4 tons of limestone may be required to bring the pH of the hectare–furrow slice of this soil to a pH of 6.5. How do you explain this difference?

3. What is the simplest means of determining soil pH and what are the principles involved?

4. Two acid soils have the same pH (5.0). It takes more than 4 tons of limestone to bring the pH of soil A to pH 7 and only 2 tons to do the same for an equal quantity of soil B. How do you account for this difference?

5. The iron analyses for an arid region soil showed an abundance of this element, yet a peach crop growing on the soil showed evidence of iron deficiency. What is a likely explanation?

6. An arid region soil, when it was first cleared for cropping, showed a pH of about 8.0. After a few years of irrigation, crop productivity declined, the soil aggregation tended to break down, and the pH went up to nearly 10. What is the likely explanation for this situation?

7. What physical and chemical treatment(s) would you suggest to bring the soil described in question 10 back to its original state of productivity?

8. In the process of "eradication" of salts from saline soils, where does the salt go when it is leached away?

9. Assume you find the pH of your soil before planting in the spring to be 5.0. You want to bring the pH to above 6.5 by the time the vegetable seeds are planted one month later. Which liming material would you likely choose and why?

10. A gardening "expert" in a weekly radio program recommended the application of low rates of ground dolomitic limestone at bimonthly intervals during the growing season to avoid raising the pH too rapidly. Do you agree with his/her advice? Why or why not?

11. Which of the following limestones will ultimately neutralize the most acidity? Limestone A: 32% CaO, 15% MgO; or limestone B: CaO_3 equivalent = 97%.

12. From the analyses of limestones A and B above, what can you say about the presence or absence of dolomite in each of these materials? Explain.

13. A neighbor complained when his azaleas were adversely affected by a generous application of limestone to the lawn immediately surrounding the azaleas. To what do you ascribe this difficulty? How would you remedy it?

14. A farmer has to choose between two finely ground calcitic limestones. Limestone A has a neutralizing power ($CaCO_3$ equivalent) of 88 and costs $14/Mg spread on the field. Limestone B has an analysis of 60% $CaCO_3$ and 30% $MgCO_3$ and costs $16/Mg spread on the field. Which is the better buy? Show calculations.

15. Following the application of 4 Mg/ha limestone to a given field, the soybeans grown on the field showed evidence of manganese deficiency. Explain why this occurred.

References

ADAMS, FRED (Ed.). 1984. *Soil Acidity and Liming*, 2nd ed. (Madison, WI: Amer. Soc. Agron.).

ADAMS, F., and Z. F. LUND. 1966. "Effect of chemical activity of soil solution aluminum on cotton root penetration of acid subsoils," *Soil Sci.*, **101**:193–98.

BALESDENT, J., G. H. WAGNES, and A. MARIOTTI. 1988. "Soil organic matter turnover in long-term field experiments as revealed by carbon-13 natural abundance," *Soil Sci. Soc. Amer. J.*, **52**:118–24.

BARBER, S. A. 1984. "Liming material and practices" in F. Adams (Ed.), *Soil Acidity and Liming*, 2nd ed. (Madison, WI: Amer. Soc. Agron.).

BRESLER, E., B. L. McNEAL, and D. L. CARTER. 1982. *Saline and Sodic Soils: Principles—Dynamics—Modeling* (Berlin: Springer-Verlag).

CARTER, D. L. 1981. "Salinity and plant productivity," in *CRC Handbook Series in Nutrition and Food* (Boca Raton, FL: CRC Press).

GARDNER, B. R., and R. L. ROTH. 1984. "Applying nitrogen in irrigation waters" in R. D. Hauck (Ed.), *Nitrogen in Crop Production* (Madison, WI: Amer. Soc. Agron.).

HARGROVE, W. L., and G. W. THOMAS. 1981. "Effect of organic matter on exchangeable aluminum and plant growth in acid soils," pp. 151–66 in *Chemistry in the Soil Environment*, ASA Special Publication No. 40 (Madison, WI: Amer. Soc. Agron. and Soil Sci. Soc. Amer.).

JENNY, H., et al. 1968. "Interplay of soil organic matter and soil fertility with state factors and soil properties," in *Organic Matter and Soil Fertility*, Pontificiae Academiae Scientiarum Scripta Varia 32 (New York: Wiley).

LATHWELL, D. J., and W. S. REID. 1984. "Crop response to lime in the Northeastern United States," in F. Adams (Ed.), *Soil Acidity and Liming*, 2nd ed. (Madison, WI: Amer, Soc. Agron.).

MEHLICH, A. 1964. "Influence of adsorbed hydroxyl and sulfate on neutralization of soil acidity," *Soil Sci. Soc. Amer. Proc.*, **28**:492–96.

NOBEL, A. D., M. E. SUMMER, and A. K. ALVA. 1988. "The pH dependency of aluminum phyto-toxicity alleviation by calcium sulfate," *Soil Sci. Soc. Amer. J.*, **52**:1398–1402.

PEECH, M. 1961. "Lime requirements vs. soil pH curves for soils of New York State," mimeographed (Ithaca, NY: Agronomy, Cornell University).

RHODES, J. D., and D. L. CORWIN. 1984. "Monitoring soil salinity," *J. Soil and Water Cons.*, **39**:172–75.

RICHARDS, L. A. (Ed.). 1947. *Diagnosis and Improvement of Saline and Alkali Soils* (Riverside, CA: U.S. Regional Salinity Lab.).

ROBERTSON, W. K., M. C. LUTRICK, and T. L. YUAN. 1982. "Heavy applications of liquid-digested sludge on three Ultisols: I. Effects on soil chemistry," *J. Environ. Qual.*, **11**:278–82.

SCHOLLENBERGER, C. J., and R. M. SALTER. 1943. "A chart for evaluating agricultural limestone," *J. Amer. Soc. Agron.*, **35**:995–96.

Organisms of the Soil

Plant and animal residues are continually being deposited on the surface of the soil. This is most noticeable in temperate regions as trees, shrubs, and grasses drop their leaves at the end of growing season. Each fall the soil of temperate regions is covered with these leaves to a depth of 15 cm or more. Yet, by the following summer, most of these residues seem to have disappeared. The soil surface may show a few leaf petiole fragments, but little evidence of the mass of material that was deposited the previous fall.

More careful examination of the underlying soil layers, including sophisticated soil analyses, will reveal that much of the residues of previous years has been incorporated into the soil while a portion remains as soil humus. Actually most of the organic residues have been devoured by soil organisms,[1] seen and unseen. In the process, the residues have been decomposed, yielding carbon dioxide and water, or have been degraded and synthesized into compounds that comprise humus. Also, essential nutrient elements held in the organic residues have been released in forms that are available for growing plants. The overall changes that have occurred are of vital importance to all living creatures.

This chapter deals with those soil organisms (animal and plant) that are responsible for the degradation and synthesis of organic materials in soils. We will be concerned about their activities rather than their scientific classification. Consequently only very broad and simple categories of organisms will be considered (Figure 9.1). Most soil organisms belong to plant life (flora). Yet the role of animals (fauna) is not to be minimized, especially in the early stages of organic matter breakdown.

Most soil organisms, plant and animal, are so small that they can be seen only with the aid of a microscope (microorganisms). But the activities of some larger organisms (macroorganisms), ranging in size from that of the larger rodents to ants and snails, have significant effects on soil physical properties. All are of significance in the soil biological processes that are so critical for plant and animal life.

[1] For a review of soil microbiology, see Paul and Clark (1989).

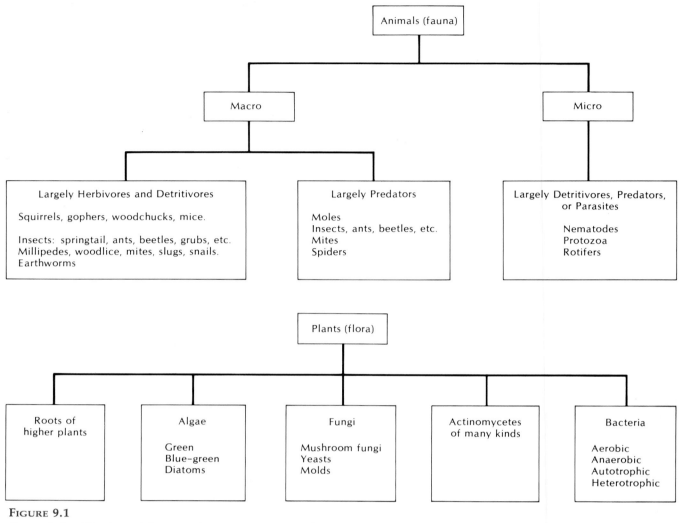

FIGURE 9.1

General classification of the more important groups of organisms commonly present in soils. Some animals subsist on plants (herbivores and detritivores), some consume other animals (predators), and some live off plants or other animals but do not consume them (parasites).

9.1

Organisms in Action

The activities of soil flora and fauna are intimately related. The relationships are shown in Figure 9.2, which illustrates how various soil organisms are involved in the degradation of higher plant tissue. First, note that live plants are subject to attack by some soil organisms known as *herbivores*. Examples are parasitic nematodes and insect larvae that attack plant roots; soil-borne termites, ants, and beetle larvae; and animals like woodchucks and mice that devour aboveground plant parts.

PRIMARY CONSUMERS

As soon as a leaf, a stalk, or a piece of bark drops to the ground, it is subject to coordinated attack by microflora and by *detritivores*, animals that live on

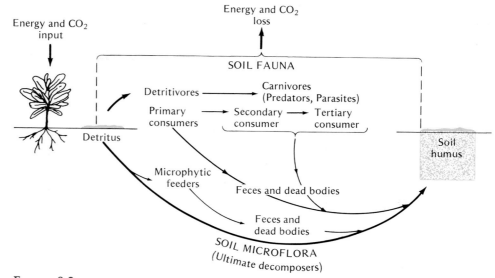

FIGURE 9.2

Diagram of the general pathway for the breakdown of higher plant tissue. Because they capture energy and carbon dioxide, the higher plants are known as *primary producers*. When the debris from dead plants (detritus) falls to the soil surface, it is attacked by the soil fauna and microflora, which are among the primary, secondary, or tertiary *consumers*. These organisms release energy and carbon dioxide and produce humus. Note that 80–90% of the total soil metabolism is due to the microflora.

dead and decaying plant tissues (Figure 9.2). These animals, which include mites, woodlice, and earthworms chew or tear holes in the tissue, opening it up to more rapid attack by the microflora. Together with microflora, these plant-eating animals (herbivores) are termed *primary consumers*.

While the actions of the microflora are mostly chemical, those of the fauna are both physical and chemical. The animals chew the plant parts and move them from one place to another on the soil surface and even into the soil. Earthworms incorporate the plant residues into the mineral soil by passing them through their bodies along with mineral soil particles. They literally eat their way through the soil. Larger animals, such as gophers, moles, prairie dogs, and rats, also burrow into the soil and bring about considerable soil mixing and granulation.

SECONDARY AND TERTIARY CONSUMERS

The primary consumers are themselves food sources for the predators and parasites that exist in the soil. Predators and some parasites are *secondary consumers*; the group includes small forms of plant life such as bacteria, fungi, algae, and lichens. Examples of carnivores are centipedes, which consume other animals such as small insects, spiders, nematodes, and snails, and certain moles that feed primarily on earthworms. Examples of microphytic feeders, which use the microflora as sources of food, are some termites, nematodes, and protozoa (Figure 9.2).

Moving farther up the food chain, the secondary consumers are prey for still other carnivores, called *tertiary consumers*. For example, ants consume centipedes, spiders, mites, and scorpions, which themselves can prey on primary or other secondary consumers. Again, the microflora are intimately involved in the decomposition of organic material associated with the fauna.

In addition to their direct attack on plant tissue, they are active within the digestive tracts of some animals. The microflora also attack the finely shredded organic material in animal feces and later decompose the bodies of dead animals. For this reason they are referred to as the *ultimate decomposers* (see Figure 9.2).

9.2

Organism Numbers, Biomass, and Metabolic Activity

Soil organism numbers are influenced by many factors, including climate, vegetation, and the physical and chemical characteristics of soils. The species composition in an arid desert will certainly be different from that in a humid forest area which, in turn, would be quite different from that in a cultivated field. Acid soils are populated by quite different species from those in alkaline soils. Likewise, species diversification and numbers are different in a tropical rainforest from those in a cool temperate area.

Nonetheless there are a few generalizations that can be made about soil organisms. For example, forested areas support a more diverse fauna than do grasslands, although the total weight of organisms and fauna activity per hectare are higher in the grasslands (see Table 9.1). Cultivated fields are generally lower than undisturbed native lands in numbers and biomass of soil organisms, especially the fauna.

COMPARATIVE ORGANISM ACTIVITY

The activities of specific groups of soil organisms are commonly identified by (a) their numbers in the soil, (b) their weight per unit volume or area of soil (biomass), and (c) their metabolic activity. The numbers and biomass of groups of organisms that commonly occur in soils are shown in Table 9.2.

TABLE 9.1

Biomass of Groups of Soil Animals Under Grassland and Forest Cover

The mass and in turn the metabolism are greatest under grasslands. The spruce, with low-base-containing leaves, encourages acid conditions and slow organic matter decomposition.

Group of organisms	Grassland meadow	Forest Oak	Forest Spruce
Herbivores	17.4	11.2	11.3
Detritivores			
Large	137.5	66.0	1.0
Small	25.0	1.8	1.6
Predators	9.6	0.9	1.2
Total	189.5	79.9	15.1

Biomass (g/m^2)[a]

Data from Macfadyen (1963).

[a] Depth about 15 cm.

TABLE 9.2

Relative Number and Biomass of Soil Flora and Fauna Commonly Found in Surface Soils[a]

Since the metabolic activity is generally related to the biomass, it is obvious that the microflora and earthworms dominate the life of most soils.

Organisms	Number per square meter	Number per gram	Biomass[b] (kg/HFS)	Biomass[b] (lb/AFS)
Microflora				
Bacteria	10^{13}–10^{14}	10^8–10^9	450–4500	400–4000
Actinomycetes	10^{12}–10^{13}	10^7–10^8	450–4500	400–4000
Fungi	10^{10}–10^{11}	10^5–10^6	1,120–11,200	800–8000
Algae	10^9–10^{10}	10^4–10^5	56–560	50–500
Microfauna				
Protozoa	10^9–10^{10}	10^4–10^5	17–170	15–150
Nematoda	10^6–10^7	10–10^2	11–110	10–100
Other fauna	10^3–10^5		17–170	15–150
Earthworms	30–300		110–1100	100–1000

[a] Generally considered 15 cm (6 in.) deep, but in some cases (e.g., earthworms) a greater depth is used.

[b] The biomass values are on a live-weight basis. Dry weights are about 20–25% of these values. HFS = hectare-furrow slice, AFS = acre-furrow slice.

Although the relative metabolic activities are not shown, they are generally related to the biomass of the organisms.

As might be expected, the microorganisms are the most numerous and have the highest biomass. Together with the earthworms, the microflora also monopolize the biological activity in soils. It is estimated that 60–80% of the total soil metabolism is due to the microflora. For these reasons, major attention will be given to the microflora along with earthworms, protozoa, and other microanimals.

Before leaving the macro soil fauna, however, recognition should be given to their importance in soil formation and management. Rodents pulverize, mix, and granulate soil and incorporate organic materials into lower horizons. They provide large channels through which water and air can move freely. The medium-sized detritivores translocate and partially digest organic residues and leave their excrement for microfloral degradation. By living in the soil, many animals favorably affect its physical condition. Others use the soil as a habitat for destructive action against higher plants. In any case, all these organisms significantly impact both the soil and growing plants.

SOURCE OF ENERGY AND CARBON

Soil organisms may be classified on the basis of their source of energy and carbon as either *autotrophic* or *heterotrophic*. The heterotrophic soil organisms obtain their energy and carbon from the breakdown of organic soil materials. These organisms are far more numerous than the autotrophs and are responsible for most of the general-purpose decay. They include the soil fauna, most bacteria, and the fungi and actinomycetes.

The autotrophs obtain their energy from sources other than the breakdown of organic materials, for example, from solar energy (photoautotrophs) and from the oxidation of inorganic elements such as nitrogen, sulfur, and iron (chemautotrophs).

FIGURE 9.3
Earthworms are perhaps the most signficant soil macroorganism in humid temperate systems, especially in relation to their effect on the physical condition of soils.

9.3

Earthworms[2]

Earthworms are probably the most important soil macroanimals (Figure 9.3). Of the more than 1800 species known worldwide, *Lumbricus terrestris*, a deep-boring reddish organism, and *Allolobophora caliginosa*, a shallow-boring pale pink organism with a length up to 25 cm, are the two most common in Europe and in the eastern and central United States. In the tropics and semitropics still other types are prevalent, some small and others surprisingly large, up to 3 m long.

It is rather interesting that *Lumbricus terrestris* is not native to America. As forests and prairies were put under cultivation, this European worm rapidly replaced the native types, which could not withstand the change in soil environment brought about by tilling the soil. Virgin lands, however, still retain at least part of their native populations.

INFLUENCE ON SOIL FERTILITY AND PRODUCTIVITY

Earthworms are important in many ways, especially in the upper 15–35 cm of soil. They ingest organic matter as well as soil. As these materials pass through the earthworm's body, they are subjected to digestive enzymes as well as to a grinding action within the animal. The weight of the material passing through their bodies (casts) each day may equal the weight of the earthworm. In the tropics, as much as 250 Mg/ha (110 tons/acre) of casts may be produced annually. Although figures only one tenth of those values are more common in the cultivated soils of temperate regions, the casts are evidence of extensive earthworm activities.

Compared to the soil itself, the casts are definitely higher in bacteria, organic matter, and available plant nutrients (Table 9.3). The rank growth of grass around earthworm casts is evidence of the favorable effect of earthworms on soil productivity.

[2] For an excellent review of earthworms, see Lee (1985).

TABLE 9.3

Comparative Characteristics of Earthworm Casts and Soils

Average of Six Nigerian Soils.

Characteristic	Earthworm casts	Soils
Silt and clay (%)	38.8	22.2
Bulk density (Mg/m^3)	1.11	1.28
Structural stability[a]	849	65
Cation exchange capacity (cmol/kg)	13.8	3.5
Exchangeable Ca^{2+} (cmol/kg)	8.9	2.0
Exchangeable K^+ (cmol/kg)	0.6	0.2
Soluble P (ppm)	17.8	6.1
Total N (%)	0.33	0.12

From de Vleeschauwer and Lal (1981).

[a] Numbers of raindrops required to destroy structural aggregates.

OTHER EFFECTS

Earthworms are important in other ways. The holes left in the soil serve to increase aeration and drainage, an important consideration in crop production and soil development. Moreover, the worms mix and granulate the soil by dragging into their burrows quantities of undecomposed organic matter such as leaves and grass, which they use as food. In some cases the accumulation is surprisingly large. This action is more important in uncultivated soils (including those where reduced tillage is practiced) than in plowed land, where organic matter may be turned under in quantity. Without a doubt, earthworms increase both the size and stability of the soil aggregates.

FACTORS AFFECTING EARTHWORM ACTIVITY

Earthworms prefer a well-aerated but moist habitat. Therefore they are found mostly in medium-textured upland soils where the moisture capacity is high rather than in droughty sands or poorly drained lowlands. Earthworms must have organic matter as a source of food. Consequently, they thrive where farm manure or plant residues have been added to the soil. A few species are reasonably tolerant of low pH, but most earthworms thrive best where the soil is not too acid.

Soil temperature affects earthworm numbers and their distribution in the soil profile. For example, a temperature of about 10 °C (50 °F) appears optimum for *Lumbricus terrestris*. The temperature sensitivity and soil moisture requirements probably account for the maximum earthworm activity noted in spring and autumn in temperate regions.

Some earthworms burrow deeply into the profile, thereby avoiding unfavorable moisture and temperature conditions. Penetration as deep as 1–2 m is not uncommon. Unfortunately, in soils not insulated by residue mulches, a sudden heavy frost in the fall may kill many earthworms before they can move lower in the profile. Soil cover is important in maintaining a high earthworm population where sudden frosts are common.

Because of their sensitivity to soil and other environmental factors, the numbers of earthworms vary widely in different soils. In very acid forest soils (Spodosols), an average of fewer than one organism per square meter is common. In contrast, more than 500 per square meter have been found on rich grassland soils (Mollisols). The numbers commonly found in arable soils range from 30 to 300 per square meter to a depth of 15 cm (Table 9.2), equivalent to from 300,000 to 3 million per hectare–furrow slice. The biomass or live weight for this number would range from perhaps 110 to 1100 kg/ha (100–1000 lb/acre).

9.4

Termites[3]

Termites, or "white ants," are major contributors to the breakdown of organic material in or at the surface of soils. There are about 2000 species of termites. They are found in about two thirds of the land areas of the world but are most prominent in the grasslands (savannahs) and forests of tropical and subtropical areas (both humid and arid). In tropical and subtropical areas termites supplement and even surpass the activities of earthworms.

Termite mud nests or mounds characterize open savannah areas in Africa, Latin America, Australia, and Asia (Figure 9.4). These insects have a very complex social life in these honeycombed mounds, which serve essentially as their "cities." In building these "cities," termites transport soil from lower layers to and above the surface soil level, thereby bringing about extensive mixing of soil materials and the plant residues they use as food.

The quantities of materials termites deposit often compare favorably with the surface castings of earthworms, amounting to tens or even hundreds of metric tons per hectare at any one time. Their mounds may be 6 m or more in height and may extend to an even greater depth into the soil in some desert areas. Each mound provides a home for 1 million or more termites. The mounds are abandoned after 10–20 years and can be broken down if the land is to be used for crop production.

[3] For an excellent discussion of termites and ants in the tropics, see Lal (1987).

FIGURE 9.4

Termite mounds in cultivated fields in Africa. A high mound in a cornfield (left). Dissected profile of another mound showing the size and depth of the underground matrix (right). [Photos courtesy Dr. R. Lal, The Ohio State University]

The effect of termites on soil productivity is generally less beneficial than that of earthworms because the digestive processes of termites, aided by microorganisms in their gut, are generally more efficient than those of earthworms. Termite deposits commonly have a lower organic matter content than the surrounding undisturbed topsoil. This is partly due to the fact that these deposits are comprised largely of low organic subsoil material brought to the surface as the mounds were being formed.

Crop growth in soil in areas where termite mounds have existed is often poor, not only because of low nutrient content in the surface layers but also because of the greater compactness of some of the mound material, the particles of which were cemented together by the termites as the mound was constructed.

Termites are significant factors in the formation of soils of tropical and subtropical areas. They also have both positive and negative effects on current land use in these areas. Termites accelerate the decay of dead trees and grasses but also disrupt crop production by the rapid development of their nests or mounds.

9.5

Ants

Although ants have less widespread influence on soil properties than earthworms and termites, locally they have notable effects. For example, some ant species are known for their exceptional ability to break down woody materials. These organisms, known for their highly organized societies, are especially active in some tropical areas of South America and Africa, both forest and arid grasslands. Some ants produce conspicuous mounds while others have underground nests. There is considerable turnover in the soil associated with the mounds and nests because subsoil is brought to the surface. Even so, probably no more than about 1% of the soil areas, even in the tropics, are affected directly by ants.

9.6

Soil Microanimals

Of the abundant microscopic animal life in soils, three groups are of some importance: nematodes, protozoa, and rotifers.

Nematodes

Nematodes—commonly called threadworms or eelworms—are found in almost all soils, often in surprisingly large numbers (Table 9.2). In size, they are almost wholly microscopic, seldom being large enough to be seen with the naked eye (Figure 9.5).

FIGURE 9.5
A nematode commonly found in soil (magnified about 120 times). More than 1000 species of soil nematodes are known. Most nematodes feed on bacteria and fungi, but plant parasitic species can be very detrimental to plant growth. [Courtesy William F. Mai, Cornell University]

The most numerous and varied of the nematodes are those that live on decaying organic matter (saprophytes) or are predatory on other nematodes, bacteria, algae, protozoa, and the like. But some nematodes, especially those of the genus *Heterodera*, can infest the roots of practically all plant species. Heavy infestations result in serious damage, especially to vegetable crops. Nematode-control measures are essential in these cases.

Protozoa

Protozoa are probably the simplest form of animal life and are the most varied and numerous of microanimals in soils (Table 9.2). Although one-celled organisms, they are considerably larger than bacteria, having a diameter range from less than 5 to greater than 100 μm, and are of a distinctly more complex organization. Soil protozoa include amoeba, ciliates (Figure 9.6a), and flagellates.

More than 250 species have been isolated in soils; sometimes as many as 40 or 50 of such groups occur in a single sample of soil. The live weight of protozoa in soils ranges from 15 to 175 kg per hectare–furrow slice (Table 9.2). A considerable number of serious animal and human diseases are attributed to protozoan infections.

Protozoa generally thrive best in moist well-drained soils and are most numerous in surface horizons. Some protozoa are predators on soil bacteria and other microflora, especially in the area immediately around the plant roots (the rhizosphere). Overall, however, the protozoa generally are not sufficiently abundant in soils to be a major factor in organic matter decay and nutrient release.

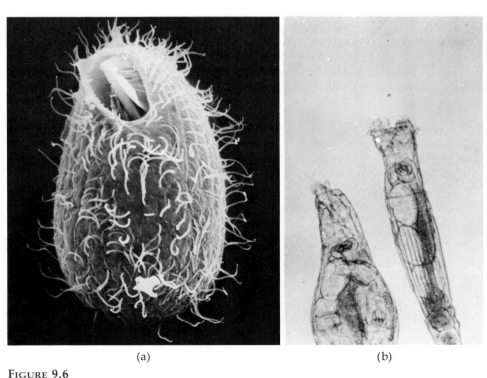

(a)   (b)

Figure 9.6

Two microanimals typical of those found in soils. (a) Scanning electron micrograph of a ciliated protozoan, *Glaucoma scintillaus*. (b) Photomicrograph of two species of rotifer, *Rotaria rotatoria* (stubby one at left) and *Philodina acuticornus* (slender one at right). [(a) Courtesy J. O. Corliss, University of Maryland; (b) courtesy F. A. Lewis and M. A. Stirewalt, Biomedical Research Institute, Rockville, MD]

ROTIFERS

This third group of soil microanimals (Figure 9.6b), of which about 100 species have been described, thrives under moist conditions, especially in swampy land where their numbers may be great. Just how important rotifers are in soils is unknown, but their activities are likely confined largely to peat bogs and wet areas of mineral soils.

9.7

Roots of Higher Plants

Higher plants are the primary producers of organic matter and storers of the sun's energy (Figure 9.2). Their roots grow and die in the soil and in so doing supply the soil fauna and microflora with food and energy. In addition, the living roots physically modify the soils as they push through cracks and make new openings of their own (Figure 9.7). Tiny initial channels are increased in size as the roots swell and grow. By removing moisture from the soil, plant roots bring about further physical stresses that stimulate soil aggregation. The roots also support myriad microorganisms with their exudates, which further stabilize these aggregates. As they later decompose,

(a)

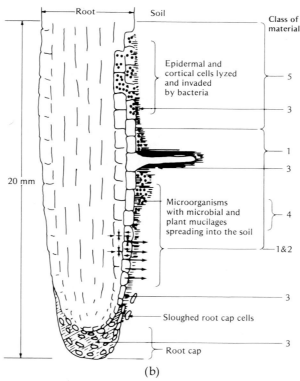

(b)

1. Simple *exudates*, which leak from plant cells to soil.
2. *Secretions*, simple compounds released by metabolic processes.
3. Plant *mucilages*, more complex organic compounds originating in root cells or from bacterial degradation.
4. *Mucigel*, a gelatinous layer composed of mucilages and soil particles intermixed.
5. *Lyzates*, compounds released through digestion of cells by bacteria.

FIGURE 9.7
(a) Photograph of a root tip illustrating how roots penetrate soil and emphasizing the root cells through which nutrients and water move into and up the plant. (b) Diagram of a root showing the origins of organic materials in the rhizosphere. [(a) From Chino (1976), used with permission of Japanese Society of Soil Science and Plant Nutrition, Tokyo; (b) redrawn from Rovira et al. (1979), used with permission of Academic Press, London]

the roots provide materials for the creation of humus not only in the top few inches but also to greater soil depths.

AMOUNT OF ORGANIC TISSUE ADDED

At crop harvest time, the mass of roots remaining in the soil is commonly 15–40% that of the aboveground crop. If an average figure of 25% is used, good crops of oats, corn, and sugarcane would be expected to leave about 2500, 4500, and 8500 kg/ha of root residues, respectively. The importance of root residues in helping to maintain soil organic matter in arable soils is often overlooked.

RHIZOSPHERE

Living roots of higher plants also affect the nutrition of soil microbes. Roots withdraw soluble nutrients from the soil solution directly but also release substances that influence nutrient availability. Significant quantities of organic compounds are exuded, secreted, or otherwise released at the surface of young roots (Figure 9.7). Organic acids so excreted can solubilize plant nutrients. Amino acids and other simple carbon-containing compounds stimulate microflora in the *rhizosphere* or zone immediately surrounding these young roots. The number of organisms in the rhizosphere may be as much as 100 times greater than elsewhere in the soil (Figure 9.8). The absorptive surfaces of the root hairs lie within the zone of this increased nutrient availability.

Also surrounding young roots are organic compounds called mucilages, which are released or sloughed off from plant roots. These compounds, when intermixed with clay particles, form a gelatinous layer called *mucigel*, which surrounds the roots (Figure 9.7). The mucigel layer may facilitate contact with the surrounding soil, especially during periods of moisture stress when the root proper may shrink in size and lose direct contact with the soil (see Figure 5.23).

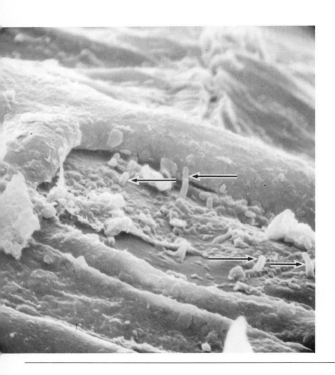

FIGURE 9.8
Soil microorganisms are found closely associated with the rhizosphere of wheat roots. Note the bacteria colonies (arrows) on the root surface in this scanning electron micrograph (magnified 3900 times). [Courtesy L. F. Elliott, USDA Agricultural Research Service, Pullman, WA]

The chemical and physical characteristics of compounds in the rhizosphere are partially responsible for plant roots being classified as soil organisms. The influence of roots on soil properties justifies their classification.

9.8

Soil Algae

Most algae are chlorophyll-bearing organisms and, like higher plants, are capable of performing photosynthesis. To do this, they need light and therefore are found mostly very near the surface of the soil. Algae perform best under moist to wet conditions.

Several hundred species of algae have been isolated from soils, the most prominent being the same throughout the world. Soil algae are divided into four general groups: blue-green, green, yellow-green, and diatoms. The green algae are most evident in nonflooded soils, especially if the pH is low. Grassland seems especially favorable for the blue-green forms, whereas in old gardens diatoms often are numerous. Both the green and the blue-green algae outnumber the diatoms.

A common range in algal population is 1–10 billion per square meter 15 cm deep (10,000–100,000 per gram). The mass of live algae in soils may range from 50–600 kg per hectare–furrow slice (Table 9.2).

Blue-green algae are especially numerous in rice soils, and when such lands are flooded, appreciable amounts of atmospheric nitrogen are fixed or changed to a combined form by these organisms. Blue-green algae, growing within the leaves of the aquatic fern Azolla, are also able to fix nitrogen in quantities significant for rice production (see Section 11.16).

9.9

Soil Fungi

Although the influence of fungi on soil is by no means entirely understood, they are known to play a very important part in the transformations of soil constituents. Over 690 species have been identified in soils, representing 170 genera. Like the bacteria and actinomycetes, fungi contain no chlorophyll. Their biomass in soils will generally range from 1000 to 10,000 kg per hectare-furrow slice (Table 9.2). Consequently, fungi must depend for their energy and carbon on the organic matter of the soil.

For convenience of discussion, fungi may be divided into three groups: yeasts, molds, and mushroom fungi. Only the last two are considered important in soils; yeasts are rare in a soil habitat.

MOLDS

The molds (Figure 9.9) are distinctly filamentous, microscopic, or semimacroscopic fungi, and they play an infinitely more important role in the soil than the mushroom fungi, approaching or even exceeding at times the influence of bacteria. Molds will develop vigorously in acid, neutral, or alkaline soils, some being favored, rather than harmed, by lowered pH (see Figure 8.12). Consequently, molds are noticeably abundant in acid surface soils, where bacteria and actinomycetes offer only mild competition. The low-pH tolerance of molds is especially important in decomposing the organic residues in acid forest soils.

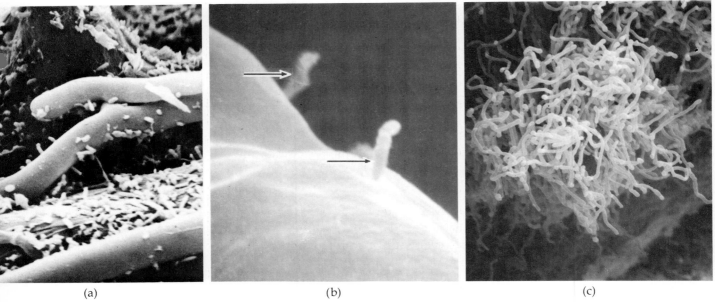

(a)   (b)   (c)

FIGURE 9.9
Scanning electron micrographs of fungi, bacteria, and actinomycetes. (a) Fungal hyphae associated with much smaller rod-shaped bacteria. (b) Rod-shaped bacteria attached to a plant root hair. (c) Actinomycete threads. [(a) Courtesy R. Campbell, University of Bristol, used with permission of American Phytopathological Society; (b, c) courtesy Maureen Petersen, University of Florida]

Many genera of molds are found in soils, four of the most common being *Penicillium*, *Mucor*, *Fusarium*, and *Aspergillus*. All of the common species occur in most soils, conditions determining which are dominant. The complexity of the organic compounds being attacked seems to determine the particular mold or molds that prevail. Their numbers fluctuate greatly with soil conditions; perhaps 100,000 to 1 million individuals per gram of dry soil (10–100 billion per square meter 15 cm deep) represents a more or less normal range in population.

MUSHROOM FUNGI

Mushroom fungi are found in forests and grasslands where there are ample moisture and organic residues. Some of these mushrooms are edible, and their production is "domesticated." They are grown in caves and specially designed houses, with composted organic materials and manures providing their source of food.

The aboveground fruiting body of most mushrooms is only a small part of the total organism. An extensive mass of fine filaments (hyphae) is found in the underlying soil or organic residue. Although mushrooms are not as widely distributed as the molds, these fungi are of significance locally, especially in the breakdown of woody tissue.

ACTIVITIES OF FUNGI

In their ability to decompose organic residues, fungi are the most versatile and perhaps the most persistent of any group. Cellulose, starch, gums, and lignin, as well as the more easily affected proteins and sugars, succumb to fungal attack. In affecting the processes of humus formation and aggregate stabilization, molds are more important than bacteria. They are especially active in acid forest soils but play a significant role in all soils.

Fungi function more efficiently than bacteria in that they transform into their tissues a larger proportion of decaying plant residues. Up to 50% of the substances decomposed by molds may become organism tissue, compared to about 20% for bacteria. Moreover, soil fertility depends in no small degree on molds, since they continue to decompose complex organic materials after bacteria and actinomycetes have essentially ceased to function.

MYCORRHIZAE[4]

An economically important, mutually beneficial (symbiotic) association between numerous fungi and the roots of higher plants is called *mycorrhizae*, a term meaning "fungus root." This association, first noted on certain forest tree species, now is known to be widespread and to affect most plant species, including many agronomic crops. The association is of great practical significance because it markedly increases the availability to plants of several essential nutrients, especially from infertile soils. Apparently, the symbiotic association provides the fungi with sugars and other organic exudates for use as food. In return, the fungi provide an enhanced availability of several essential nutrients, including phosphorus, zinc, copper, calcium, magnesium, manganese, and iron.

There are two types of mycorrhizal associations of considerable practical importance, *ectomycorrhiza* and *endomycorrhiza*. The ectomycorrhiza group includes hundreds of different fungal species associated primarily with trees, such as pine, birch, hemlock, beech, oak, spruce, and fir. These fungi, stimulated by root exudates, cover the surface of feeder roots with a fungal mantle. The hyphae of the fungi penetrate the roots and develop around the cells of the cortex but do not penetrate these cell walls (Figure 9.10).

The endomycorrhiza group, the most important of which are called *vesicular arbuscular* (VA) *mycorrhizae*, penetrates the root cell walls, enters the root cells, and forms hyphal masses within the cells. This group is perhaps the most common and most widespread of mycorrhizae, with some 89 identified species of fungi in soils from the tropics to the Arctic forming VA mycorrhizal associations. The roots of most agronomic crops, including corn, cotton, wheat, potatoes, beans, alfalfa, sugarcane, cassava, and dryland rice, have VA mycorrhizal associations as do most fruits and vegetables such as apples, grapes, and citrus. Many trees, including maple, yellow poplar, redwood and such important tree crops as apple, cacao, coffee, and rubber also have VA mycorrhizae.

The root cortical cell walls of host plants are penetrated by the hyphae of VA mycorrhizae. Inside the plant cell highly branched, small structures known as *arbuscules* are formed by the fungi. These structures are considered to be the sites of transfer of mineral nutrients from the fungi to the host plants (Figure 9.10). Other structures, called *vesicles*, serve as storage organs for the plant nutrients and other products.

The increase in nutrient availability provided by mycorrhizae is thought to result from the nutrient-absorbing surface provided by the fine filamentous hyphae of the fungi. The surface area of mycorrhizal infiltrated roots has been calculated to be as much as 10 times that of the uninfested roots. Also, the soil volume from which nutrients are absorbed is greater for mycorrhizal associations. The very fine fungal hyphae extend up to 8 cm

[4] For excellent presentations on mycorrhizae, see Menge (1981) and Harley and Smith (1983).

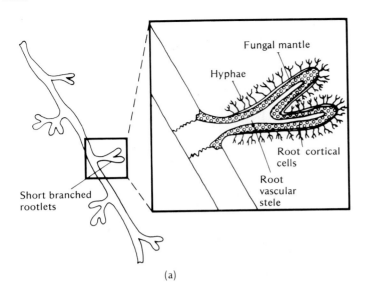

(a)

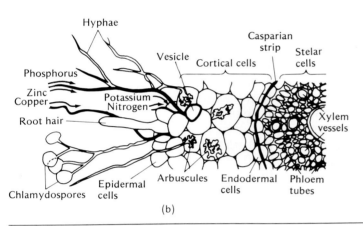

(b)

FIGURE 9.10
Diagram of ectomycorrhiza and vesicular arbuscular (VA) mycorrhiza association with plant roots. (a) The ectomycorrhiza fungi association produces short branched rootlets that are covered with a fungal mantle, the hyphae of which extend out into the soil and between the plant cells but do not penetrate the cells. (b) In contrast, the VA mycorrhizae penetrate not only between cells but into certain cells as well. Within these cells, the fungi form structures known as *arbuscules* and *vesicles*. The former transfer nutrients to the plant, and the latter store these nutrients. In both types of association, the host plant provides sugars and other food for the fungi and receives in return essential mineral nutrients that the fungi absorb from the soil. [Redrawn from Menge (1981)]

into soil surrounding the roots, thereby increasing the absorption of nutrients, such as phosphorus, zinc and copper, which do not diffuse readily to the roots (Table 9.4). During drought, water stress in mycorrhizal-infested plants is less than in uninfested plants. Also, VA mycorrhizae are thought to be responsible for the transfer of phosphorus from one plant to nearby

TABLE 9.4
Effect of Inoculation with Mycorrhiza and of Added Phosphorus on the Content of Different Elements in the Shoots of Corn

Expressed in micrograms per plant.

Element in plant	No phosphorus		25 mg/kg phosphorus added	
	No mycorrhiza	Mycorrhiza	No mycorrhiza	Mycorrhiza
P	750	1,340	2,970	5,910
K	6,000	9,700	17,500	19,900
Ca	1,200	1,600	2,700	3,500
Mg	430	630	990	1,750
Zn	28	95	48	169
Cu	7	14	12	30
Mn	72	101	159	238
Fe	80	147	161	277

From Lambert et al. (1979).

(a)   (b)   (c)

FIGURE 9.11

The effect of mycorrhizae on availability of phosphorus to a pasture legume, *Pueraria phaseoloides*: (a) no treatment, (b) rock phosphate, (c) rock phosphate plus mycorrhizae. [Courtesy Dr. Fritz Kramer, CIAT, Cali, Colombia]

neighboring species, especially in grasslands. Research findings of recent years have brought to light the significant role mycorrhizae play in helping plants absorb nutrients from relatively infertile soils (Figure 9.11).

9.10

Soil Actinomycetes

Actinomycetes resemble molds in that they are filamentous, often profusely branched. Their mycelial threads are smaller, however, than those of fungi. Actinomycetes are similar to bacteria in that they are unicellular and of about the same diameter. They break up into spores that closely resemble bacteria. Although historically they were often classified with the fungi, actinomycetes, like bacteria, have no nuclear membrane. Consequently, they are sometimes called *thread* or *filamentous bacteria* (Figure 9.9).

Actinomycetes develop best in moist, well-aerated soil. However, in times of drought, they remain active to a degree not usually exhibited by either bacteria or molds. Actinomycetes are generally rather sensitive to acid soil conditions, their optimum development occurring at pH values between 6.0 and 7.5 (see Figure 8.12). This sensitivity to soil reaction is used to control potato scab, an actinomycete disease of wide distribution. By lowering the soil pH with sulfur, the disease can be controlled (see Section 8.9).

NUMBERS AND ACTIVITIES OF ACTINOMYCETES

Actinomycete numbers in soil exceed those of all organisms except bacteria, their numbers sometimes reaching hundreds of millions, about one tenth the population of bacteria. In actual live weight, actinomycetes often exceed bacteria, a biomass of more than 4500 kg per hectare–furrow slice being present in some soils. These organisms are especially numerous in soils high in humus, such as old meadows or pastures, where the acidity is not too great. The aroma of freshly plowed land so noticeable at certain times of the year is probably owing to actinomycetes as well as to certain molds.

Actinomycetes undoubtedly are of great importance in the decomposition of soil organic matter and the liberation of its nutrients. Apparently, they reduce to simpler forms even the more resistant compounds, such as cellulose, chitin, and phospholipids. An abundance of actinomycetes in soils long under sod is an indication of their ability to attack complex compounds.

9.11

Soil Bacteria

CHARACTERISTICS

Bacteria are single-cell organisms, one of the simplest and smallest forms of life known. Their ability to increase rapidly in numbers is extremely important in soils. Their capacity for rapid reproduction allows bacteria to adjust their activities quickly in response to changes in their environment.

Bacteria are very small; the larger individuals seldom exceed 4–5 μm (0.004–0.005 mm) in length, and the smaller ones approach the size of an average clay particle. The shape of bacteria varies in that they may be nearly round, rod-like, or spiral. In the soil, the rod-shaped organisms seem to predominate (Figure 9.9b).

BACTERIAL POPULATIONS IN SOILS

As with other soil organisms, the numbers of bacteria are extremely variable. But the numbers are high, ranging from a few billion to 3 trillion organisms in each kilogram of soil. A biomass of a few hundred kilograms to 5 Mg (metric tons) live weight in a hectare–furrow slice of fertile soil is commonly encountered (2.2 tons/acre–furrow slice) (see Table 9.2).

In the soil, bacteria exist as mats, clumps, and filaments called *colonies*. Many soil bacteria in addition to their vegetative (growing) phase are able to produce spores or similar resistant bodies (a resting stage). This latter capability is important as it allows the organisms more readily to survive unfavorable conditions.

SOURCE OF ENERGY

Soil bacteria are either *autotrophic* or *heterotrophic* (see Section 9.2). The autotrophs obtain their energy from the oxidation of inorganic substances such as ammonium, sulfur, and iron and most of their carbon from carbon dioxide. They are not very numerous but play vital roles in controlling nutrient availability to higher plants. Most soil bacteria are heterotrophic— that is, both their energy and their carbon come directly from organic matter. Heterotrophic bacteria, along with fungi and actinomycetes, account for the general breakdown of organic matter in soils.

IMPORTANCE OF BACTERIA

Bacteria as a group, almost without exception, participate vigorously in all of the organic transactions so vital if a soil is to support higher plants. They not only rival but often exceed both fungi and actinomycetes in this regard.

Bacteria also hold near monopolies in several basic enzymatic transformations. For example, different autotrophic bacteria have the ability to oxidize (or reduce) selected chemical elements in soils. Thus, through *nitrogen oxidation (nitrification)*, selected bacteria oxidize ammonium compounds to the nitrite form, and others complete the oxidation to the nitrate form. Because nitrate is the form preferentially absorbed by most plants, this oxidation has implications for plant nutrition. Other bacteria are responsible for *sulfur oxidation*, which yields sulfate ions that can be readily taken up by plants. Also, bacterial oxidation and reduction of inorganic ions such as iron and manganese not only influence the availability of these elements, but also help determine the color of the soils.

A second critical process that depends largely on bacteria is *nitrogen fixation*—the biochemical combination of atmospheric nitrogen with hydrogen to form ammonia which is then incorporated into organic nitrogen compounds usable by higher plants. The process can be carried out by bacteria in soils independent of plants, but the amount of nitrogen fixed is much greater if the bacteria are intimately associated with plant roots, much as are the mycorrhizal fungi (see Section 9.9).

9.12

Conditions Affecting Growth of Soil Bacteria

Many conditions of the soil affect the growth of bacteria. Among the most important are the supplies of oxygen and moisture, the temperature, the amount and nature of the soil organic matter, the pH, and the amount of exchangeable calcium present. These effects are briefly outlined below.

1. Oxygen requirements
   a. Some bacteria use mostly oxygen gas (aerobic).
   b. Some bacteria use mostly combined oxygen (anaerobic).
   c. Some bacteria use either free or combined oxygen (facultative).
   d. All three of the above types usually function simultaneously in a soil.
2. Moisture relationships
   a. Optimum moisture level for higher plants (moisture potential −0.1 to −1 bar) is usually best for most bacteria.
   b. The moisture content affects oxygen supply (see 1 above).
3. Suitable temperature range
   a. Bacterial activity generally is greatest at 20–40 °C (about 70–100 °F).
   b. Ordinary soil temperature extremes seldom kill bacteria and often only temporarily suppress activity.
4. Organic matter requirements
   a. Organic matter is used as an energy source by the majority of bacteria (heterotrophic).
   b. Autotrophic bacteria do not require organic matter as an energy source.
5. Exchangeable calcium and pH relationships
   a. High calcium concentration and pH values from 6 to 8 generally are best for most bacteria.
   b. Calcium and pH values determine the specific bacteria present.
   c. Certain bacteria function at very low pH (<3.0) and a few at high pH values.
   d. Exchangeable calcium seems to be more important than pH.

9.13

Injurious Effects of Soil Organisms on Higher Plants

SOIL FAUNA

It already has been suggested that certain of the soil fauna are injurious to higher plants. For instance, some rodents and moles may severely damage crops. Snails and slugs in some climates are dreaded pests, especially of vegetables. Ants, because they transfer aphids on certain plants, must be held in check by gardeners. Also, most plant roots are infested with nematodes, sometimes so seriously as to make the successful growth of certain crops both difficult and expensive.

MICROFLORA AND PLANT DISEASES

In general, the microflora exert the most devastating effects on higher plants. Although bacteria and actinomycetes contribute their quota of plant diseases, fungi are responsible for most of the common soil-borne diseases of crop plants. Included are wilts, damping off, root rots, and clubroot of cabbage and similar crops. In short, disease infestations occur in great variety and are induced by many different organisms.

Soils are easily infested with disease organisms. These are transferred to the soil through farm implements, plants, and even through manure from animals that were fed infected plants. Erosion of soil also can carry diseases from one field to another. Once a soil is infested, it is apt to remain so for a long time.

DISEASE CONTROL BY SOIL MANAGEMENT

Prevention is the best defense against soil-borne diseases. Strict quarantine systems will restrict the transfer of seed-borne pathogens from one farm to another. Elimination of the host crop from the infested field will help in cases where no alternate hosts are available. In some cases, crop rotations and tillage practices can be used to help control a disease.

Regulation of the pH is effective in controlling some diseases. Keeping the pH low can control potato scab (see Figure 9.12) while raising the pH to 7.0 and above helps control clubfoot of cabbage, a fungus disease.

FIGURE 9.12
Soil-borne pathogens damage roots as well as other belowground organs. This potato has been attacked by the potato scab actinomycete, which may be present in soils with pH above about 5.0. [Courtesy Dr. F. E. Manzer, University of Maine]

Wet, cold soils favor some seed rots and seedling diseases known as damping off. Good drainage and ridging help control these diseases. Steam or chemical sterilization is a practical method of treating greenhouse soils for a number of diseases. The breeding of plants immune to particular diseases has been successful in the case of a number of other crops.

COMPETITION FOR NUTRIENTS

Another way in which soil organisms may be detrimental to higher plants, at least temporarily, is by competition for available nutrients. Soil organisms can quickly absorb essential nutrients into their own bodies, so that the more slowly growing higher plants can use only what is left. Nitrogen is the element for which competition is usually greatest, although similar competition occurs for phosphorus, potassium, calcium, and even the micronutrients. This subject is considered in greater detail in Chapter 10.

OTHER DETRIMENTAL EFFECTS

Under conditions of poor drainage, active soil microflora may deplete the already limited soil oxygen supply. The shortage of oxygen may affect plants adversely in two ways. First, the plant roots require a certain minimum amount of oxygen for normal growth and nutrient uptake. Second, oxidized forms of several elements, including nitrogen, sulfur, iron, and manganese, will be chemically reduced by further microbial action. In the cases of nitrogen and sulfur, some of the reduced forms are gaseous and may be lost to the atmosphere. In soils that are quite acid, the reduction of iron and manganese may produce soluble forms of these elements in toxic quantities. Thus, nutrient deficiencies and toxicities, both microbiologically induced, can result from the same basic set of conditions.

9.14

Competition Among Soil Microorganisms

In addition to competition between microorganisms and higher plants, there exists in soils an intense intermicrobial rivalry for food. When fresh organic matter is added, the vigorous heterotrophic soil organisms (bacteria, fungi, and actinomycetes) compete with each other for this source of food. The bacteria dominate initially because they reproduce rapidly and prefer simple compounds. As these simple compounds are broken down, the fungi, and particularly the actinomycetes, become more competitive. Undoubtedly, such competition for food is the rule and not the exception in soils (Figure 9.13).

FIGURE 9.13
Organisms compete with each other in the soils. The growth of a fungus, *Fusarium*, was rapid when this organism was grown alone in a soil (left), but when a certain bacterium, *Agrobacterium*, was also introduced, the fungal growth did not appear. [Courtesy M. A. Alexander, Cornell University]

ANTIBIOTICS PRODUCED IN SOILS

There is another type of microbial conflict just as intense and even more deadly than rivalry over food. To increase their ability to compete, certain soil bacteria, fungi, and actinomycetes can produce antibiotics. Because these substances are detrimental to organisms other than those that produce them, antibiotics have revolutionized the treatment of certain human and animal diseases. Many antibiotics now on the market, such as penicillin, streptomycin, and aureomycin, can be produced by organisms found in soils. Thus, the soil harbors not only disease organisms but also organisms that are the source of lifesaving drugs, the discovery of which marked an epochal advance in medical science.

9.15

Effects of Agricultural Practice on Soil Organisms

Changes in environment affect both the number and kinds of soil organisms. Clearing forested or grassland areas for cultivation drastically changes the soil environment. First, the quantity and quality of plant residues (food for the organisms) is markedly reduced. Second, the number of species of higher plants is reduced. Monoculture or even common crop rotations provide a much narrower range of original plant materials than nature provides in forests or grasslands.

While agricultural practices have different effects on different organisms, a few generalizations can be made. Figure 9.14 illustrates some principles relating to the total soil organism diversity and population. For example, some agricultural practices generally reduce the species diversity as well as the total organism population (owing, for example, to erosion and monoculture). Other practices, such as adding lime, fertilizers, and manures to an infertile soil, generally will increase the activities of the microflora. Pesticides (especially the fumigants) can sharply reduce organism numbers, at least on a temporary basis. However, monoculture cropping systems, which generally reduce species numbers and diversity, may actually increase the organism count of certain species.

Most changes in agricultural technology have ecological effects on soil organisms that can affect higher plants and animals, including humans. The effects of pesticides, both positive and negative, provide evidence of this.

Changes in diversity and overall population density of soil organisms

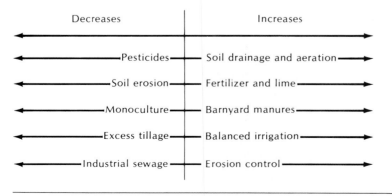

FIGURE 9.14
A simple diagrammatic illustration of the general effects of major agricultural practices on the diversity and population densities of soil organisms. Note that practices promoting sustained crop production and the protection and improvement of soils generally favor both diversity and overall organism population numbers.

For this reason, the ecological side effects of modern technology must be carefully scrutinized.

9.16

Benefits of Soil Organisms

In their influence on crop production, the soil fauna and flora are indispensable. Of their many beneficial effects on higher plants, only the most important can be mentioned here.

ORGANIC MATTER DECOMPOSITION

Perhaps the most significant contribution of the soil fauna and flora to higher plants is their ability to decompose organic matter. By decomposition plant residues are broken down, releasing organically held nutrients for use by plants. Nitrogen is a prime example. At the same time the stability of soil aggregates is enhanced not only by the slimy intermediate products of decay but by the more resistant portion, humus. Plants naturally profit from these beneficial chemical and physical effects.

INORGANIC TRANSFORMATIONS

The transformation of inorganic compounds is of great significance to higher plants. The presence in soils of nitrates, sulfates, and, to a lesser degree, phosphate ions is due primarily to the action of microorganisms. Organically bound forms of nitrogen, sulfur, and phosphorus are converted by the microbes into plant-available forms.

Likewise, the availability of the other essential elements such as iron and manganese is determined largely by microbial action. In well-drained soils, these elements are oxidized by autotrophic organisms to their higher valence states, in which forms their solubilities are quite low. This keeps the greater portion of iron and manganese, even under fairly acid conditions, in insoluble and nontoxic forms. If such oxidation did not occur, plant growth would be jeopardized because of toxic quantities of these elements in solution.

NITROGEN FIXATION

The fixation of elemental nitrogen, which cannot be used directly by higher plants, into compounds usable by plants is one of the most important microbial processes in soils (see Chapter 11). Although some blue-green algae and certain actinomycetes can fix nitrogen, bacteria are most important in the capture of gaseous nitrogen. The *nodule organisms*, especially those of legumes, *free-fixing bacteria* of several kinds, and some actinomycetes are most noted for their ability to fix nitrogen. Worldwide, enormous quantities of nitrogen are fixed annually into forms usable by higher plants.

It is obvious that the organisms of the soil must have energy and nutrients if they are to function efficiently. To obtain these, soil organisms break down organic matter, aid in the production of humus, and leave behind compounds that are useful to higher plants. These biotic features and their practical significance are considered in the next chapter, which deals with soil organic matter.

9.17

Conclusion

Soil organisms are vital to the cycle of life in earth. They incorporate plant and animal residues into the soil and digest them, returning carbon dioxide to the atmosphere where it can be recycled through higher plants. Simultaneously, they create humus, the organic constituent so vital to good physical and chemical soil conditions. During decomposition, soil organisms release essential plant nutrients in inorganic forms that can be absorbed by plant roots or leached from the soil. They even help control soil color through oxidation–reduction reactions they stimulate.

Animals, particularly earthworms, are responsible for mechanically incorporating residues into the soil, and they leave open channels through which water and air can flow. The microflora (plants) are responsible for most of the organic decay. Fungi, actinomycetes, and bacteria and major general-purpose decay organisms, while bacteria and algae play special roles in providing essential elements for higher plants. Competition among soil microbes and between these organisms and higher plants is responsible for some nutrient deficiencies. Microbial requirements are factors in determining the success of most soil management systems.

Study Questions

1. At which stage in the breakdown of organic residues (primary, secondary, or tertiary) do the microflora become active? Explain.
2. Contrast the diversity of soil organisms in a native forest area with an adjacent grassland area. What would happen to the diversity and organism numbers if the land were cleared and cultivated in each of these areas? Explain
3. If you wanted to dig for worms for a fishing trip, describe the soil, cover, and moisture conditions where you would most likely find the worms.
4. Explain why earthworms are said to be the most important soil macroanimal.
5. Termite mounds are commonly lower in organic matter than the surrounding soil, whereas earthworm casts are commonly higher. Explain the difference.
6. Nematodes generally are most important to agriculture than the other microanimals. Explain.
7. Contrast the numbers of microorganisms in the rhizosphere with those in the surrounding soil. What accounts for the difference?
8. Which class of microflora would you expect to be most active under the following conditions and why?

a. In fresh barnyard manure added to a soil
b. In rice paddies
c. In very acid soils
d. In the decomposition of the most resistant organic compounds
e. As an agent for most soil-borne diseases
f. Where atmospheric nitrogen is being fixed (two)
g. Where photosynthesis is occurring

9. You notice marked differences in the growth of pine trees in two different sandy soil areas and suspect differences in the uptake of phosphorus and other nutrient elements. Which of the microflora is most likely to account for these differences? Explain.
10. Contrast the importance to humans of heterotrophic and autotrophic organisms.
11. What are the ways in which microorganisms affect each other and affect higher plants?
12. Which group of microorganisms is concerned with oxidation and reduction of inorganic ions? Of what importance are these transformations?
13. Of what importance and how widespread are mycorrhizal associations?

References

Chino, M. 1976. "Electron microprobe analysis of zinc and other elements within and around rice root growth in flooded soils," *Soil Sci. Plant Nutr. J.*, **22**:449.

De Vleeschauwer, D., and R. Lal. 1981. "Properties of worm casts under secondary tropical forest regrowth," *Soil Sci.*, **132**:175–81.

Harley, J. L., and S. E. Smith. 1983. *Mycorrhizal Sym-*

biosis (New York: Academic Press).

JEX, G. W., et al. 1985. "Humidity-induced increase in water repellency in some sandy soils," *Soil Sci. Soc. Amer. J.*, **49**:1177–82.

LAL, R. 1987. *Tropical Ecology and Physical Edaphology* (New York: Wiley).

LAMBERT, D. H., D. E. BAKER, and H. COLE, Jr. 1979. "The role of mycorrhizae in the interactions of phosphorus with zinc, copper, and other elements," *Soil. Sci. Soc. Amer. J.*, **43**:976–80.

LEE, K. E. 1985. *Earthworms, Their Ecology and Relationships with Soils and Land Use* (New York: Academic Press).

MACFADYEN, A. 1963. In J. Doeksen and J. van der Drift (Eds.), *Soil Organisms* (Amsterdam: North-Holland Publ. Co.).

MENGE, J. A. 1981. "Mycorrhizae agriculture technologies," pp. 383–424 in *Background Papers for Innovative Biological Technologies for Lesser Developed Countries*, paper No. 9, Office of Technology Assessment Workshop, Nov. 24–25, 1980 (Washington, DC: U.S. Government Printing Office).

PAUL E. A., and F. E. CLARK. 1989 *Soil Microbiology and Biochemistry* (New York: Academic Press).

ROVIRA, A. D., R. C. FOSTER, and J. K. MARTIN. 1979. "Origin, nature and nomenclature of the organic materials in the rhizosphere," in J. L. Harley and R. S. Russell (Eds.), *The Soil-Root Interface* (New York: Academic Press).

YOST R. S., and R. L. FOX. 1982. "Influence of mycorrhizae on the mineral contents of cowpea and soybean growth in an Oxisol," *Agron. J.*, **74**:475–81.

10

Soil Organic Matter and Organic Soils

Organic matter[1] influences physical and chemical properties of soils far out of proportion to the small quantities present. It commonly accounts for as much as one third or more of the cation exchange capacity of surface soils and is responsible, perhaps more than any other single factor, for the stability of soil aggregates. Furthermore, organic matter supplies energy and body-building constituents for most of the microorganisms whose general activities were considered in chapter 9.

10.1

Sources of Soil Organic Matter

The original source of the soil organic matter is plant tissue. Under natural conditions, the tops and roots of trees, shrubs, grasses, and other native plants annually supply large quantities of organic residues. Even with harvested crops, one tenth to one third of the plant tops commonly fall to the soil surface and remain there or are incorporated into the soil. Naturally, all of the roots remain in the soil. As these organic materials are decomposed and digested by soil organisms, they become part of the underlying soil by infiltration or by actual physical incorporation. Accordingly, the residues of higher plants provide food for soil organisms, which in turn create stable compounds that help maintain the soil organic levels.

Animals usually are considered secondary sources of organic matter. As they attack the original plant tissues, they contribute waste products and leave their own bodies as their life cycles are consummated. Certain forms of animal life, especially the earthworms, termites, and ants, also play an important role in the translocation of soil and plant residues.

[1] For a review of the subject, see Stevenson (1982a) and Tate (1987).

10.2

Composition of Plant Residues

The moisture content of plant residues is high, varying from 60 to 90%, with 75% a representative figure (Figure 10.1). On a weight basis, the dry matter is mostly carbon and oxygen, with less than 10% each of hydrogen and inorganic elements (ash); however, on an elemental basis (number of atoms of the elements), hydrogen predominates. In representative plant residues, there are 8 hydrogen atoms for every 3.7 carbon atoms and 2.5 oxygen atoms. These three elements dominate the bulk of organic tissue in the soil.

Even though other elements are present only in small quantities, they play a vital role in plant nutrition and in meeting microorganism body requirements. The essential elements such as nitrogen, sulfur, phosphorus, potassium, calcium, and magnesium are particularly significant. These will be discussed later in more detail.

The actual organic compounds in plant tissue are many and varied; however, they can be grouped into a small number of classes, as shown in Figure 10.1. Representative percentages as well as ranges common in plant materials are shown.

GENERAL COMPOSITION OF COMPOUNDS

The carbohydrates, which range in complexity from simple sugars and starches to cellulose, are usually the most prominent of the organic compounds found in plants. Lignins, which are complex compounds with "ring"-type structures, are found in older plant tissue and especially woody tissues. They are very resistant to decomposition. Fats and oils, which are somewhat more complex than carbohydrates and less so than lignins, are found primarily in seeds.

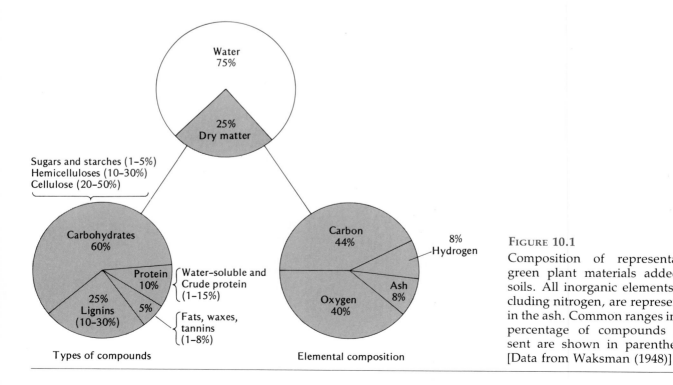

FIGURE 10.1

Composition of representative green plant materials added to soils. All inorganic elements, including nitrogen, are represented in the ash. Common ranges in the percentage of compounds present are shown in parentheses. [Data from Waksman (1948)]

Proteins contain—in addition to carbon, oxygen, and hydrogen—nitrogen and smaller amounts of other essential elements such as sulfur, manganese, copper, and iron. Proteins are primary sources of these essential elements. The simple proteins are decomposed easily while the more complex crude proteins are resistant to breakdown.

10.3

Decomposition of Organic Compounds

RATE OF DECOMPOSITION

Organic compounds vary greatly in their rate of decomposition. They may be listed in terms of ease of decomposition as follows.

1. Sugars, starches, and simple proteins   rapid decomposition
2. Crude proteins
3. Hemicelluloses
4. Cellulose
5. Fats, waxes, etc.
6. Lignins   very slow decomposition

All organic compounds usually begin to decompose simultaneously when fresh plant tissue is added to a soil. The sugars and simple proteins decompose most readily; at the other extreme, lignins are the most resistant to breakdown.

When organic tissue is added to soil, three general reactions take place.

1. The bulk of the material undergoes enzymatic oxidation with carbon dioxide, water, energy, and heat as the major products.
2. The essential elements such as nitrogen, phosphorus, and sulfur are released and/or immobilized by a series of specific reactions relatively unique for each element.
3. Compounds very resistant to microbial action are formed either through modification of compounds in the original plant tissue or by microbial synthesis. (Collectively, these resistant compounds comprise soil humus.)

Each of these reactions has great practical significance as discussions that follow will indicate.

DECOMPOSITION—AN OXIDATION PROCESS

In a well-aerated soil, all of the organic compounds found in plant residues are subject to oxidation. Since the oxidizable fraction of plant materials is composed largely of carbon and hydrogen, the oxidation of the organic compounds in soil may be represented as follows.

$$-(C, 4H) \; + 2\,O_2 \xrightarrow[\text{oxidation}]{\text{enzymatic}} CO_2 + 2\,H_2O + \text{energy}$$

Carbon- and hydrogen-containing compounds

Many intermediate steps are involved in this overall reaction, and it is accompanied by important side reactions that involve elements other than carbon and hydrogen. Even so, this basic reaction accounts for most of the decomposition of organic matter in the soil as well as the oxygen consumption and release of carbon dioxide.

BREAKDOWN OF PROTEINS

The plant proteins also succumb to microbial decay, yielding not only carbon dioxide and water but also amino acids such as glycine (CH_2NH_2COOH) and cysteine ($CH_2HSCHNH_2COOH$). In turn, these nitrogen and sulfur compounds are further broken down, eventually yielding simple inorganic ions such as NH_4^+, NO_3^- and SO_4^{2-}.

EXAMPLE OF ORGANIC DECAY

The process of organic decay in time sequence is illustrated in Figure 10.2. First, assume a situation where no readily decomposable materials are present in the soil and where the microbial numbers and activity are low. When an abundance of fresh, decomposable tissue is added, the organisms promptly attack the easily decomposable compounds such as sugars, starches, and cellulose, releasing carbon dioxide and water. Simultaneously, the number of soil microorganisms suddenly increases dramatically. Soon microbial activity is at its peak, at which point energy is being liberated

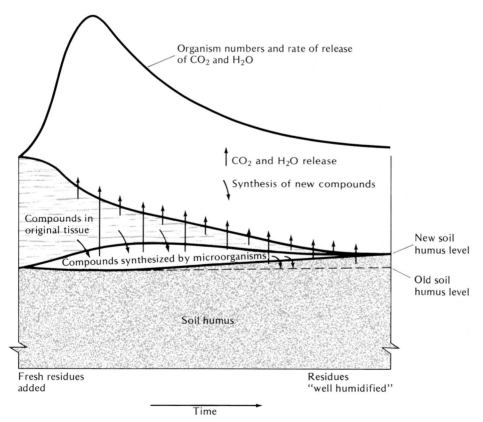

FIGURE 10.2

Diagrammatic illustration of the general changes that occur when fresh plant residues are added to soil. The time required for the process will depend on the nature of both the residue and the soil. While the rapid initial increase and subsequent decline in microorganism numbers and in CO_2 and H_2O are most obvious, the modest increase in soil humus level should not be overlooked.

rapidly, and carbon dioxide is being formed in large quantities as are new organic compounds synthesized by the microbes. General-purpose decay bacteria, fungi, and actinomycetes are fully active and are decomposing and synthesizing at the same time. The microbial tissue may at times account for as much as one third of the organic fraction of a soil. At the height of microbial activity, even the original soil organic matter is subject to some breakdown.

As soon as the easily decomposed food is exhausted, microorganism numbers begin to decline. When the microbial cells die, their bodies are devoured by living microbes, with the continued evolution of carbon dioxide and water. As further reduction in available food and energy occurs, microbial activity continues to decline, and the general-purpose soil organisms again sink back into comparative quiescence. This is associated with a release of simple products such as nitrates and sulfates. The organic matter now remaining is mostly a dark, heterogeneous colloidal mass of resistant and newly synthesized compounds, usually referred to as *humus*.

The reactions just described are nothing more than a process of enzymatic digestion such as that which occurs when foods enter the human digestive system. The products of these enzymatic activities in soils include: (a) energy appropriated by the microorganisms or liberated as heat; (b) carbon dioxide and other simple end products; and (c) humus. The products will be considered in order.

10.4

Energy of Soil Organic Matter

The quantity of energy soil organisms use for digesting plant residues and humus is remarkably high. For example, the application of 20 Mg (metric tons) of farm manure containing 5000 kg of dry matter embodies about 105 million kilojoules (kJ) [25 million kilocalories (kcal)] of latent energy. This is equivalent to the energy in more than 3 Mg of anthracite coal. The potential energy of a hectare–furrow slice of a soil with 4% organic material is about 1675 million kJ (400 million kcal), equivalent to the heat value of 50 Mg of anthracite coal or 225 barrels of oil. If all this energy equivalent were changed to heat, it would be enough to raise the temperature of that amount of soil (20% moisture) by more than 600 °C.

In the digestion process, most of the energy released is lost as heat, with only a small amount being used for activities of the organisms. The elevated temperature of a compost pile is a practical illustration of the dissipated heat.

RATE OF ENERGY LOSS FROM SOILS

Certain estimates made at the Rothamsted Experiment Station in England give some idea of the rate of energy dissipation from soils (Russell and Russell, 1950). It was calculated that the *annual* heat loss from a hectare of an untreated, low-producing soil was equivalent to the heat value of nearly a metric ton of anthracite coal. The comparable figure on a productive soil receiving liberal supplies of farmyard manure was more than 12 Mg/ha of coal annually. If the entire annual heat loss from these plots took place at one time and all the heat was absorbed by the soil, a temperature rise of 12 °C could be expected for the poor soil and 145 °C for the more productive soil. The magnitude of such energy loss is surprising even for the poorer plot.

10.5

The Carbon Cycle

As organic matter decays, carbon dioxide is among the immediate break-down products. This is a reminder that carbon is a common constituent of all organic matter and is involved in essentially all life processes. Consequently, the transformations of this element, termed the *carbon cycle*, are in reality a *biocycle* or "cycle of life" that makes possible the continuity of life on earth. These changes are diagrammed in Figure 10.3. Note that humus and carbon dioxide are relatively stable components of this cycle.

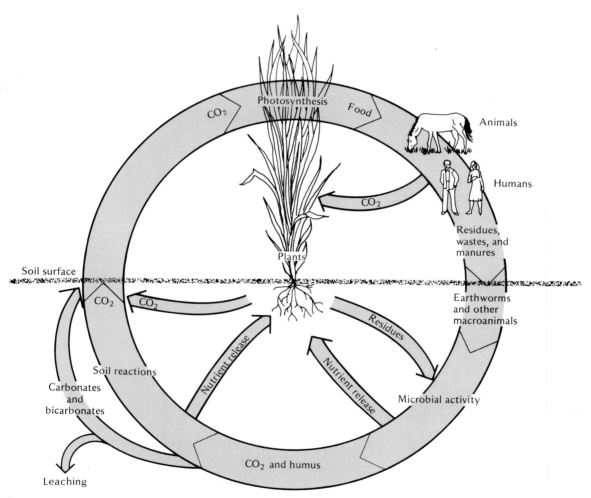

FIGURE 10.3

Transformation of carbon commonly spoken of as the *carbon cycle*, the *biocycle*, or *life cycle*. Plants assimilate CO_2 from the atmosphere into organic compounds using energy from the sun. Man and other higher animals obtain energy and body tissue from plant products and return wastes and residues to the soil. Macro- and microorganisms digest these organic materials, releasing nutrients for plants and leaving CO_2 and humus as relatively stable products. Carbonates and bicarbonates of Ca, Mg, K, etc., are removed in leaching, but eventually the carbon returns to the cycle in the form of CO_2. The total CO_2 is released to the atmosphere where it is again available for plant assimilation. This biocycling illustrates how carbon is the focal point of *energy* transformations and makes possible the continuity of life on Earth.

RELEASE OF CARBON DIOXIDE

Through the process of photosynthesis, carbon dioxide is assimilated by higher plants and converted into numerous organic compounds, such as those described in Section 10.3. As these organic compounds reach the soil in plant residues, they are digested, and carbon dioxide is evolved. Microbial activity is the main soil source of carbon dioxide, although appreciable amounts come from the respiration of rapidly growing plant roots, and small amounts are brought down dissolved in rain water. Under optimum conditions, more than 100 kg/ha of carbon dioxide can be evolved per day, with 25–30 kg being more common. Much of the carbon dioxide of the soil ultimately escapes to the atmosphere, where it can again be used by plants, thus completing the cycle.

A much smaller amount of carbon dioxide reacts in the soil, producing carbonic acid (H_2CO_3) and the carbonates and bicarbonates of calcium, potassium, magnesium, and other base-forming cations. The bicarbonates are readily soluble and can be removed in drainage or used by higher plants.

OTHER CARBON PRODUCTS OF DECAY

The decay of organic matter results in other carbon products. Small quantities of elemental carbon are found in soil. Under highly anaerobic conditions, methane (CH_4) and carbon bisulfide (CS_2) may be produced in small amounts. But of all the simple carbon products, carbon dioxide is by far the most abundant and, in terms of human welfare, by far the most important.

The carbon cycle is all-inclusive because it involves not only the soil and its teeming fauna and flora and higher plants of every description but also all animal life, including humans. Its failure to function properly would mean disaster to all.

10.6

Simple Decomposition Products

As the enzymatic changes of the soil organic matter proceed, simple inorganic products emerge. Along with carbon dioxide, the most important are those containing the essential plant nutrients, especially nitrogen, sulfur, and phosphorus. For example, as proteins are attacked by microbes, amino acids appear. These, in turn, are broken down to produce first ammonium compounds and sulfides and finally nitrates and sulfates. Similar breakdowns of other organic compounds release inorganic phosphates as well as cations such as Ca^{2+}, Mg^{2+}, and K^+. The overall process that produces these inorganic (mineral) forms is called *mineralization*.

Fortunately, most of the inorganic ions released by decomposition are readily available to higher plants and microorganisms. Although the nitrates and sulfates are subject to leaching, the phosphates tend to be retained in insoluble calcium, iron, and aluminum compounds. In all cases, the decay of organic tissue is an important source of these essential elements.

As organic matter decays, cations such as Ca^{2+}, K^+, Na^+, and Mg^{2+} are released to the soil solution and are subject to adsorption by negatively charged soil colloids. They can later be released for plant uptake, and some will be removed by leaching waters.

10.7

Humus—Genesis and Nature

Since the material we refer to as "humus" is a heterogenous mixture of complex organic compounds, our knowledge of its nature and formation is incomplete.[2] Research suggests, however, that the complex compounds that make up humus are not merely degraded plant materials. The bulk of these compounds have resulted from two general types of biochemical reactions: *decomposition* and *synthesis*.

Decomposition occurs as chemicals in the plant residues are broken down or drastically modified by soil organisms. Even lignin is split and degraded, many of its structural units being destroyed. Other simpler organic compounds that also result from the breakdown may take part immediately in the second of the humus-forming processes, biochemical synthesis. These simpler chemicals are metabolized into new compounds in the body tissue of the soil microorganisms. The new compounds are subject to further modification and synthesis as the microbial tissue is subsequently attacked by other soil microorganisms.

Additional synthetic reactions involve such breakdown products of lignin as the phenols and quinones. These decomposition products, present initially as separate molecules called monomers, are enzymatically stimulated to join together into polymers. By this process of polymerization (linkage) *polyphenols* and *polyquinones* are formed. These high molecular weight compounds interact with nitrogen-containing amino compounds and give rise to a significant component of resistant humus. The formation of these polymers is encouraged by the presence of colloidal clays.

Studies suggest that there are two general groups of compounds that collectively make up humus, the *humic* group and the *nonhumic* group. These will now be considered.

HUMIC GROUP

The humic substances make up about 60–80% of the soil organic matter. They are comprised of the most complex materials, which are also the most resistant to microbial attack. Humic substances are characterized by aromatic, ring-type structures that include *polyphenols* (numerous phenolic compounds tied together) and *polyquinones*, which are even more complex. They are formed as just described, by decomposition, synthesis, and polymerization.

The humic substances have no sharply defined physical or chemical properties (e.g., melting points), which are characteristic of the more simple nonhumic compounds. They are amorphous, dark in color, and have high to very high molecular weights, varying from a few hundred to several thousand.

Solubility Groupings. On the basis of resistance to degradation and of solubility in acids and alkalis, humic substances have been classified into three chemical groupings: (a) *fulvic acid*, lowest in molecular weight and lightest in color, soluble in both acid and alkali, and most susceptible to

---

[2] For discussions of humus formation, see Haynes (1986), Stevenson (1982b), and Frimmel and Christman (1988).

microbial attack; (b) *humic acid*, medium in molecular weight and color, soluble in alkali but insoluble in acid, and intermediate in resistance to degradation; and (c) *humin*, highest in molecular weight, darkest in color, insoluble in both acid and alkali, and most resistant to microbial attack.

It should be emphasized that even fulvic acid, the most easily degraded, is still quite stable in the soil and is more resistant to microbial attack than most freshly applied crop residues. Depending on the environment, it may take 15–50 years to destroy fulvic acid-type compounds in the soil; hundreds of years are required to destroy humic acid. Despite differences in chemical and physical makeup, the three humic groups have some similarities with regard to properties, such as the ability to absorb and release cations. Consequently, they will all be considered under the general term "humic materials."

NONHUMIC GROUP

The nonhumic group comprises about 20–30% of the organic matter in soils. Nonhumic substances are less complex and less resistant to microbial attack than those of the humic group. They are comprised of specific organic compounds with definite physical and chemical properties. Some of these nonhumic materials have been only modified by microbial action while others were synthesized as breakdown occurred.

Included among the nonhumic substances are *polysaccharides*, polymers that have sugar-like structures and a general formula of $C_n(H_2O)_m$, where n and m are variable. Polysaccharides are especially effective in enhancing soil aggregate stability. Also included are *polyuronides*, which are not found in plants but have been synthesized by the soil microbes and held as part of the organism body tissue. On the death of the organism, the polyuronides, as mentioned earlier, may be subject to further microbial attack or can interact with other organic materials in the soil.

Some simpler compounds are part of the nonhumic group. For example, organic acids and some protein-like materials are included. Although these simpler materials are not present in large quantities, their presence is critical, especially as they affect the availability of some plant nutrients such as nitrogen and iron.

Figure 10.4 shows the approximate proportions of organic carbon from plant residues that are found in humic and nonhumic compounds 1 year after the residues were added to the soil. Much of the carbon has returned to the atmosphere as carbon dioxide, but nearly one third remains in the soil either as live organisms (5%) or as humus.

GROUP INTERACTIONS

It should not be inferred that the humic and nonhumic groups react independently. Just the opposite is true—they are continually interacting and are commonly linked together. Even as the respective compounds are forming, reactions occur among members of these two groups. For example, proteins and other nitrogen compounds react with a wide variety of organic compounds in both groups, including the humic acids and polysaccharides. These reactions are essential for nitrogen conservation in soils because in the resultant compounds the protein nitrogen is somehow protected from microbial attack. Without this protection, the proteins would succumb easily to decomposition and hydrolysis, releasing the nitrogen as soluble NH_4^+ and NO_3^- forms, which are rapidly lost from the soil.

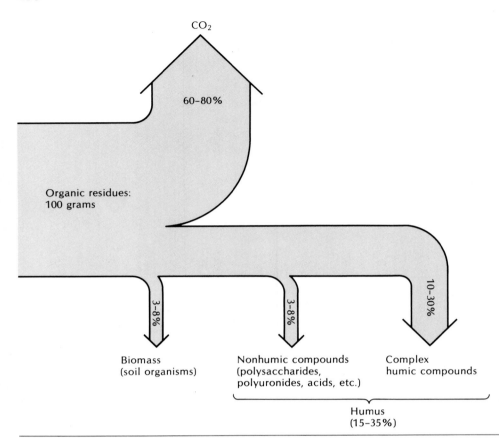

FIGURE 10.4
Diagram to illustrate the disposition of 100 g of organic residues 1 year after they were incorporated into the soil. More than two thirds of the residues have been oxidized to CO_2 and less than one third remains in the soil, some in the bodies of soil organisms but a larger component as humus. [Estimates summarized from several sources]

CLAY–HUMUS COMBINATIONS[3]

Another means of stabilizing soil nitrogen is through interaction with some layer silicate clays. These clays are known to attract and hold substances such as amino acids, peptides, and proteins, forming complexes that protect the nitrogen-containing compounds from microbial degradation. For example, vermiculite forms interlayer complexes that strongly resist decomposition. Thus, clay joins the humic polymers and other organic materials such as polysaccharides in protecting relatively simple nitrogen compounds from microbial attack.

Specific protein–clay linkages are not the only important reactions that occur between organic and inorganic compounds in soils. In some cases, clays seem to catalyze some oxidation and polymerization reactions even in the absence of microorganisms. Also various organomineral complexes involve linkage of phenols or organic acids with silicate clays or iron and aluminum oxides. More than half the organic matter in soil is likely associated in some definite way with clay and other inorganic constituents. These associations account in part for the high organic matter content of clay soils.

STABILITY OF HUMUS

Humus is subject to continual microbial attack. Without the annual addition of plant residues, microbial action results in a reduction in soil organic

[3] For detailed discussions of this subject, see Huang and Schnitzer (1986) and Loll and Bollag (1983).

matter levels. Even so, humus is more resistant to decay than are most compounds found in plant residues. Each year newly synthesized polymers join those formed in previous months, years, and decades, or even earlier. Studies have shown that some organic carbon converted to humus hundreds of years ago is still present in soils, evidence that some of the humic materials are extremely resistant to microbial attack. This resistance is important in maintaining organic matter levels in soils and also helps protect nitrogen and other essential nutrients that are bound in the resistant humic complex.

HUMUS DEFINED

From the previous discussion, two facts are obvious: (a) Humus is a mixture of complex compounds, not a single material. (b) The compounds that make up humus have been synthesized by microorganisms in the soil from products of the breakdown or alteration of the original plant tissue. These two facts lead to the following definition.

Humus is a complex and rather resistant mixture of brown or dark brown amorphous and colloidal organic substances that results from microbial decomposition and synthesis and has chemical and physical properties of great significance to soils and plants.

10.8

Humus—Colloidal Characteristics

In Chapter 7, the colloidal nature of humus was discussed and the following major characteristics of this important soil constituent were mentioned.

1. The tiny colloidal humus particles (micelles) are composed of carbon, hydrogen, and oxygen (probably in the form of polyphenols, polyquinones, polyuronides, and polysaccharides).
2. The surface area of humus colloids per unit mass is very high, generally exceeding that of silicate clays.
3. The colloidal surfaces of humus are negatively charged, the sources of the charge being hydroxy (—OH), carboxylic (—COOH), or phenolic (⬡—OH) groups. The extent of the negative charge is pH dependent (high at high pH values).
4. At high pH values the cation exchange capacity of humus on a mass basis (150–300 cmol/kg) far exceeds that of most silicate clays.
5. The water-holding capacity of humus on a mass basis is 4–5 times that of the silicate clays.
6. Humus has a very favorable effect on aggregate formation and stability.
7. The black color of humus tends to distinguish it from most of the other colloidal constituents in soils.
8. Cation exchange reactions with humus are qualitatively similar to those occurring with silicate clays.

The humic micelles, like particles of clay, carry a swarm of adsorbed cations (Ca^{2+}, H^+, Mg^{2+}, K^+, Na^+, etc.). Thus, colloidal humus may be represented by the same illustrative formula used for clay.

$$Ca^{2+}$$
$$Al^{3+}$$
$$H^+ \quad \boxed{Micelle}$$
$$M^+$$

and the same reactions serve to illustrate cation exchange in both (see Section 7.11). As before, M represents other metallic cations, such as potassium, magnesium, and sodium.

EFFECT OF HUMUS ON NUTRIENT AVAILABILITY

Humus enhances mineral breakdown and, in turn, nutrient availability in two ways. First, humic acids can attack the minerals and bring about their decomposition, thereby releasing essential base-forming cations.

$$KAlSi_3O_8 + \quad H^+\boxed{Micelle} \rightarrow HAlSi_3O_8 + \quad K^+\boxed{Micelle}$$

Microcline   Humic acid   Acid silicate   Adsorbed K

The potassium is changed from a molecular to an adsorbed state, which is more readily available to higher plants.

The second mechanism for increasing the availability of some cations is through the formation of stable organomineral complexes with these ions. For example, polysaccharides and fulvic acids form such complexes with metallic ions such as Fe^{3+}, Cu^{2+}, Zn^{2+}, and Mn^{2+}. The cations are attracted from the minerals in which they are found and held in complex form by the organic molecule. Later they may be taken up by plants or may take part in the synthesis of clay and other inorganic constituents.

These examples are reminders of the beneficial interreactions of clay and humus. It is most fortunate that these two soil constituents tend to reinforce each other in many soil–plant relationships.

10.9

Direct Influences of Organic Compounds on Higher Plants

It was once thought that organic matter as such is directly absorbed by higher plants. Although this opinion was later discarded, we now know that higher plants can absorb certain organic nitrogen compounds. For example, some amino acids, such as alanine and glycine, can be absorbed directly. However, in normal soils such substances are available in quantities far too small to satisfy plant needs for nitrogen.

The uptake by plants of vanillic acid and other phenol carboxylic acids has been established by the use of radioactive carbon, but the significance of these acids in practical agriculture is not known. The beneficial effects of an exceedingly small absorption of organic compounds might be explained by the presence of growth-promoting substances. In fact, various growth-promoting compounds such as vitamins, amino acids, auxins, and gibberellins are formed as organic matter decays. These may at times stimulate both higher plants and microorganisms.

On the other hand, some soil organic compounds are harmful; for example, dihydroxystearic acid, which is toxic to higher plants, has been isolated from soils. However, such "toxic matter" may be a product of unfavorable soil conditions that will disappear when the conditions are corrected. Apparently, good drainage and tillage, lime, and fertilizers reduce the probability of organic toxicity.

10.10

Influence of Soil Organic Matter on Soil Properties

As background for discussion of the practical maintenance of soil organic matter, it is useful to keep in mind the remarkable effects of this all-important constituent on a number of soil properties.

1. Effect on soil color—brown to black
2. Influence on physical properties
   a. Granulation encouraged
   b. Plasticity, cohesion, and so on, reduced
   c. Water-holding capacity increased
3. High cation exchange capacity
   a. 2–30 times as great as mineral colloids (mass basis)
   b. Accounts for 20–90% of the adsorbing power of mineral soils
4. Supply and availability of nutrients
   a. Easily replaceable cations present on humus colloids
   b. Nitrogen, phosphorus, sulfur, and micronutrients held in organic forms
   c. Release of elements from minerals by acid humus

10.11

Carbon/Nitrogen Ratio

Reference has been made to the close relationship existing between the organic matter and the nitrogen contents of soils. Since carbon makes up a large and rather definite proportion of this organic matter, it is not surprising that the carbon to nitrogen (C/N) ratio of soils also is fairly constant. This fact is important in controlling the available nitrogen, total organic matter, and the rate of organic decay, and in developing sound soil management schemes.

RATIO IN SOILS

The C/N ratio in the organic matter of the furrow slice (upper 15 cm) of arable soils commonly ranges from 8:1 to 15:1, the median being between 10:1 and 12:1. In a given climatic region, little variation is found in this ratio, at least in similarly managed soils. The variations that do occur seem to be correlated generally with climatic conditions. For instance, the C/N ratio tends to be lower in soils of arid regions than in those of humid regions when annual temperatures are about the same. The ratio is also lower in warmer regions than in cooler ones if the rainfalls are about equal. Also, C/N is lower for subsoils, in general, than for the corresponding surface layers.

RATIO IN PLANTS AND MICROBES

The C/N ratio in plant material varies, ranging from 20:1 to 30:1 in legumes and farm manure, to 100:1 in certain strawy residues, and to as high as 400:1 in sawdust. In the bodies of microorganisms the C/N ratio is not only more constant but also much lower, ordinarily falling between 4:1 and 9:1.

Among microorganisms bacterial tissue in general is somewhat richer in protein than fungi and, consequently, has a lower ratio.

It is apparent, therefore, that most organic residues entering the soil contain large amounts of carbon and comparatively small amounts of total nitrogen. The C/N ratio is high, and the C/N values for soils are between those of higher plants and microbes.

10.12

Significance of Carbon/Nitrogen Ratio

The C/N ratio in soil organic matter is important for two major reasons: (a) Keen competition among microorganisms for available soil nitrogen occurs when residues having a high C/N ratio are added to soils. (b) Because the C/N ratio is relatively constant in soils, the maintenance of carbon— and, hence, soil organic matter—is constrained by the soil nitrogen level. The significance of the C/N ratio is apparent in a practical example of the influence of highly carbonaceous material on the availability of nitrogen.

PRACTICAL EXAMPLE

Assume that a cultivated soil in a condition favoring vigorous nitrification is examined. Nitrates are present in moderate amounts, and the C/N ratio is relatively low. The general-purpose decay organisms are at a low level of activity, as evidenced by low carbon dioxide production (Figure 10.5).

Now suppose that large quantities of organic residues with a high C/N ratio (50:1) are incorporated in this soil under conditions supporting vigorous digestion. A change quickly occurs. The heterotrophic flora—bacteria, fungi, and actinomycetes—become active and multiply rapidly, producing carbon dioxide in large quantities. Under these conditions, nitrate nitrogen essentially disappears from the soil because of the persistent microbial demand for this element to build microbe tissues. For the time being, then, little or no mineral nitrogen (NH_4^+ or NO_3^-) is available to higher plants. As decay occurs, the C/N ratio of the remaining plant material decreases because carbon is being lost and nitrogen conserved.

This condition persists until the activities of the decay organisms gradually subside because they lack easily oxidizable carbon. Their numbers decrease, carbon dioxide formation drops off, nitrogen demand by microbes becomes less acute, and the release of nitrates (nitrification) can proceed. Again, nitrates appear in quantity, and the original conditions prevail except that, for the time being, the soil is somewhat richer both in nitrogen and humus. This sequence of events, an important phase of the carbon cycle, is shown in Figure 10.5.

Very important

REASON FOR C/N RATIO CONSTANCY

As the decomposition processes continue, both carbon and nitrogen are lost—the carbon as carbon dioxide and the nitrogen as inorganic compounds that are leached or absorbed by plants. In time the percentage of the total nitrogen being removed equals the percentage of the total carbon being lost. At this point, the C/N ratio in the topsoil, commonly between 10:1 and 12:1 in humid regions (somewhat lower in arid areas), becomes more or less constant, always being somewhat greater than the ratios characterizing microbial tissue. The ratio in some cultivated, well-weathered tropical soils may be somewhat higher than for soils in temperate zones.

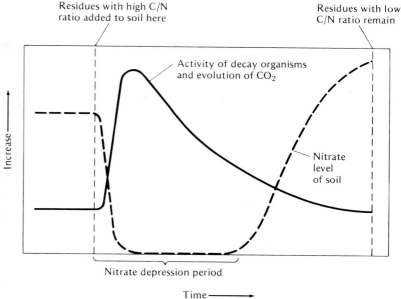

FIGURE 10.5

Cyclical relationship between the stage of decay of organic residues and the presence of nitrate nitrogen in soil. As long as the C/N ratio is high, the general-purpose decay organisms are dominant and the nitrifiers are more or less inactive. During the period of nitrate depression that results, higher plants can obtain little nitrogen from the soil. The length of this period will depend upon a number of factors, of which the C/N ratio is most important.

PERIOD OF NITRATE DEPRESSION

The time interval of nitrate depression (Figure 10.5) may be only a week or so, or it may last throughout the growing season. The rate of decay of organic residues will lengthen or shorten the period, as the case may be. For example, the greater the quantity of decomposable residues applied, the longer nitrification (nitrate formation) will be blocked. And the lower the C/N ratio of the residues applied, the more rapidly the cycle will run its course. Hence, alfalfa and clover residues interfere less with nitrification and yield nitrogen more quickly than if oats or wheat straw were incorporated in the soil. Also, mature residues, whether legume or nonlegume, generally have a much higher C/N ratio than do younger succulent materials (Figure 10.6). These facts are of practical significance and should be considered when organic residues are added to a soil.

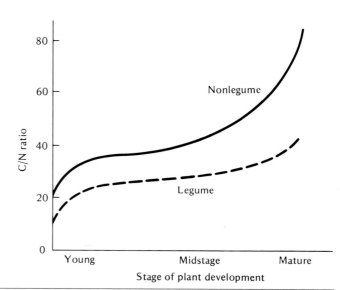

FIGURE 10.6

The C/N ratio of organic residues added to soil will depend upon the maturity of the plants turned under. The older the plants, the larger will be the C/N ratio and the longer will be the period of nitrate suppression. Obviously, leguminous tissue has a distinct advantage over nonlegumes since it promotes a more rapid organic turnover in soils.

CARBON/NITROGEN RATIO AND ORGANIC MATTER LEVEL

Since the C/N ratio of soils in a given region is reasonably constant, the level of soil nitrogen definitely influences the level of organic carbon and vice versa. And since a rather definite ratio (about 1:1.7) exists between the organic carbon and the soil humus, the amount of organic matter that can be maintained in any soil is largely dependent on the amount of organic nitrogen present. The ratio between nitrogen and organic matter is, thus, also rather constant. A value for the organic matter/nitrogen ratio of 20:1 is commonly used for average soils. The practical significance of this relatively constant ratio is that a soil's organic matter content cannot be increased without simultaneously increasing its organic nitrogen content and vice versa. This should be kept in mind while reading the sections that follow.

10.13

Amount of Organic Matter and Nitrogen in Soils

The amount of organic matter[4] (O.M.) in mineral soils varies widely; mineral surface soils contain from a trace to 20–30% of organic matter. The ranges and representative organic matter and nitrogen contents of different soil orders are shown in Table 10.1. The relatively wide range of organic matter encountered in these soils, even in comparatively localized areas, is immediately apparent. Among the mineral soils, Aridisols (dry soils) are generally the lowest in organic matter and Mollisols the highest. Naturally, the organic soils (Histosols) are dominated by this constituent.

The data in the Table 10.1 are for surface soils only. The organic matter contents of the subsoils are generally much lower (Figure 10.7). Since organic residues in both cultivated and virgin soils are incorporated in or deposited on the surface, the organic matter accumulates in the upper layers. Also, note that poorly drained soils, such as those in boggy areas, are commonly higher in organic matter than their well-drained counterparts.

TABLE 10.1

Suggested Ranges in Nitrogen and Organic Matter Contents Commonly Found in the Upper 15 cm of Eight Different Soil Orders and Representative Levels of These Two Constituents for Each Order

Soil order	Organic matter (%)		Nitrogen (%)	
	Range	Representative	Range	Representative
Alfisols	0.8–6.5	3.0	0.04–0.35	0.14
Aridisols	0.2–1.7	1.0	0.01–0.10	0.06
Histosols	20–98	80.0	0.60–3.00	2.40
Mollisols	1.5–6.5	4.0	0.07–0.30	0.18
Oxisols	1.5–5.0	3.0	0.07–0.25	0.13
Spodosols	1.5–5.0	3.5	0.06–0.20	0.14
Vertisols	1.5–3.0	2.0	0.06–0.16	0.10
Ultisols	1.5–4.0	2.0	0.07–0.20	0.09

[4]No figures can be given for the humus content of mineral soils because there is no satisfactory method for its determination. Likewise, there is no satisfactory method of measuring the exact quantity of soil organic matter. The usual procedure is to measure the organic carbon content, which can be done accurately. This figure multiplied by the factor 1.7 will give the approximate amount of organic matter present.

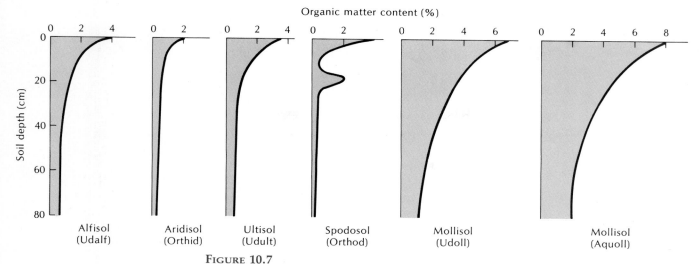

FIGURE 10.7
Distribution of organic matter in representative soils of five soils orders. Note the high organic content in the Mollisols and the very low organic content in the dry soil (Aridisol). Also note that the Mollisol with restricted drainage (Aquoll) has the highest organic matter level.

ORGANIC MATTER/NITROGEN RATIO

Data in Table 10.1 illustrate the close correlation between the organic matter and nitrogen contents of soils, which was mentioned in the preceding section. The average organic matter content of mineral soils is about 20 times that of nitrogen although the ratio varies somewhat among the soil orders. This O.M./N ratio of 20:1 is of considerable value in making rough calculations involving these two soil constituents.

10.14

Factors Affecting Soil Organic Matter and Nitrogen

INFLUENCE OF CLIMATE

Climatic conditions, especially *temperature* and *rainfall*, exert a dominant influence on the amounts of nitrogen and organic matter found in soils (see Jenny, 1941). As one moves from warmer to cooler areas in the United States, the organic matter and nitrogen of comparable soils tend to increase. At the same time, the C/N ratio increases somewhat. In general, the decomposition of organic matter is accelerated in warm climates; a lower rate of decay is the rule in cool regions. Within belts of uniform moisture conditions and comparable vegetation, the average total organic matter and nitrogen increase from two to three times for each 10 °C decline in mean annual temperature.

The situation is well illustrated by conditions in the Mississippi Valley region (Figure 10.8). Here the northern Great Plains soils contain considerably greater amounts of total organic matter and nitrogen than do those to the south. High temperatures also result in rapid organic matter decomposition in the irrigated soils of southern desert regions. Rates of release of nutrients are accelerated, but the residual organic matter accumulation is lower than in cooler soils. The influence of temperature on the rapidity of decay and disappearance of organic material is apparent.

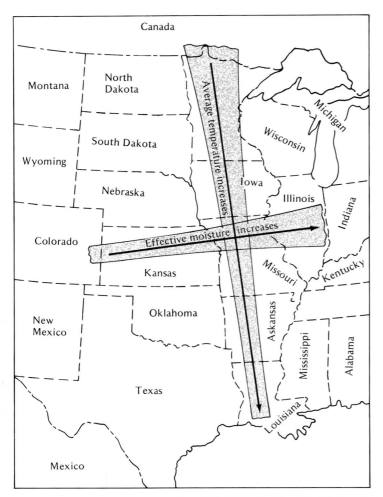

FIGURE 10.8
Influence of the average annual temperature and the effective moisture on the relative organic matter contents of grassland soils of the Midwest. Of course, the soils must be more or less comparable in all respects except for climatic differences. Note that higher temperatures yield soils lower in organic matter. The effect of increasing moisture is exactly opposite, favoring a higher level of this constituent. These climatic factors affect forest soils in much the same way.

Soil moisture also exerts a positive influence on the accumulation of organic matter and nitrogen in soils. Ordinarily, under comparable conditions, the nitrogen and organic matter increase as the effective moisture becomes greater (Figure 10.8). The C/N ratio also increases, especially in grasslands areas. The explanation lies mostly in the scantier vegetation of drier regions. Although a correlation of increasing organic matter with rainfall exists, remember that the level of organic matter in any one soil is influenced by temperature and other factors as well as by precipitation. Climatic influences never work independently.

INFLUENCE OF NATURAL VEGETATION

It is difficult to differentiate between the effects of climate and vegetation on organic matter and nitrogen contents of soil. Grasslands generally dominate the subhumid and semiarid areas, whereas trees are dominant in humid regions. In climatic zones where the natural vegetation includes both forests and grasslands, the total organic matter is higher in soils developed under grasslands than those under forests (Figure 10.9). Apparently, the nature of the grassland organic residues and their mode of decomposition encourage a reduced rate of decay and thus a higher organic level than is found under forests.

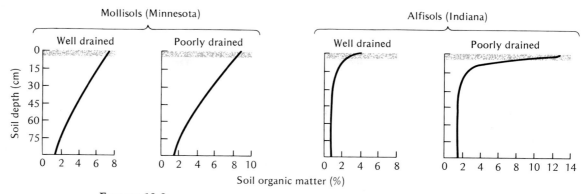

FIGURE 10.9

Distribution of organic matter in four soil profiles. Note that the Mollisols (soils developed under grasslands) have a higher organic matter content in the profiles as a whole than the soils developed under forest vegetation (Alfisols). Poor drainage results in a higher organic matter content, particularly in the surface horizon.

EFFECTS OF TEXTURE, DRAINAGE, AND OTHER FACTORS

Beside the two broader aspects—temperature and moisture—just discussed, numerous local relationships are involved in the amount of organic matter retained in soils. In the first place, the *texture* of the soil, other factors being constant, influences the percentage of humus and nitrogen present. Soils high in clay and silt are generally higher in organic matter than are coarser-textured soils (Figure 10.10). The rate of organic residues returned to the soil is also usually higher in the finer-textured soils, and the rate of oxidation may be somewhat slower than in sandy soils. Also, as pointed out in Section 10.7, there are interactions between organic materials and clays. Through the organomineral complexes that result, soils high in clay are able to protect the protein nitrogen and, in turn, the organic matter from degradation. This results in higher organic matter contents.

Again, *poorly drained* soils, because of their high moisture content and relatively poor aeration, are generally much higher in organic matter and

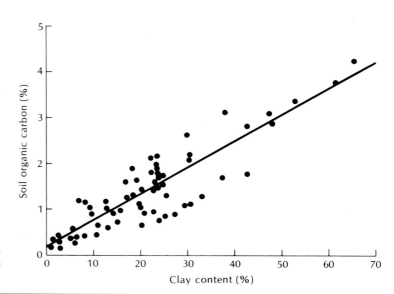

FIGURE 10.10

The effect of clay content on the soil organic carbon levels in the southern Great Plains area of the United States. [From Nichols (1984)]

nitrogen than their better drained equivalents (Figure 10.9). For instance, soils lying along streams are often quite high in organic matter, owing in part to their poor drainage.

INFLUENCE OF CROPPING AND TILLAGE

A very marked change in the soil organic matter content occurs when a virgin soil, developed under either forest or prairie, is brought under cultivation. The changes are illustrated in Figure 10.11, which shows the general decline in organic matter of grassland soils with time of cultivation. It is common to find cropped land much lower in both nitrogen and organic matter than comparable virgin areas. The decrease is not surprising because in nature all the organic matter produced by the vegetation is returned to the soil, whereas in cultivated areas much of the plant material is removed for human or animal food and relatively less finds its way back to the land. Also, soil tillage breaks up the organic residues and brings them into direct contact with soil organisms, thereby increasing the rate of decomposition.

Modern conservation tillage practices help maintain high surface soil organic matter levels (Figure 10.12). Compared to conventional tillage, these practices leave a higher proportion of the residues on or near the soil surface. This technique protects the soil from erosion and also discourages the rapid decay of the crop residues.

INFLUENCE OF ROTATIONS, RESIDUES, AND PLANT NUTRIENTS

The nature and sequence of crops as well as the use of lime, fertilizers, and manures influence soil organic matter levels. The relationships are illustrated by the graphs in Figure 10.13 showing data from the famous Morrow plots at the University of Illinois. Different crop sequences and manure, lime, and fertilizer treatments were initiated on these plots in 1876. Figure 10.13 shows the organic matter content on selected plots since 1903. A study of this figure verifies the following conclusions.

1. A rotation of corn, oats, and clovers resulted in a higher soil organic matter level than did continuous corn. The nitrogen fixed by the clover

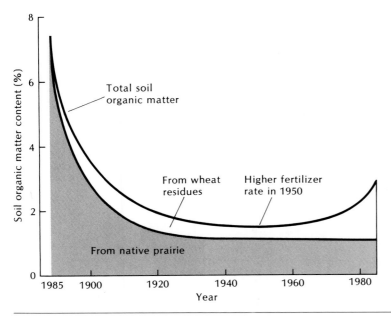

FIGURE 10.11

Changes in soil organic matter when a Udollic Ochraqualfs with prairie vegetation was plowed and cropped annually to wheat. Note the very rapid decline during the first 30 years and leveling off thereafter. Also note that 60 years after the cropping started most of the organic matter remaining is the very stable forms found in the original prairie soil when it was first plowed. When a higher rate of fertilizer was applied starting in 1950, wheat yields increased, and the increased residues resulted in an increase in soil organic matter. [Drawn from data in Balesdent et al. (1988)]

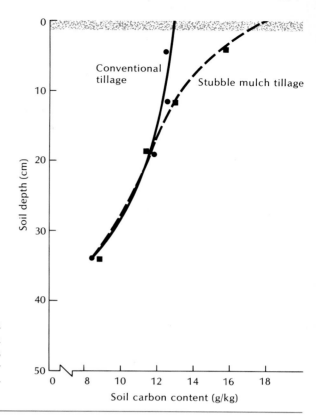

FIGURE 10.12
Effect of 44 years of conventional and stubble mulch tillage on the organic carbon content of a Mollisol (Typic Haploxeroll) in a semiarid region of Oregon. The stubble mulch tillage leaves much of the crop residues on or near the soil surface, thereby reducing its rate of decomposition. [Drawn from data of Rasmussen and Rhode (1988)]

and the higher crop residues from this rotation likely accounted for the higher level of organic matter.

2. The application of manure, lime, and phosphorus helped maintain much higher organic matter levels, especially where a rotation of crops

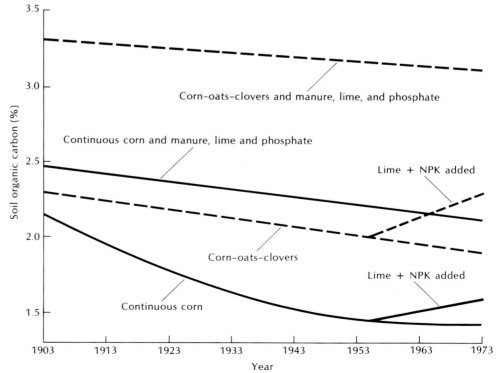

FIGURE 10.13
The soil organic carbon contents of selected treatments of the famous Morrow plots at the University of Illinois. Since the treatments were started in 1876, they had been under way for 27 years in 1903 when the first measurements shown were recorded. Note that the addition of manure, lime, and phosphate along with crop rotation has helped maintain higher organic carbon levels. Also note the application of lime, nitrogen, phosphate, and potassium (LNPK) in 1955 to previously untreated plots had resulted in significant organic carbon increases by 1973. [From Odell et al. (1984)]

was followed. Again, the increased return of organic matter through the added manure and increased crop residues likely accounts for this high soil organic matter level.

3. The application of lime and fertilizers (LNPK) to previously unfertilized and manured plots starting in 1955 resulted in increased soil organic matter levels. This increase was probably also due to the increased returns to the soil of crop residues coming from the higher crop yields that resulted from the lime and fertilizer applications.

The results on these world-famous plots remind us of the importance of high soil productivity in the maintenance of soil organic matter. Soils that are kept highly productive by supplemental applications of fertilizers, lime, and manure and by the choice of high-yielding crop varieties are likely to have more organic matter than comparable less productive soils. The amounts of root and top residues returned to the soil are invariably dependent on the level of soil productivity. Nevertheless, the most productive cultivated soil will likely be considerably lower in organic matter than soil in a nearby virgin area.

10.15

Regulation of Soil Organic Matter in Mineral Soils

A few principles governing organic matter regulation should be emphasized. First, no attempt should be made to try to maintain higher soil organic matter levels than the soil–plant–climate control mechanisms dictate. It would be foolhardy, for example, to try to maintain as high a level of organic matter in Texas soils as is found in Minnesota.

Second, because of the linkage between soil nitrogen and organic matter, management practices must maintain adequate nitrogen inputs as a requisite for adequate organic matter levels. Accordingly, the inclusion of legumes in the crop rotation and judicious use of nitrogen-containing fertilizers to enhance high soil productivity are two desirable practices. At the same time, steps must be taken to minimize the loss of nitrogen by leaching, erosion, or volatilization (see Section 11.20).

Third, the soil must receive a continuous supply of organic materials if the soil organic level is to be maintained (see Section 10.14). Animal manures, composts, organic wastes, and crop residues are the primary sources of these organic materials. In addition, some cover crops or crops grown specifically to supply organic materials are turned into the soil as *green manures*. Although the total organic additions needed to maintain soil organic matter will vary from one situation to another, research in the Midwest and Northwest suggests that additions of 5–6 Mg/ha per year are required for this purpose.

Fourth, attempts should be made to minimize the constraints to crop production. Moderate applications of lime and fertilizers should help remove constraints from chemical toxicities and deficiencies. Vigorously growing plants leave root and top residues on which an adequate organic level depends.

Fifth, mechanical stirring of the soil by tillage should be only that needed to control weeds and to maintain adequate soil aeration. Conservation tillage practices that minimize tillage (see Section 15.12) should be followed to the degree feasible. These practices leave much of the residues on or near the soil surface and thereby slow down the rate of residue decay. In time, conservation tillage could lead to higher organic matter levels.

FIGURE 10.14
How the difficulty and expense of maintaining the organic matter of cultivated soils increase as the support level is raised. While the curves are quite similar, their positions in relation to the percentage of organic matter possible differ with *texture* [compare (a) with (c)] and *climate* [compare (b) with (c)]. Other factors also are involved, especially the type of crop rotation employed. In practice the average amount of organic matter in a cultivated soil should be maintained as high as is economically feasible.

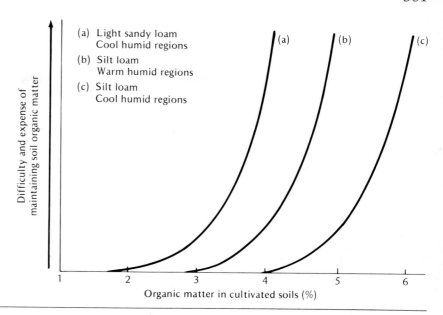

(a) Light sandy loam
   Cool humid regions
(b) Silt loam
   Warm humid regions
(c) Silt loam
   Cool humid regions

Difficulty and expense of maintaining soil organic matter

Organic matter in cultivated soils (%)

Sixth and last, there should be no hesitation about taking the land out of cultivation and letting it return to its natural vegetation state where such a move is appropriate. Large areas of land are under cultivation today that never should have been cleared.

ECONOMY IN HUMUS MAINTENANCE

Since the rate at which carbon is lost from the soil increases rapidly as the organic content is raised, maintenance of the humus at a high level is not only difficult but also expensive (Figure 10.14). It is, therefore, unwise to hold the organic matter above a level consistent with crop yields that have the best return. Just what this level should be depends on climatic environments, soil conditions, and the particular crops grown and their sequence. Obviously, organic matter contents should be higher in North Dakota than in central Kansas, where the temperature is higher, or in northern Montana, where the effective rainfall is lower. In any event, the soil organic matter and nitrogen should always be maintained at as high a level as is economically feasible.

10.16

Organic Soils (Histosols)

Discussion up to this point has focused on organic matter in predominately mineral soils. Attention will now be given to those soils whose properties are dominated by organic matter. These soils are termed *organic* soils and are classified in the *Histosols* order in *Soil Taxonomy*.

Histosols are by no means as extensive as mineral soils, occupying less than 1% of the world's land area. However, when properly drained and managed, Histosols are among the world's most productive soils, especially for high-value vegetable and floriculture crops.

Technically, all organic soils must have at least 20% organic matter (12% C) by weight. This minimum level increases with increased clay content until it reaches 30% if as much as 60% of the mineral matter is clay. Soils with these and higher organic matter contents have distinct physical and chemical properties.

10.17

Genesis of Organic Deposits

Organic deposits accumulate in marshes, bogs, and swamps, which are habitats for water-loving plants such as pondweeds, cattails, sedges, reeds, mosses, shrubs, and even some trees. For generations, the residues of these plants have sunk into the water, which by reducing oxygen availability has inhibited their oxidation and, consequently, has acted as a partial preservative.

LAYERING OF PEAT BEDS

As one generation of plants follows another, layer after layer of organic residue is deposited in the swamp or marsh (Figure 10.15). The constitution of these successive layers changes as time goes on because a sequence of different plant life occurs. The succession is by no means regular or definite, as a slight change in climate or water level may alter the sequence entirely. The profile of an organic deposit is, therefore, characterized by layers that differ in their degree of decomposition and in the nature of the original plant tissue.

10.18

Distribution, Nature, and Use of Peat Accumulations

Peat deposits are found all over the world. Although they are most extensive in areas with cold climates, peat beds are found even in the tropics. Of approximately 300 million ha of peatlands worldwide, about 7.5 million ha are found in the United States. Three quarters of this peatland is in glaciated areas where peat accumulation was encouraged because drainage of some lands was impeded by the standing glaciers. In Wisconsin, Minnesota, New York, and Michigan, materials for large areas of productive organic soils were laid down during glaciation. Other peatlands (about one quarter) are found in near-coastal areas such as the Everglades in Florida, the tule-reed beds of California, and similar water-endowed areas in Louisiana and Texas.

TYPES OF PEAT

Based on the nature of the parent materials, peat deposits are grouped into three general types: *fibrous, woody,* and *sedimentary.* Fibrous peats are the accumulation of fibrous materials from a variety of plants including sedges, sphagnum and other mosses, grasses, and cattails. These peats have unusually high water-holding capacities. As they decay, fibrous materials can make satisfactory field soils, although some of them (e.g., moss peats) may be quite acid because of their low ash content.

Woody peats develop from residues of trees and are most commonly (but not always) at the surface of the organic accumulation. Cultivated woody peats have a characteristic granular structure and are brown to black in color,

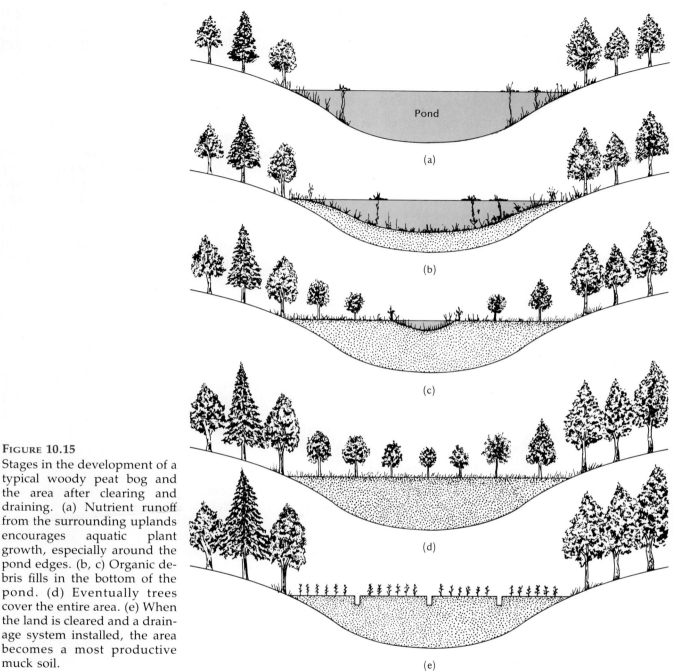

Pond

(a)

(b)

(c)

(d)

(e)

FIGURE 10.15
Stages in the development of a typical woody peat bog and the area after clearing and draining. (a) Nutrient runoff from the surrounding uplands encourages aquatic plant growth, especially around the pond edges. (b, c) Organic debris fills in the bottom of the pond. (d) Eventually trees cover the entire area. (e) When the land is cleared and a drainage system installed, the area becomes a most productive muck soil.

according to the degree of humification. The water-holding capacity of woody peat is somewhat lower than that of fibrous peat, making it less desirable for greenhouse and nursery use. However, woody residues produce a superior field soil, much prized for vegetable production. In the United States such woody peat is confined mostly to Wisconsin, Michigan, and New York.

Sedimentary peat is generally undesirable as a soil. Formed from mixtures of various materials deposited in deep water, this material is highly colloidal and compact and also rubbery when wet. On drying, however, sedimentary peat resists rewetting and remains in a hard, lumpy condition, making it unsatisfactory for use in growing plants. Fortunately, in most cases it occurs far down in the profile and is unnoticed unless it interferes with drainage of the bog area.

USES OF PEAT

In the United States, the most extensive use of peat is as a field soil, especially for vegetable and flower production. Peat products are also used widely for potted plants, and for home flower and vegetable gardens, and as a mulch on home gardens and lawns as well as for commercial greenhouse production. Fiber pots made of compressed peats are used to start plants, which can later be transplanted to soil without being removed from the pot. Peat utilization is a profitable business in Europe and North America.

Peat deposits are used extensively for fuel, especially in the Soviet Union where energy for some power stations is derived from peat. The use of peat for domestic fuel is declining but is still prevalent in remote areas of the Soviet Union and elsewhere in Europe.

10.19

Classification of Organic Soils

Two types of classification are employed for organic soils, one using traditional field terminology and the other the terminology of the *Soil Taxonomy* classification system. In each case, considerable emphasis is placed on the stage of breakdown of original plant materials.

PEAT VERSUS MUCK

As a practical matter, organic soils are commonly differentiated on the basis of their state of decomposition. Those deposits that are slightly decayed or nondecayed are termed *peat*; those that are markedly decomposed are called *muck*. In peat soils, one can identify the kinds of plants in the original deposit, especially in the upper horizons, whereas mucks are generally decomposed to the point that the original plant parts cannot be identified.

SOIL TAXONOMY—ORDER

In the *Soil Taxonomy* system, organic soils are identified as the order *Histosols* (see Section 3.7). Four suborders have been established. Three of the suborders, *Fibrists*, *Hemists*, and *Saprists*, are of agricultural importance. Their major characteristics are shown in Table 10.2. Note that the degree of decomposition of the original plant materials is the dominant feature dis-

TABLE 10.2

The Names and Major Characteristics of the Three Suborders of Histosols of Greatest Agricultural Significance

Suborder	Degree of organic decay	Bulk density (Mg/m³)	Common color	Water-holding capacity
Fibrist	Low	<0.1	Brown to yellow	High
Hemist	Intermediate	0.1–0.2	Intermediate	Intermediate
Saprist	High	>0.2	Dark grey to black	Low

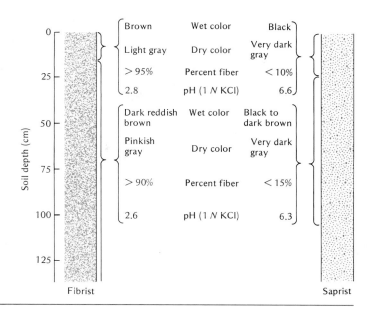

Figure 10.16

Two organic soil profiles showing distinguishing characteristics of each. (a) An uncultivated soil of the Fibrist suborder with a high percentage of undecomposed fiber and high acidity. (b) In contrast, a cultivated soil of the Saprist suborder with only 10–20% undecomposed fibers and reasonably low acidity. Both soils are found in Minnesota. [Modified from Farnham and Finney (1966)]

tinguishing among these suborders, the Fibrists having the least decayed materials, the Saprists the most, and the Hemists being intermediate in this regard.

Drawings of organic soil profiles representative of the Saprist and Fibrist suborders are shown in Figure 10.16. The Fibrist soil has a high percentage of undecomposed fibrous materials and is very acid. The cultivated Saprist soil has a nearly neutral pH, low undecomposed fiber percentage, and is darker in color.

10.20

Physical Characteristics of Field Peat Soils

Color

A typical cultivated Histosol is dark brown to intensely black in color even though it may have developed from materials that were gray, brown, or reddish brown. The changes that the organic matter undergoes as it decomposes seem to be similar to those occurring in the organic residues of mineral soils as they are broken down.

Bulk Density

Histosols are light in weight when dry. The bulk density of a peat surface soil is only 0.20–0.30 Mg/m^3 compared to 1.25–1.45 Mg/m^3 for mineral surface soils. Thus, a hectare–furrow slice of a dry Histosol weighs only 15–20% as much as a dry mineral soil.

Water-Holding Capacity

A third important property of a Histosol is its high water-holding capacity on a mass basis. While a dry mineral soil will adsorb and hold from one fifth to two fifths its weight of water, a well-humified organic soil will retain two to four times its dry weight of moisture. Undecayed or only slightly decomposed moss or sedge peat has an even greater water-holding capacity, being able to hold water to the extent of 12, 15, or even 20 times its dry weight. This explains in part the value of moss and sedge peat in greenhouse and nursery operations.

Unfortunately, Histosols in the field do not greatly surpass mineral soils in their capacity to supply plants with water for two reasons. First, their amounts of *unavailable* water are much higher proportionately than those of mineral soils. Second, the light weight of organic soils means that a given *volume* of organic soil may not hold much more water than the same volume of a mineral soil. For these two reasons a given volume of peat soil at optimum moisture will supply only slightly more water to plants than a comparable mineral soil.

STRUCTURE

A fourth outstanding characteristic of a typical woody or fibrous organic soil is its almost invariably loose physical condition. While humified organic matter is largely colloidal and possesses high adsorptive powers, its cohesion and plasticity are rather low. Most Histosols are, therefore, porous, open, and easy to cultivate. These characteristics make this type of soil especially desirable for vegetable production. However, during dry periods, light, loose peat, whose granular structure has been destroyed by cultivation, may erode badly in a high wind and result in extensive crop damage. Peat also may ignite when dry. Such a fire often is difficult to extinguish and may continue for several years.

10.21

Chemical Characteristics of Histosols

The colloidal nature of organic soil material already has been emphasized (see Section 7.2). The same graphic formula employed for mineral soils can be used to represent the colloidal complex of organic soils.

$$
\begin{array}{l}
Ca^{2+} \\
H^+ \\
Mg^{2+}
\end{array}
\boxed{\text{Micelle}}
$$

The micelle is humus, and the base-forming cations are adsorbed in the same order of magnitude as in mineral soils: $Ca^{2+} > Mg^{2+} > K^+$ or Na^+ (see Section 7.3).

CATION EXCHANGE

On a mass basis, maximum cation exchange capacities of organic colloids are two to ten times higher than those for the inorganic colloids. But on a volume basis, as shown in Table 10.3, most of this difference disappears. In fact, a cubic meter of vermiculite has about twice the cation exchange capacity of an equal volume of organic colloid. Under acid conditions, this difference is even more pronounced since the negative charges on the organic colloids are pH dependent in contrast to the permanent charges on 2:1-type clay colloids. However, fine-grained micas, kaolinite, and hydrous oxide clays have lower capacities than the humus, even on a volume basis.

TABLE 10.3

Representative Maximum Cation Exchange Capacities of an Organic Colloid and Several Inorganic Colloids

On a weight basis, the CEC of humus greatly exceeds that of the minerals, but on a volume basis both vermiculite and smectites have higher capacities

	Cation exchange capacity	
Colloid	Weight basis (cmol/kg)	Volume basis (cmol/L)
Humus (organic)	300	75
Vermiculite	120	150
Smectite	90	113
Fine-grained micas	25	31
Kaolinite	5	6
Hydrous oxides	3	4

Because plants, whether in the field or the greenhouse, absorb their nutrients from a given volume rather than mass of soil, the volume basis for evaluating cation exchange capacity appears to be more appropriate. In some deep, artificially drained peat soils, however, deep root penetration is encouraged, resulting in a large volume of soil from which nutrients can be absorbed.

FORMS OF NUTRIENT ELEMENTS

In mineral soils, most of the essential nutrients are held in relatively unavailable forms, with only small amounts being in exchangeable and soil solution forms. As the data in Table 10.4 illustrate, the opposite is true for organic soils, at least for base-forming cations. High percentages of calcium, magnesium, and potassium are held in exchangeable form by the organic colloids. Thus, most of these soil-held cations are available for plant growth and for soil microorganisms.

pH AND BUFFERING RELATIONS

On a mass basis, Histosols have a higher buffering capacity against pH change than do mineral soils; but on a volume basis, the differences tend to disappear. The plow layer of an organic soil may be more responsive to pH change from additions of a given rate of limestone than would some mineral soils.

TABLE 10.4

Percent of Three Cations Commonly Found in Exchangeable Forms in Organic and Mineral Soils

The remainder is held in the solid framework of the soils.

	Percent in exchangeable form	
Cation	Mineral soil	Organic soil
Calcium	25	80
Magnesium	7	70
Potassium	0.5	35

TABLE 10.5
Cation Exchange Data for Representative Organic and Inorganic Soils

Exchange characteristics	Weight basis (cmol/kg)		Volume basis (cmol/L)	
	Organic soil	Mineral soil	Organic soil	Mineral soil
Exchangeable Ca	150	8	38	10
Other exchangeable bases, M	40	3	10	4
Exchangeable H and Al	60	5	15	6
Cation exchange capacity	250	16	63	20
Percentage base saturation	76	69	76	69
pH	5.0–5.2	5.6–5.8	5.0–5.2	5.6–5.8

It is important to note that the pH of a Histosol at a given percentage base saturation is generally lower than that of a representative mineral soil. Data in Table 10.5 illustrate this point. The representative organic soil has a higher percentage base saturation than the mineral soil, yet the pH of the organic soil is lower. Note also that the difference in cation exchange capacity between the organic and mineral soils is much less when they are compared on a volume basis.

NITROGEN AND ORGANIC MATTER

Histosols are variable chemically, but they have some characteristics in common. A representative analysis of the surface layer of an arable organic soil is given in Table 10.6, along with a similar analysis of a representative humid region mineral soil.

The high nitrogen and organic contents of organic soils are self-evident and need no further emphasis. Note, however, that Histosols have a C/N ratio of about 20:1 compared to 12:1 for a representative cropped mineral soil. Even so Histosols show vigorous nitrification (nitrate release) in spite of their high C/N ratio. Apparently some of the carbon in peats is very resistant to microbial attack and is not readily usable by general-purpose decay organisms. Consequently, these organisms are not excessively encouraged, and they do not tie up the nitrates.

PHOSPHORUS AND POTASSIUM

Both the phosphorus and potassium contents of peat are low, the latter exceedingly so in comparison with a mineral soil (Table 10.6). To grow crops on an organic soil, therefore, requires applying phosphorus as well as potassium in large amounts. Unlike mineral soils, organic soils do not fix

TABLE 10.6
Analyses for Representative Peat and Mineral Surface Soils

Constituent	Peat		Mineral	
	g/100 g soil	kg/HFS	g/100 g soil	kg/HFS
Organic matter	80.00	440,000	4.0	88,000
Nitrogen (N)	2.50	13,750	0.15	3,300
Phosphorus (P)	0.15	825	0.05	1,100
Potassium (K)	0.10	550	1.70	37,400
Calcium (Ca)	2.00	11,000	0.40	8,800
Magnesium (Mg)	0.30	1,650	0.30	6,600
Sulfur (S)	0.60	3,300	0.04	880

phosphorus and potassium. In particular, phosphorus solubility is affected relatively little by pH in organic soils.

CALCIUM AND pH

The high calcium content of a hectare–furrow slice of organic soils is apparently due to the high content of this element in the seepage water that enters the swamp from the surrounding uplands. The calcium is less subject to leaching than in mineral soils and is adsorbed by the organic materials. Even with reasonable levels of exchangeable calcium, however, most Histosols are distinctly acid, often very markedly so.

MAGNESIUM AND SULFUR

The amount of magnesium in a hectare–furrow slice of an organic soil is usually much less than in a mineral soil (Table 10.6). Organic soils, long intensively cropped, may develop a magnesium deficiency unless fertilizers or limestone containing this constituent have been used.

The abundance of sulfur in Histosols is expected because plant tissue always contains considerable sulfur. Consequently, organic deposits are comparatively high in this constituent. When sulfur oxidation is vigorous, sulfur is abundant enough to meet most crop plant needs.

10.22

Management of Histosols

The productivity of organic soils depends as much on proper management as that of mineral soils. When the land is first cleared and drained, it often is used as a pasture for a few years or is devoted to nonintensive crops such as rice, oats, or rye. These require a minimum of seasonal care but will provide some return while the organic debris is being broken down. Heavy tillage implements are often used to help with this process. After a few years, the land is ready for more intensive cash-crop production.

DRAINAGE AND WATER TABLE

Water levels for organic soils should be maintained between 45 and 75 cm from the soil surface. This assures a ready water supply for vegetables and other shallow-rooted crops grown on these soils. It also keeps water close enough to the surface to minimize surface dryness and thereby reduce wind erosion.

Another benefit of keeping reasonably high water tables is to reduce the rate of oxidation of the organic matter and the associated subsidence of the soil. Subsidence can be as high as 5 cm a year if the soil is allowed to dry to a depth of a meter or more. Even with water tables of about 50 cm, the annual subsidence rate is commonly 1 cm a year. This means that the water table should be kept no lower than necessary to assure adequate root aeration.

STRUCTURAL MANAGEMENT

Plowing is not ordinarily necessary because peat is porous and open unless it contains considerable silt and clay. In fact, a cultivated Histosol may need packing rather than loosening. For this reason, a roller or packer is an important implement in the management of such land. Cultivation tends to

FIGURE 10.17
Windbreaks, such as these in Michigan, help protect valuable muck land from blowing. The unprotected field in the lower left has been wetted by sprinkler to prevent its blowing. [Courtesy USDA Soil Conservation Service]

destroy the original granular structure, leaving the soil in a powdery condition when dry. It is then susceptible to wind erosion, a serious problem in some sections (Figure 10.17).

USE OF LIME

Lime, which often must be used on mineral soils, usually is less necessary on Histosols unless they have developed in regions low in calcium in the surrounding uplands. On acid mucks containing appreciable quantities of inorganic matter, however, the situation is quite different. The highly acid conditions result in the dissolution of iron, aluminum, and manganese in toxic quantities. Under these conditions, large amounts of lime may be necessary to obtain normal plant growth—but overliming should be avoided. At high pH levels, micronutrients such as zinc, copper, and manganese are bound by the organic materials, which leads to deficiencies of these elements.

COMMERCIAL FERTILIZERS

Of much more widespread importance than lime are commercial fertilizers. As organic soils are very low in phosphorus and potassium, these elements must be added. Except on freshly cleared peats, decay and nitrate release (nitrification) are frequently too slow to meet the crop demands for nitrogen. Consequently, this element is needed in large amounts along with the phosphorus and potassium, especially where vegetables are being grown.

MICRONUTRIENTS

Organic soils need not only potassium, phosphorus, and nitrogen but often also need some of the micronutrient elements such as copper, zinc, and manganese. Boron deficiencies also are becoming evident.

In management terms, one is encouraged to think of organic soils as distinctly and even radically different from most mineral soils. In many respects, this is true. Yet, fundamentally, the same types of change occur in the two groups; nutrients become available in much the same way, and their management is based on the same principles of fertility and water management.

10.23

Peat and Container-Grown Plants

For many years, plants have been grown in containers of various kinds. Researchers have long used potted plants for their experiments, and homeowners use them for various purposes, especially for indoor flower enjoyment. Recently, there has been large-scale expansion of commercial production of plants in pots, primarily for ornamentals but also to a lesser extent for food crops.

Root growth in containers presents special problems, primarily physical. Closed pots frequently become water-logged because of the difficulty in supplying exactly the proper amount of water. Overwatering causes water-logging. Most containers have holes at the bottom to drain out excess water, but the water drains out of holes at the bottom of the pot only when the soil at the bottom is saturated. Under these conditions, if the potting materials are mostly mineral soil, there are few air-filled pore spaces and the root zone can rapidly become anaerobic. The problem is intensified because the low soil volume the plant has available makes frequent watering necessary.

Thus management of plants in containers is much different than management of in-field plants. One solution for the problems has been substitution of mixtures of coarse organic and inorganic particles for field soils in containers. Originally these mixtures were termed "potted composts" and contained some field soil and compost. More recently, commercial peat and wood products (sawdust, bark, chips) have become the dominant organic components, and vermiculite or perlite (both geological materials), porous volcanic rock (scoria), or sands are the most important inorganic constituents. These materials retain water within the coarse particles themselves, but have large air-filled pores between the particles through which oxygen can be supplied to the plant roots, even immediately after watering. An added advantage is the low bulk density of the materials, which lowers shipping and handling costs, for ornamentals particularly. The versatility of peat and similar organic materials is illustrated by this example.

10.24

Conclusion

The importance of organic matter in enhancing the usefulness of soils for agricultural purposes cannot be overemphasized. It supplies essential nutrients and has unexcelled capacity to hold water and adsorb cations. But the indirect effects of organic matter on soil structure, soil water, soil aeration, and soil temperature likely exceed in importance its function as a direct supplier of water and nutrients. Organic matter also functions as a source of food for soil microbes and thereby helps enhance and control their activities.

The maintenance of soil organic matter in mineral soils is perhaps the most important challenge to modern and traditional agriculture alike. By encouraging high crop yields, abundant residues (including nitrogen) can be returned to the soil either directly or through feed-consuming animals. Also the rate of destruction of soil organic matter can be minimized by restricting soil tillage and by keeping at least part of the crop residues at or near the surface.

The highly productive organic soils (Histosols) share some properties with the organic matter in mineral soils. However, the wetland conditions under which these soils formed have resulted in certain unique properties that set them apart from their inorganic counterparts. Practical considerations suggest that soil volume rather than soil mass be used to compare the management needs of organic soils with those of mineral soils. For organic soils the water table should be kept as near the surface as good crop growth will permit to prevent wind erosion as well as the rapid decomposition of the peat and the subsequent subsidence of the soil.

Study Questions

1. Assume you mix a mass of tree leaves into a flower garden bed. List the five groups of organic compounds likely to be found in these residues and indicate which ones are (a) a major source of nitrogen for plant nutrition; (b) the major source of carbon dioxide during the first few weeks of decomposition; and (c) the group that, when modified by microbes, provides compounds that are included among the so-called humic materials in soils.

2. A farmer incorporates wheat straw into the upper few inches of a reasonably productive soil and then plants a corn crop. Very soon the corn crop shows evidence of nitrogen deficiency, which is not displayed by an adjacent corn field where no wheat straw was incorporated. How do you account for this difference? Would you expect the same difference if the farmer had turned under a legume crop such as alfalfa? Explain.

3. Why is the carbon cycle sometimes referred to as the "Cycle of Life"?

4. A home gardening "expert" in a weekly radio broadcast explains that humus is merely a mixture of compounds that originally were found in plants and that have resisted microbial decay. Do you agree or disagree with the expert? Explain.

5. Distinguish between humic and nonhumic organic compounds in soils and indicate which are the major sources for the following properties of soil organic matter: (a) high cation exchange capacity; (b) aggregate formation and stability; (c) high molecular weight; (d) persistence of organic matter in soils; and (e) high content of polyphenols and polyquinones.

6. Simple sugars are subject to ready decomposition by soil organisms and much of these organic materials disappears in a matter of weeks. For what period of time have the more resistant humic compounds remained in the soil?

7. Explain how humus enhances the processes of soil formation and of nutrient availability.

8. Soils in Minnesota are known to be higher in organic matter than soils developed from comparable parent materials in Kansas. In turn, soils in Kansas contain more organic matter then those in New Mexico. What accounts for the difference in each case?

9. Soil and crop management practices that ignore the maintenance of soil nitrogen are not likely to help maintain soil organic matter levels. What is the reason for this?

10. What was the effect on soil organic matter of clearing areas in the United States with Mollisols and Alfisols of their native vegetation and then cropping those land areas? Why did the change take place?

11. What would you expect has been the effect of irrigation and crop production on the organic matter level of an Aridisol in Arizona? Explain.

12. The application of lime and fertilizer effected an increase in the soil organic matter content of a continuous corn plot that had previously received no fertilizer or lime for a period of 69 years. Explain how these chemicals, which contain no organic matter, could bring about this change.

13. Describe how the organic materials from which Histosols form are deposited and indicate where these materials are most prevalent in the United States and in the world.

14. The cation exchange and water-holding capacities of Histosols are known to be much higher than those of adjacent mineral soils. Yet an acre–furrow slice of a Histosol may provide little more exchangeable nutrients and water to plants than the mineral soils. Explain.

15. Contrast the physical and chemical properties of Histosols and mineral soils.

References

BALESDENT, J., G. H. WAGNER, and A. MARIOTTI. 1988. "Soil organic matter turnover in long-term field experiments as revealed by carbon-13 natural abundance," *Soil Sci. Soc. Amer. J.*, **52**:118–24.

FARNHAM, R. S., and H. R. FINNEY. 1966. "Classification of organic soils," *Adv. Agron.*, **17**:115–62.

FRIMMEL, F. H., and R. F. CHRISTMAN. 1988. *Humic Substances and Their Role in the Environment* (New York: Wiley).

HAYNES, R. J. 1986. "The decomposition process: mineralization, immobilization, humus formation and degradation" in R. J. Haynes (Ed.), *Mineral Nitrogen in the Plant–Soil System*, (New York: Academic Press).

HUANG, P. M., and M. SCHNITZER (Eds.). 1986. *Interactions of Soil Minerals with Natural Organics and Microbes*, SSSA Special Publication No. 17 (Madison, WI: Soil Sci. Soc. Amer.)

JENNY, H. 1941. *Factors of Soil Formation* (New York: McGraw-Hill).

LOLL, M. J., and J. M. BOLLAG. 1983. "Protein transformation in soil," *Adv. Agron.*, **36**:352–82.

LYON, T. L., H. O. BUCKMAN, and N. C. BRADY. 1952. *The Nature and Properties of Soils*, 5th ed. (New York: Macmillan), p. 171.

MCKENZIE, W. E. 1974. "Criteria used in soil taxonomy to classify organic soils," in *Histosols: Their Characteristics, Classification and Use*, SSSA Special Publication No. 6 (Madison, WI: Soil Sci. Soc. Amer.), pp. 1–10.

NICHOLS. 1984. "Relation of organic carbon to soil properties and climate in the southern Great Plains," *Soil Sci. Soc. Amer. J.*, **48**:1382–84.

ODELL, R. T., S. W. MELSTED, and V. M. WALKER. 1984. "Changes in organic carbon and nitrogen of Morrow plots under different treatments 1904–1973," *Soil Sci. Soc. Amer. J.*, **137**:160–71.

PARENT, L. E., J. A. MILLETTE, and G. R. MEHUYS. 1982. "Subsidence and erosion of a Histosol," *Soil Sci. Soc. Amer. J.*, **46**:404–08.

RASMUSSEN, P. E., and C. R. RHODE. 1988. "Long-term tillage and nitrogen fertilization effects on organic nitrogen and carbon in a semi-arid soil," *Soil Sci. Soc. Amer. J.*, **52**:1114–17.

RUSSELL, E. J., and E. W. RUSSELL. 1950. *Soil Conditions and Plant Growth* (London: Longmans, Green), p. 194.

STEVENSON, F. J. 1982a. "Origin and distribution of nitrogen in soil," in F. J. Stevenson (Ed.), *Nitrogen in Agricultural Soils* (Madison, WI: Amer. Soc. Agron., Crop Sci. Soc. Amer., and Soil Sci. Soc. Amer.), pp. 1–42.

STEVENSON, F. J. 1982b. *Humus Chemistry—Genesis, Composition Reactions* (New York: Wiley).

TATE, R. L. III. 1987. *Soil Organic Matter: Biological and Ecological Effects* (New York: Wiley).

WAKSMAN, S. A. 1948. *Humus* (Baltimore: Williams & Wilkins), p. 95.

11

Nitrogen and Sulfur Economy of Soils

Of the various essential elements, nitrogen probably has been subjected to the most study, and for many good reasons still receives much attention.[1] The amount of this element in available forms in the soil is small, while the quantity withdrawn annually by crops is comparatively large. When there is too much nitrogen in readily soluble forms, it is lost in drainage and may become a water pollutant. Nitrogen can be added to the soil by some microbes that "fix" it from the atmosphere, and can then be released back to the atmosphere by still other organisms. Nitrogen can acidify the soil as it is oxidized. Most soil nitrogen is unavailable to higher plants. All in all, nitrogen is an important nutrient element that must be conserved and carefully managed.

11.1

Influence of Nitrogen on Plant Development

FAVORABLE EFFECTS

Nitrogen is an integral component of many compounds, including chlorophyll and enzymes, essential for plant growth processes. It is an essential component of amino acids and related proteins, which are critical not only as building blocks for plant tissue but also in the cell nuclei and protoplasm in which hereditary control is vested. Nitrogen is essential for carbohydrate use within plants and stimulates root growth and development as well as the uptake of other nutrients.

---

[1] Nitrogen in soils and in crop production is covered extensively in the reviews edited by Stevenson (1982) and Hauck (1984).

Plants respond quickly to applications of nitrogen. This element encourages aboveground vegetative growth and gives a deep green color to the leaves. It increases the plumpness of cereal grains and tends to produce succulence, a quality particularly desirable in such crops as lettuce and radishes. Nitrogen deficiency is evident when the older leaves of plants turn yellow or yellowish green and tend to drop off.

OVERSUPPLY

When too much nitrogen is applied, excess vegetative growth occurs, and the plants lodge (fall over) with the slightest wind. Crop maturity is delayed, and the plants are more susceptible to disease and insect pests.

However, not all plants are harmed by large amounts of nitrogen. Many crops, such as the grasses and vegetables, need nitrogen in quantity for optimum development. Detrimental effects to the vegetable crops and grasses should not result unless excessive quantities of nitrogen are applied or nitrate levels in the foliage become toxic to humans or other animals. At the same time, potential adverse effects on the environment must be considered.

11.2
Origin and Distribution of Nitrogen

The nitrogen contents of surface mineral soils normally range from 0.02 to 0.5%, a value of about 0.15% being representative. A hectare of such a soil would likely contain about 3.3 Mg of nitrogen while the air above that hectare of soil would contain nearly 300,000 Mg of the element. Obviously, the atmosphere (80% N) is a seemingly limitless source of nitrogen although it is not readily usable by plants in elemental form.

Most of the soil nitrogen is in organic form. Proteins and other organic nitrogen compounds are associated with humus and with some silicate clays, which protect them from rapid microbial breakdown. Only about 2–3% of organic nitrogen is mineralized each year under normal conditions. Ammonium ions tightly bound (fixed) by clay may account for up to 8% of the nitrogen in surface soils and up to about 40% in subsoils. This clay-fixed nitrogen is only slowly available to plants.

The quantity of nitrogen in the readily available nitrate and ammonium forms is seldom more than 1–2% of the total soil nitrogen, except where large amounts of chemical fertilizers have been applied. This is fortunate because these soluble forms are easily lost from soils through leaching and volatilization. Only enough usable nitrogen is needed to supply the daily requirements of the growing plants.

11.3
The Nitrogen Cycle

The interaction among the various forms of nitrogen in soils, plants, and animals and nitrogen in the atmosphere constitute the *nitrogen cycle* (Figure 11.1). It has attracted scientific study for years, and its practical significance is beyond question.

The nitrogen income of arable soils is derived from such materials as commercial fertilizers, crop residues, green and farm manures, and ammonium and nitrate salts brought down by precipitation. In addition, certain

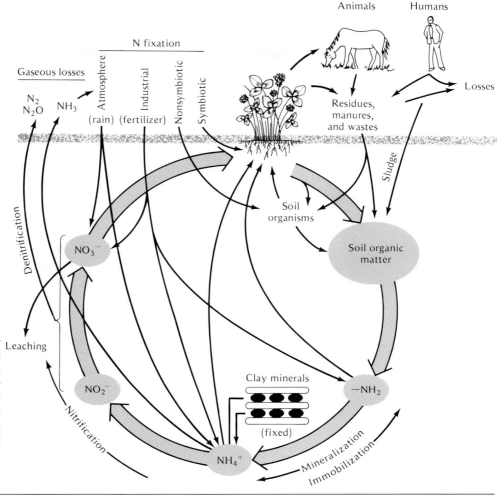

FIGURE 11.1
The nitrogen cycle, emphasizing the primary biological transformations (large shaded circle), the means of replenishment, and the losses from the soil. Chemical fertilizers from industrial nitrogen fixation are an increasingly important source of this element, but excessive use has serious environmental implications.

microorganisms can combine atmospheric nitrogen into compounds usable by plants. Nitrogen depletion results from crop removal, drainage, erosion, and loss in gaseous forms.

Much of the nitrogen added to the soil undergoes many transformations before it is removed. The nitrogen in organic combination is subjected to simplification, first to simple amino compounds[2] (R—NH_2), then to ammonium (NH_4^+) ion, and finally to nitrate (NO_3^-). Even then, nitrogen is allowed no rest because it is either appropriated by microorganisms and higher plants, removed in drainage, or lost by denitrification and volatilization. The cyclic transfer goes on and on. The mobility of nitrogen is remarkable, rivaling that of carbon.

MAJOR DIVISIONS OF THE NITROGEN CYCLE

At any one time, the great bulk of the nitrogen in a humid zone soil is in organic combinations protected from loss but largely unavailable to higher

[2] Amino acids such as glycine (CH_2NH_2COOH) and alanine (CH_3CHNH_2COOH) are examples of these simpler compounds. The R in the type formula represents the part of the organic molecule with which the amino group (NH_2) is associated. For example, for glycine, $R =$ —CH_2COOH.

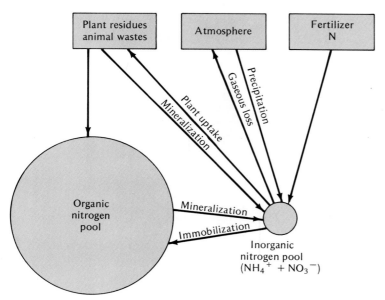

FIGURE 11.2
Two major pools of soil nitrogen, their primary sources and processes of transfer.

plants. For this reason, much scientific effort has been devoted to the study of organic nitrogen—how it is stabilized and how it may be released to forms usable by plants. The process of tying up nitrogen in organic forms is called *immobilization*; its slow release—specifically, organic to inorganic conversion—is called *mineralization* (Figure 11.2).

11.4
Immobilization and Mineralization

The process of *immobilization* is simply the conversion of inorganic nitrogen ions (NO_3^- and NH_4^+) into organic forms. It occurs most commonly when plant and animal residues, especially those low in nitrogen, are added to soils. As microorganisms attack the residues, they absorb the inorganic nitrogen ions and convert them to organic tissue where, in effect, the nitrogen is immobilized. When the organisms die, some of the organic nitrogen in their bodies may be converted into forms that make up the humus complex, and some may be released as NO_3^- and NH_4^+ ions.

The release of organically bound nitrogen to inorganic mineral forms (NH_4^+ and NO_3^-) is termed *mineralization*. Although a whole series of reactions are involved, the net effect can be visualized quite simply. Heterogeneous soil organisms simplify and hydrolyze the organic nitrogen compounds, ultimately producing the NH_4^+ and NO_3^- ions. The enzymatic process may be indicated as follows, using an amino compound (RNH_2) as an example of the organic nitrogen source.

$$\text{Mineralization} \longrightarrow$$

$$\mathbf{RNH_2} \underset{-H_2O}{\overset{+H_2O}{\rightleftharpoons}} \text{ROH} + \mathbf{NH_4^+} \underset{-O_2}{\overset{+O_2}{\rightleftharpoons}} \mathbf{NO_2^-} + 4\,H^+ \underset{-[O]}{\overset{+[O]}{\rightleftharpoons}} \mathbf{NO_3^-}$$

$$\longleftarrow \text{Immobilization}$$

Experiments with isotopically tagged nitrogen (^{15}N) have shown that only 2–3% of the immobilized nitrogen is mineralized annually. Even so, the release of nitrogen to inorganic forms has long supplied a significant portion of crop needs and may be about 60 kg N/ha per year for a representative surface soil in humid temperate regions. In arid zone soils, the quantities mineralized are less, except in irrigated areas that have received large inputs of plant residues.

11.5

Fate of Ammonium Compounds

The ammonium nitrogen (as shown in Figure 11.1) may move in five directions. First, considerable amounts are appropriated by soil micro-organisms who incorporate it into organic forms in their bodies. Second, higher plants are able to use this form of nitrogen, young plants being especially capable in this respect. Some plants, such as blueberries and some lowland rices, even prefer ammonium nitrogen to the nitrate form.

Third, ammonium ions are subject to interlayer fixation (binding) by vermiculite and, to a certain extent, by fine-grained micas and organic matter (Section 11.6). In these fixed forms, the nitrogen is not subject to rapid oxidation, although in time it may become available.

Fourth, some ammonia gas can be lost directly to the atmosphere (volatilized), especially in alkaline soils (Section 11.7). Such losses are significant when large quantities of ammonia are added to the soil in fertilizers. Finally, when plant and animal syntheses temporarily are satisfied, the remaining ammonium nitrogen may go in a fifth direction. It is readily oxidized by certain special-purpose forms of bacteria, first to nitrites and then to nitrates. The process, called *nitrification*, will be discussed in Section 11.8.

11.6

Ammonia Fixation

In agricultural soils, 5–20% of the total nitrogen is found as fixed ammonium ion, with 10% a representative figure. Both organic and inorganic soil fractions have the ability to bind or "fix" the ammonia.[3] However, different mechanisms and compounds are involved in these two types of fixation, and they will be considered separately.

FIXATION BY CLAY MINERALS

Several clay minerals with a 2:1-type structure have the capacity to "fix" both ammonium and potassium ions. Vermiculite has the greatest capacity, followed by fine-grained micas and some smectites. Ammonium and potassium ions are apparently just the right size to fit into the "cavities" between the crystal units of these minerals, thereby becoming trapped, or fixed, as a rigid part of the crystal (shown for K^+ in Figure 12.16). The ions are held in a

[3] This chemical fixation of ammonia is the entrapment or other strong binding of NH_4^+ ions by organic matter and certain silicate clays. This type of fixation (by which K^+ ions are similarly bound), is not to be confused with the very beneficial biological fixation of atmospheric nitrogen gas into compounds usable by plants (see Section 11.12).

TABLE 11.1

Total Nitrogen Levels of A and B Horizons of Four Cultivated Virginia Soils and the Percentage of the Nitrogen Present as Nonexchangeable or Fixed NH_4^+

Note the higher percentage of fixation in the B horizon.

	Total N (mg/kg)		Nitrogen fixed as NH_4^+ (%)	
Soil great group (series)	*A Horizon*	*B Horizon*	*A Horizon*	*B Horizon*
Hapludults (Bojac)	812	516	5	18
Paleudults (Dothan)	503	336	5	14
Hapludults (Groseclose)	1792	458	3	17
Hapludults (Elioak)	1110	383	6	26

From Baethgen and Alley (1987).

nonexchangeable form, from which they are released slowly to higher plants and microorganisms. The relationship of the various forms of ammonium might be represented as follows.

$$NH_4^+ \quad \rightleftharpoons \quad \boxed{\text{Micelle}}\,NH_4^+ \quad \rightleftharpoons \quad \boxed{\text{Micelle} \quad NH_4^+}$$

(in soil solution) (exchangeable) (fixed)

Ammonium fixation by clay minerals is generally greater in subsoils than in topsoil because of the higher clay content of subsoils (Table 11.1). In some cases fixation may be considered an advantage because it is a means of conserving soil nitrogen. In other cases the rate of release of the fixed ammonium is too slow to be of much practical value.

FIXATION BY ORGANIC MATTER

Anhydrous ammonia (NH_3), or other fertilizers that contain free ammonia or that form it when added to the soil, can react with soil organic matter to form compounds that resist decomposition. In this sense, the ammonia can be said to be chemically "fixed" or bound by the organic matter. The exact mechanisms by which this type of fixation occurs is not known, but specific chemical reactions with components of humus have been established. The reactions take place most readily in the presence of oxygen and at high pH values.

The practical significance of ammonium ion fixation by organic matter depends on the circumstances. In organic soils with a high fixing capacity, the reaction could result in a serious loss of available nitrogen and would dictate the use of fertilizers other than those that supply free ammonia. In normal practice on mineral soils, however, organic fixation should not be too disadvantageous since the fixed ammonia is subject to subsequent slow release by mineralization.

11.7

Ammonia Volatilization

Ammonia present in soils from manure and residue breakdown and from fertilizers that contain or produce it can be lost in significant quantities, especially from sandy soils and from alkaline or calcareous soils. This loss is most serious when the materials are applied to the soil surface. Surface applications not only minimize opportunities for the ammonia to react with soil colloids, but temperatures of surface soil are usually high, which enhances ammonia volatilization. Incorporating the manure and fertilizers into the top few centimeters of soil can reduce ammonia loss by 25–75%.

Loss of gaseous ammonia from nitrogen fertilizers applied to the surface of fish ponds and paddy soils can be appreciable even on slightly acid soils. The applied fertilizer stimulates algal growth in the paddy water; the algae extract carbon dioxide from the water, just as higher plants extract this gas from the atmosphere. The reduction in carbon dioxide causes a marked increase in paddy water pH, levels above 9.0 not being uncommon. At these pH values, ammonia is released from ammonium compounds and volatilizes directly into the atmosphere. As with upland soils, this loss can be reduced significantly if the fertilizer is placed below the soil surface.

11.8
Nitrification

Nitrification is a process of enzymatic oxidation of ammonia to nitrates by certain microorganisms in the soil. It takes place in two coordinated steps. The first step is the production of nitrite (NO_2^-) ions by one group of organisms (*Nitrosomonas*), apparently followed immediately by further oxidation to the nitrate (NO_3^-) form by another organism (*Nitrobacter*). The enzymatic oxidation is represented very simply as follows.

Step 1

$$\overbrace{NH_4^+ \xrightarrow{[O]} HONH_2 \xrightarrow{-2\,H} \tfrac{1}{2}\,HONNOH \xrightarrow{[O]} NO_2^- + H^+ + \text{energy}}^{\textit{Nitrosomonas}}$$

Ammonium   Hydroxylamine   Hyponitrite   Nitrite

Step 2

$$\overbrace{NO_2^- \xrightarrow{[O]} NO_3^- + \text{energy}}^{\textit{Nitrobacter}}$$

Under most conditions favoring the two reactions, the second transformation is thought to follow the first so closely as to prevent any great accumulation of the nitrite. This is fortunate because nitrites are generally toxic to higher plants and mammals.

SOIL CONDITIONS AFFECTING NITRIFICATION

The nitrifying bacteria are much more sensitive to their environment than most heterotrophic organisms. Soil conditions that influence the population and activity of nitrifiers and, consequently, the vigor of nitrification will receive brief consideration.

Ammonia Level. Nitrification can take place only if there is a source of ammonia to be oxidized. Factors such as high C/N ratio of residues, which prevents the release of ammonia, also prevent nitrification. However, if the ammonia is present at too high a level, it also constrains nitrification. Heavy localized applications to alkaline soils of anhydrous ammonia or urea, which forms ammonia through hydrolysis, appear to be toxic to *Nitrobacter*, resulting in the accumulation of toxic levels of nitrite ions.

Aeration. Soil aeration and good soil drainage are needed to provide the oxygen for the nitrification process. Plowing and modest cultivation may promote nitrification. Rates of nitrification are generally somewhat slower under minimum tillage than where plowing and some cultivation are practiced.

Source of Carbon. The nitrifiers (autotrophs) use carbon dioxide and bicarbonate ions as sources of carbon for their body tissues. In normal soils, the carbon dioxide and bicarbonate levels are sufficient to provide this carbon.

Temperature and Moisture. The temperature most favorable for nitrification is 25–35 °C (77–95 °F). Consequently, nitrification is slow in cool soils in the spring, and the soil's ability to provide nitrates for plants is curtailed. Under these conditions, fertilizer nitrate must be provided directly where needed. Nitrification rates decline at temperatures above 35 °C and essentially cease at temperatures above 50 °C (122 °F).

Nitrification is also retarded by both very low and very high moisture conditions. The optimum moisture for higher plants is also optimum for nitrification; however, appreciable nitrification occurs when the soil moisture is at or even below the wilting coefficient.

Exchangeable Base-Forming Cations and pH. Nitrification proceeds most rapidly where there is an abundance of exchangeable base-forming cations. The lack of these cations accounts in part for the slow nitrification in acid mineral soils and thus for the seeming sensitivity of the organisms to a low pH. However, within reasonable limits, acidity itself seems to have less influence on nitrification when adequate base-forming cations are present. This is especially true of peat soils. Even at pH values below 5, these soils may show remarkable accumulations of nitrates.

Fertilizers. Nitrifying organisms seem to have nutrient requirements not too different from those of higher plants. Consequently, nitrification may be stimulated by applying phosphorus or potassium-containing fertilizers if they are needed to provide an appropriate nutrient balance. As noted in Section 11.5 however, applications of large quantities of ammonium nitrogen to strongly alkaline soils should be avoided to prevent loss of ammonia gas to the atmosphere and to alleviate any negative effects of the NH_4^+ ion on nitrification.

Pesticides. Nitrifying organisms are quite sensitive to some pesticides. If used at high rates many pesticides inhibit nitrification almost completely and others slow the process down. Most studies suggest, however, that at ordinary field rates of application the majority of pesticides have only minimal effect on nitrification. This subject is covered in greater detail in Chapter 19.

11.9

Fate of Nitrate Nitrogen

The nitrate nitrogen of the soil, whether added in fertilizers or formed by nitrification, may go in four directions (Figure 11.1). It may (a) be incorporated into microorganisms; (b) be assimilated into higher plants; (c) be lost in drainage; or (d) escape from the soil as a gas.

USE BY SOIL ORGANISMS AND PLANTS

Both plants and soil microorganisms readily assimilate nitrate nitrogen. However, if microbes have a ready food supply (for example, carbonaceous organic residues), they use nitrates more rapidly than higher plants (see Section 10.12), which will suffer if supplemental fertilizer nitrogen is not applied.

LEACHING AND GASEOUS LOSS

Negatively charged nitrate ions are not adsorbed by the negatively charged colloids that dominate most soils. Consequently, they are subject to ready leaching from the soil, and move downward freely with the water. The quantity of nitrate nitrogen lost in drainage water depends on the climate and cultural conditions. Loss is low in unirrigated arid and semiarid regions and high in humid areas and where irrigation water is applied. Where modest fertilizer applications are made, usually no more than 5–10% of the nitrogen will be lost by leaching.

Heavy nitrogen fertilization, especially for vegetables and other cash crops grown on coarse-textured soils, may increase nitrate loss by leaching as may the use of high levels of nitrogen fertilizers applied in the fall before a spring crop is planted. Such losses can be further accentuated by increased use of conservation tillage systems, which, by keeping water on the land, can increase leaching and concomitant loss of nitrates.

Where leaching water from areas receiving higher fertilizer nitrogen rates have been combined with concentrated leachings from huge feedlots, the nitrate concentration in drainage water may rise to levels that could be harmful to humans, livestock, and other animals. The need to take steps to minimize nitrate leaching is obvious. This subject is given further consideration in Section 18.8.

Gaseous loss of nitrogen from soils may occur when nitrates are reduced to nitrogen oxide compounds and to elemental nitrogen. This process, termed *denitrification*, is carried out by soil microorganisms and, less commonly, by purely chemical reactions. Both mechanisms for denitrification will be discussed briefly.

11.10

Denitrification

REDUCTION BY ORGANISMS

Biochemical reduction of nitrate nitrogen to gaseous compounds is a widespread phenomenon. The microorganisms involved are common facultative anaerobic forms. The exact mechanisms by which the reductions take place are not known. However, the five-valent nitrogen in nitrate is reduced stepwise primarily to the zero-valent elemental nitrogen as follows.

$$2\ NO_3^- \xrightarrow{-2[O]} 2\ NO_2^- \xrightarrow{-2[O]} 2\ NO \xrightarrow{-[O]} N_2O \xrightarrow{-[O]} N_2$$

| Nitrate ions (+5) | Nitrite ions (+3) | Nitric oxide (+2) | Nitrous oxide (+1) | Elemental nitrogen (0) |

Each step in the reaction is triggered by a specific reductase enzyme. Note, however, that the reaction can stop at any stage and the gaseous products of

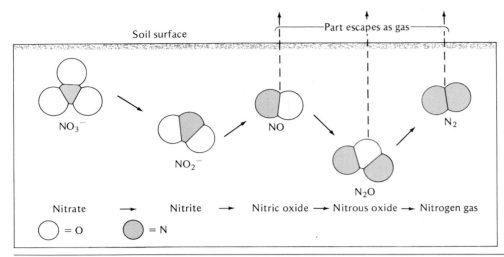

Part escapes as gas

NO₃⁻

NO₂⁻

NO

N₂O

N₂

Nitrate ⟶ Nitrite ⟶ Nitric oxide ⟶ Nitrous oxide ⟶ Nitrogen gas

◯ = O   ● = N

FIGURE 11.3

Illustration of the reductive nature of the denitrification process. Organisms take an oxygen from nitrate ion and leave the nitrite ion, which in turn loses an oxygen and produces nitric oxide, and so on.

that stage (NO, N_2O, and N_2) can be released to the atmosphere (Figure 11.3). The oxygen atoms become incorporated into the bodies of the anaerobic bacteria.

Under field conditions, nitrous oxide (N_2O) and elemental nitrogen (N) are lost in the largest quantities with nitrous oxide dominant if ample nitrite ion (NO_2^-) is present and if the soil is not too low in oxygen (Figure 11.4). Nitric oxide (NO) loss is generally not great and apparently occurs most readily under acid conditions.

It might be mentioned that nitrous oxide (N_2O) can also be formed in small quantities by certain *Nitrosomonas* bacteria during ammonia oxidation in aerobic soils. While this observation is of some scientific interest, it is not thought to be of great practical importance.

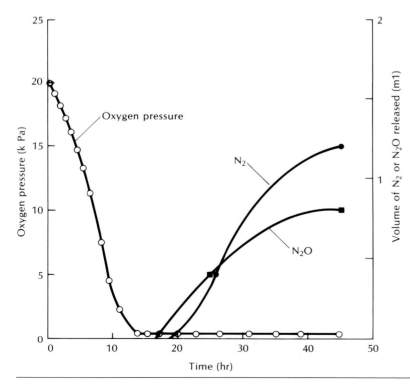

FIGURE 11.4

The reduction in oxygen pressure at the center of a soil aggregate following a rain and the subsequent denitrification, with the release of N_2O and N_2 gases. This loss can take place at the center of the soil aggregate even though the surrounding soil is not saturated with water. [From Leffelaar (1986)]

CHEMICAL REDUCTION

There are nonmicrobial processes by which nitrogen may be reduced in soils to gaseous forms. For instance, nitrite ions in a slightly acid solution will evolve nitrogen gas when brought in contact with certain ammonium salts, with simple amino compounds such as urea, and even with lignins, phenols, and carbohydrates. The following reaction suggests what may happen to urea.

$$2\ HNO_2\ +\ CO(NH_2)_2\ \rightarrow\ CO_2 \uparrow\ +\ 3\ H_2O\ +\ 2\ N_2 \uparrow$$

Nitrous acid;   Urea
nitrite ion

The reaction that results in this gaseous loss is strictly chemical and does not require either the presence of microorganisms or adverse soil conditions. Its practical importance, however, is not great.

QUANTITY OF NITROGEN LOST THROUGH DENITRIFICATION

As might be expected, the exact magnitude of the denitrification losses will depend on the cultural and soil conditions. In well-drained humid-region soils, denitrification losses may be no more than 5–15 kg N/ha per year if heavy rainfall does not provide temporary periods of poor aeration (Table 11.2). Where drainage is restricted and where large amounts of ammonia and urea fertilizer are applied, substantial losses of nitrogen might be expected, 30–60 kg/ha per year not being uncommon.

In flooded soils such as those used for rice culture, losses by denitrification may be very high. Often 60–70% of the applied fertilizer nitrogen is volatilized as oxides of nitrogen or elemental nitrogen. The process by which the loss takes place is shown in Figure 11.5. By incorporating the fertilizer into the anaerobic zone of the soil, and largely preventing the formation of nitrates by nitrification, losses can be reduced. In some cases, fertilizer-use efficiency has been doubled by deep placement of the fertilizer.

ATMOSPHERIC POLLUTION

The loss of nitrous oxide (N_2O) from the soil into the atmosphere is thought to have some harmful effects on the environment. As the nitrous oxide

TABLE 11.2

Fate of Applied Nitrogen in Two Experiments, One in Ohio and the Other in a Less Humid Area in Oklahoma

In the more humid area (Ohio) loss by leaching and volatilization was much higher than in the drier area (Oklahoma).

Fate of applied N	Sorghum-Sudan grass[a] (Oklahoma) (%)	Corn[b] (Ohio) (%)
Plant uptake	60	29
Soil organic matter	33	21
Leached	0	33
Gaseous loss[c]	7	17
	100	100

[a] Calculated from 3 years' data from eight soils (Smith et al., 1982).
[b] Calculated from 3 years' data (Chichester and Smith, 1978).
[c] Unaccounted for and presumed lost by volatilization.

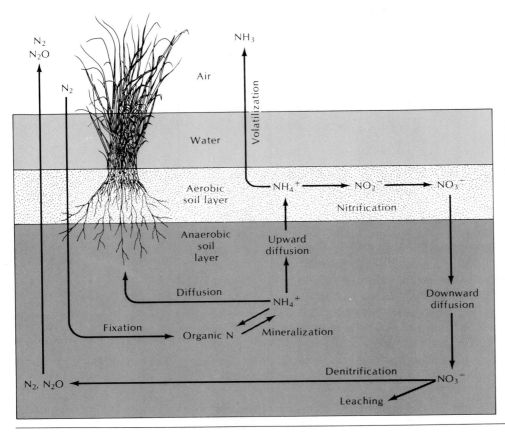

FIGURE 11.5
Nitrification–denitrification reaction and kinetics of the related processes controlling nitrogen loss from the aerobic–anaerobic layer of a flooded soil system. Nitrates, which form in the thin aerobic soil layer just below the soil–water interface, diffuse into the anaerobic (reduced) soil layer below and are denitrified to the N_2 and N_2O gaseous forms, which are lost to the atmosphere. Placing the urea or ammonium-containing fertilizers in the anaerobic layer prevents the oxidation of ammonium ions to nitrates, thereby greatly reducing the loss. [From Patrick (1982)]

moves up into the stratosphere, products of its oxidation unite with water to form nitric acid (HNO_3), which is a major constituent of so-called acid rain. The nitrous oxide may also participate in reactions that result in the destruction of ozone (O_3), a gas that helps shield the Earth from harmful ultraviolet radiation from the sun. If this ozone shield were to be removed, the Earth's surface would be warmed, and serious ecological consequences would result. While there are other much larger sources of nitrous oxide such as tropical forests and woodlands and, in industrialized areas, the exhaust fumes of gasoline engines, nitrous oxide loss from agricultural soils is considerable and may have implications far beyond agriculture.

11.11
Nitrification Inhibitors[4]

Losses of nitrogen through volatilization and leaching are generally higher when considerable nitrate nitrogen is present in the soil over long periods of time. The losses are high when nitrogen fertilizers are applied in the fall for the succeeding crop or even in the spring on wet soils. To minimize these losses, it is desirable to keep much of the applied nitrogen in the ammonium form rather than the nitrate form.

In recent years chemicals have been found that can inhibit the nitrification process. Two of the more effective are nitrapyrin, sold under the brand name of N-Serve, and etridiazol, sold commercially as Dwell (Figure 11.6).

---

[4] For reviews of this subject, see ASA (1980).

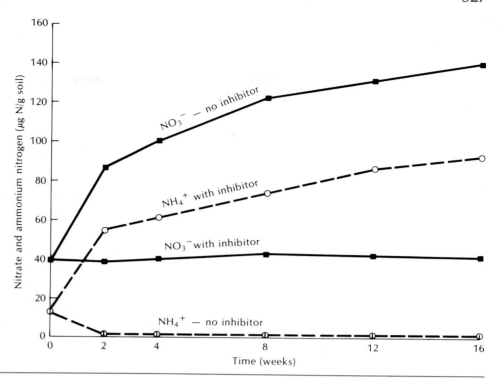

FIGURE 11.6
The effect of nitrification inhibitors on nitrate and ammonium ion levels in a sludge-compost amended Hagerstown silty clay loam (Typic Hapludalf fine, mixed Mesic). With no inhibitor, nitrate levels went up very rapidly while ammonium levels went down. As NH_4^+ ions were formed, apparently, they were being oxidized to nitrates. With the inhibitor present nitrates remained essentially unchanged, but NH_4^+ ions increased significantly since they were not being oxidized. Inhibitor data are an average of figures obtained with nitrapyrin and etridiazol. [Data from McClung et al. (1983)]

Both compounds inhibit oxidation of ammonia to nitrates. Nitrapyrin has been most widely used. It is relatively inexpensive and is effective at rates of 0.5 kg/ha or less. Usually nitrapyrin is applied with anhydrous ammonia or incorporated into dry fertilizer such as urea.

Response to nitrapyrin is generally best in situations where nitrate leaching or denitrification losses may be great. Thus, significant responses to the use of nitrapyrin have been noted on sandy soils (reduced leaching) and on imperfectly drained soils (reduced denitrification).

Nitrapyrin's effectiveness also is determined by soil temperatures. The inhibitor is quite readily hydrolyzed as temperatures increase from 10 °C to 20–30 °C and so does not persist very long in warm soils.

SULFUR-COATED UREA

Another means of reducing gaseous loss of nitrogen is by coating urea pellets with elemental sulfur. These sulfur-coated urea products have been found to be effective in increasing plant use of the nitrogen, especially on sandy soils and on some tropical soils, including rice paddies. The sulfur slows down the release of urea to the soil solution, and consequently limits the hydrolysis of urea to ammonia and the oxidation of ammonia to nitrate; thus the sulfur coating reduces the availability of nitrate for the denitrification process. Unfortunately, however, sulfur adds significantly (25–50%) to the cost of urea fertilizer. Moreover, sulfur increases soil acidity because oxidation of the sulfur in the soil produces sulfuric acid (see Section 11.25). Although this may be satisfactory for alkaline soils, it can cause problems in soils that are already quite acid. Sulfur-coated urea will likely be economical only where high yield increases are obtained and/or where high crop values prevail. Other "slow release" nitrogen fertilizers, either resin-coated or low-solubility compounds, suffer from similar practical limitations.

11.12

Biological Nitrogen Fixation

Biological nitrogen fixation is the biochemical process by which elemental nitrogen is combined into organic forms. It is carried out by a number of organisms including several species of bacteria (not all of which are associated with legumes), a few actinomycetes, and blue-green algae (cyanobacteria).

The quantity of nitrogen fixed globally each year is enormous, having been estimated at about 175 million Mg (metric tons) (Table 11.3). This exceeds the amount applied in chemical fertilizers. It is no wonder that biological nitrogen fixation is said to be, next to photosynthesis, the most important biological process on earth.

THE MECHANISM

Although biological nitrogen fixation is accomplished by a number of different organisms, a common mechanism appears to be involved. The overall effect of the process is to reduce nitrogen gas to ammonia.

$$N_2 + 6\,H + 6\,e^- \xrightarrow[\text{(Fe, Mo)}]{\text{(nitrogenase)}} 2\,NH_3$$

Ammonia in turn is combined with organic acids to form amino acids and ultimately proteins.

$$NH_3 + \text{organic acids} \rightarrow \text{amino acids} \rightarrow \text{proteins}$$

The site of nitrogen reduction is the enzyme *nitrogenase*, a two-protein complex consisting of a larger ion, a molybdenum-containing member, and a smaller companion containing iron. Nitrogenase is indeed of great significance to humankind.

FIXATION SYSTEMS

Biological nitrogen fixation occurs through a number of organism systems with or without direct association with higher plants (Table 11.4). Included are the following.

TABLE 11.3
Global Biological Nitrogen Fixation from Different Sources

	Area (10^6 ha)	Nitrogen fixed per year	
Use		Rate (kg/ha)	Total fixed (10^6 Mg)
Arable (cropped)			
Legumes	250	140	35
Nonlegumes	1,150	8	9
Permanent meadows and grassland	3,000	15	45
Forest and woodland	4,100	10	40
Unused	4,900	2	10
Ice covered	1,500	0	0
Total land	14,900	1	139
Sea	36,100	1	36
Grand total	51,000		175

From Burns and Hardy (1975).

TABLE 11.4
Information on Different Systems of Biological Nitrogen Fixation

N-fixing systems	Organisms involved	Plants involved	Site of fixation
Symbiotic			
Obligatory			
Legumes	Bacteria (*Rhizobium*)	Legumes	Root nodules
Nonlegumes (angiosperms)	Actinomycetes (*Frankia*)	Nonlegumes (angiosperms)	Root nodules
Associative			
Morphological involvement	Blue-green algae, bacteria	Various higher plants and microorganisms	Leaf and root nodules, lichens
Nonmorphological involvement	Blue-green algae, bacteria	Various higher plants and microorganisms	Rhizosphere (root environment) Phyllosphere (leaf environment)
Nonsymbiotic	Blue-green algae, bacteria	Not involved with plants	Soil, water independent of plants

Modified from Burns and Hardy (1975).

1. Symbiotic systems, nodule-forming (obligatory)
   a. With legumes
   b. With nonlegumes
2. Symbiotic systems, nonnodule forming (associative)
3. Nonsymbiotic systems

Although the legume symbiotic systems have received the most attention historically, recent findings suggest that the other systems are also very important worldwide and may even rival the legume-associated systems as suppliers of biological nitrogen to the soil. Each major system will be discussed briefly.

11.13

Symbiotic Fixation with Legumes

The symbiosis, or living together, of legumes and bacteria of the genus *Rhizobium* provides the major biological source of fixed nitrogen in agricultural soils. *Rhizobium* bacteria invade the root hairs and the cortical cells, ultimately inducing the formation of nodules that serve as a home for the organisms (Figure 11.7). The host plant supplies the bacteria with carbohydrates for energy, and the bacteria reciprocate by supplying the plant with fixed nitrogen compounds, an association that is mutually beneficial.

ORGANISMS INVOLVED

There is considerable specificity between *Rhizobium* species and the host plants they will invade. A given *Rhizobium* species will inoculate some legumes but not others. This specificity of interaction is the basis for classifying rhizobia and their host plants into seven so-called cross-inoculation groups (Table 11.5). Legumes that can be inoculated by a given *Rhizobium* species are included in the same cross-inoculation group. Thus, *Rhizobium*

(a)   (b)   (c)

FIGURE 11.7
Photos illustrating soybean nodules. In (a) the nodules are seen on the roots of the
soybean plant, and a closeup (b) shows a few of the nodules associated with the
roots. A scanning electron micrograph shows a single plant cell within the nodule
stuffed with the bacterium *Rhizobium japonicum*. [(c) Courtesy W. J. Brill, University
of Wisconsin]

trifolii inoculates *Trifolium* species (most clovers), *Rhizobium phaseoli* inoculates *Phaseolus vulgaris* (dry beans), and so on.

In areas where a given legume has been growing regularly, the appropriate species of *Rhizobium* may well be present in the soil. All too often, however, the natural *Rhizobium* population in the soil is too low or the strain of the *Rhizobium* species present is not effective. In such circumstances,

TABLE 11.5
Cross-Inoculation Groups of Legumes and Associated Rhizobia

Group	Rhizobium species	Legume
Alfalfa	R. meliloti	Melilotus (certain clovers), Medicago (alfalfa), Trigonella (fenugreek)
Clover	R. trifolii	Trifolium spp. (clovers)
Soybean	R. japonicum	Glycine max (soybeans)
Lupini	R. lupini	Lupinus (lupines), Ornithopus spp. (serradella)
Bean	R. phaseoli	Phaseolus vulgaris (dry bean), Phaseolus coccineus (runner bean)
Peas and vetch	R. leguminosarum	Pisum (peas), Vicia (vetch), Lathyrus (sweet pea), Lens spp. (lentil)
Cowpea miscellany	Various	Vigna (cowpea), Lespedeza (lespedeza), Arachis (peanut), Stylosanthes (stylo), Desmodium (desmodium), Cajanus (pigeon pea), Crotolaria (crotalaria), Pueraria (kudzu), Acacia

TABLE 11.6

Average Yield of Soybeans as Affected by the Application of different Strains of Inoculant Containing *Rhizobium japonicum* (1975–76)

The inoculant was applied in the row on a soil very low in organism numbers.

Inoculant strain	Inoculum population (10⁷ MPN/g)[a]	Yield of soybeans (kg/ha)
No. 110 (single strain)	2.8	3565
No. 138 (single strain)	6.3	3468
Commercial[b]	2.3	3660
Control	—	1139

From Bezdicek et al. (1978).

[a] MPN = most probable number.
[b] Specific strain not known.

special cultures of *Rhizobium* are applied, either by coating the legume seeds with the culture or by applying the inoculant directly to the soil. Effective and competitive strains of *Rhizobium*, which are available commercially, often give significant yield increases (Table 11.6). In the United States, most of the inoculant sold is used for soybeans, although some is used for forage legumes.

QUANTITY OF NITROGEN FIXED

The rate of biological fixation of nitrogen is greatly dependent on soil and climatic conditions. The legume–rhizobium associations generally function best on soils that are not too acid and are well supplied with essential nutrients. However, high levels of available nitrogen, whether from the soil or added in fertilizers, tend to depress biological nitrogen fixation (Figure 11.8). Apparently, symbiotic nitrogen fixation tends to operate only when the nitrogen is needed by the plant.

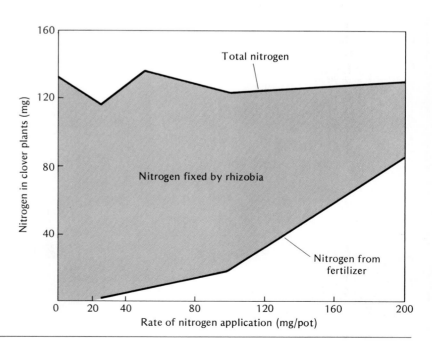

FIGURE 11.8
Influence of added inorganic nitrogen on the total nitrogen in clover plants, the proportion supplied by the fertilizer and that fixed by the rhizobium organisms associated with the clover roots. Increasing the rate of nitrogen application decreased the amount of nitrogen fixed by the organisms in this greenhouse experiment. [From Walker et al. (1956)]

TABLE 11.7
Typical Levels of Nitrogen Fixation from Different Systems

Crop or plant	Associated organism	Typical levels of nitrogen fixation (kg N/ha per yr)
Symbiotic		
Legumes (nodulated)	Bacteria (*Rhizobium*)	
Alfalfa (*Medicago sativa*)		150–250
Clover (*Trifolium pratense* L.)		100–150
Soybean (*Glycine max* L.)		50–150
Cowpea (*Vigna unquiculata*)		50–100
Lupine (*Lupinus*)		50–100
Vetch (*Vicia vilbosa*)		50–125
Bean (*Phaseolus vulgaris*)		30–50
Nonlegumes (nodulated)		
Alders (*Alnus*)	Actinomycetes (*Frankia*)	50–150
Species of *Gunnera*	Blue-green algae (*Nostoc*)	10–20
Nonlegumes (nonnodulated)		
Pangola grass (*Degetaria decumbens*)	Bacteria (*Azospirillum*)	5–30
Bahia grass (*Paspalum notatum*)	Bacteria (*Azotobacter*)	5–30
Azolla	Blue-green algae (*Anabaena*)	150–300
Nonsymbiotic	Bacteria (*Azotobacter, Clostridium*)	5–20
	Blue-green algae (various)	10–50

From experiments around the world, it is possible to suggest typical levels of nitrogen fixation, not only by legume-associated organisms but also by other mechanisms. Such levels are shown in Table 11.7. Although these levels would not pertain to any specific ecological condition, they do indicate the relative levels of fixation that occur. It is important to note that the legume associations fix large quantities of nitrogen.

EFFECT ON SOIL NITROGEN LEVEL

The data in Table 11.7 show that symbiotic nitrogen fixation is definitely beneficial to agriculture and forestry. This source of nitrogen is not available to the cereals, vegetables, and other crops lacking symbiotic associations. Even so, however, the amount of nitrogen fixed symbiotically is seldom enough to satisfy plant needs. For some leguminous crops such as beans, peas, and soybeans, most of the nitrogen must come from the soil. Consequently it should not be assumed that the symbiotic systems consistently increase soil nitrogen. Only in cases where the soil is low in available nitrogen would this likely be true. Legumes and other such "symbiotic" crops are, therefore, nitrogen *savers* for the soil rather than nitrogen contributors. Exceptions to that general rule may be when essentially all the plant material is incorporated into the soil, such as is the case with green manures.

11.14

Fate of Nitrogen Fixed by Legume Bacteria

The nitrogen fixed by *Rhizobium* and the nodule organisms goes in three directions. First, it is used directly by the host plant, which thereby benefits greatly from the symbiosis described in Section 11.13. Second, as the nitrogen passes from plant roots and nodules into the soil, some of it is

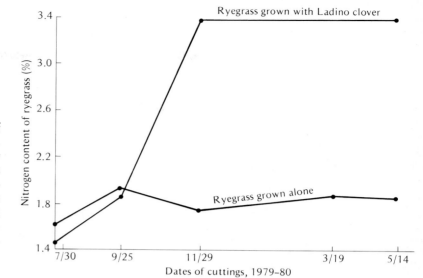

FIGURE 11.9
Nitrogen content of five field cuttings of ryegrass grown alone or with Ladino clover. For the first two harvests, nitrogen fixed by the clover was not available to the ryegrass and the nitrogen content of the ryegrass forage was low. In subsequent harvests, the fixed nitrogen apparently was available and was taken up by the ryegrass. This was probably due to the mineralization of dead Ladino clover root tissue. [From Broadbent et al. (1982)]

mineralized and becomes available almost immediately as ammonium or nitrate compounds. This nitrogen is available to any crop grown in association with the legume. The vigorous development of a grass in a legume–grass mixture is evidence of this rapid release (Figure 11.9).

The third pathway for the fixed nitrogen is to be incorporated into the bodies of the general-purpose decay organisms and, finally, into the soil organic matter. This process of immobilization was discussed in Section 11.4.

11.15

Symbiotic Fixation with Nodule-Forming Nonlegumes

About 160 species from 13 genera of nonlegumes are known to develop nodules and to accommodate symbiotic nitrogen fixation (Bond, 1977). Included are several important groups of angiosperms, listed in Table 11.8. The roots of these plants, which are present in some forested areas and

TABLE 11.8
Number and Distribution of Major Actinomycete-Nodulated Nonlegume Angiosperms

In comparison there are about 13,000 legume species.

Genus	Family	Species[a] nodulated	Geographic distribution
Alnus	Betulaceae	33/35	Cool regions of the northern hemisphere
Ceanothus	Rhamnaceae	31/35	North America
Myrica	Myricaceae	26/35	Many tropical, subtropical and temperate regions
Casuarina	Casuarinaceae	24/25	Tropics and subtropics
Elaeagnus	Elaeagnaceae	16/45	Asia, Europe, N. America
Coriaria	Coriariaceae	13/15	Mediteranean to Japan, New Zealand, Chile to Mexico

Selected from Torrey (1978).

[a] Number of species nodulated/total number of species in genus.

wetlands, are inoculated by soil actinomycetes of the genus *Frankia*. The actinomycetes invade the root hairs, and distinctive nodules form.

The rates of nitrogen fixation per hectare compare favorably with those of the legume–*Rhizobium* complexes (Table 11.7). On a worldwide basis, the total nitrogen fixed in this way may even exceed that caused by some legumes used in agriculture. While the rates of cycling of the fixed nitrogen may not be quite as high as those associated with agricultural crops, the total quantity fixed places the actinomycete-induced fixation on a par with *Rhizobium* fixation.

Certain blue-green algae are known to develop nitrogen-fixing symbiotic relations with green plants. One involves nodule formation on the stems of *Gunnera*, an angiosperm common in marshy areas of the southern hemisphere. Algae of the genus *Nostoc* are involved, and the rate of fixation is typically 10–20 kg N/ha per year (Table 11.7).

11.16

Symbiotic Nitrogen Fixation Without Nodules

Recent studies have called attention to several significant nonlegume symbiotic nitrogen-fixing systems that do not create nodules. Among the most significant are those involving blue-green algae. One system of considerable practical importance is the *Azolla–Anabaena* complex, which flourishes in certain rice paddies of tropical and semitropical areas. The *Anabaena* blue-green algae inhabit cavities in the leaves of the floating fern *Azolla* and fix quantities of nitrogen comparable to those of the better *Rhizobium*–legume complexes (Table 11.7).

A more widespread but less intense nitrogen-fixing phenomenon is that which occurs in the *rhizosphere*, or root environment, of nonlegumes and especially of grasses. The organisms responsible are bacteria, especially those of the *Azospirillum* and *Azotobacter* genera (Table 11.7). The organisms use root exudates as sources of energy for their nitrogen-fixing activities.

There are differences of opinion as to the rates of fixation that occur in the rhizosphere. Relatively high values have been observed in association with certain tropical grasses, but rates of 5–30 kg N/ha per year are thought to be more typical. Even so, because of the vast areas of tropical grasslands, the total quantities of nitrogen fixed by rhizosphere organisms are likely very high (Table 11.3).

11.17

Nonsymbiotic Nitrogen Fixation

There exist in soils and water certain free-living microorganisms that are able to fix nitrogen. Because these organisms are not directly associated with higher plants, the transformation is referred to as *nonsymbiotic* or *free-living*.

FIXATION BY HETEROTROPHS

Several different groups of bacteria and blue-green algae are able to fix nitrogen nonsymbiotically. In upland mineral soils, the major fixation is brought about by species of two genera of heterotrophic aerobic bacteria: *Azotobacter* (in temperate zone soils) and *Beijerinckia* (in tropical soils). Certain anaerobic bacteria of the genus *Clostridium* also are able to fix nitrogen. Because of pockets of low oxygen supply in most soils, even when they are in the best tilth, aerobic and anaerobic bacteria probably work side by side in many agricultural soils.

The amount of nitrogen fixed by these heterotrophs varies greatly with the pH, soil nitrogen level, and the sources of organic matter available to the organisms for energy. In any case, under normal agricultural conditions the rates of nitrogen fixation by these organisms are thought to be much lower than those associated with legumes, being in the range of 3–15 kg N/ha per year (Table 11.7).

FIXATION BY AUTOTROPHS

Among the autotrophs able to fix nitrogen are certain photosynthetic bacteria and blue-green algae (Havelka et al., 1982). In the presence of light, these organisms are able to fix both carbon dioxide and nitrogen simultaneously. The contribution of the photosynthetic bacteria is uncertain but that of blue-green algae is thought to be of some significance, especially in wetland areas and in rice paddies. In some cases, the blue-green algae have been found to fix sufficient nitrogen for moderate rice yields, but normal levels may be no more than 20–30 kg N/ha per year. Nitrogen fixation by blue-green algae in upland soils also occurs, but the level is much lower than is found under wetland conditions.

11.18

Addition of Nitrogen to Soil in Precipitation

The atmosphere contains ammonia and other nitrogen compounds released from the soil and plants as well as from the combustion of coal and petroleum products. Nitrates exist, too, in small quantities, resulting from the oxidation of atmospheric nitrogen oxides and from electrical discharges (lightning) in the atmosphere. Another source of nitrogen compounds is the exhaust from automobile and truck engines, which contributes a considerable amount to the atmosphere, especially downwind from large cities. These atmosphere-borne nitrogen compounds are added to the soil through rain and snow. Although the rates of addition per hectare are typically small, the total quantity of nitrogen added is not insignificant and may be tens of kilograms per hectare near highly polluted areas.

The quantity of ammonia and nitrates in precipitation varies markedly with location and with season. The quantities are greater in comparable areas in the tropics than in humid temperate regions and greater in the latter than under semiarid temperate climates. Rainfall additions of nitrogen compounds are highest near cities and industrial areas and near huge animal feedlots (Table 11.9). There is special concern for the depositions of nitrates and other nitrogen oxides from these areas of concentration because they are associated with increased soil acidity. The environmental impact of ''acid'' rain and its influence on vegetation (forests and crops) as well as on lakes tends to overshadow the beneficial effects of precipitation-supplied nitrates as nutrients for growing plants.

The ammonium nitrogen added to the soil in precipitation is generally greater in amount than nitrogen in the nitrate form, although the quantities added in ammonium form show wide variations. The range in total nitrogen added annually is 1–20 kg N/ha. A figure of 5–8 kg N/ha would be typical

TABLE 11.9

Amounts of Nitrogen Brought Down in Precipitation Annually in Different Parts of the United States

| Areas in the United States | Range in annual deposition (kg/ha) | | Total nitrogen (kg/ha) |
	Nitrate nitrogen	Ammonium nitrogen[a]	
Industrialized Northeast	4.3–7.4	8.6–14.8	12.9–22.2
Borders of NE industrialized areas	2.8–4.1	5.6– 8.2	8.4–12.3
Open areas in West	0.4–0.6	0.8– 1.2	1.2– 1.8

U.S./Canada work group #2 (1982).

[a] Calculated as twice the nitrate deposition, the approximate ratio from numerous measurements, although it may be incorrect at any specific site.

for temperate regions. This annual acquisition of nitrogen in a readily available form to each hectare of land affords some aid in the maintenance of soil fertility.

11.19

Reactions of Nitrogen Fertilizers

The worldwide use of nitrogen-containing fertilizers has expanded greatly in recent years. As a result, nitrogen in the soil solution, particularly in the localized soil zones to which the fertilizers have been applied, often is dominated by fertilizer-applied materials. While the ammonium and nitrate ions coming from fertilizers react in a similar way to comparable ions released by microbial breakdown of organic materials, their high concentration in the local zones of application and tendency to acidify the soil deserve special attention.

HIGH CONCENTRATION

The localized concentration of anhydrous ammonia (NH_3), ammonium-containing salts, and urea (which hydrolyzes to ammonia) stimulates several reactions. The fixation of ammonium ions by clays and organic matter is enhanced. High localized levels of ammonia inhibit the second stage of nitrification, resulting in the undesirable accumulation of nitrite ions (see Section 11.8). In alkaline soils, a high concentration of ammonium ion can result in the release of some ammonia gas directly to the atmosphere.

The addition of large amounts of nitrogen-containing fertilizers may affect the microbial processes of free fixation and gaseous nitrogen loss. In general, fixation by free-living organisms is depressed by high levels of mineral nitrogen. Gaseous losses, on the other hand, are often encouraged by abundant nitrates. Heavy nitrogen fertilization followed by nitrification thus tends to increase losses of nitrogen from the soil.

In most soil situations, the effects of higher localized concentrations of fertilizer materials on nitrogen transformations are not too serious. Economic considerations often counterbalance the biochemical disadvantage of localized concentration. In any case, it can be assumed that fertilizer nitrogen will react in soils in a manner very similar to that of nitrogen released by biological transformations.

SOIL ACIDITY

Ammonium-containing fertilizers and those that form ammonia upon reacting in the soil can increase soil acidity (see Section 16.9 for a more thorough discussion of this). The process of nitrification (see the equations in Section 11.8) releases hydrogen ions that become adsorbed on the soil colloids. For best crop growth in humid regions, continued and substantial use of acid-forming fertilizers must be accompanied by applications of lime.

The nitrate component of fertilizers does not increase soil acidity. In fact, nitrate fertilizers containing metallic cations in the molecule (for example, calcium or sodium nitrate) have a slight alkalizing effect.

11.20

Practical Management of Soil Nitrogen

The problem of nitrogen control is twofold: (a) the maintenance of an adequate nitrogen supply in the soil and (b) the regulation of the soluble forms of nitrogen to ensure a ready availability to meet crop demands.

NITROGEN BALANCE SHEET

Major gains and losses of available soil nitrogen are diagrammed in Figure 11.10. While the relative additions and losses by various mechanisms will vary greatly from soil to soil, the principles illustrated in the diagram are valid.

The major loss of nitrogen from most soils is that removed in crop plants. A good crop of wheat or cotton may remove only 100 kg N/ha, nearly half of which may be returned to the soil in the stalks or straw. A bumper silage corn crop, in contrast, may contain over 250 kg N/ha, and a good annual yield of alfalfa or well-fertilized grass hay removes more than 300 kg N/ha. It is obvious that modern yield levels require nutrient inputs far in excess of those of a generation ago.

Erosion, leaching, and volatilization losses are determined to a large degree by water management practices. Their magnitudes are so dependent upon specific situations that generalizations are difficult. However, soil and crop management practices that give optimum crop yields and, in turn, crop residues will likely hold these sources of nitrogen loss to a satisfactory minimum.

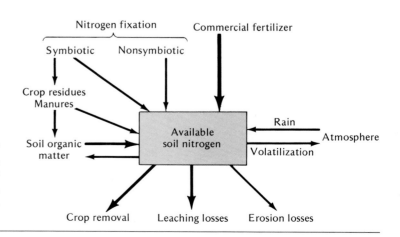

FIGURE 11.10

Major gains and losses of available soil nitrogen. The widths of the arrows indicate roughly the magnitude of the losses and the additions often encountered. It should be emphasized that the diagram represents average conditions only and that much variability is to be expected in the actual and relative quantities of nitrogen involved.

MEETING THE DEFICIT

In practice, nitrogen deficits are met from four sources: crop residues, farm manure, legumes, and commercial fertilizers. On dairy farms and beef feed lots, much of the deficit will be met by the first three methods, fertilizers being used as a supplementary source. Mountain ranges in the western part of the United States depend on some symbiotic fixation plus animal manures and nonsymbiotic fixers.

On most other types of farms, fertilizers will play a major role. Where vegetables and other cash crops with high nutrient requirements are grown, the nitrogen deficit will be met almost entirely with commercial fertilizers. Even with general field crops, modern yield levels can be maintained only through the extensive use of fertilizers.

REGULATION OF SOLUBLE FORMS

Regulation of the levels of soluble forms after nitrogen enters the soil is not easy. Availability at the proper time and in suitable amounts, with a minimum of loss, is the ideal. Frequent application of nitrogen fertilizer in small amounts is one way to accomplish this purpose, although application costs place some limitations on this practice.

Soils in any given climate or farming system tend to assume what may be called a *normal* or *equilibrium content* of nitrogen. Consequently, under ordinary methods of cropping and manuring, any attempt to raise permanently the nitrogen content to a materially higher level than this normal will be attended by unnecessary waste caused by drainage and other losses. Generally the best results are obtained by keeping the nitrogen suitably active by the use of legumes and other organic materials with a low C/N ratio and by the applications of lime and commercial fertilizers.

There are some exceptions to a soil remaining at its equilibrium nitrogen content. If a soil that is naturally low in organic matter and nitrogen is heavily fertilized over a period of time, and if crop residues are returned to the soil, the soil nitrogen level will likely increase. Such a situation is common where soils of dry areas (Aridisols) are irrigated and cropped to high fertilizer-requiring vegetables or alfalfa. To some degree then the equilibrium level of soil nitrogen is governed by the agricultural system employed.

The recommendation concerning the soils' equilibrium level of nitrogen is the same as that already made for soil organic matter (see Section 10.15), and it is known to be both economical and effective. In short, the practical problem is to supply adequate nitrogen to the soil, to keep it mobile and, at the same time, to protect it from excessive losses caused by leaching, volatilization and erosion.

11.21

Importance of Sulfur

Sulfur has long been recognized as essential for plant and animal growth.[5] It is known to be indispensable for many reactions in every living cell. Sulfur is a constituent of the amino acids methionine, cysteine, and cystine, deficiencies of which result in serious human malnutrition. The vitamins biotin, thiamine, and B_1 contain sulfur, and the structure of proteins is determined

---

[5] For a discussion of sulfur and agriculture, see Tabatabai (1986).

to a considerable extent by sulfur groups. The properties of certain protein enzymes, such as those concerned with photosynthesis and nitrogen fixation, are thought to be attributable to the type of sulfur linkages present. As with the other essential elements, sulfur plays a unique role in plant and animal metabolism.

DEFICIENCIES OF SULFUR

Only in recent years have deficiencies of sulfur become common. In the past, sulfur-bearing fertilizers such as ordinary superphosphate (which contains $CaSO_4$) and ammonium sulfate were important sources of sulfur. In addition, atmospheric sulfur dioxide, a by-product of the combustion of sulfur-rich coals and residual fuel oils, supplied large quantities of this element to both plants and soils. Thus by seemingly incidental means, the sulfur needs of crops in the past were largely satisfied, especially in areas near industrial centers (Table 11.10).

In recent years, increased demand for high-analysis fertilizers has forced manufacturers to use alternatives to ordinary superphosphate and ammonium sulfate. As a result, many sulfur-free fertilizers such as triple superphosphate and diammonium phosphate are on the market, and the average sulfur content of fertilizers has decreased. Even sulfur-containing pesticides, so commonly used a few years ago, have been largely replaced by organic materials free of sulfur.

The replacement of wood and coal for domestic heating by natural gas, electricity, and low-sulfur fuel oil also has affected the amount and distribution of sulfur dioxide in the atmosphere. Intensified efforts in the United States to reduce air pollution in and around cities and industrial areas will likely further reduce the quantity of sulfur in the air. The recognition that clean air is a primary goal will necessitate finding alternative means of supplying sulfur for plant growth.

Coupled with the reductions in the supply of sulfur to soils and plants is the increased removal of this element in harvested crops. Yields have grown markedly higher during the past 20–25 years, and much of the sulfur used by the crops has not been returned. The quantity of sulfur thus removed is about the same as that of phosphorus. It is not surprising, therefore, that increased attention is being given to sulfur.

AREAS OF DEFICIENCY

Sulfur deficiencies have been reported in most areas of the world but are more prevalent in areas where soil parent materials are low in sulfur, where

TABLE 11.10
Annual Deposition of Sulfur in Precipitation at Locations in Rural and Urban Areas Compared with Sites Near Industrial Areas

State	Sulfur deposition (kg/ha)	
	Rural and urban	Near industry
Alabama	3–7	7–17
Florida	2–10	34
Indiana	22–37	142
New York	10–20	35–76
North Carolina	3–15	11–43
Texas	3–8	10–14
Wisconsin	16 (rural)	168

Selected data from Olsen and Rehm (1986).

extreme weathering and leaching has removed this element, or where there is little replenishment of sulfur from the atmosphere. In many tropical countries, one or more of these conditions prevail and sulfur-deficient areas are common.

In the United States, deficiencies of sulfur are most common in the Southeast, the Northwest, California, and the Great Plains. In the Northeast and in other areas with heavy industry and large cities, sulfur deficiencies do not yet seem to be widespread.

Crops vary in their sulfur requirements. Legume crops such as alfalfa, the clovers, and soybeans have high sulfur requirements, as do cotton, sorghum, sugar beets, cabbage, turnips, and onions. Forests, grasses, and cereals generally have lower sulfur requirements, although wheat in the northwestern states is often quite responsive to sulfur applications.

11.22

Natural Sources of Sulfur

There are three major natural sources from which plants can be supplied with available sulfur: (a) organic matter; (b) soil minerals; and (c) sulfur gases in the atmosphere. These will be considered in order.

ORGANIC MATTER

In humid-region surface soils, most of the sulfur is present in organic forms, with amounts up to 90% and more of the total sulfur in this form being not uncommon. As is the case for nitrogen, the exact organic sulfur combinations in the organic matter are not known; however, the bulk of the sulfur is known to be found in proteins and in specific amino acid forms such as cysteine, cystine, and methionine. These materials are bound with the humus and clay fractions, which protects them from microbial attack. Over time, however, the microorganisms can simplify the bound sulfur into soluble inorganic forms.

In arid and semiarid regions, the proportion of organic sulfur will not be so high, especially in the subsoils. This is because of the lower organic matter level of these soils and of the presence of gypsum ($CaSO_4 \cdot 2H_2O$), which supplies inorganic sulfur. In the lower horizon of arid and semiarid soils, organic sulfur may comprise far less than half the sulfur present.

SOIL MINERALS

The inorganic forms of sulfur are not as plentiful as the organic forms, but they include the soluble and available compounds on which plants and microbes depend.

Sulfur is held in several mineral forms in soils, with the sulfide and sulfate minerals being most widespread. The sulfates are most easily solubilized, and the SO_4^{2-} ion is easily assimilated by plants. Sulfate minerals are most common in regions of low rainfall where they accumulate in the lower horizons of some Mollisols and Aridisols (Figure 3.12). They also accumulate even in the surface of saline soils of arid and semiarid regions.

Sulfides that are found in some humid region soils with restricted drainage must be oxidized to the sulfate form before the sulfur can be assimilated by plants. When these soils are drained, the oxidation can occur and ample available sulfur is released.

Another mineral source of sulfur is the clay fraction of some soils high in

Fe, Al oxides and kaolinite. These clays are able to adsorb some of the sulfur and it can later be slowly released from this bound form by anion exchange, especially at low pH.

ATMOSPHERIC SULFUR[6]

The soil is one of several natural sources of sulfur-containing gases that are being added to the atmosphere. In recent years, industrial areas around the world have become the primary source of atmospheric sulfur. The combustion of fuels, especially high-sulfur coal, releases sulfur dioxide and other sulfur gases and solids into the atmosphere. Some of these materials are oxidized in the atmosphere to sulfates, forming sulfuric acid by reacting with water and sulfate salts such as calcium and magnesium sulfate. Some of the solids and gases return to the earth as dry particles and gases, *dry deposition*, whereas some are brought down with precipitation, *wet deposition*.

The industrialized northeastern states have the highest deposition of sulfur, ranging from 30–75 kg S/ha per year. As one moves away from the industrialized plant sources, this level declines to figures one fourth to one half those cited. In open areas of the Western United States, away from industrial cities and smelting plants, the level drops to 2–5 kg S/ha per year. Thus, the effect of atmospheric deposition greatly depends on location and industrial concentration (Figure 11.11).

Atmospheric sulfur becomes part of the soil–plant system in three ways. The wet deposition materials, which are usually high in sulfuric acid (H_2SO_4), are absorbed by soils. Part of the dry deposition is also absorbed directly by soils, while some is absorbed directly by plants. The quantity plants can absorb directly is variable, but, in some cases, 25–35% of the plant sulfur can come from this source even if available soil sulfate is adequate. In sulfur deficient soils, about half of what plants absorb can come from the atmosphere if it is well-supplied with sulfur-containing materials.

Recent environmental concerns about high sulfur levels in the atmosphere

---

[6] For a discussion of this topic, see NAS (1983).

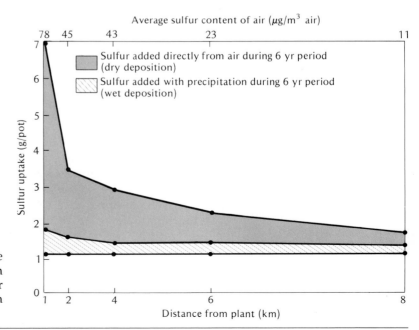

FIGURE 11.11
Sulfur added to soils as affected by distance from an oil-burning industrial plant in Sweden. Note the rapid dropoff in sulfur added directly from the air. [From Johannson (1960)]

and the resulting "acid rain" (see Section 11.25) are of great practical significance to agriculture. On the one hand, efforts to reduce atmospheric sulfur are welcomed in areas adversely affected by air pollution resulting from urban and industrial wastes. Toxic effects of high sulfur levels on vegetation are noted over wide areas downwind from the pollution source. But in other areas, where atmospheric sulfur levels are not sufficiently high to be toxic to plants, drastic reductions in atmospheric sulfur would result in deficiencies of this element for optimum crop growth. In these cases, the sulfur deficiency would need to be met by increasing the sulfur content of fertilizers, thereby increasing crop production costs. These matters are discussed further in Chapter 19.

Forms of Sulfur

This discussion has identified three major forms of sulfur in soils and fertilizers: *sulfides, sulfates,* and *organic forms.* To these must be added the fourth important form, *elemental sulfur,* the starting point of most of the man-made chemical sulfur compounds. The following sections will show the relationship among these forms.

11.23

The Sulfur Cycle

The major transformations that sulfur undergoes in soils are shown in Figure 11.12. The inner circle shows the relationships among the four major forms of this element in soils and in fertilizers. The outer portions show the most important sources of sulfur and how this element is lost from the system.

Some similarity between the sulfur and nitrogen cycles is evident (compare Figures 11.1 and 11.12). In each case, the atmosphere is an important source of the element in question. Each is held largely in the organic fraction of the soil, and each depends to a considerable extent on microbial action for its various transformations.

Figure 11.12 should be referred to frequently as a more detailed examination of sulfur in plants and soils is begun.

11.24

Role of Sulfur Compounds in Soils

Mineralization

Sulfur acts much like nitrogen as it is absorbed by plants and microorganisms and moves through the sulfur cycle. The organic forms of sulfur must be mineralized by soil organisms if the sulfur is to be used by plants. The rate at which mineralization occurs depends on the same environmental factors that affect nitrogen mineralization, including moisture, aeration, temperature, and pH. When conditions are proper for general microbial activity, sulfur mineralization occurs. The mineralization reaction might be expressed as

$$\text{organic sulfur} \rightarrow \text{decay products} \xrightarrow{O_2} SO_4^{2-} + H^+$$

organic sulfur
(proteins and
other organic
combinations)

decay products
(H_2S and other
sulfides are
simple examples)

SO_4^{2-}
Sulfates

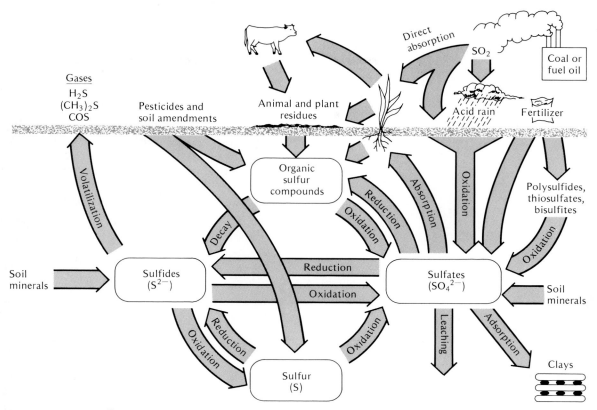

FIGURE 11.12

The sulfur cycle, showing some of the transformations that occur as this element changes form in soils, plants, and animals. It is well to keep in mind that except for certain soils in arid areas, the great bulk of the sulfur is in the form of organic compounds.

IMMOBILIZATION

Immobilization of inorganic forms of sulfur occurs when low-sulfur, energy-rich organic materials are added to soils not plentifully supplied with inorganic sulfur. The mechanism is thought to be the same as for nitrogen. The energy-rich material stimulates microbial growth, and the inorganic sulfate is synthesized into microbial tissue. Only when the microbial activity subsides does the inorganic sulfate form again appear in the soil solution.

These facts suggest that, like nitrogen, sulfur in soil organic matter may be associated with organic carbon in a reasonably constant ratio. The ratio among carbon, nitrogen, and sulfur for a number of soils on three different continents is given in Table 11.11. A C/N/S ratio of 130:10:1.3 is reasonably

TABLE 11.11

Mean Carbon/Nitrogen/Sulfur Ratios in a Variety of Soils

Location	Description and number of soils	C/N/S ratio
North Scotland	Agricultural, noncalcareous (40)	147:10:1.4
Minnesota	Mollisols (6)	114:10:1.55
Minnesota	Spodosols (24)	132:10:1.22
Oregon	Agricultural, varied (16)	145:10:1.01
Eastern Australia	Acid soils (128)	152:10:1.21
Eastern Australia	Alkaline soils (27)	140:10:1.52

From Whitehead (1964).

representative. Undoubtedly, there is a definite relationship among the amounts of these elements, a relationship that is in accord with their reactions in soils.

VOLATILIZATION

During the microbial breakdown of organic materials, several sulfur-containing gases are formed, including hydrogen sulfide (H_2S), carbon disulfide (CS_2), carbonyl sulfide (COS), and mercaptan (CH_3SH). All are more prominent in anaerobic soils. Hydrogen sulfide is commonly produced in waterlogged soils by reduction of sulfates by anaerobic bacteria. Most of the others are formed from the microbial decomposition of sulfur-containing amino acids. Although these gases can be adsorbed by soil colloids, some escape to the atmosphere where they undergo chemical changes and eventually return to the soil.

11.25

Sulfur Oxidation and Reduction

THE PROCESS

During the microbial decomposition of organic sulfur compounds, sulfides are formed along with other incompletely oxidized substances such as elemental sulfur, thiosulfates, and polythionates. These reduced substances are subject to oxidation, just as are the ammonium compounds formed when nitrogenous materials are decomposed. The oxidation reactions may be illustrated with hydrogen sulfide and elemental sulfur.

$$H_2S + 2\,O_2 \rightarrow H_2SO_4 \rightarrow 2\,H^+ + SO_4^{2-}$$

$$2\,S + 3\,O_2 + 2\,H_2O \rightarrow 2\,H_2SO_4 \rightarrow 4\,H^+ + 2\,SO_4^{2-}$$

The oxidation of some sulfur compounds, such as sulfites (SO_3^{2-}) and sulfides (S^{2-}), can occur by strict chemical reactions. However, most of the sulfur oxidation occurring in soils is thought to be *biochemical* in nature. Biochemical sulfur oxidation is accomplished by a number of autotrophic bacteria, including five species of the genus *Thiobacillus*.

Since the environmental requirements and tolerances of these five species vary considerably, sulfur oxidation can occur over a wide range of soil conditions—for example, at pH values from less than 2 to higher than 9. This is in contrast to nitrification, the comparable nitrogen oxidation process, which requires a rather narrow pH range closer to neutral.

Like nitrate ions, sulfate ions tend to be unstable in anaerobic environments. They are reduced to sulfide ions by a number of bacteria of two genera, *Desulfovibrio* (five species) and *Desulfotomaculum* (three species). The organisms use the combined oxygen in sulfate to oxidize organic materials. A representative reaction showing this reduction is

$$2\,R\!-\!CH_2OH + SO_4^{2-} \rightarrow 2\,R\!-\!COOH + 2\,H_2O + S^{2-}$$
Organic alcohol   Sulfate   Organic acid   Sulfide

In poorly drained soils, the sulfide ion will react immediately with iron or manganese, which in anaerobic conditions would be present in the reduced forms. This reaction might be expressed as

$$Fe^{2+} + S^{2-} \rightarrow \quad FeS$$
Iron sulfide

$$Mn^{2+} + S^{2-} \rightarrow \quad MnS$$
Manganese sulfide

Sulfide ions will also undergo hydrolysis to form gaseous hydrogen sulfide, which is the cause of the rotten egg smell of swampy or marshy areas. Sulfur reduction may take place with sulfur-containing ions other than sulfates. For example, sulfites (SO_3^{2-}), thiosulfates ($S_2O_3^{2-}$), and elemental sulfur (S) are rather easily reduced to the sulfide form by bacteria and other organisms.

The oxidation and reduction of inorganic sulfur compounds determine to a considerable extent the quantity of sulfate, the essential nutrient form, present in soils at any one time. The state of sulfur oxidation also determines to a marked degree the acidity of a soil.

SULFUR OXIDATION AND ACIDITY

As the oxidation reactions for H_2S and S show, oxidizing sulfur is an acidifying process. For every sulfur atom oxidized, two hydrogen ions are formed. By adding sulfur this acidifying reaction is utilized to reduce the extreme alkalinity of alkali soils of arid regions and to reduce the pH of soils for the control of diseases such as potato scab (Figure 9.12). The acid-forming reactions must also be taken into consideration when choosing a fertilizer, since some materials contain elemental sulfur and might lower soil pH unfavorably.

The addition of atmospheric sulfur to soils through precipitation increases soil acidity.[7] The pH of this so-called "acid rain" may be 4 or even lower compared to about 5.6 for "natural" precipitation (see Figure 11.13). These low pH values harm plants and wildlife as well as people. Fortunately, steps are being taken to reduce excess sulfur- and nitrogen-containing acids in the atmosphere.

EXTREME SOIL ACIDITY

The acidifying effect of sulfur oxidation can bring about extremely acid soil conditions, for example, when coastal land under brackish water or sea-water is drained and put under cultivation. During past periods of submergence, sulfates in the water have been reduced to sulfides, generally as iron and manganese sulfides, in which form they are stabilized. During periods of partial drying, some elemental sulfur has been formed by partial oxidation of the sulfides. The sulfide and elemental sulfur content in these submerged areas is hundreds of times greater than would be found in comparable upland soils.

When high-sulfide lands are drained, the sulfides and/or elemental sulfur are quickly oxidized, forming sulfuric acid. The soil pH may drop as low as 1.5, a level unknown in normal upland soils. Obviously, plant growth cannot occur under these conditions. Furthermore, the quantity of limestone needed to neutralize the acidity is so high as to make this remedy completely uneconomical.

---

[7] Note that some of this acidity results from nitrogen oxides as they react in the atmosphere to form nitric and nitrous acids. Sulfur and nitrogen are jointly responsible for this problem.

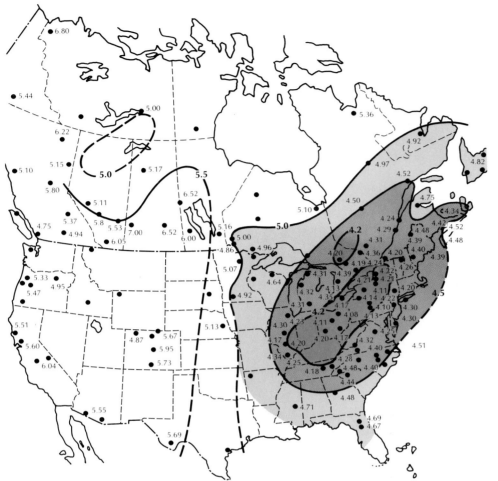

FIGURE 11.13
Annual mean value of pH in precipitation weighted by the amount of precipitation in the United States and Canada for 1980. [From U.S./Canada Work Group No. 2 (1982)]

Sizeable areas of high-sulfur soils (classified as sulfaquepts and sulfaquents in *Soil Taxonomy* and commonly called *acid sulfate soils* or *cat-clays*) are found in southeast Asia and along the Atlantic coasts of South America and Africa. They also occur in tideland areas along the coasts of several other areas, including the Netherlands and the southeastern part and west coast of the United States. So long as these soils are kept submerged, the soil pH does not drop prohibitively. Consequently, production of paddy rice is sometimes possible under these conditions.

11.26

Sulfate Retention and Exchange

Sulfate ion is the form in which plants absorb most of their sulfur from soils. Since this ion is quite soluble, it would be readily leached from the soil, especially in humid regions, were it not for its adsorption by the soil colloids. As was pointed out in Chapter 7, most soils have some anion exchange capacity associated with iron and aluminum oxide clays and, to a limited extent, with 1:1-type silicate clays. The sulfate ion is attracted by the positive charges that characterize acid soils containing these clays. It also reacts directly with hydroxy groups associated with these clays. A general equation indicates how this may occur.

$$
\begin{array}{c}
\mid \\
\text{Al} \\
\end{array}
\text{HO} \diagup\diagdown \text{OH} + \text{KHSO}_4 \rightleftharpoons \text{HO} \diagup\diagdown \text{SO}_4\text{—K} + \text{H}_2\text{O}
$$

This reaction to the right is favored in acid soils. If the soil pH were increased and hydroxide ions added, the reaction would be driven to the left, and sulfate ions would be released. Accordingly, this is an ion exchange reaction, the hydroxide being exchanged for an KSO_4^- anion.

Most soils will retain some sulfate, although the quantity held is generally small and the strength of sulfate retention is low compared to that of phosphate. Sulfate retention is generally higher in the subsoil than the topsoil, since iron and aluminum oxides are more prominent in subsoils. Soils of the southeastern United States tend to be higher in sulfate adsorption than elsewhere in the country because of their higher content of iron and aluminum oxides and 1:1-type silicate clays, especially in the subsoils. This is fortunate since the topsoils in this area are quite low in sulfur.

Sulfur adsorption and exchange by soils are important since these processes can provide a supply of available sulfur for plants. Also, sulfur adsorption helps the soil hold an otherwise mobile element that could easily be lost by leaching.

11.27

Sulfur and Soil Fertility Maintenance

Figure 11.14 depicts the major gains and losses of sulfur from soils. The problem of maintaining adequate quantities of sulfur for crops is becoming increasingly important. Although this element is added to soils through adsorption from the atmosphere and as an incidental component of many fertilizers, each of these sources are apt to decline in the future. Even though chances for widespread sulfur deficiencies are generally less than for nitrogen, phosphorus, and potassium, increasing crop removal of sulfur makes it essential that farmers be on guard to prevent deficiencies of this element.

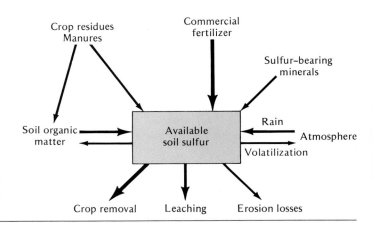

FIGURE 11.14

Major gains and losses of available soil sulfur. This represents average conditions from which considerable variation occurs in the field.

Crop residues and farmyard manures can help replenish the sulfur removed in crops, but greater and greater dependence must be placed on fertilizer additions. Regular applications of sulfur-containing materials now are necessary for good crop yields in large areas far removed from industrial plants. There will certainly be an increased necessity for the use of sulfur in the future.

11.28
Conclusion

Sulfur and nitrogen have much in common in soils. Both are held by soil colloids in forms that become slowly available. Both are found in proteins and other organic substances as part of the soil organic matter. Their release as inorganic ions (SO_4^{2-}, NH_4^+, and NO_3^-), in which form they are available to higher plants, is accomplished by soil microorganisms. Anaerobic soil organisms are able to change these elements into gaseous forms, which are then released to the atmosphere to be joined by similar gases released from industrial and urbanized areas and from automobiles and trucks. The gases are then deposited on plants, soils, and other surface objects in forms that are popularly termed "acid rain." This has serious consequences both for agriculture and for society in general.

There are some significant differences between nitrogen and sulfur. A much higher proportion of soil sulfur as compared to soil nitrogen is found in inorganic compounds, especially in some drier areas where gypsum ($CaSO_4 \cdot 2H_2O$) is abundant in the subsoil. Some soil organisms have the ability to fix elemental nitrogen gas into compounds usable by plants. Also, nitrogen, which is removed in much larger quantities by plants, must be replenished regularly by organic residues, manure, or chemical fertilizers.

In rural areas away from cities and industrial plants, sulfur deficiencies in plants are increasingly common. As a result of increased removal of sulfur by crops and reductions in "incidental" applications of sulfur to soils, this element will likely join nitrogen as a regular intentional constituent of chemical fertilizers.

Study Questions

1. A farmer is experiencing difficulty in harvesting wheat because the crop tends to fall over prior to harvest. There appears to be no disease or insect problems. Would you advise heavier nitrogen fertilizer applications to alleviate this problem? Explain.
2. The atmosphere is said to be an important source of soluble nutrients for both sulfur and nitrogen. Compare the amounts of each element coming to the soil from the atmosphere as well as the mechanisms by which each element is made available to plants.
3. Would you expect the process of nitrification to be occurring vigorously a few days after a heavy application of straw was made and worked into a soil? Why or why not?
4. When barnyard manure is incorporated into a moist alkaline surface soil, major products of the decomposition process are ammonium compounds. What happens to the NH_4^+ ions? Explain.
5. Assume you are a salesperson and one of your products is a nitrification inhibitor. What specific soil conditions would you seek if you wanted to demonstrate the biological and economic potential of your product?
6. There is much concern today about the contamination of air and water by the use of nitrogen fertilizers. What relationship, if any, do the processes of nitrification and denitrification have to these concerns? Explain.

7. A study has shown that the worldwide amount of nitrogen reaching the soil from legume–*Rhizobium* association makes up only a fraction of the total nitrogen that is added to the soil. What are the other major sources of biologically fixed nitrogen, and what factors influence the amounts of combined nitrogen they supply?

8. Nitrogen is "fixed" from the atmosphere and also is "fixed" by vermiculite clays and humus. Differentiate between these two processes and indicate the role in each played by soil microbes.

9. Assume you sent some seeds of a high-yielding soybean variety to a friend in the highlands of East Africa where this leguminous crop has never been grown before but where temperature and moisture conditions should be satisfactory. The crop did very poorly even though no disease or insect problems were apparent. What is the likely difficulty, and what could you do to alleviate it? Explain.

10. You have read in the newspaper about "acid rain." Where would you go if you wanted to sample it? What characteristics would you expect it to have? What are its advantages and disadvantages to agriculture?

11. Sulfur deficiencies are being experienced in areas in the southwestern, northwestern, and even midwestern parts of the United States, areas in which such deficiencies were not widespread 20 years ago. What are the reasons for these increases in areas of sulfur deficiency?

12. How would you compare nitrogen and sulfur with respect to (a) total amount in (1) soils and (2) the atmosphere; (b) relative amounts in organic and inorganic forms; (c) reactions with clays (silicate and hydrous oxides); and (d) dependence on soil microorganisms for their mobility in soils?

13. Would you be surprised if sulfur deficiencies occurred in well-weathered soils of Africa in areas far removed from cities and industry? Explain.

References

ASA. 1980. *Nitrification Inhibitors—Potentials and Limitations*, ASA Special Publication No. 38 (Madison, WI: Amer. Soc. Agron., Soil Sci. Soc. Amer.)

BAETHGEN, W. E., and M. M. ALLEY. 1987. "Nonexchangeable ammonium nitrogen contribution to plant available nitrogen," *Soil Sci. Soc. Amer. J.,* **50**:110–15.

BEZDICEK, D. F., et al. 1972. "Evaluation of peat and granular inoculum for soybean yield and N fixation under irrigation," *Agron. J.,* **76**:865–68.

BOND, G. 1977. "Some reflections on *Alnus*-type root nodules," in W. Newton, J. R. Postgate, and C. Rodriguez-Barrueco (Eds.), *Recent Developments in Nitrogen Fixation* (New York: Academic Press).

BURNS, R. C., and R. W. F. HARDY. 1975. *Nitrogen Fixation in Bacteria and Higher Plants* (Berlin: Springer-Verlag).

BROADBENT, F. E., T. NAKASHIMA, and G. Y. CHANG. 1982. "Estimation of nitrogen fixation by isotope dilution in field and greenhouse experiments," *Agron. J.,* **74**:625–28.

CHICHESTER, F. W., and S. J. SMITH. 1978. "Disposition of ^{15}N-labeled fertilizer nitrate applied during corn culture in field lysimeters," *J. Environ. Qual.,* **7**:227–33.

HAUCK, R. D. (Ed.). 1984. *Nitrogen in Crop Production* (Madison, WI: Amer. Soc. Agron., Crop Sci. Soc. Amer., Soil Sci. Soc. Amer.).

HAVELKA, U. D., M. G. BOYLE, and R. W. F. HARDY. 1982. "Biological nitrogen fixation," in F. J. Stevenson (Ed.), *Nitrogen in Agricultural Soils*, Agronomy Series No. 22 (Madison, WI: Amer. Soc. Agron., Crop Sci. Soc. Amer., Soil Sci. Soc. Amer.).

JOHANNSON, O. 1960. "On sulfur problems in Swedish agriculture," *Kgl. Lanabr. Ann.,* **25**: 57–169.

LEFFELAAR, P. A. 1986. "Dynamics of partial anaerobiosis, denitrification and water in a soil aggregate," *Soil Sci.,* **142**:352–66.

McCLUNG, G., D. C. WOLF, and J. E. FOSS. 1983. "Nitrification inhibition by nitrapyrin and etridiazol in soils amended with sewage sludge compost," *Soil Sci. Soc. Amer. J.,* **47**:75–80.

NAS. 1983. *Acid Deposition: Atmospheric Processes in Eastern North America* (Washington, DC: National Academy Press).

OLSEN, R. A., and G. W. REHM. 1986. "Sulfur in precipitation and irrigation waters and its effect on soils and plants," in M. A. Tabatabai (Ed.), *Sulfur in Agriculture*, Agronomy Series No. 27 (Madison, WI: Amer. Soc. Agron., Crop Sci. Soc. Amer., Soil Sci. Soc. Amer.).

PATRICK, W. H., Jr. 1982. "Nitrogen transformations in submerged soils," in F. J. Stevenson (Ed.), *Nitrogen in Agricultural Soils*, Agronomy Series No. 22 (Madison, WI: Amer. Soc. Agron., Crop Sci. Soc. Amer., Soil Sci. Soc. Amer.).

SMITH, S. J., et al. 1982. "Disposition of fertilizer nitrates applied to sorghum–Sudan grass in the Southern Plains," *J. Environ. Qual.,* **11**:341–44.

STEVENSON, F. J. (Ed.). 1982. *Nitrogen in Agricultural Soils*, Agronomy Series No. 22 (Madison, WI: Amer. Soc. Agron., Crop Sci. Soc. Amer., Soil Sci. Soc. Amer.).

TABATABAI, J. S. (Ed.). 1986. *Sulfur in Agriculture*, Agronomy Series No. 27 (Madison, WI: Amer. Soc. Agron., Crop Sci. Soc. Amer., Soil Sci. Soc. Amer.).

TERMAN, G. L. 1978 "Atmospheric sulphur—the agronomic aspects," *Tech. Bull.* 23 (Washington, DC: The Sulphur Institute).

Torrey, J. G. 1978. "Nitrogen fixation by actinomycete-induced angiosperms," *BioScience,* **28**:586–92.

U.S./Canada Work Group #2. 1982 *Atmospheric Sciences and Analysis,* Final Report, J. L. Ferguson and L. Machta, Co-chairmen (Washington, DC: Environmental Protection Agency).

Walker, T. W., et al. 1956. "Fate of labeled nitrate and ammonium nitrogen when applied to grass and clover grown separately and together," *Soil Sci.,* **81**:339–52.

Whitehead, D. C. 1964. "Soil and plant-nutrition aspects of the sulfur cycle," *Soils and Fertilizers,* **27**:1–8.

Phosphorus and Potassium

Next to nitrogen, phosphorus and potassium are most critical essential elements in influencing plant growth and production throughout the world. Unlike nitrogen, these elements are not supplied through biochemical fixation but must come from other sources to meet plant requirements. The sources include (a) commercial fertilizer; (b) animal manures; (c) plant residues, including green manures; (d) human, industrial, and domestic wastes; and (e) native compounds of potassium and phosphorus, both organic and inorganic, already present in the soil. The first four sources will be considered in later chapters; in this chapter we will examine the ways and means of utilizing the body of soil as a source of these mineral elements.

12.1

Importance of Phosphorus

Phosphorus is essential for plant growth.[1] It is a component of adenosine diphosphate (ADP) and adenosine triphosphate (ATP), the two compounds involved in most significant energy transformations in plants. ATP, synthesized from ADP through both respiration and photosynthesis, contains a high-energy phosphate group that drives most biochemical processes requiring energy. For example, the uptake of some nutrients and their transport within the plant, as well as the synthesis of new molecules, are energy-using processes that ATP helps to implement.

Phosphorus also plays a critical role in the life cycle of plants. It is an essential component of deoxyribonucleic acid (DNA), the seat of genetic inheritance in plants as well as animals, and of the various forms of ribonucleic acid (RNA) needed for protein synthesis. Obviously, phosphorus is essential for numerous metabolic processes.

[1] For a review of the significance of this element, see Khasawneh et al. (1980).

Among the more significant functions and qualities of plants on which phosphorus has an important effect are

1. Photosynthesis.
2. Nitrogen fixation.
3. Crop maturation: flowering and fruiting, including seed formation.
4. Root development, particularly of the lateral and fibrous rootlets.
5. Strength of straw in cereal crops, thus helping to prevent lodging.
6. Improvement of crop quality, especially of forages and of vegetables.

12.2

The Phosphorus Cycle

The biological cycling of phosphorus from the soil to higher plants and return is illustrated in Figure 12.1. Note that the plant roots absorb soluble inorganic and, to a lesser extent, organic phosphorus compounds and translocate them to aboveground plant parts. The phosphorus in the plants is returned to the soil either in crop residues or in human and animal wastes. Microorganisms decompose the residues and temporarily tie up at least part of the phosphorus in their body tissue. Some of it becomes associated with the soil organic matter where it is subject to future release. Some is very slowly converted to the soluble forms that plant roots can absorb, thereby starting a repeat of the cycle.

In most soils, the amount of phosphorus in the available form at any one time is very low, seldom exceeding about 0.01% of the total phosphorus in the soil. Thus available phosphorus levels must be supplemented on most soils by adding chemical fertilizers. Unfortunately, much of such added phosphorus is converted to the less available secondary mineral forms, from which it, too, is released very slowly and becomes useful to plants only over a period of years. The problem of maintaining phosphorus in an available form will receive our attention first.

12.3

The Phosphorus Problem

The phosphorus problem is threefold. First, the total phosphorus level of soils is low, commonly no more than one tenth to one fourth that of nitrogen, and one twentieth that of potassium. The phosphorus content of soils ranges from 200 to 2000 kg P per hectare–furrow slice (HFS) with an average of about 1000 kg P/HFS. Second, the native phosphorus compounds are mostly unavailable for plant uptake, some being highly insoluble. Third, when soluble sources of phosphorus such as those in fertilizers and manures are added to soils, they are fixed[2] or are changed to unavailable forms and in time react further to become highly insoluble forms.

---

[2] Note that the term fixation as applied to phosphorus has the same general meaning as the chemical fixation of potassium or ammonium ions, that is, the chemical being fixed is bound, entrapped, or otherwise held tightly by soil solids in a form that is relatively unavailable to plants. In contrast, the fixation of gaseous nitrogen refers to the biological conversion of N_2 gas to combined forms that plants can utilize.

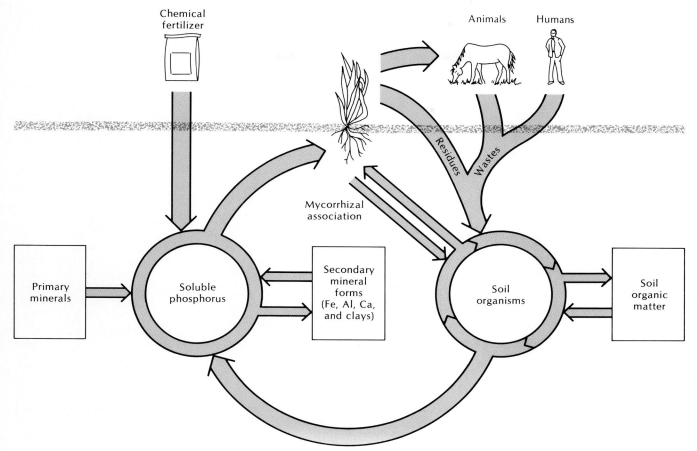

Figure 12.1
The phosphorus cycle in soils. At any one time 98–99% of the phosphorus is associated with primary or secondary minerals and soil organic matter. Some 1–2% is in microbial tissue, and only 0.01% exists as soluble phosphorus. Nonetheless, the circle that shows the interaction among plants, microbes, and soluble phosphorus is most significant to higher plants since it shows how soluble phosphorus is supplied. Chemical fertilizers are increasingly important as sources of soluble phosphorus. Note that there is no significant loss of phosphorus by leaching or in gaseous forms.

Fertilizer practices in many areas illustrate the problem of phosphorus unavailability. To obtain high yields, farmers commonly apply more phosphorus in fertilizers than is removed by the crops. For highly fertilized vegetable crops, phosphorus application rates are often double or triple the crop removal rates. For the United States overall, phosphorus fertilizer additions have generally exceeded crop removal (Figure 12.2). Research has quantified this inefficiency of use, showing less than 15% of fertilizer-applied phosphorus is normally taken up by the crop during the year the fertilizer is applied.

Some long-term benefits accrue from practices that add more phosphorus than the crops remove. In time, soil phosphorus levels are increased, often to high enough levels to reduce significantly future requirements for phosphorus fertilizers. Consequently, a buildup of phosphorus reserves has some practical implications.

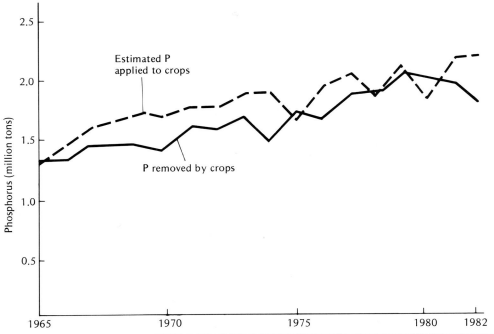

Figure 12.2
Estimates of fertilizer phosphorus applied in the United States from 1965 to 1982 compared to the removal of this element by crops. [From Douglas (1982)]

12.4

Phosphorus Compounds in Soils

Both inorganic and organic forms of phosphorus occur in soils[3] and both are important to plants as sources of this element. The relative amounts in the two forms vary greatly from soil to soil, but it is not at all uncommon for more than half the phosphorus to be in the organic form. Data in Table 12.1 give some idea of the relative proportions of organic and inorganic phosphorus in mineral soils. The organic fraction generally constitutes 20–80% of the total. Despite the variation that occurs, it is evident that consideration of soil phosphorus would not be complete without some attention to both forms (Figure 12.3).

INORGANIC COMPOUNDS

Most inorganic phosphorus compounds in soils fall into one of two groups: (a) those containing calcium and (b) those containing iron and aluminum. The calcium compounds of most importance are listed in Table 12.2. The apatite minerals are the most insoluble and unavailable of the group. They may be found in even the more weathered soils, especially in their lower horizons. The persistence of apatite minerals is an indication of the extreme insolubility and consequent unavailability of the phosphorus contained therein.

The simpler compounds of calcium, such as mono- and dicalcium phosphates, are readily available for plant growth. Except on recently fertilized soils, however, these compounds are present only in extremely small quantities because they easily revert to the more insoluble forms.

Much less is known of the exact constitution of the iron and aluminum

[3] For a review of soil phosphorus, see Olsen and Khasawneh (1980) and Stevenson (1986).

TABLE 12.1

Total Phosphorus Content of Soils from Different Areas and the Percentage of Total Phosphorus in the Organic Form

Soils	Number of samples	Total P (mg/kg)	Organic fraction (%)
Western Oregon soils			
Hill soils	4	357	66
Old valley-filling soils	4	1479	30
Recent valley soils	3	848	26
Iowa soils			
Mollisols	2	613	42
Alfisols	2	574	37
Alfisols	2	495	53
Arizona soils			
Surface soils	19	703	36
Subsoils	5	125	34
Australia soils			
Spodosol	1	398	65
Vertisol	1	362	86
Mollisol	1	505	75
Hawaii soils			
Hydrandept	1	4700	37
Haplustoll	1	2250	21
Gibbsiorthox	1	1414	19
Gibbsiorthox (Subsoil)	1	2575	7

Data for Oregon, Iowa, and Arizona from sources quoted by Brady (1974); Australia from Fares et al. (1974); Hawaii from Soltanpour et al. (1988).

phosphates contained in soils. The compounds involved are probably hydroxy phosphates such as strengite ($FePO_4 \cdot 2H_2O$) and variscite ($AlPO_4 \cdot 2H_2O$). Such hydroxy phosphates are most stable in acid soils and are quite insoluble.

Many investigators have shown that phosphates react with certain silicate minerals such as kaolinite. There is some uncertainty, however, as to the exact form in which this phosphorus is held in the soil. Most evidence indicates that the phosphates may actually encourage some crystal breakdown of the silicate clays and that the resulting insoluble products are iron or aluminum phosphates, such as those described in the preceding paragraph.

ORGANIC PHOSPHORUS COMPOUNDS

There has been relatively less work done on the organic phosphorus compounds in soils, even though this fraction often comprises more than half of the total soil phosphorus. As a consequence, the nature of most of the organic-bound phosphorus in soils is not known. However, three main groups of organic phosphorus compounds found in plants are also present in soils. These are (a) inositol phosphates—phosphate esters of a sugar-like compound, inositol ($C_6H_6(OH)_6$); (b) nucleic acids; and (c) phospholipids. While other organic phosphorus compounds are present in soils, the identity and amounts present are not known.

Inositol phosphates are the most abundant of the known organic phosphorus compounds, making up 10–50% of the total. They are thought to be of microbial origin. Inositol phosphates tend to be quite stable in acid and alkaline conditions and interact with the higher molecular weight humic

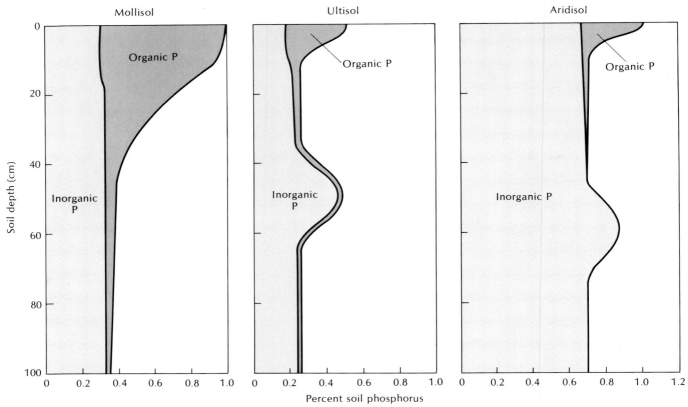

Figure 12.3

Phosphorus content of representative soils of three soil orders. Organic phosphorus makes up a high proportion of the soil phosphorus in the Mollisol and a much lower proportion in the others. The Aridisol has high phosphorus content because little of the inorganic phosphorus compounds has been leached from the soil. The higher content in the subsoil is due to fixation by Fe, Al oxides (Ultisol) and by calcium compounds (Aridisol).

compounds. These properties may account for their relative abundance in soils.

Nucleic acids, exemplified by ribonucleic acid (RNA) and deoxyribonucleic acid (DNA), are phosphorus-containing compounds in soils. Once thought to comprise about half the soil organic phosphorus, nucleic acids are now known to account for no more than about 1–5%.

The third known form of organic phosphorus is phospholipids, fat-like compounds of microbial origin. They make up only about 0.2–2.5% of the organic phosphorus.

It is obvious that half or more of the organic phosphorus is present in unknown forms. However, our ignorance of the specific compounds does not detract from the importance of these compounds as suppliers of phosphorus through microbial breakdown. This will be discussed after consideration of inorganic forms.[4]

[4] Although plants are able to absorb some organic phosphorus compounds directly, the level of such absorption is thought to be very low compared to that of inorganic phosphates.

12.5

Factors Controlling Availability of Inorganic Soil Phosphorus

The availability of inorganic phosphorus is largely determined by (a) soil pH; (b) soluble iron, aluminum, and manganese; (c) presence of iron-, aluminum-, and manganese-containing minerals; (d) available calcium and calcium minerals; (e) amount and decomposition of organic matter; and (f) activities of microorganisms. The first four factors are interrelated since soil pH drastically influences the reaction of phosphorus with the different ions and minerals.

12.6

pH and Phosphate Ions

The availability of phosphorus to plants is determined to no small degree by the ionic form of this element. The ionic form in turn is determined by the pH of the solution in which the ion is found (Figure 12.4). Thus, in highly acid solutions only $H_2PO_4^-$ ions are present. If the pH is increased, first HPO_4^{2-} ions and finally PO_4^{3-} ions dominate. This situation is represented by the equation

$$H_2PO_4^- \underset{H^+}{\overset{OH^-}{\rightleftharpoons}} H_2O + HPO_4^{2-} \underset{H^+}{\overset{OH^-}{\rightleftharpoons}} H_2O + PO_4^{3-}$$
(very acid solutions)   (very alkaline solutions)

At intermediate pH levels two of the phosphate ions may be present simultaneously. Thus, in solutions at pH 7.0, both $H_2PO_4^-$ and HPO_4^{2-} ions are found. The $H_2PO_4^-$ ion is somewhat more available to plants than is the HPO_4^{2-} ion. In soils, however, this relationship is complicated by the presence or absence of other compounds or ions at different pH levels. For

Figure 12.4
Relationship between solution pH and the relative concentrations of three soluble forms of phosphate. In the pH range common for soils, the $H_2PO_4^-$ ions predominate. The $H_2PO_4^-$ and HPO_4^{2-} ions are most commonly taken up by plants.

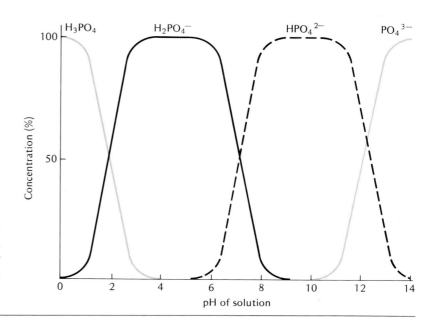

example, the presence of soluble iron and aluminum under very acid conditions, or calcium at high pH values, will markedly affect the availability of the phosphorus. The effect of these ions in acid soils will be discussed first.

12.7

Inorganic Phosphorus Availability in Acid Soils

Assume that you are dealing with either a nutrient solution or an organic soil very low in inorganic matter. Assume also that these media are acid in reaction but that they are low in iron, aluminum, and manganese. The $H_2PO_4^-$ ions, which would dominate under these conditions, would be readily available for plant growth. Normal phosphate absorption by plants would be expected so long as the pH was not too low.

PRECIPITATION BY IRON, ALUMINUM, AND MANGANESE IONS

If the same degree of acidity should exist in a normal mineral soil, however, quite different results would be expected (Figure 12.5). Some soluble iron, aluminum, and manganese are usually found in strongly acid mineral soils. Reaction with the $H_2PO_4^-$ ions would immediately occur, resulting in the formation of insoluble hydroxy phosphates. This chemical precipitation may be represented as follows, using the aluminum cation as an example.

$$Al^{3+} + H_2PO_4^- + 2\,H_2O \rightleftharpoons 2\,H^+ + Al(OH)_2H_2PO_4$$
$$\text{(soluble)} \qquad\qquad\qquad\qquad \text{(insoluble)}$$

In most strongly acid soils the concentration of the iron and aluminum ions greatly exceeds that of the $H_2PO_4^-$ ions. Consequently, the reaction moves to the right, forming the insoluble phosphate. This leaves only minute quantities of the $H_2PO_4^-$ ion immediately available for plants under these conditions.

An interesting series of reactions occur with iron, aluminum and man-

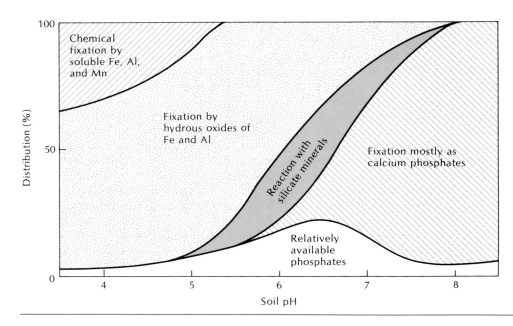

Figure 12.5
Inorganic fixation of added phosphates at various soil pH values. Average conditions are postulated, and it is not to be inferred that any particular soil would have exactly this distribution. The actual proportion of the phosphorus remaining in an available form will depend upon contact with the soil, time for reaction, and other factors. It should be kept in mind that some of the added phosphorus may be changed to an organic form in which it would be temporarily unavailable.

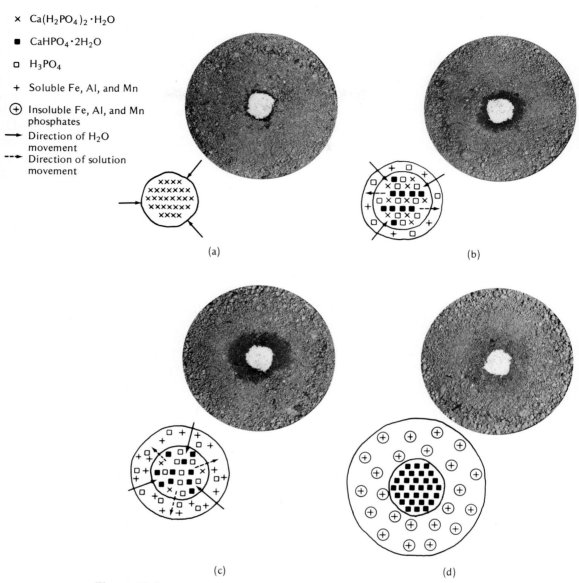

× Ca(H$_2$PO$_4$)$_2$·H$_2$O

■ CaHPO$_4$·2H$_2$O

□ H$_3$PO$_4$

+ Soluble Fe, Al, and Mn

⊕ Insoluble Fe, Al, and Mn phosphates

→ Direction of H$_2$O movement

--→ Direction of solution movement

(a)

(b)

(c)

(d)

Figure 12.6
Reaction of Ca(H$_2$PO$_4$)$_2$·H$_2$O granules with moist soil. (a) The granule has just been added to the soil and is beginning to absorb water from it. (b) In the moistened granule H$_3$PO$_4$ and CaHPO$_4$·2H$_2$O are being formed, and H$_3$PO$_4$ begins to move out into the soil as more soil water is being absorbed. (c) The H$_3$PO$_4$-laden solution moves into the soil, dissolving and displacing Fe, Al, and Mn and leaving insoluble CaHPO$_4$·2H$_2$O in the granule. (d) The Fe, Al, and Mn ions have reacted with the phosphate to form insoluble compounds, which, along with the residue of CaHPO$_4$·2H$_2$O, are the primary reaction products. [Photos courtesy G. L. Terman and National Plant Food Institute, Washington, DC]

ganese compounds when fertilizers containing soluble Ca(H$_2$PO$_4$)$_2$·H$_2$O are added to soils, even those of relatively high pH (Figure 12.6). The Ca(H$_2$PO$_4$)$_2$·H$_2$O in the fertilizer granules attracts water from the soil, resulting in the following reaction.

$$Ca(H_2PO_4)_2·H_2O + H_2O \rightarrow CaHPO_4·2H_2O + H_3PO_4$$

As more water is attracted, a H_3PO_4-laden solution with a pH of about 1.4 moves outward from the granule. This solution is sufficiently acid to dissolve and displace large quantities of iron, aluminum, and manganese. In acid soils these ions promptly react with the phosphate to form complex compounds, which later revert to the hydroxy phosphates of iron, aluminum, and manganese. In neutral to alkaline soils equally insoluble calcium phosphates are formed. Consequently, the immediate products of the addition to soils of a water-soluble compound [$Ca(H_2PO_4)_2 \cdot H_2O$] are a group of insoluble iron, aluminum, managanese, and calcium compounds. Fortunately, the phosphorus in these freshly precipitated compounds remains available, to a degree, for plant uptake. But when these freshly precipitated compounds are allowed to "age" or to revert to more insoluble forms, availability to plants is greatly reduced.

REACTION WITH HYDROUS OXIDES

The $H_2PO_4^-$ ion reacts not only with iron, aluminum, and manganese ions but even more extensively with insoluble hydrous oxides of these elements, such as gibbsite ($Al_2O_3 \cdot 3H_2O$) and goethite ($Fe_2O_3 \cdot 3H_2O$) (Figure 12.5). The compounds resulting from these reactions are likely to be hydroxy phosphates, just as in the case of the chemical precipitation described above. Their formation can be illustrated by means of the following equation if the hydrous oxide of aluminum is represented as aluminum hydroxide.

$$\begin{matrix} HO \\ \quad \searrow \\ HO \nearrow \end{matrix} Al\text{—}OH + H_2PO_4^- \rightleftharpoons \begin{matrix} HO \\ \quad \searrow \\ HO \nearrow \end{matrix} Al\text{—}H_2PO_4 + OH^-$$

$$\text{(soluble)} \qquad \text{(insoluble)}$$

By means of this and similar reactions the formation of several basic phosphate minerals containing either iron or aluminum or both is thought to occur. Since several such compounds are possible, fixation of phosphorus by this mechanism probably takes place over a relatively wide pH range (Figure 12.5). Furthermore, the large quantities of hydrous Fe, Al oxides present in most soils make possible the fixation of extremely large amounts of phosphorus by this means.

   Therefore, as both of the fixation equations show, the acid condition that assures the presence of the readily available $H_2PO_4^-$ ion in mineral soils also results in conditions conducive to the vigorous *fixation* or precipitation of the phosphorus by iron, aluminum, and manganese compounds (Figure 12.6).

FIXATION BY SILICATE CLAYS

A third means of fixation of phosphorus under moderately acid conditions involves silicate minerals such as kaolinite. Although there is some doubt about the actual mechanisms involved, the overall effect is essentially the same as when phosphorus is fixed by simpler iron and aluminum compounds. Some scientists visualize the fixation of phosphates by silicate minerals as a surface reaction between exposed —OH groups on the mineral crystal and the $H_2PO_4^-$ ions. Other investigators have evidence that aluminum and iron ions are actually removed from the edges of the silicate crystals and then form hydroxy phosphates of the same general formula as those already discussed. This type of reaction might be expressed as follows.

$$[Al] \quad + H_2PO_4^- + 2\,H_2O \rightleftharpoons 2\,H^+ + Al(OH)_2H_2PO_4$$
$$\text{(in silicate crystal)} \qquad\qquad\qquad \text{(insoluble)}$$

Although phosphates react with different ions and compounds in acid soils, apparently similar insoluble iron and aluminum compounds are formed in each case. Major differences from soil to soil are probably due to differences in rate of phosphate precipitation and in the surface area of the reaction products once the reactions have occurred. We will discuss the effect of surface area later (Section 12.10).

ANION EXCHANGE

In addition to taking part in reactions that yield insoluble precipitates and hydroxy phosphates, phosphorus can participate in simple anion exchange reactions. As was pointed out in Section 7.14, certain positively charged colloidal particles are the seat of exchange between two anions. This can be illustrated with $>AlOH_2^+$ to represent the positively charged particle.

$$\boxed{>AlOH_2^+}OH^- + H_2PO_4^- \rightleftharpoons \boxed{>AlOH_2^+}H_2PO_4^- + OH^-$$

(positively charged particle)   (positively charged particle)

Note that the reaction is reversible; that is, the OH^- ion can replace the $H_2PO_4^-$ ion and remove it from the colloidal surface to the soil solution. This reaction shows how anion exchange takes place and illustrates the importance of liming acid soils in helping to maintain a higher level of available phosphorus.

12.8

Inorganic Phosphorus Availability at High pH Values

The availability of phosphorus in alkaline soils[5] is determined largely by the solubility of the calcium compounds in which the phosphorus is found. If an $H_2PO_4^-$-containing fertilizer such as concentrated superphosphate is added to an alkaline soil (e.g., at pH = 8.0), the $H_2PO_4^-$ ion quickly reacts to form less soluble compounds. Although intermediate compounds are formed, tricalcium phosphate $[Ca_3(PO_4)_2]$ is the most significant of these products. The reaction involving $Ca(H_2PO_4)_2 \cdot H_2O$ and $CaCO_3$ in the soil can be shown as follows.

$$Ca(H_2PO_4)_2 \cdot H_2O + 2\,CaCO_3 \rightarrow Ca_3(PO_4)_2 + 2\,CO_2 + 2\,H_2O$$

The solubility of the compounds and, in turn, the availability to plants of the phosphorus they contain decrease as the phosphorus changes from the $H_2PO_4^-$ ion to tricalcium phosphate $[Ca_3(PO_4)_2]$. Although this compound is quite insoluble, it may be converted further in the soil to even more insoluble compounds. Hydroxy, oxy, carbonate, and even fluorapatite compounds, such as those shown in Table 12.2, may be formed if conditions are favorable and if sufficient time is allowed. These compounds are thousands of times more insoluble than freshly formed tricalcium phosphate.

The reversion to insoluble calcium phosphates may also occur in soils of the eastern United States (Spodisols, Ultisols, and Alfisols) that have been heavily limed. The problem is much more serious in soils of the western

[5] See Sample et al. (1980) for a discussion of this subject.

TABLE 12.2
Inorganic Calcium Compounds of Phosphorus Often Found in Soils

Listed in order of increasing solubility.

Compound	Formula
Fluorapatite	$[3Ca_3(PO_4)_2] \cdot CaF_2$
Carbonate apatite	$[3Ca_3(PO_4)_2] \cdot CaCO_3$
Hydroxy apatite	$[3Ca_3(PO_4)_2] \cdot Ca(OH)_2$
Oxy apatite	$[3Ca_3(PO_4)_2] \cdot CaO$
Tricalcium phosphate	$Ca_3(PO_4)_2$
Octacalcium phosphate	$Ca_8H_2(PO_4)_6 \cdot 5H_2O$
Dicalcium phosphate	$CaHPO_4 \cdot 2H_2O$
Monocalcium phosphate	$Ca(H_2PO_4)_2$

states, however, because of the widespread natural presence of excess $CaCO_3$. The problem of utilizing phosphates in alkaline soils of arid areas (e.g., Aridisols) is therefore fully as serious as it is on highly acid Spodisols and Ultisols.

12.9

pH for Maximum Inorganic Phosphorus Availability

With insolubility of phosphorus occurring at both extremes of the soil pH range (Figure 12.5), the question arises as to the range in soil reaction in which minimum fixation occurs. The basic iron and aluminum phosphates have a minimum solubility around pH 3–4. At higher pH values some of the phosphorus is released and the fixing capacity somewhat reduced. Even at pH 6.5, however, much of the phosphorus is probably still chemically combined with iron and aluminum. As the pH approaches 6, precipitation as calcium compounds begins; at pH 6.5 the formation of insoluble calcium salts is a factor in rendering the phosphorus unavailable. Above pH 7.0, even more insoluble compounds, such as the apatites, are formed.

These facts seem to indicate that maximum phosphate availability to plants is obtained when the soil pH is maintained in the 6.0–7.0 range (Figure 12.5). Even in this range, however, the fact should be emphasized that phosphate availability may still be very low and that added soluble phosphates are readily fixed by soils. The low recovery (perhaps 10–15%) by plants of phosphates added to field mineral soils in a given season is partially due to this fixation. A much higher recovery would be expected in organic soils and in many potted mixes where calcium, iron, and aluminum concentrations are not high.

12.10

Availability and Surface Area of Phosphates

When soluble phosphates are added to soils, insoluble phosphates are formed with calcium, or iron and aluminum. However, the total surface area of these phosphate-containing particles is reasonably high; consequently the availability of the phosphorus is appreciable. Thus, even though the water-soluble phosphorus may be precipitated in the soil in a matter of a few days, the freshly precipitated compounds will release much of their phosphorus to growing plants.

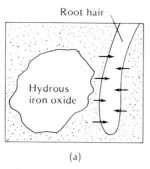
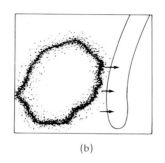
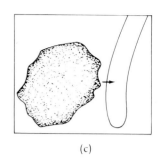

(a)   (b)   (c)

Figure 12.7
How relatively soluble phosphates are rendered unavailable by compounds such as hydrous oxides of Fe and Al. (a) The situation just after application of a soluble phosphate. The root hair and the hydrous iron oxide particle are surrounded by soluble phosphates. (b) Within a very short time most of the soluble phosphate has reacted with the surface of the iron oxide crystal. The phosphorus is still fairly readily available to the plant roots since most of it is located at the surface of the particle where exudates from the plant can encourage exchange. (c) In time the phosphorus penetrates the crystal and only a small portion is found near the surface. Under these conditions its availability is low.

EFFECTS OF AGING

With time, changes take place in the reaction products of soluble phosphates and soils. These changes generally result in a reduction in surface area of the phosphates and a similar reduction in their availability. An increase in the crystal size of precipitated phosphates occurs in time. This decreases their surface area. Also, the phosphorus held by calcium carbonate, or by iron or aluminum oxide particles, penetrates into the particle itself (Figure 12.7). This leaves less of the phosphorus near the surface where it can be made available to growing plants. By these processes of aging, phosphate availability is reduced. Consequently, the supply of available phosphorus to plants is determined not only by the kinds of compounds that form but also by their surface areas.

RELATION TO SOIL TEXTURE

Most of the compounds with which phosphorus reacts are in the finer soil fractions. As a consequence phosphorus fixation tends to be more pronounced in clay soils than in the coarser textured ones. This fact is borne out by the data in Figure 12.8, which clearly illustrates the tendency of clays to reduce phosphate availability.

Figure 12.8
The effect of clay content on the percent recovery of fertilizer phosphorus by seven crops grown on four calcareous soils. Soils with higher clay contents tended to fix the phosphorus in forms not readily available to the crop plants. [From Olsen et al. (1977); used with permission of World Phosphate Industry Association, Paris, France]

TABLE 12.3

Comparative Rates of Phosphorus Fixation of Several Soils Varying the Content and Kinds of Clays

Fe, Al oxides (especially amorphous types) fix largest quantities and silicate clays least.

| Soil great group | Location | Clay | | P fixation (kg P/2 × 10⁶ kg soil) |
		Percent	Type	
Haplustoll	Hawaii	70%	Silicate clays	0
Gibbsihumox	Hawaii	70%	Crystal Fe,Al oxides	1800
Hydrandept	Hawaii	70%	Amorphous Fe,Al oxides	5600
Paleudult	Peru	6%	Fe,Al oxides	200
Hapludult	North Carolina	38%	Fe,Al oxides	680

Data quoted from several sources by Sanchez and Uehara (1980).

12.11

Phosphorus-Fixing Power of Soils

In light of the discussion in the preceding section, it is interesting to note the actual quantities of phosphorus that soils are capable of fixing. Data from selected Oxisols, Ultisols, and Mollisols in Table 12.3 show fixation capacity. Note that soils high in clay fix more phosphorus, especially if the clays are primarily Fe, Al oxides and if they are amorphous rather than crystalline. The Mollisol (Haplustoll) apparently was very low in Fe, Al oxides and fixed essentially no phosphorus.

The high capacity of soil for fixing phosphorus explains why much fertilizer-supplied phosphorus is quickly rendered unavailable for crops. Fortunately, over a period of years plants are able to absorb at least some of these fixed materials.

12.12

Organic Matter, Microbes, and Available Phosphorus

Phosphorus held in organic form can be mineralized and immobilized by the same general processes pertinent for nitrogen and sulfur. The following reaction illustrates this point.

$$\text{organic P forms} \underset{\text{microbes}}{\overset{\text{microbes}}{\rightleftharpoons}} H_2PO_4^- \overset{Fe^{3+},\ Al^{3+},\ Ca^{2+}}{\rightleftharpoons} \underset{\text{Fixed phosphorus}}{\text{Fe, Al, Ca phosphates}}$$

(Immobilization ← / Mineralization →)

Soluble phosphorus compounds are released as organic residues and humus are decomposed. The resulting soluble inorganic phosphate ion ($H_2PO_4^-$) is subject to uptake by plants or to fixation into insoluble forms. Should organic residues low in phosphorus but high in other nutrients be added to a soil, rapid microbial activity would take place and available $H_2PO_4^-$ in the soil solution would temporarily disappear, just as was the case for soluble NH_4^+, NO_3^- and SO_4^{2-} ions.

Organic matter influences phosphorus availability in two other ways. First, known sources of organic phosphorus (the nucleic acids) are adsorbed

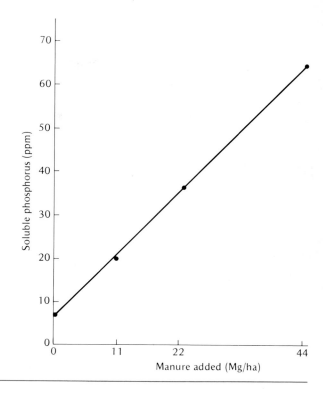

Figure 12.9
The effect of added organic matter (manure) on the soluble phosphorus level of soil at pH 7.2. As the manure decomposed, organic acids were released that formed stable complexes with iron and aluminum compounds and also affected the solubility of calcium phosphates. [Data from El-Baruni and Olsen (1979)]

by humic compounds as well as by silicate clays. The adsorptive reactions probably protects the organic phosphorus from microbial attack. Secondly, specific organic compounds form complexes with iron and aluminum ions and hydrous oxides, thereby preventing these materials from reacting with phosphates. Manure is known to influence the availability of inorganic phosphorus compounds (Figure 12.9).

12.13

Intensity and Quantity Factors

It is important to be aware of the different forms in which phosphorus is found in soils. But even more important is knowledge of a soil's ability to maintain sufficiently high levels of phosphorus in the soil solution to assure satisfactory plant growth. The concentration of phosphorus in the soil solution is a measure of the *intensity (I) factor* of phosphorus nutrition. If this factor is maintained at about 0.2 mg/kg or above, maximum yield of most crop plants will occur (Table 12.4).

As plant roots take up phosphorus from the soil solution, it is at least partially replenished by release of phosphorus from the so-called labile forms, a pool of soil solids that readily exchange phosphorus with the soil solution (see Figure 12.10). This source of soil solution replenishment is known as the *quantity (Q) factor* of phosphorus nutrition. Although the Q factor involves all solid forms of phosphorus that releases the element to the solution, it is made up primarily of the slowly available forms shown in Figure 12.10. These are freshly precipitated iron, aluminum, manganese, and calcium compounds and surface-adsorbed phosphates that have not yet penetrated the particles on which they are held.

As shown in Figure 12.11, the Q level needed to provide a given intensity (I)—or soil solution phosphorus level—will vary from soil to soil. Tropical

TABLE 12.4
**Concentration of Phosphorus in Soil Solution
That Provided 95% of Maximum Yield of
Several Crops in Hawaii**

Crop	Soil	Approximate P in soil solution (mg/kg)
Cassava	Halii	0.005
Peanut	Halii	0.01
Corn	Halii	0.05
Soybean	Halii	0.20
Cabbage	Kula	0.04
Tomato	Kula	0.20
Head lettuce	Kula	0.30

From Fox (1981).

clays high in iron and aluminum need a high Q level to assure an I level sufficient for normal growth. In contrast, a sandy soil with low clay and iron and aluminum contents provides a higher intensity (I) with a given quantity (Q) of phosphorus.

When plant roots remove a given quantity of phosphorus from the soil solution (reduce the I factor), the level of I is drastically reduced in the sandy soil and may be reduced only slightly in the clay soil. The latter is thus said to be more highly buffered with respect to phosphorus than the sandy soil. The *potential buffer capacity* (PBC) of a soil is given by

$$PBC = \frac{\Delta Q}{\Delta I}$$

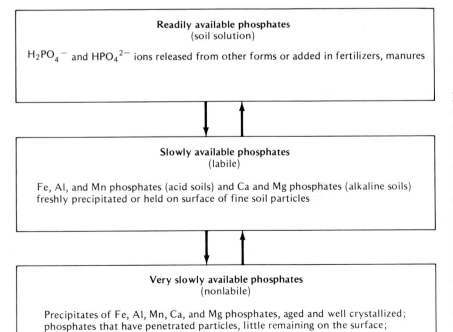

Figure 12.10
Classification of phosphorus compounds in soils in three major groups. Fertilizer phosphates are generally in the readily available phosphate (soil solution) group, but are quickly converted to the slowly available (labile) forms. These can be utilized by plants at first, but upon aging are rendered less available and are then classed as very slowly available (nonlabile). At any one time perhaps 80–90% of the soil phosphorus is in very slowly available forms. Most of the remainder is in the slowly available form; perhaps less than 1% would be expected to be readily available. Note that phosphorus moves from one form to another, although the movement is slow.

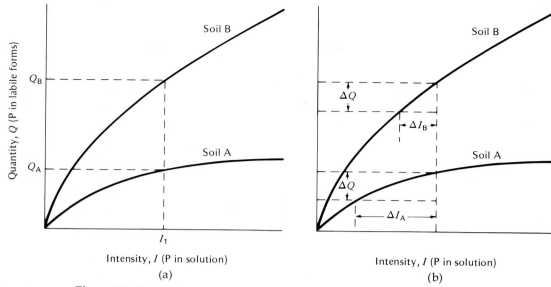

Figure 12.11
Relationship between the intensity factor (P in solution) and the quantity factor (P in solid labile forms in quilibrium with P in solution) on two soils. Soil A may be a sandy soil low in phosphorus-binding compounds, while soil B is a heavier textured soil high in Fe, Al, Mn, Mg, or Ca that holds the phosphorus in relatively unavailable forms. (a) To provide a given level of P in the soil solution (intensity), a much higher quantity must be present in soil B than in soil A. (b) However, when a given quantity (ΔQ) of phosphorus is removed from the soil by a crop, the decrease in intensity is much greater for soil A (ΔI_a) than for soil B (ΔI_b). Soil B is more highly buffered than soil A.

ΔQ is the change in the quantity (Q factor), and ΔI is the change in the intensity (I factor).

The Q/I relations of phosphorus in soils and the potential buffer capacity are important in determining the fertilizer phosphate requirements for high yields. We shall see in Section 12.21 that this same general relationship holds for potassium.

12.14

Practical Control of Phosphorus Availability

From a practical standpoint, the problem of phosphorus utilization is thorny. The inefficient use by plants of applied phosphates has long been known. Most crops do not take up more than about 10–15% of the phosphorus added in fertilizers during the year the fertilizer is applied. This is due not only to the tendency of the soil to fix the added phosphorus but also to the slow rate of movement of this element to plant roots in the soil. Phosphate is essentially an immobile nutrient.

Continued application of phosphate fertilizers tends in time to increase the level of this nutrient in the soil and particularly its level in the labile forms that can release phosphorus to the soil solution. Thus, even though much of the phosphorus added in fertilizer is not used during the year of application, it may provide an important source in future years.

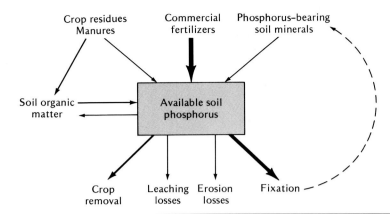

Figure 12.12
How the available phosphorus level in a soil is depleted and replenished. Note that the two main features are the addition of phosphate fertilizers and the fixation of the element in insoluble forms. It should be remembered that the amount of available phosphorus in the soil at any one time is relatively small, especially when compared to the amounts of calcium, magnesium, and potassium.

LIMING AND PLACEMENT OF FERTILIZERS

The small amount of control that can be exerted over phosphate availability seems to be associated with *liming*, *fertilizer placement*, and *organic matter maintenance*. By holding the pH of soils between 6.0 and 7.0, the phosphate fixation can be kept at a minimum (Figure 12.5). Phosphate fertilizers are commonly placed in localized bands to prevent rapid reaction with the soil. In addition, phosphatic fertilizers are quite often granulated to retard still more their contact with the soil. The effective utilization of phosphorus in combination with animal manures is evidence of the importance of organic matter in increasing the availability of this element. Finally, soluble phosphate-containing fertilizers can be successfully applied to the foliage of some vegetable crops and fruit trees.

In spite of these precautions, a major portion of the added phosphates still reverts to less available forms (Figure 12.12). Remember, however, that the reverted phosphorus is not lost from the soil and through the years is slowly made available to growing plants. This can become an important factor, especially in soils that have been heavily phosphated for years.

In summary, maintaining sufficient available phosphorus in a soil largely narrows down to a twofold program: (a) the addition of phosphorus-containing fertilizers and (b) the regulation, to some degree, of the fixation in the soil of both the added and the native phosphates.

12.15

Potassium—The Third "Fertilizer" Element[6]

The history of fertilizer usage in the United States shows that nitrogen and phosphorus received most of the attention when commercial fertilizers first appeared on the market. Although the role played by potassium in plant nutrition has long been known, the importance of potassium fertilization has received full recognition only in comparatively recent years.

The reasons that a widespread deficiency of this element did not develop earlier are at least twofold. First, the supply of available potassium was so high in many soils that it took many years of cropping for a serious depletion to appear. Second, even in soils having insufficient potassium for optimum

[6] For further information on this topic, see Mengel and Kirby (1982) and Munson (1985).

crop yields, production was more drastically limited by the lack of nitrogen and phosphorus. As the use of nitrogen and phosphorus fertilizers expanded, crop yields increased and so did the removal of soil potassium. As a consequence, the drain on soil potassium has been greatly increased. This, coupled with considerable loss by leaching, has raised the demand for potassium in commercial fertilizers.

12.16

Effects of Potassium on Plant Growth

Potassium plays many essential roles in plants. It is an activator of dozens of enzymes responsible for such plant processes as energy metabolism, starch synthesis, nitrate reduction, and sugar degradation. Potassium is extremely mobile within the plant and helps regulate the opening and closing of stomates in the leaves and the uptake of water by root cells.

Potassium is essential for photosynthesis, for protein synthesis, for starch formation, and for the translocation of sugars. This element is important in grain formation, and is absolutely necessary for tuber development. All root crops generally respond to applications of potassium. As with phosphorus, it may be present in large quantities in the soil and yet exert no harmful effect on the crop.

Potassium increases crop resistance to certain diseases and, by encouraging strong root and stem systems, helps to prevent the undesirable "lodging" of plants that is sometimes caused by excessive nitrogen. Potassium delays maturity, thereby working against undue ripening influences phosphorus can exert. In a general way, potassium exerts a balancing effect on the effects of both nitrogen and phosphorus; consequently it is especially important in a multinutrient fertilizer.

Note that sodium has been found partially to take the place of potassium in the nutrition of certain plants. Where there is a deficiency of potassium, native soil sodium or that added in such fertilizers as sodium nitrate, may be useful.

12.17

The Potassium Cycle

Figure 12.13 shows the major forms in which potassium is held in soils and the changes it undergoes as it is cycled through the soil and plant systems. The original sources of potassium are the primary minerals, such as the micas and potassium feldspar (microcline). As these minerals weather, their rigid lattice structures become more pliable. For example, potassium held between the 2:1-type crystal layers of mica is in time made more available, first through nonexchangeable but slowly available forms and finally through the readily exchangeable and the soil solution forms from which it is absorbed by plant roots.

At any one time most of the potassium is in the primary mineral and nonexchangeable forms. To maintain the levels of exchangeable and soil solution potassium high enough to encourage good crop production, chemical fertilizers are used. The sections that follow give greater details on the reactions involved in the potassium cycle.

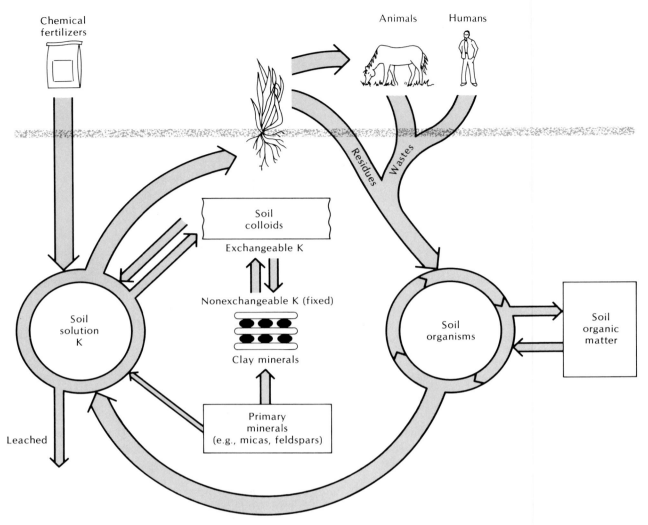

Figure 12.13
Major components of the potassium cycle. Emphasis is placed on the biological cycling of soil solution potassium to plants, humans, and animals and return through soil organisms to the soil solution. Primary and secondary minerals are original sources of the element, and organic matter holds and releases some potassium. Interaction with soil colloids to give exchangeable and nonexchangeable (fixed) forms of potassium is shown. The bulk of the potassium is held in the primary and secondary (clay) minerals.

12.18

The Potassium Problem

AVAILABILITY OF POTASSIUM

In contrast to phosphorus, potassium is found in comparatively high levels in most mineral soils, except those of a sandy nature. In fact, the total quantity of this element is generally greater than that of any other major nutrient element. Amounts as great as 35,000–50,000 kg K/HFS (31,000–45,000 lb/AFS) are not at all uncommon (see Table 1.3).

Yet the quantity of potassium held in an easily exchangeable condition at any one time often is very small. Most of this element is held rigidly as part of the primary minerals or is fixed in forms that are at best only moderately available to plants. Therefore, the situation in respect to potassium utilization parallels that of phosphorus and nitrogen in at least one way. A very large proportion of all three of these elements in the soil is insoluble and relatively unavailable to growing plants.

LEACHING LOSSES

Unlike the situation with respect to phosphorus, however, much potassium is lost by leaching. Drainage waters from soils receiving liberal fertilizer applications usually have considerable quantities of potassium. A representative humid region soil receiving only moderate rates of fertilizer, the annual loss of potassium by leaching is usually about 35 kg/ha (31 lb/acre). However, since considerable potassium is adsorbed by soil colloids, leaching losses of this element normally do not result in yield losses except on very sandy soils.

CROP REMOVAL

Potassium removal by crops is high, often being three to four times that of phosphorus and equaling that of nitrogen. The removal of 140–180 kg/ha (125–156 lb/acre) of potassium by a 60 Mg/ha silage corn crop is not at all unusual.

   This loss of potassium is made even more critical by the tendency of plants to take up soluble potassium far in excess of their needs if sufficiently large quantities are present. This tendency is termed *luxury consumption*, because the excess potassium absorbed does not increase crop yields to any extent.

EXAMPLES OF LUXURY CONSUMPTION

The principles involved in luxury consumption are shown by the graph of Figure 12.14. For many crops there is a direct relationship between the available potassium (soil plus fertilizer) and the removal of this element by the plants. However, only a certain amount of potassium is needed for optimum yields; this is termed *required potassium*. All potassium taken up

Figure 12.14
General relationship between the potassium content of plants and the available soil potassium. If excess quantities of potash fertilizers are applied to a soil, the plants will absorb potassium in excess of that required for optimum yields. This luxury consumption may be wasteful, especially if the crops are completely removed from the soil.

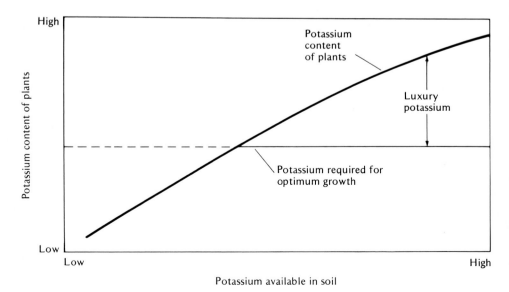

by the plant above this critical required level is considered a *luxury*. The removal of luxury potassium by plants is decidedly wasteful.

Wasteful luxury consumption occurs especially with forage crops. For example, large applications of potassium fertilizers are made during the first year of a three- or four-year perennial hay crop with the exception that this one application would supply subsequent years' needs. Unfortunately, much of the added potassium is likely to be absorbed wastefully by the first crop of hay, leaving too little of the added potassium for subsequent crops.

In summary, then, the problem of potassium is at least threefold: (a) a very large proportion of the element is relatively unavailable to higher plants; (b) it is subject to wasteful leaching losses; and (c) the removal of potassium by crop plants is high, especially when luxury quantities of this element are supplied. With these ideas as a background, the various forms of potassium in soils and their availabilities will now be considered.

12.19

Forms and Availability of Potassium in Soils

The various forms of potassium in soils can be classified in three general groups: (a) *unavailable*, (b) *readily available*, and (c) *slowly available*. Although most of the soil potassium is in the first of these three forms, from an immediate practical standpoint the latter two are of greater significance.

The relationship among these three general categories is shown diagrammatically in Figure 12.15. This figure should be referred to as the different forms are discussed.

RELATIVELY UNAVAILABLE FORMS

Some 90–98% of all soil potassium in a mineral soil is in relatively unavailable forms (Figure 12.15). The compounds containing most of this form of potassium are the feldspars and the micas. These minerals are quite resistant to weathering and supply relatively small quantities of potassium during a given growing season. However, their cumulative release of potassium over a period of years undoubtedly is of some importance. This release is enhanced by the solvent action of carbonic acid and of stronger organic and

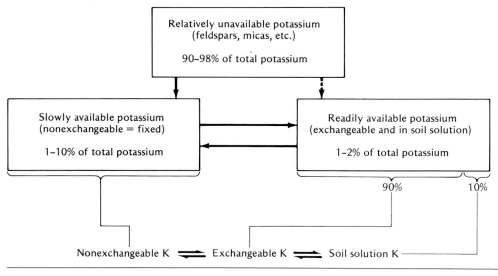

Figure 12.15
Relative proportions of the total soil potassium in unavailable, slowly available, and readily available forms. Only 1–2% is rated as readily available. Of this, approximately 90% is exchangeable, and only 10% appears in the soil solution at any time. [Modified from Attoe and Truog (1945)]

inorganic acids, as well as by the presence of acid clays and humus (see Section 2.4).

READILY AVAILABLE FORMS

Only 1–2% of the total soil potassium is readily available. Available potassium exists in soils in two forms: (a) in the soil solution and (b) as exchangeable potassium adsorbed on the soil colloidal surfaces. Although most of this available potassium is in the exchangeable form (approximately 90%), soil solution potassium is most readily absorbed by higher plants. Unfortunately, potassium in soil solution is subject to considerable leaching loss.

As represented in Figure 12.15, the two form of readily available potassium are in dynamic equilibrium. Such a situation is of extreme practical importance. When plants absorb soil potassium from the soil solution, exchangeable potassium immediately moves into the soil solution until the equilibrium is again established. When water-soluble fertilizers are added to the soil, the equilibrium reverses—soil solution potassium moves onto the exchange complex. The exchangeable potassium can be seen as an important buffer mechanism for soil solution potassium.

SLOWLY AVAILABLE FORMS

In the presence of vermiculite, smectite, and other 2:1-type minerals, the K^+ ions as well as NH_4^+ ions in the soil solution (or added as fertilizers) not only become adsorbed but also may become definitely "fixed" by the soil colloids (Figure 12.16). The potassium (and ammonium) ions are just the right size to fit between layers in the crystals of these normally expanding clays and become an integral part of the crystal. These ions cannot be replaced by ordinary exchange methods and consequently are referred to as *nonexchangeable ions*. As such, the ions are no longer readily available to higher plants.

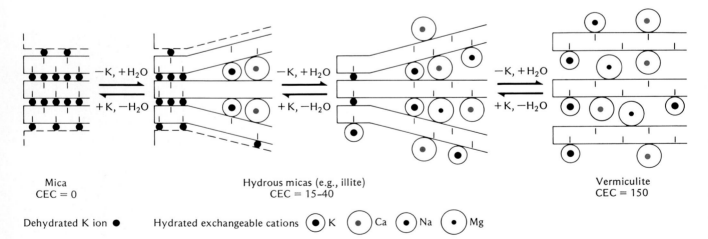

Mica
CEC = 0

Hydrous micas (e.g., illite)
CEC = 15–40

Vermiculite
CEC = 150

Dehydrated K ion ●   Hydrated exchangeable cations (●)K (•)Ca (●)Na (•)Mg

Figure 12.16
Diagrams to illustrate the release of potassium from primary micas to fine-grained mica (illite) and then to vermiculite and the fixation of exchangeable potassium by reversing these release reactions. Note that the dehydrated K^+ ion is much smaller than the hydrated ions of Na^+, Ca^{2+}, Mg^{2+}, etc. Thus, when potassium is added to a soil containing 2:1-type minerals such as vermiculite, the reaction may go to the left and K^+ ions will be held tightly (fixed) in between layers within the crystal, giving fine-grained mica structure. [Modified from McLean (1978)] Note that NH_4^+ ions are fixed in a similar manner.

Nonexchangeable ions are in equilibrium, however, with the more available forms and consequently act as an extremely important reservoir of slowly available nutrients. The entire equilibrium may be represented for potassium as follows.

$$\text{nonexchangeable K} \underset{}{\overset{\text{slow}}{\rightleftharpoons}} \text{exchangeable K} \underset{}{\overset{\text{rapid}}{\rightleftharpoons}} \text{soil solution K}$$

The importance of this equilibrium to practical agriculture should not be overlooked. It is of special value in the conservation of added potassium and also is a means by which reserve potassium can eventually be released for plant use.

RELEASE OF FIXED POTASSIUM

The quantity of nonexchangeable or "fixed" potassium in some soils is quite large. Because of the equilibrium just discussed, the fixed potassium in such soils is continually released to the exchangeable form in amounts large enough to be of great practical importance. The data in Table 12.5 indicate the magnitude of the release of nonexchangeable potassium from certain soils. In these soils, the potassium removed by crops was supplied largely from nonexchangeable forms.

12.20

Factors Affecting Potassium Fixation in Soils

Four soil conditions markedly affect the amounts of potassium fixed: (a) the nature of the soil colloids, (b) wetting and drying, (c) freezing and thawing, and (d) the presence of excess lime.

EFFECTS OF TYPE OF CLAY AND MOISTURE

The ability of the various soil colloids to fix potassium varies widely. Kaolinite and other 1:1-type clays fix little potassium. On the other hand, clays of

TABLE 12.5
Potassium Removal by Very Intensive Cropping and the Amount of This Element Coming from the Nonexchangeable Form

Soil	Total K used by crops		Percent derived from nonexchangeable form
	kg/ha	lb/acre	
Wisconsin soils[a]			
Carrington silt loam	133	119	75
Spencer silt loam	66	59	80
Plainfield sand	99	88	25
Mississippi soils[b]			
Robinsonville fine silty loam	121	108	33
Houston clay	64	57	47
Ruston sandy loam	47	42	24

[a] Average of six consecutive cuttings of Ladino clover, from Evans and Attoe (1948).
[b] Average of eight consecutive crops of millet, from Gholston and Hoover (1948).

the 2:1 type, such as vermiculite, fine-grained mica (illite), and smectite, fix potassium very readily and in large quantities. Even silt-sized fractions of some micaceous minerals fix and subsequently release potassium.

Potassium (and ammonium) ions are attracted between layers in the negatively charged clay crystals. The tendency for fixation is greatest in minerals where the major source of negative charge is in the silica (tetrahedral) sheet. Consequently, vermiculite has a greater fixing capacity than montmorillonite (see Table 7.5 for formulas for these minerals).

Alternate wetting and drying, and freezing and thawing, has been shown to result in the fixation of potassium in nonexchangeable forms as well as its ultimate release to the soil solution. Although the practical importance of these physical conditions is recognized, the mechanisms are not well understood.

INFLUENCE OF pH

Applications of lime sometimes result in an increase in potassium fixation in soils (Figure 12.17). This is not surprising since in strongly acid soils the tightly held hydrogen and hydroxy aluminum ions are able to keep the potassium ions from close association with the colloidal surfaces, which reduces their susceptibility to fixation. As the pH increases, the hydrogen and hydroxy aluminum ions are removed or neutralized, making it easier for the potassium ions to move closer to the colloidal surfaces where they become susceptible to fixation.

Liming may have adverse effects on potassium availability in other ways. For example, in soils where the negative charge is pH dependent, liming increases the cation exchange capacity, which results in an increased potassium adsorption by the soil colloids and a decrease in the potassium level in the soil solution (Figure 12.18). Furthermore, high calcium levels in the soil solution may reduce potassium uptake by the plant. Finally, potassium deficiency has been noted in soils with excess calcium carbonate. Potassium fixation as well as cation ratios may be responsible for these adverse effects.

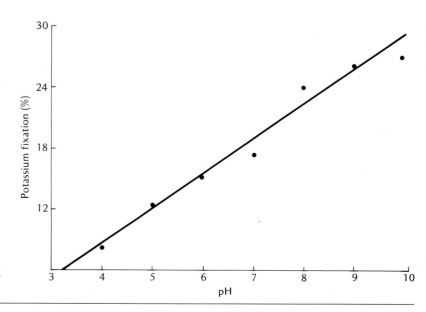

Figure 12.17

The effect of pH on the fixation of potassium soils in India. [From Grewal and Kanwar (1976)]

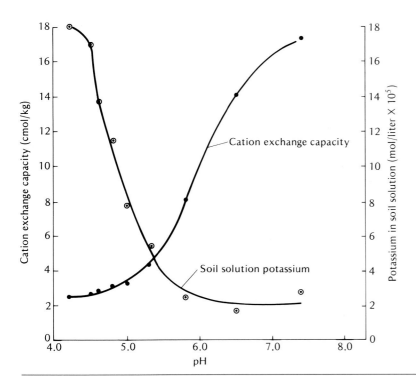

Figure 12.18
The influence of increased pH resulting from lime additions on the pH-dependent cation exchange capacity of a soil and the level of potassium in the soil solution. As the cation exchange capacity increases, some of the soil solution potassium is attracted to the adsorbing colloids. [Data from Magdoff and Bartlett (1980)]

12.21

Intensity and Quantity Factors

As was the case for phosphorus, the power of soils to supply potassium to plant roots needs to be described and measured. There are marked differences between the compounds in which these two elements are found in soils and in the rates at which they are made available to plants. But the concepts of *intensity* and *quantity* discussed in Section 12.13 in relation to phosphorus are also useful in describing and measuring potassium-supplying power.

The intensity factor is a measure of the potassium that is immediately available to the root—the potassium in the soil solution. Since the absorption of the potassium ion by plant roots is affected by the activity in the soil solution of other cations, particularly of calcium and magnesium, some authorities prefer to use the ratio

$$\frac{[K^+]}{\sqrt{[Ca^{2+}] + [Mg^{2+}]}}$$

rather than the potassium concentration alone to indicate the intensity factor.

The quantity factor is a measure of the capacity of the soil to maintain the level of potassium in the soil solution over the period of time the crop is being produced. This capacity is due mainly to the exchangeable potassium, although some nonexchangeable forms release sufficient potassium during a growing season to provide a notable portion of the crop needs. The main point to remember, however, is that the quantity factor designates the potassium sources capable of helping to replenish the soil solution (intensity) level of potassium.

The concept of the buffer capacity of the soil, which indicates how the potassium level in the soil solution (*intensity*) varies with the amount of labile form of this element (*quantity*), is useful in explaining differences in potassium-supplying power of soils. The principle is the same as that shown for phosphorus in Section 12.13. Clay-textured soils that are high in the quantity factor are well buffered, whereas soils that are sandy are apt to be poorly buffered.

12.22

Practical Implications in Respect to Potassium

FREQUENCY OF APPLICATION

Frequent light applications of potassium have some advantages over heavier and less frequent ones. Such a conclusion is based on the luxury consumption of potassium by some crops, the ease with which this element is lost from the soil solution by leaching, and the fact that excess potassium is subject to fixation. Although the fixation has definite conserving features, these in most cases tend to be outweighed by the disadvantages of leaching and luxury consumption.

POTASSIUM-SUPPLYING POWER OF SOILS

A second very important suggestion is that full advantage should be taken of the potash-supplying power of soils. The idea that each kilogram of potassium removed by plants or through leaching must be returned in fertilizers may not always be correct. In some soils the large quantities of moderately available forms of potassium already present can be utilized. More often, however, slowly available forms are not found in significant quantities, and supplementary additions are necessary. Moreover, the importance of lime in reducing leaching losses of potassium should not be overlooked as a means of effectively utilizing the power of soils to furnish this element.

Soils of arid zones commonly can supply adequate potassium for many years, even under irrigation. However, continued crop removal can deplete even these soils. Also, deep-rooted plants such as cotton may depend on the subsoil for much of their potassium. Increasing the availability of this element at depths below the plow layer is difficult.

POTASSIUM LOSSES AND GAINS

The problem of maintaining soil potassium is outlined diagrammatically in Figure 12.19. Crop removal of potassium generally exceeds that of the other essential elements, with the possible exception of nitrogen. Annual losses of

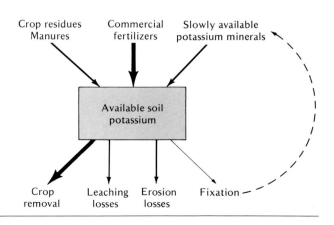

Figure 12.19
Gains and losses of *available* soil potassium under average field conditions. The approximate magnitude of the changes is represented by the width of the arrows. For any specific case the actual amounts of potassium added or lost undoubtedly may vary considerably from this representation. As was the case with nitrogen and phosphorus, commercial fertilizers are important in meeting crop demands.

potassium from plant removal as great as 100 kg/ha or more are not uncommon, particularly if the crop is a legume and is cut several times for hay. As might be expected, therefore, the return of crop residues and manures is very important in maintaining soil potassium. For example, 10 Mg of average animal manure supplies about 40 kg of potassium, fully equal to the amount of nitrogen thus supplied.

The annual losses of available potassium by leaching and erosion greatly exceed losses of nitrogen and phosphorus. Potassium losses through leaching are generally not as great, however, as the corresponding losses of available calcium and magnesium. In contrast, the total potassium removal by erosion generally exceeds that of any other major nutrient element. Such losses of soil minerals are indeed serious.

INCREASED USE OF POTASSIUM FERTILIZERS

In the past, potassium in fertilizers were added only to supplement potassium returned in crop residues or released from slowly available (labile) forms. While these sources are still very important, fertilizers are used increasingly to supply much of the potassium needed for crop production. This is true especially in cash-crop areas and in regions where sandy soils are prominent. Even in some finer textured soils, the release of potassium from mineral form is much too slow to support maximum crop yields. Consequently, increased usage of commercial potassium fertilizers must be expected if yields are to be increased or even maintained.

12.23
Conclusion

The availability of phosphorus to plant roots has a double constraint: the low total phosphorus level in soils and the small percentage of this level that is present in available forms. Furthermore, even when soluble phosphates are added to soils, they are quickly fixed into insoluble forms that in time become quite unavailable to growing plants. In acid soils, the phosphorus is fixed primarily by iron, aluminum, and manganese; in alkaline soils, by calcium and magnesium. This fixation greatly reduces the efficiency of phosphate fertilizers so that little of the added phosphorus can be taken up by plants. In time, however, this fixed phosphorus can build up and can serve as a reserve pool for plant absorption.

Potassium is generally abundant in soils, but it, too, is present mostly in forms that are relatively unavailable for plant absorption. Fortunately, however, some soils contain considerable nonexchangeable but slowly available forms of this element. Over time this potassium can be released to exchangeable and soil solution forms that can be quickly absorbed by plant roots. The reserve of this element is important because the crop requirements for potassium are high, three to five times that of phosphorus and equal to that of nitrogen.

Study Questions

1. Compare soil phosphorus and soil nitrogen with respect to the total content in soils, distribution between organic and inorganic forms, and the role of microbes in determining their availability.

2. The same amount of a soluble phosphorus fertilizer was added to two field soils, (a) an Entisol with a sandy surface soil and (b) an Ultisol with a clay surface soil. A crop of corn was grown on each soil. In which case would you expect more phosphorus in the harvested crop? Explain.

3. Assume you mix some finely chopped wheat straw with soil and place it in a greenhouse pot. Tomatoes that were transplanted to the pot soon showed phosphorus deficiency, which was not the case in a companion pot to which no wheat straw had been added. How do you account for this when you consider that some phosphorus was supplied by the wheat straw?

4. An examination of a soil in Maine on which potatoes had been grown each year for the past 15 years showed a higher phosphorus content in the surface layer than a nearby virgin soil on which no crops had been grown. What is a logical explanation for this?

5. Contrast the fixation of elemental nitrogen (N_2), NH_4^+ ions, $H_2PO_4^-$ ions, and K^+ ions and indicate the advantages and disadvantages of each.

6. Would you expect more or less phosphorus to be fixed by an Oxisol as compared to a Vertisol? Explain.

7. What is the fate of the $H_2PO_4^-$ added in fertilizers to an Aridisol?

8. What is the fate of a heavy application of a potassium-containing fertilizer added to a clay loam soil (high in vermiculite) on which alfalfa, a forage crop, is being grown?

9. What is meant by the intensity factor and the quantity factor in relation to the supply of potassium to plant roots?

10. Which is likely to have the higher buffering capacity with respect to potassium supply, a sand or a clay loam soil? Explain.

11. Thirty years ago the total amount of phosphorus applied in fertilizer in the United States exceeded that for potassium. The opposite is true today. How do you account for this?

12. In the spring a certain soil showed the following soil test: soil solution K = 20 kg/HFS; exchangeable K = 200 kg/HFS. After two crops of alfalfa had been grown on the field, a second test showed: soil solution = 15 kg/HFS and exchangeable K = 150 kg/HFS. The total potassium removed in the alfalfa crops was 250 kg K/ha. Explain why there was not a greater reduction in the soil solution and exchangeable K levels?

References

ATTOE, O. J., and E. TRUOG. 1945. "Exchangeable and acid soluble potassium as regards availability and reciprocal relationships," *Soil Sci. Soc. Amer. Proc.*, **10**:81–6.

BRADY, N. C. 1974. *The Nature and Properties of Soils*, 8th ed. (New York: Macmillan).

DOUGLAS, J. R. 1982. "U.S. fertilizer situation–1990," A paper presented at The Fertilizer Institute Seventh World Fertilizer Conference, Sept. 12–14, 1982, San Francisco, CA.

EL-BARUNI, B., and S. R. OLSEN. 1979. "Effect of manure on solubility of phosphorus in calcareous soils," *Soil Sci.*, **112**:219–25.

EVANS, C. E., and O. J. ATTOE. 1948. "Potassium supplying power of virgin and cropped soils," *Soil Sci.*, **66**:323–34.

FARES, F., et al. 1974. "Quantitative survey of organic phosphorus in different soil types," *Phosphorus in Agriculture*, **63**: 25–40.

FOX, R. L. 1981. "External phosphorus requirements of crops," in *Chemistry in the Soil Environment*, ASA Special Publication No. 40 (Madison, WI: Amer. Soc. Agron. and Soil Sci. Soc. Amer.), pp. 223–39.

GHOLSTON, L. E., and C. D. HOOVER. 1948. "The release of exchangeable and nonexchangeable potassium from several Mississippi and Alabama soil upon continuous cropping," *Soil Sci. Soc. Amer. Proc.*, **13**:116–21.

GREWAL, J. S., and J. S. KANWAR. 1976. *Potassium and Ammonium Fixation in Indian Soils* (Review), (New Delhi, India: Indian Council for Agricultural Research).

KHASAWNEH, F. E., et al. (Eds). 1980. *The Role of Phosphorus in Agriculture* (Madison, WI: Amer. Soc. Agron., Crop Sci. Soc. Amer., Soil Sci. Soc. Amer.).

MAGDOFF, F. R., and R. J. BARTLETT. 1980. "Effect of liming acid soils on potassium availability," *Soil Sci.*, **129**:12–14.

MCLEAN, E. O. 1978. "Influence of clay content and clay composition on potassium availability," in G. S. Sekhon (Ed.), *Potassium in Soils and Crops* (New Delhi, India: Potash Research Institute of India), pp. 1–19.

MENGEL, K., and E. A. KIRKBY, 1982. *Principles of Plant Nutrition* (Bern, Switzerland: International Potash Institute).

MUNSON, R. D. (Ed.). 1985. *Potassium in Agriculture*, (Madison, WI: Amer. Soc. Agron., Crop Sci. Soc.

Amer., Soil Sci. Soc. Amer.).

OLSEN, S. R., R. A. BOWMAN, and F. S. WATANABE. 1977. "Behavior of phosphorus in the soil and interactions with other nutrients," *Phosphorus in Agriculture*, **70**:31–46.

OLSEN, S. R., and F. E. KHASAWNEH. 1980. "Use and limitations of physical-chemical criteria for assessing the status of phosphorus in soils," in F. E. Khasawneh et al. (Eds.), *The Role of Phosphorus in Agriculture* (Madison, WI: Amer. Soc. Agron., Crop Sci. Soc. Amer., Soil Sci. Soc. Amer.).

SAMPLE, E. C., et al. 1980. "Reactions of phosphate fertilizers in soils," in F. E. Khasawneh et al. (Eds.), *The Role of Phosphorus in Agriculture* (Madison, WI: Amer. Soc. Agron., Crop Sci. Soc. Amer., Soil Sci. Soc. Amer.).

SANCHEZ, P. A., and G. UEHARA. 1980. "Management considerations for acid soils with high phosphorus fixation," in F. E. Khasawneh et al. (Eds.), *The Role of Phosphorus in Agriculture* (Madison, WI: Amer. Soc. Agron., Crop Sci. Soc. Amer., Soil Sci. Soc. Amer.).

SOLTANPOUR, P. N., R. L. FOX, and R. C. JONES. 1988. "A quick method to extract organic phosphorus from soils", *Soil Sci. Soc. Amer. J.*, **51**:255–56.

STEVENSON, F. J. 1986. *Cycles of Soil Carbon, Nitrogen, Phosphorus, Sulfur, and Micronutrients* (New York: Wiley).

Micronutrient Elements

Of the seventeen elements known to be essential for plant growth, eight are required in such small quantities that they are called *micronutrients*[1] or *trace elements*. These are iron, manganese, zinc, copper, boron, molybdenum, cobalt, and chlorine.

Other elements, such as silicon, vanadium, nickel, and sodium, appear to be helpful for the growth of certain species. Still others, such as chromium, tin, iodine, and fluorine, have been shown to be essential for animal growth but are apparently not required by plants. As better techniques of experimentation are developed and as more pure salts are made available, it is likely that these and other elements may be added to the list of essential nutrients.

There are several reasons for the widespread concern for micronutrients.

1. Improved crop varieties and macronutrient fertilizer practices have greatly increased crop production and thereby the micronutrient removal.
2. The trend toward high-analysis fertilizers has reduced the use of impure salts, which formerly contained some micronutrients.
3. Increased knowledge of plant nutrition has helped in the diagnosis of trace-element deficiencies that formerly might have gone unnoticed.

13.1

Deficiency Versus Toxicity

All micronutrients are required in very small quantities (Figure 13.1). In fact, they are harmful when present in the soil in larger amounts than can be tolerated by plants or by animals consuming the plants (see Figure 1.11). Accordingly, the range of concentration of these elements for optimal plant growth is not too great. Molybdenum, for example, may be beneficial if

---

[1] For review articles on this subject, see Nicholas and Egan (1975) and Stevenson (1986).

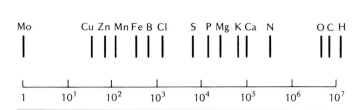

Figure 13.1
Relative numbers of atoms of the essential elements in alfalfa at bloom stage, expressed logarithmically. Note that there are more than 10 million hydrogen atoms for each molybdenum atom. Even so, normal plant growth would not occur without molybdenum. Cobalt is generally present in even smaller quantities in plants than is molybdenum. [Modified from Viets (1956)]

added at rates as low as 35–70 g/ha (0.5–1.0 oz/acre), whereas application rates of 3–4 kg/ha of available molybdenum may adversely affect some plants. More importantly, at these higher rates, molybdenum concentrates in the plant tissue to levels that are highly toxic to animals consuming the plants. Even natural soil levels of molybdenum may result in toxic levels of this element in plants.

Although somewhat larger amounts of the other micronutrients are required and can be tolerated by plants, control of the quantities added is essential, especially for maintaining nutrient balance. The concepts of deficiency, adequacy, and toxicity for several micronutrients are illustrated in Figure 13.2.

13.2

Role of Micronutrients

While there is considerable variation in the specific roles of the various micronutrients[2] in plant and microbial growth processes, a common one is participation in enzyme systems (Table 13.1). For example, copper, iron, and molybdenum are capable of acting as "electron carriers" in enzyme systems that bring about oxidation–reduction reactions in plants. Apparently, such reactions, which are essential to plant development and reproduction, require the presence of these micronutrients. Zinc and manganese also function in enzyme systems necessary for important reactions in plant metabolism.

Molybdenum and manganese are essential for certain nitrogen transformations in microorganisms as well as in plants. Molybdenum is a component of the enzyme *nitrogenase*, which is essential for the process of nitrogen fixation, both symbiotic and nonsymbiotic. It is also present in the enzyme *nitrate reductase*, which is responsible for the reduction of nitrates in soils and plants.

[2] For a review of the role of individual micronutrients, see Mengel and Kirkby (1982) and Marschner (1986).

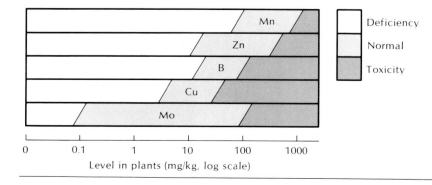

Figure 13.2
Deficiency, normal, and toxicity levels in plants for five micronutrients. Note that the range is shown on a logarithm base and that the upper limit for manganese is about 10,000 times the lower range for molybdenum. [Levels taken from Allaway (1968)]

TABLE 13.1
Functions of Several Micronutrients in Higher Plants

Micronutrient	Functions in higher plants
Zinc	Present in several dehydrogenase, proteinase, and peptidase enzymes; promotes growth hormones and starch formation; promotes seed maturation and production.
Iron	Present in several peroxidase, catalase, and cytochrome oxidase enzymes; found in ferredoxin, which participates in oxidation–reduction reactions (e.g., NO_3^- and SO_4^{2-} reduction, N fixation); important in chlorophyll formation.
Copper	Present in laccase and several other oxidase enzymes; important in photosynthesis, protein and carbohydrate metabolism, and probably nitrogen fixation.
Manganese	Activates decarboxylase, dehydrogenase, and oxidase enzymes; important in photosynthesis, nitrogen metabolism, and nitrogen assimilation.
Boron	Activates certain dehydrogenase enzymes; facilitates sugar translocation and synthesis of nucleic acids and plant hormones; essential for cell division and development.
Molybdenum	Present in nitrogenase (nitrogen fixation) and nitrate reductase enzymes; essential for nitrogen fixation and nitrogen assimilation.
Cobalt	Essential for nitrogen fixation; found in vitamin B_{12}.

Zinc plays a role in protein synthesis, in the formation of some growth hormones and in the reproductive process of certain plants. Copper is involved in both photosynthesis and respiration and in the use of iron. Copper also stimulates lignification of all plant cell walls. A boron deficiency decreases the rate of water absorption, root growth, and translocation of sugars in plants. Iron is involved in chlorophyll formation and degradation and in the synthesis of proteins contained in the chloroplasts. Manganese seems to be essential for photosynthesis, respiration, and nitrogen metabolism.

The role of chlorine is still somewhat obscure; however, it is known to influence photosynthesis, and root growth also suffers if it is absent. Cobalt is essential for the symbiotic fixation of nitrogen. In addition, legumes and some other plants have a cobalt requirement independent of nitrogen fixation, although the amount required is small compared to that for the nitrogen-fixation process. It is obvious that the micronutrients play a complex role in plant metabolism.

13.3

Source of Micronutrients

Deficiencies of micronutrients are commonly related to low contents of these elements in the parent rocks or transported parent material. Similarly, toxic quantities are commonly related to abnormally large amounts in the soil-forming rocks and minerals.

INORGANIC FORMS

Sources of the eight micronutrients vary markedly from area to area. The wide range of these elements in soils and suggested contents in a representative soil is shown on Table 13.2.

All of the micronutrients have been found in varying quantities in igneous rocks. Two of them, iron and manganese, have prominent structural positions in certain of the original silicate materials. Others, such as cobalt and

TABLE 13.2

Major Natural Sources of the Eight Micronutrients and Their Suggested Contents in a Representative Humid Region Surface Soil

Element	Major forms in nature	Range (mg/kg)[a]	Representative surface soil (mg/kg)
		Analyses of soils	
Iron	Oxides, sulfides, and silicates	10,000–100,000	25,000
Manganese	Oxides, silicates, and carbonates	20–4,000	1,000
Zinc	Sulfides, carbonates, and silicates	10–300	50
Copper	Sulfides, hydroxy carbonates, and oxides	2–100	20
Boron	Borosilicates, borates	2–100	10
Molybdenum	Sulfides, oxides, and molybdates	0.2–5	2
Chlorine[b]	Chlorides	7–50	10
Cobalt	Silicates	1–40	8

[a] Equivalent to parts per million.
[b] Much higher in saline and alkaline soils.

zinc, also may occupy structural positions as minor replacements for the major constituents of silicate minerals, including clays.

As mineral decomposition and soil formation occur, the mineral forms of micronutrients are changed just as macronutrients are. Oxides and, in some cases, sulfides of elements such as iron, manganese, and zinc are formed (Table 13.2). Secondary silicates, including the clay minerals, may contain considerable quantities of iron and manganese and smaller quantities of zinc and cobalt. The micronutrient cations released as weathering occurs are subject to colloidal adsorption, just as are calcium or hydrogen ions.

Anions such as borate and molybdate may undergo adsorption or reaction in soils similar to that of phosphate ions. Chlorine, by far the most soluble of the group, is added to soils in considerable quantities each year through rainwater. Its incidental addition in fertilizers and in other ways helps prevent the deficiency of chlorine under field conditions.

ORGANIC FORMS

Organic matter is an important secondary source of some of the trace elements. They seem to be held as complex combinations by the organic colloids. Copper is especially tightly held. In the profiles of uncultivated soils, micronutrients tend to be highest in the upper layers, most of them presumably in the organic fraction. Correlations between soil organic matter and contents of copper, molybdenum, and zinc have been noted. Although the elements thus held are not always readily available to plants, their release through decomposition is undoubtedly an important fertility factor.

FORMS IN SOIL SOLUTION

The dominant forms of micronutrients that occur in the soil solution are found in Table 13.3. The specific forms present are determined largely by the pH and by soil aeration. Note that cations may be present in the form of either simple cations or hydroxy metal cations. The simple cations tend to be dominant under highly acid conditions; the more complex hydroxy metal cations form as the soil pH is increased.

Molybdenum is present mainly as MoO_4^{2-}, an anionic form that reacts at low pH in ways similar to those of phosphate ions. Although boron also may be present in anionic form at high pH levels, research suggests that undissociated boric acid (H_3BO_3) is the form that is dominant in the soil solution and is absorbed by higher plants.

TABLE 13.3

Forms of Micronutrients Dominant in the Soil Solution

Micronutrient	Dominant soil solution forms
Iron	Fe^{2+}, $Fe(OH)_2^+$, $Fe(OH)^{2+}$, Fe^{3+}
Manganese	Mn^{2+}
Zinc	Zn^{2+}, $Zn(OH)^+$
Copper	Cu^{2+}, $Cu(OH)^+$
Molybdenum	MoO_4^{2-}, $HMoO_4^-$
Boron	H_3BO_3, $H_2BO_3^-$
Cobalt	Co^{2+}
Chlorine	Cl^-

From data in Lindsay (1972).

The cycling of macronutrients through the soil–plant–animal system is illustrated in Figure 13.3. Note the various means by which these elements are supplied to the soil solution and, in turn, to growing plants. This figure should be kept in mind as attention is turned to micronutrient availability.

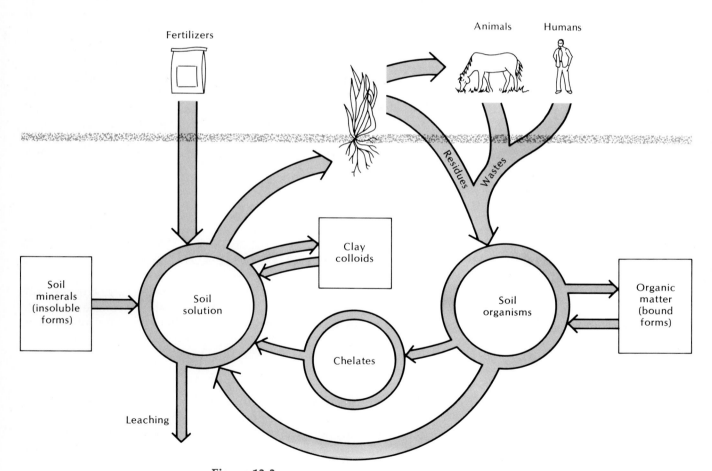

Figure 13.3

Micronutrient cycle in soil showing the major pathways of these elements as they are taken up by plants and are recycled through the soil. Although all micronutrients may not follow each of the pathways, most are involved in the major components of the cycle. The formation of chelates, which keep these elements in soluble forms, is a unique feature of this cycle.

13.4

Conditions Conducive to Micronutrient Deficiency

Micronutrients are most apt to limit crop growth in (a) highly leached acid sandy soils; (b) organic soils; (c) soils very high in pH; and (d) soils that have been very intensively cropped and heavily fertilized with macronutrients only.

Strongly leached acid sandy soils are low in most micronutrients for the same reasons they are deficient in most macronutrients. Their parent materials were originally deficient in the elements, and acid leaching has removed much of the small quantity of micronutrients originally present. In the case of molybdenum, acid soil conditions also have a markedly depressing effect on availability.

The micronutrient contents of organic soils depend on the extent of the washing or leaching of these elements into the bog area as the deposits were formed. In most cases this rate of movement was too slow to produce deposits as high in micronutrients as are the surrounding mineral soils. Intensive cropping of organic soils and their ability to bind certain elements, notably copper, also accentuate micronutrient deficiencies. When much of the harvested crops, especially vegetables, is removed from the land, micro- as well as macronutrients must eventually be supplied in the form of fertilizers if good crop yields are to be maintained. Intensive cropping of mineral soils heavily fertilized with macronutrients also can hasten the onset of micronutrient shortage, especially if the soils are coarse in texture.

The soil pH, especially in well-aerated soils, has a decided influence on the availability of all the micronutrients except chlorine. Under very acid conditions, molybdenum is rendered unavailable; at high pH values, all the micronutrient cations are unfavorably affected. Overliming or a naturally high pH is associated with deficiencies of iron, manganese, zinc, copper, and even boron. Such conditions occur naturally in many of the calcareous soils of arid regions.

13.5

Factors Affecting Availability of Micronutrient Cations

Each of the five micronutrient cations (iron, manganese, zinc, copper and cobalt) is influenced in a characteristic way by the soil environment. However, certain soil factors have the same general effects on the availability of all five cations.

SOIL pH

The micronutrient cations are most soluble and available under acid conditions. In very acid soils, there is a relative abundance of the ions of iron, manganese, zinc, and copper (Figure 13.4). In fact, under acid conditions, the concentration of one or more of these elements often is sufficiently high to be toxic to common plants. As indicated in Chapter 8, one of the primary reasons for liming acid soils is to reduce the concentrations of these ions.

As the pH is increased, the ionic forms of the micronutrient cations are changed first to the hydroxy ions of the elements and, finally, to the insoluble hydroxides or oxides. The following example uses the ferric ion as typical of the group.

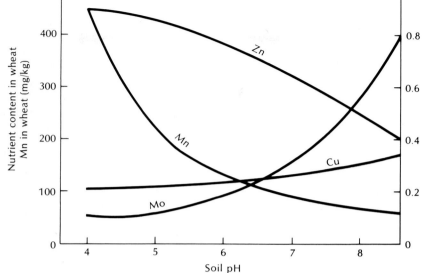

Figure 13.4

Effect of soil pH on the concentrations of manganese, zinc, copper, and molybdenum in wheat plants. The soils were from different countries around the world. The molybdenum levels are extremely low, but increase with increasing pH. Manganese and zinc levels decrease as the pH rises, while copper is little affected. [Redrawn from Sillanpaa (1982)]

$$Fe^{3+} \xrightarrow{\ OH^- \ } Fe(OH)^{2+} \xrightarrow{\ OH^- \ } Fe(OH)_2^+ \xrightarrow{\ OH^- \ } Fe(OH)_3$$

Simple cation   Hydroxy metal cations   Hydroxide
(soluble)   (soluble)   (insoluble)

All of the hydroxides of the micronutrient cations are insoluble, some more so than others. The exact pH at which precipitation occurs varies from element to element and even between oxidation states of a given element. For example, the higher valent states of iron and manganese form hydroxides that are much more insoluble than their lower valent counterparts. In any case, the principle is the same: At low pH values, the solubility of micronutrient cations is at a maximum, and as the pH is raised, their solubility and availability to plants decrease. The desirability of maintaining an intermediate soil pH, perhaps slightly below 7, is obvious.

Zinc availability is reduced drastically by magnesium-containing limestone. This may be caused by interactions between the two cations in the soil or in the plant.

OXIDATION STATE AND pH

Three of the trace element cations are found in soils in more than one valence state. These are iron, manganese, and copper. The lower valent states are encouraged by conditions of low oxygen supply and relatively higher moisture level. They are responsible for the subdued subsoil colors, grays and blues in poorly drained soils in contrast to the bright reds, browns, and yellows of well-drained mineral soils.

The changes from one valence state to another are, in most cases, brought about by microorganisms and organic matter. In some cases, the organisms may obtain their energy directly from the inorganic reaction. For example, the oxidation of manganese from the bivalent manganous form, Mn^{2+}, to a tetravalent form (MnO_2) can be carried on by certain bacteria and fungi. In other areas, organic compounds formed by the microbes may be responsible for the oxidation or reduction. In general, high pH favors oxidation, whereas acid conditions are more conducive to reduction.

At pH values common in soils, the oxidized states of iron, manganese,

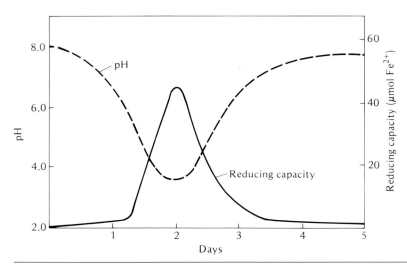

Figure 13.5
Response of one variety of sunflowers to iron deficiency. When the plant became stressed owing to iron deficiency, plant exudates lowered the pH and increased the reducing capacity immediately around the roots. Iron is solubilized and taken up by the plant, the stress is alleviated, and conditions return to normal. [From Marschner et al. (1974) as reported by Olsen et al. (1981); used with permission of *American Scientist*]

and copper are generally much less soluble than are the reduced states. The hydroxides (or hydrous oxides) of these high-valence forms precipitate even at low pH values and are extremely insoluble. For example, the hydroxide of trivalent ferric iron precipitates at pH values of 3.0 to 4.0, whereas ferrous hydroxide does not precipitate until a pH of 6.0 or higher is reached.

The interaction of soil acidity and aeration in determining micronutrient availability is of great practical importance. The micronutrient cations are somewhat more available under conditions of restricted drainage, flooded soils generally showing higher availabilities than well-aerated soils. Very acid soils that are poorly drained often supply toxic quantities of iron and manganese. Such toxicity is less apt to occur under well-drained conditions.

At the high end of the soil pH range, good drainage and aeration often have the opposite effect. Well-oxidized calcareous soils are sometimes deficient in available iron, zinc, or manganese even though adequate total quantities of these trace elements are present. The hydroxides of the high-valence forms of these elements are too insoluble to supply the ions needed for plant growth. In contrast to the other micronutrients, at high soil pH values, molybdenum availability may be so high that the levels of this nutrient in the plants are toxic to the animals eating them.

There are marked differences in the sensitivity of different plant varieties to iron deficiency in soils with high pH. This is apparently caused by differences in each plant's ability to solubilize iron immediately around the roots. Efficient varieties respond to iron stress by acidifying the immediate vicinity of the roots and by excreting compounds capable of reducing the iron to a more soluble form, with a resultant increase in availability (Figure 13.5).

OTHER INORGANIC REACTIONS

Micronutrient cations interact with silicate clays in two ways. First, they may be involved in cation exchange reactions much like those of calcium or hydrogen. Second, they may be more tightly bound or fixed to certain silicate clays, especially the 2:1 type. Zinc, manganese, cobalt, and iron ions are found as elements in the crystal structure of silicate clays. Depending on the conditions, they may be released from the clays or fixed by them in a manner similar to that by which potassium is fixed. The fixation may be serious in the case of cobalt, and sometimes zinc, because each element is present in soil in such a small quantity (Table 13.2).

The application of large quantities of phosphate fertilizers can adversely affect the supply of some of the micronutrients. The uptake of both iron and zinc may be reduced in the presence of excess phosphates. From a practical standpoint, phosphate fertilizers should be used in only those quantities required for good plant growth.

Lime-induced chlorosis (iron deficiency) in fruit trees is encouraged by the presence of the bicarbonate ion. Bicarbonate-containing irrigation waters enhance the level of this ion in some soils. The chlorosis apparently results from iron deficiency in soils with high pH because the bicarbonate ion interferes in some way with iron metabolism.

ORGANIC COMBINATIONS

Each of the five micronutrient cations may be held in organic combination. Microorganisms also assimilate the cations, which are apparently required for many microbial transformations. The organic compounds in which these trace elements are combined vary considerably, but they include proteins, amino acids, and constituents of humus, including the humic acids (see Section 10.7) and acids such as citric and tartaric. Among the most important are the *organic complexes*, combinations of the metallic cations and certain organic groups. These complexes may protect the micronutrients from certain harmful reactions, such as the precipitation of iron by phosphates and vice versa (see Section 12.12). Soluble complexes increase micronutrient availability while insoluble ones decrease the availability. These complexes, called *chelates*, are considered in the next section.

13.6

Chelates

Chelate, a term derived from a Greek word meaning "claw," is applied to compounds in which certain metallic cations are complexed or bound to an organic molecule or anion. These organic molecules may be synthesized by plant roots and released to the surrounding soil, may be present in the soil humus, or may be synthetic compounds added to the soil to enhance micronutrient availability. In complexed form, the cations are protected from reaction with inorganic soil constituents that would make them unavailable for uptake by plants. Iron, zinc, copper, and manganese are among the cations that form chelate complexes. An example of an iron chelate ring structure is shown in Figure 13.6

The effect of chelation can be illustrated with iron. In the absence of

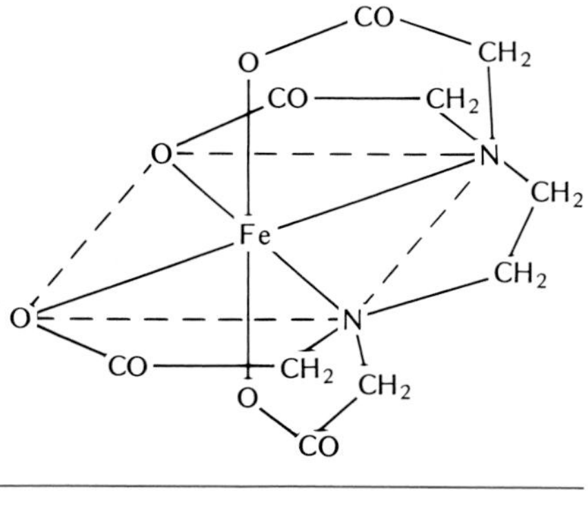

Figure 13.6
Structural formula for a common iron chelate, ferric ethylenediaminetetraacetate (FeEDTA). The iron is protected and yet can be utilized by plants.

chelation, when an inorganic iron salt such as ferric sulfate is added to a calcareous soil, most of the iron is quickly rendered unavailable by reaction with hydroxide.

$$Fe^{3+} + 3\,OH^- \rightleftharpoons FeOOH + H_2O$$
(available) (unavailable)

In contrast, if the iron is added in the form of an iron chelate, such as FeEDDHA (see Table 13.4), the iron remains in the chelate form, which is available for uptake by plants.

$$FeEDDHA + 3\,OH^- \rightleftharpoons FeOOH + EDDHA^{3-} + H_2O$$
(available) (unavailable)

The mechanism by which micronutrients from chelates are absorbed by plants is still unclear. The organic chelating agents are not absorbed normally by growing plants. It appears that the primary role the chelate plays is to hold the metallic cations near the root surface until direct absorption of the free cations can take place. Once the micronutrient cations are inside the plant, other organic chelates (such as citrates) may be "carriers" of these cations to different parts of the plant.

STABILITY OF CHELATES

Some of the major synthetic chelating agents are listed in Table 13.4. They vary in their stability and suitability as sources of micronutrients. Except in soils of very high pH, iron chelates tend to be more stable than those of copper and zinc, which are, in turn, more stable than those of manganese. Accordingly, in this example iron is more strongly attracted by the chelating agents than the other micronutrients. Consequently, if a zinc chelate is added to a soil with significant quantities of available iron, the following reaction may occur.

$$Zn\ chelate + Fe^{2+} \rightleftharpoons Fe\ chelate + Zn^{2+}$$

Since the iron chelate is more stable than its zinc counterpart, the reaction goes to the right and the released zinc ion is subject to reaction with the soil. It is obvious that an added metal chelate must be stable within the soil if it is to have lasting advantage.

TABLE 13.4

Common Chemical Names, Formulas, and Abbreviations for Major Chelating Agents

Name	Formula	Abbreviation
Ethylenediaminetetraacetic acid	$C_{10}H_{16}O_8N_2$	EDTA
Diethylenetriaminepentaacetic acid	$C_{14}H_{23}O_{10}N_3$	DTPA
Cyclohexanediaminetetraacetic acid	$C_{14}H_{22}O_8N_2$	CDTA
Ethylenediaminedi(o-hydroxyphenylacetic acid)	$C_{18}H_{20}O_6N_2$	EDDHA
Hydroxyethylethylenediaminetriacetic acid	$C_{10}H_{18}O_7N_2$	HEDTA
Nitrilotriacetic acid	$C_6H_9O_6N$	NTA
Ethyleneglycol-bis(2-aminoethyl ether)tetraacetic acid	$C_{14}H_{24}O_{10}N_2$	EGTA
Citric acid	$C_6H_8O_7$	CIT
Oxalic acid	$C_2H_2O_4$	OX
Pyrophosphoric acid	$H_4P_2O_7$	P_2O_7
Triphosphoric acid	$H_5P_3O_{10}$	P_3O_{10}

From Norvell (1972).

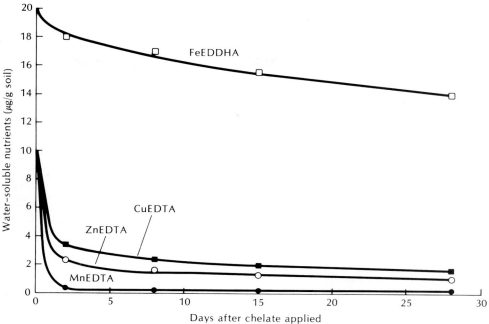

Figure 13.7
Average reduction in water solubility of four chelated micronutrients when incubated with four calcareous soils with pH higher than 8. The iron chelate was most stable in these soils, that with manganese the least. Because of the small quantities needed by plants, even the copper and zinc chelates would likely provide adequate nutrients for plant absorption. [Drawn using data from Ryan and Hariq (1983)]

It should not be inferred that only iron chelates are effective. The chelates of other micronutrients, including zinc, manganese, and copper, have been used successfully to supply these nutrients (Figure 13.7). Apparently, replacement of other micronutrients by iron in the soil is sufficiently slow to permit absorption by plants of the other added micronutrients. Also, because foliar spray and banded applications are often used to supply zinc and manganese, the possibility of reaction of these elements with iron in the soil can be reduced or eliminated.

The use of synthetic chelates in the United States is substantial in spite of the fact that they are quite expensive. They are used primarily to meet micronutrient deficiencies of citrus and other fruit trees. Although chelates may not replace the more conventional methods of supplying most micronutrients, they offer possibilities in special cases. Agricultural and chemical research will likely continue to increase the opportunities for their use.

Chelate stability also varies depending on the specific chelating agent used. For four micronutrients, the following order of stability of some of the major chelates has been calculated by Norvell (1972).

Fe: EDDHA > DTPA, CDTA > EDTA > EGTA = HEDTA > NTA

Cu: DTPA > HEDTA = CDTA > EDTA = EDDHA > EGTA = NTA

Zn: DPTA > CDTA = HEDTA = EDTA > NTA > EGTA

Mn: DPTA = CDTA > EDTA = EGTA = HEDTA > NTA

13.7

Factors Affecting Availability of Micronutrient Anions

Unlike the cations needed in trace quantities by plants, the anions seem to have relatively little in common. Chlorine, molybdenum, and boron are quite different chemically, so little similarity would be expected in their reaction in soils.

CHLORINE

Chlorine is absorbed in larger quantities by most crop plants than any of the micronutrients except iron. Most of the chlorine in soils is in the form of the chloride ion, which leaches rather freely from humid region soils. In semi-arid and arid regions, a higher concentration might be expected, with the amount reaching the point of salt toxicity in some of the poorly drained saline soils. In most well-drained areas, however, one would not expect a high chlorine content in the surface of arid-region soils.

Except where toxic quantities of chlorine are found in saline soils, there are no common situations that reduce the availability and use of this element. Accretions of chlorine from the atmosphere along with those from fertilizer salts such as potassium chloride are sufficient to meet most crop needs. However, crop responses to chloride additions have been noted and are thought to be due to reduction of diseases such as the "take all" disease of winter wheat.

BORON

The availability of boron is related to the soil pH, this element being most available in acid soils. In fact, boron is rather easily leached from acid sandy soils. When the pH is increased by liming, boron availability decreases, being lowest between pH 7 and 9. Apparently this element is fixed or bound by soil colloids as the pH increases, resulting in lime-induced boron deficiency in some cases. Note that the mechanism is likely different from that of phosphate fixation, which is high in acid soils.

Boron is also absorbed by humus, being bound more firmly by humus than by inorganic colloids. Consequently, the organic matter serves as a major reservoir for boron in many soils.

The availability of boron is impaired by long dry spells, especially following periods of optimum moisture conditions. Boron deficiencies are common in calcareous Aridisols and in neutral to alkaline soils that are high in pH.

Soluble boron is present in soils mostly in the form of boric acid (H_3BO_3). A second soluble form ($H_2BO_3^-$) is present in smaller quantities, its activity increasing from acid to neutral and alkaline conditions.

MOLYBDENUM

Soil pH is the most important factor affecting the availability and plant uptake of molybdenum. The following equation shows the forms of this element present at low and high soil pH.

$$H_2MoO_4 \underset{+H^+}{\overset{+OH^-}{\rightleftarrows}} HMoO_4^- + H_2O \underset{+H^+}{\overset{+OH^-}{\rightleftarrows}} MoO_4^{2-} + H_2O$$

At low pH values, the relatively unavailable H_2MoO_4 and $HMoO_4^-$ forms are prevalent, whereas the more readily available MoO_4^{2-} anion is dominant at pH values above 5 or 6. The MoO_4^{2-} ion is subject to adsorption by oxides of iron and possibly aluminum just as is phosphate ion, but calcium molybdate is much more soluble than its phosphate counterpart.

The liming of acid soils will usually increase the availability of molybdenum (Figure 13.4). The effect is so striking that some researchers, especially those in Australia and New Zealand, argue that the primary reason for liming very acid soils is to supply molybdenum. Furthermore, in some instances an ounce or so of molybdenum added to acid soils has given about the same increase in the yield of legumes as has the application of several tons of lime.

The use of phosphate by plants seems to complement that of molybdenum and vice versa. For this reason, molybdate salts are often applied along with phosphate carriers to molybdenum-deficient soils. This practice apparently encourages the uptake of both elements and is a convenient way to add the extremely small quantities of molybdenum required.

A second common anion, the sulfate ion, seems to have the opposite effect on plant uptake of molybdenum. Sulfate reduces molybdenum uptake, although the specific mechanism for this antagonism is not yet known.

13.8
Need for Nutrient Balance

Nutrient balance among the trace elements is as essential as macronutrient balance, but is even more difficult to maintain. Some of the plant enzyme systems that depend on micronutrients require more than one element. For example, both manganese and molybdenum are needed for the assimilation of nitrates by plants. The beneficial effects of combinations of phosphates and molybdenum have already been discussed. Apparently, some plants need zinc and phosphorus for optimum use of manganese. The use of boron and calcium depends on the proper balance between these two nutrients. A similar relationship exists between potassium and copper, and between potassium and iron, in the production of good-quality potatoes. Copper utilization is favored by adequate manganese, which in some plants is assimilated only if zinc is present in sufficient amounts. Of course, the effects of these and other nutrients will depend on the specific plant being grown, but the complexity of the situation can be seen from the examples cited.

ANTAGONISM

Some enzymatic and other biochemical reactions requiring a given micronutrient may be "poisoned" by the presence of a second trace element in toxic quantities. Examples include the following.

1. Excess copper or sulfate may adversely affect the use of molybdenum.
2. Iron deficiency is encouraged by an excess of zinc, manganese, copper, or molybdenum.
3. Excess phosphate may encourage a deficiency of zinc, iron, or copper but enhance the absorption of molybdenum.
4. Heavy nitrogen fertilization intensifies copper and zinc deficiencies.
5. Excess sodium or potassium may adversely affect manganese uptake.
6. Excess lime reduces boron uptake.
7. Excess iron, copper, or zinc may reduce the absorption of manganese.

Some of these antagonistic interactions may be used effectively in reducing toxicities of certain of the micronutrients. For example, copper toxicity of citrus groves caused by residual copper from fungicidal sprays may be reduced by adding iron and phosphate fertilizers. Sulfur additions to calcareous soils containing toxic quantities of soluble molybdenum may reduce the availability and, hence, the toxicity of molybdenum.

These examples of nutrient interactions, both beneficial and detrimental, emphasize the highly complicated nature of the biological transformations in which micronutrients are involved. The total area where unfavorable nutrient balances require special micronutrient treatment is increasing with more intensive cropping of soils.

13.9

Soil Management and Micronutrient Needs

Although the characteristics of each micronutrient are quite specific, some generalizations with respect to management practices are possible.

In seeking the cause of plant abnormalities, one should keep in mind the conditions under which micronutrient deficiencies or toxicities are likely to occur. Sandy soils, mucks, and soils having very high or very low pH values are suspect. Areas of intensive cropping and heavy macronutrient fertilization may be deficient in micronutrients.

CHANGES IN SOIL ACIDITY

In very acid soils, one might expect toxicities of iron and manganese and deficiencies of phosphorus and molybdenum. These can be corrected by liming and by appropriate fertilizer additions. Calcareous soils may have deficiencies of iron, manganese, zinc, and copper and, in a few cases, toxic quantities of molybdenum.

No specific statement can be made concerning the pH value most suitable for all the elements. However, medium-textured soils generally supply adequate quantities of micronutrients when the soil pH is held between 6 and 7. In sandy soils, a somewhat more acid reaction may be justified because the total quantity of micronutrients is low, and even at pH 6.0, some cation deficiencies may occur.

SOIL MOISTURE

Drainage and moisture control can influence micronutrient solubility in soils. Improving the drainage of acid soils will encourage the formation of the oxidized forms of iron and manganese. These are less soluble and, under acid conditions, less toxic than the reduced forms.

Moisture control at high pH values can have the opposite effect. High moisture levels maintained by irrigation may result in the chemical reduction of high-valence compounds, the oxides of which are extremely insoluble. Flooding a soil will favor the reduced forms, which are more available to growing plants. Poor drainage also increases the availability of molybdenum, in some soils to the point of producing plants with toxic levels of this element.

FERTILIZER APPLICATIONS

The most common management practice to overcome micronutrient deficiencies (and some toxicities) is the application of commercial fertilizers. Examples of fertilizer materials applied for each of the micronutrients are shown in Table 13.5. The materials are most commonly applied to the soil, although in recent years foliar sprays and even seed treatments have been used. Foliar sprays of dilute inorganic salts or organic chelates are more effective than soil treatments where high soil pH or other factors render the soil-applied nutrients unavailable. Treating seeds with small dosages (20–40 g/ha) of molybdenum has had satisfactory results on molybdenum-deficient acid soils.

The micronutrients can be applied to the soil either as separate materials or incorporated in standard macronutrient carriers. Unfortunately, the solubilities of copper, iron, manganese, and zinc can be reduced by such incorporation, but boron and molybdenum remain in reasonably soluble

TABLE 13.5
A Few Commonly Used Fertilizer Materials That Supply Micronutrients

Micronutrient	Commonly used fertilizers		Nutrient content (%)
Boron	Borax	$Na_2B_4O_7 \cdot 10H_2O$	11
	Sodium pentaborate	$Na_2B_{10}O_{16} \cdot H_2O$	18
Copper	Copper sulfate	$CuSO_4 \cdot 5H_2O$	25
Iron	Ferrous sulfate	$FeSO_4 \cdot 7H_2O$	19
	Iron chelates	NaFeEDDHA	6
Manganese	Manganese sulfate	$MnSO_4 \cdot 3H_2O$	26–28
	Manganese oxide	MnO	41–68
Molybdenum	Sodium molybdate	$Na_2MoO_4 \cdot 2H_2O$	39
	Ammonium molybdate	$(NH_4)_6Mo_7O_{24} \cdot 4H_2O$	54
Zinc	Zinc sulfate	$ZnSO_4 \cdot H_2O$	35
	Zinc oxide	ZnO	78
	Zinc chelate	$Na_2ZnEDTA$	14

Selected from Murphy and Walsh (1972).

condition. Liquid macronutrient fertilizers containing polyphosphates (see Section 16.3) encourage the formation of complexes that protect added micronutrients from adverse chemical reactions.

Economic responses to micronutrients are becoming more widespread as intensity of cropping increases. For example, responses of fruits, vegetables and field crops to zinc and iron applications have been noted in areas with neutral to alkaline soils. Even on acid soils, deficiencies of the elements have been demonstrated. Molybdenum, which has been used for some time for forage crops and for cauliflower and other vegetables, has received attention in recent years in soybean-growing areas, especially those with acid soils. Seed treatments with about 18 g Mo/ha have produced good responses. These examples, along with those from muck areas and sandy soils, where micronutrients have been used for years, illustrate the need for these elements if optimum yields are to be maintained.

Marked differences in crop needs for micronutrients make fertilization a problem where rotations are being followed. On general-crop farms, vegetables are sometimes grown in rotation with small grains and forages. If the boron fertilization is adequate for a vegetable crop such as red beets or even for alfalfa, the small-grain crop grown in the rotation is apt to show toxicity damage. These facts emphasize the need for specificity in determining crop-nutrient requirements and for care in meeting these needs (Table 13.6).

MICRONUTRIENT AVAILABILITY

Major sources of micronutrients and the general reactions that make them available to higher plants and microorganisms are summarized in Figure 13.8. Original and secondary minerals are the primary sources of these elements, but the organic forms (as they are broken down in the soil) are also important sources. The micronutrients are used by higher plants and microorganisms in important life-supporting processes. Removal of nutrients in harvested crops reduces the soluble ion pool, and if it is not replenished with chemical fertilizers, nutrient deficiencies will result. These fertilizers, although intended primarily for supplying the macronutrients, are increasingly important sources of micronutrients as well.

TABLE 13.6
Areas of Micronutrient Deficiency, Range in Recommended Rates of Application in the Deficient Areas, and Some Crops Having a High Requirement for Micronutrients

Micronutrient	Area deficient in United States[a] (10^6 ha)	Common Range in recommended application rates (kg/ha)	Crops having a high requirement
Iron	1.54	0.5–10	Blueberries, cranberries, rhododendron, peaches, grapes, nut trees
Manganese	5.26	5–30	Dates, beans, soybeans, onions, potatoes, citrus
Zinc	2.63	0.5–20	Citrus and fruit trees, soybeans, corn, beans
Copper	0.24	1–20	Citrus and fruit trees, onions, small grains
Boron	4.85	0.5–5	Alfalfa, clovers, sugar beets, cauliflower, celery, apples, other fruits
Molybdenum	0.61	0.05–1	Alfalfa, sweet clover, cauliflower, broccoli, celery

[a] Calculated from Burgess (1966).

WORLDWIDE MANAGEMENT PROBLEMS

Micronutrient deficiencies have been diagnosed in most crop-production areas of the United States and Europe. However, in some developing countries, particularly in the tropics, the extent of these deficiencies is much less well-known. Limited research suggests that there may be large areas with deficiencies of one or more of these elements. The management principles established for the economically developed countries should be helpful in alleviating the problem.

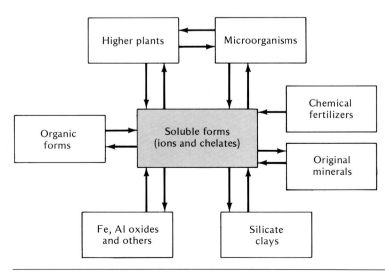

Figure 13.8
Diagram of soil sources of soluble forms of micronutrients and their utilization by plants and microorganisms.

13.10

Conclusion

Micronutrients are becoming increasingly important to world agriculture as crop removal of these essential elements increases. Soil and plant tissue tests confirm that these elements are limiting crop production over wide areas and suggest that attention to them will likely increase in the future.

In most cases, soil management practices that avoid extremes in soil pH and optimize the return of plant residues will minimize the risk of micronutrient deficiencies or toxicities. But increasingly noted are situations where micronutrient problems can be solved only by the application of micronutrient fertilizers. Such materials are becoming common components of fertilizers for field and garden use and will likely become even more necessary in the future.

Study Questions

1. Since 175–200 kg/ha of nitrogen may be needed to produce a "bumper" corn crop and only 30 g/ha of molybdenum, some have assumed that nitrogen is more essential for plant growth than molybdenum. Is this assumption correct? Explain.

2. Which of the micronutrients play a critical role in the process of nitrogen fixation?

3. Since only small quantities of micronutrients are needed for crop production, would it be wise to add large quantities of fertilizers containing these nutrients now and thereby satisfy the crop's future requirements? Explain.

4. What are some sources of chelates and of what importance are they in enhancing micronutrient availability? Under what soil conditions would they most likely be useful?

5. Peaches and other fruits are often grown on highly alkaline irrigated soils of arid regions. A common constraint is said to be iron deficiency, even though these soils are known to contain large total quantities of this element. How do you explain this situation and what are some feasible options for removing the constraint?

6. An acid soil is limed to bring the pH from 5.0 to 7.0. Unfortunately, a symptom of nutrient deficiency appears on the soybeans being grown on this soil. One person suggests molybdenum deficiency. Do you agree? If not, what are other possible explanations?

7. Cauliflower production on an acid soil has been declining. The farmer applied about 1 kg/ha of a fertilizer material containing one of the micronutrients and received a yield increase. Which of the micronutrients would it most likely have been? Explain.

8. Two Aridisols both at pH 8.0 were developed from the same general parent material, but one has restricted drainage while the other is well drained. Plants growing on the well-drained soil show iron deficiency symptoms; plants on the less well-drained site do not. What is the likely explanation for this?

9. In which three of the mineral soil orders would you most likely find micronutrient deficiencies? In which two orders would you *least* likely find such deficiencies? (Assume comparable parent materials in all cases). Give reasons for your choices. (This question should bring out some discussion.)

10. Since adequate boron is required for the production of good quality table beets, some companies purchase only beets that have been fertilized with specified amounts of this element. Unfortunately, an oat crop that follows the fertilized beets does very poorly compared to oats following unfertilized beets. Give possible explanations for this situation.

References

ALLAWAY, W. H. 1968. "Agronomic controls over the environmental cycling of trace elements," *Adv. Agron.*, 20:235–74.

ANONYMOUS. 1965. "A survey of micronutrient deficiencies in the U.S.A. and means of correcting them," *Report of the Soil Test Commission of Soil Sci. Soc. Amer.*, Madison, WI.

BERGER, K. C. 1962. "Micronutrient deficiencies in the United States," *J. Agr. and Food Chem.*, 10:178–81.

BURGESS, W. D. 1966. *The Market for Secondary and Micro-*

nutrients (New York: Allied Chemical Corp.)

ELLIOTT, H. A., M. R. LIBERATI and C. P. HUANG. 1986. "Competitive adsorption of heavy metals by soils," *J. Environ. Qual.* **15**:214–19.

GUPTA, U. C., and J. LIPSETT. 1981. "Molybdenum in soils, plants, and animals," *Adv. Agron.,* **34**:73–116.

LINDSAY, W. L. 1972. "Inorganic phase equilbria of micronutrients in soils," in J. J. Mortvedt, P. M. Giordano, and W. L. Lindsay (Eds.), *Micronutrients in Agriculture* (Madison, WI: Soil Sci. Soc. Amer.).

MARSCHNER, H., A. KALISCH, and V. ROMHELD. 1974. "Mechanism of iron uptake in different plant species," *Proc. 7th Int. Colloquium on Plant Analysis and Fertilizer Problems,* Hanover, West Germany.

MARSCHNER, H. 1986. *Mineral Nutrition of Higher Plants* (New York: Academic Press).

MENGEL, K., and E. A. KIRKBY. 1982. *Principles of Plant Nutrition* (Bern, Switzerland: Int. Potash Institute).

MORTVEDT, J. J., P. M. GIORDANO, and W. L. LINDSAY (Eds.) 1975. *Micronutrients in Agriculture* (Madison, WI: Soil Sci. Soc. Amer.).

MURPHY, L. S., and L. M. WALSH. 1972. "Correction of micronutrient deficiencies with fertilizers," in J. J. Mortvedt, P. M. Giordano, and W. L. Lindsay (Eds.), *Micronutrients in Agriculture* (Madison, WI: Soil Sci. Soc. Amer.)

NICHOLAS, D. J. D., and A. R. EGAN (Eds.). 1975. *Trace Elements in Soil–Plant–Animal Systems* (New York: Academic Press).

NORVELL, W. A. 1972. "Equilibria of metal chelates in soil solution," in J. J. Mortvedt, P. M. Giordano, and W. L. Lindsay (Eds.), *Micronutrients in Agriculture* (Madison, WI: Soil Sci. Soc. Amer.).

OLSEN, R. A., R. B. CLARK, and J. H. BENNETT. 1981. "The enhancement of soil fertility by plant roots," *Amer. Scientist,* **69**:378–84.

RYAN, J., and S. N. HARIQ. 1983. "Transformation of incubated micronutrients in calcareous soils," *Soil Sci. Soc. Amer. J.* **47**:806–10.

SILLANPAA, M. 1982. *Micronutrients and the Nutrient Status of Soils: A Global Study* (Rome, Italy: UN Food and Agriculture Organization).

STEVENSON, F. J. 1986. *Cycles of Soil Carbon, Nitrogen, Sulfur and Micronutrients* (New York: Wiley).

VIETS, F. J., Jr. 1965. "The plants' need for and use of nitrogen," in *Soil Nitrogen (Agronomy,* 10) (Madison, WI: Amer. Soc. Agron.).

WILSON, D. O., and H. M. REISENAUER. 1963. "Cobalt requirements of symbiotically grown alfalfa," *Plant and Soil,* **19**:364–73.

Losses of Soil Moisture and Their Regulation

Wise use and management of water are perhaps the most critical factors in schemes to increase food supplies. At the same time, industrial and domestic requirements for water have expanded dramatically and have created strong competition for traditional agricultural uses. This competition forces increased emphasis on efficiency of water use for agriculture.

The central role of soils and plants in the hydrological cycle[1] is illustrated in Figure 14.1. This figure also shows the interrelationship among the various uses of water. Further, it points out the need to minimize the loss of water from plant and soil surfaces in both the vapor and liquid forms.

14.1

Fate of Precipitation and Irrigation Water

The water that falls on the land or is added to a soil by irrigation moves in a number of directions (see Figure 14.1). In vegetated areas, 5–40% is usually *intercepted* by plant foliage (Table 14.1) and returns to the atmosphere by evaporation without ever reaching the soil. In some evergreen forested areas, one third to one half the precipitation is intercepted and does not reach the soil.

In level areas with friable soils, most of the added water penetrates the soil. But in rolling to hilly areas, especially if the soils are not loose and open, considerable *runoff* and *erosion* take place, thereby reducing the proportion of the water that can move into the soil. In extreme cases, as much as one fourth of the precipitation will be lost in this manner.

Once the water penetrates the soil, some of it is subject to downward percolation and eventual loss from the root zone by *drainage*. In humid areas,

---

[1] Excellent reviews of soil–water–plant relations are given in Taylor et al. (1983) and Kramer (1983).

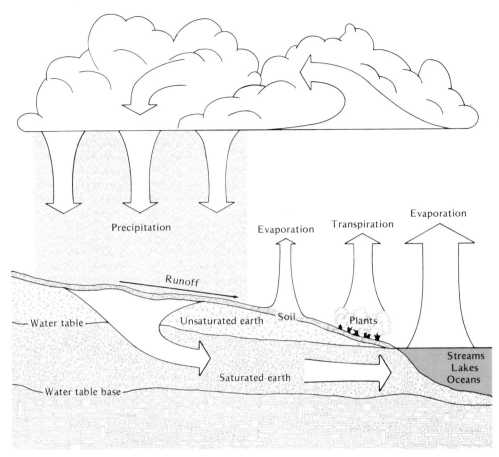

Precipitation

Evaporation   Transpiration   Evaporation

Runoff

Water table

Unsaturated earth   Soil   Plants

Streams
Lakes
Oceans

Saturated earth

Water table base

Figure 14.1
Diagram of the hydrologic cycle showing the interrelationships among soil, plants, and the atmosphere. Only water is lost from soils in evaporation and transpiration from plant surfaces. Solid particles are often carried in surface runoff water and nutrients in percolation water. [From Science of Food and Agriculture (1983)]

up to 50% of the precipitation may be lost as drainage water. However, during periods of low rainfall, some of this downward percolating water may later move back up into the plant root zone by *capillary rise*, and thereby become available for plant absorption. Such movement is of practical importance in areas with deep soils and semiarid to arid climates. Water also may reenter the plant root zone along streams where ground water moves outward from the stream.

TABLE 14.1
Percentage Interception of Precipitation by Several Crop and Tree Species at Different Locations in the United States

Note the high interception for the forest species and close-growing crops such as alfalfa.

Species	Location	Percent of precipitation intercepted by plant
Alfalfa	MO	22
Corn	MO	7
Soybeans	NJ	15
Ponderosa pine	ID	22
Douglas fir	WA	34
Maple–beech	NY	43

Crop data from Haynes (1954); forest data from Kittridge (1948).

The water remaining in the soil, sometimes referred to as *soil storage water*, is now subject to two major vapor losses. Some moves upward by capillarity and is lost by *evaporation* from the soil surface. The remainder is absorbed by plants and then moves through the roots and stems to the leaves where it is lost by *transpiration*—evaporation through the stomates of the leaf surfaces. The moisture in the atmosphere is later returned to the soil surface in precipitation or irrigation water, and the cycle starts again. Figure 14.1 should be studied to gain an understanding of the fate of water added to soil in the field.

14.2

The Soil–Plant–Atmosphere Continuum

As scientists have studied the soil–plant–atmosphere continuum (SPAC) illustrated in Figure 14.2, they have discovered that the same basic principles apply for the retention and movement of water whether it is in soil, in plants, or in the atmosphere. In Chapter 5, the free energy level of water was seen to be a major controlling factor in determining soil water behavior. The

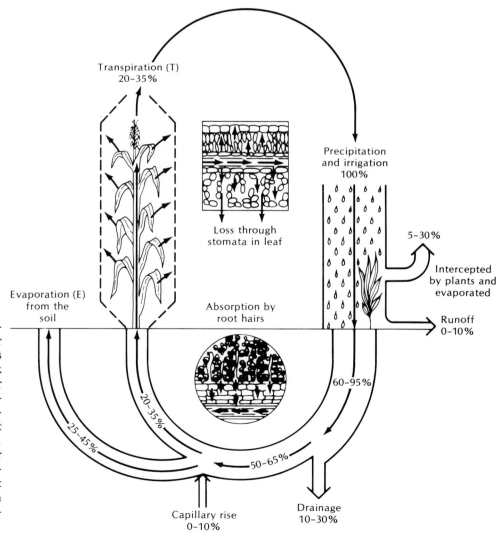

Figure 14.2
Soil–plant–atmosphere continuum (SPAC) showing water movement from soil to plants to the atmosphere and back to the soil. Water behavior through the continuum is subject to the same free energy relations covering soil water that were discussed in Chapter 5. Note the suggested ranges for partitioning of the precipitation and irrigation water as it moves through the continuum in a humid to subhumid climatic zone.

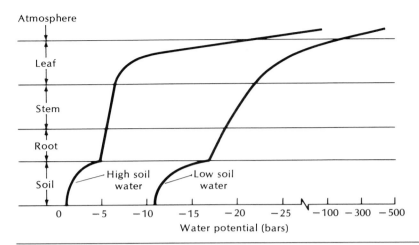

Figure 14.3
Change in moisture potential as water moves from the soil through the root, stem, and leaf to the atmosphere. Note that water potential decreases as the water moves through the system. [Adapted from Hillel (1980)]

same can be said for soil–plant and plant–atmosphere relations (see Figure 14.3). As water moves through the soil to plant roots, into the roots, across cells into stems, up the plant stems to the leaves, and is evaporated from the leaf surfaces, its tendency to move is determined by differences in free energy levels of the water or by the moisture potential.

The moisture potential must be higher in the soil than in the plant roots if water is to be absorbed from the soil. Likewise, movement up the stem to the leaf cells is in response to differences in moisture potential (see Figure 14.3).

TWO POINTS OF RESISTANCE

Changes in water potential illustrated in Figure 14.3 suggest major resistance at two points as the water moves through the SPAC: the root–soil water interface and the leaf cell–atmosphere interface. This means that two primary factors determine whether plants are well-supplied with water: (a) the rate at which water is supplied by the soil to the absorbing roots and (b) the rate at which water is evaporated from the plant leaves. Factors affecting the soil's ability to supply water were discussed in Chapter 5. Attention will now be given to the loss of water by evaporation and factors affecting this loss.

14.3

Evapotranspiration

As shown in Figure 14.2, vapor losses of water from soils occur by *evaporation* (E) at the soil surface and by *transpiration* (T) from the leaf surfaces. The combined loss resulting from these two processes, termed *evapotranspiration* (ET), is responsible for most of the water removal from soils during a crop growing period. On irrigated soils located in arid regions, for example, ET commonly accounts for the loss of 75–100 cm of water during the growing season of a crop such as alfalfa. Obviously, the phenomenon is of special significance to growing plants.

RADIANT ENERGY

Solar radiant energy provides the 2260 joules (J) (540 calories) needed to evaporate each gram of water whether from the soil (E) or from leaf surfaces (T). On a cloud-free day solar radiation is high, and evapotranspiration is encouraged. On cloudy days the solar radiation striking the soil and plant surfaces is reduced, and the evaporative potential is not as great.

Shading by plant leaves reduces evaporation (E) from a soil surface. The degree of this shading, known as *leaf area index*, significantly affects the radiant energy that reaches the soil surface and, in turn, the evaporation that takes place. Thus, newly planted row crops have a very low leaf area index, and they intercept very little solar energy. In contrast, perennial forage crops such as alfalfa and forest stands have very high leaf area indices; they shade the soil efficiently and thereby reduce markedly the soil surface evaporation.

ATMOSPHERIC VAPOR PRESSURE

Evaporation occurs when the atmospheric vapor pressure is low compared to the vapor pressure at the plant and soil surfaces. Evaporation is high from irrigated soils in arid climates (low atmospheric vapor pressure) and much lower in humid regions at comparable temperatures.

TEMPERATURE

A rise in temperature increases the vapor pressure at the leaf and soil surfaces but has much less effect on the vapor pressure of the atmosphere. As a result, on hot days there is a sharp difference in vapor pressure between leaf or soil surfaces and the atmosphere, and evaporation proceeds rapidly. Plants and especially soils may be warmer than the atmosphere on bright, clear days. This temperature difference definitely enhances the rate of evaporation.

WIND

A dry wind will continually sweep away moisture vapor from a wet surface. Hence, high winds intensify evaporation from both soils and plants. Farmers of the Great Plains of the United States dread the hot winds characteristic of that region.

SOIL MOISTURE SUPPLY

Evapotranspiration (ET) is higher where plants are grown on soils with modest moisture levels at or near the field capacity than where soil moisture is low (Table 14.2). This is to be expected because at low soil moisture levels water uptake by plants is restricted.

TABLE 14.2

Effect of Soil Moisture Level on Evapotranspiration Losses

Where the surface moisture content was kept high, total evapotranspiration losses were greater than when medium level of moisture was maintained.

Moisture condition of soil[a]	Evapotranspiration (cm)	
	Corn	Alfalfa
High	45	62
Medium	32	52

Calculated from Kelly (1954).

[a] High moisture—irrigated when upper soil layers were 50% depleted of available water. Medium moisture—irrigated when upper layers were 85% depleted of available water.

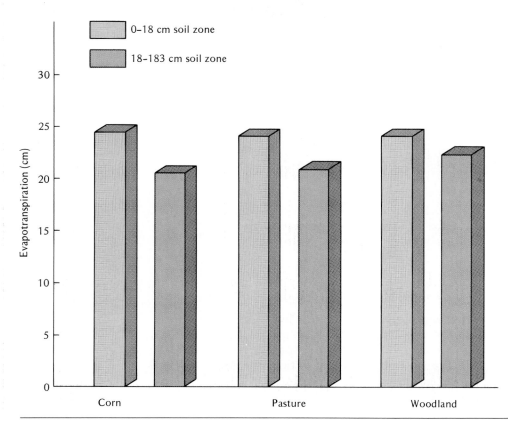

Figure 14.4
Evapotranspiration loss from the surface layer (0–18 cm) compared with loss from the subsoil (18–183 cm). Note that more than half the water lost came from the upper 18 cm and only half from the *next 165 cm* of depth. Periods of measurement: corn, May 23–September 25; pasture, April 15–August 23; woodland, May 25–September 28. [Calculated from Dreibelbis and Amerman (1965)]

In most cases, the furrow slice provides most of the water for surface evaporation (E). However, because of the large volume of the subsoil, it can provide a significant portion of the overall moisture needed for evapotranspiration (ET) (see Figure 14.4). In fact, this situation prevails in many regions having alternating moist and dry conditions during the year. Subsoil moisture stored during the moist periods is available for evapotranspiration during the dry periods. Subsoil moisture is a major source of water for crops in subhumid and semiarid areas such as the Great Plains area of the United States.

Some plant characteristics can affect ET over a growing season. The length of the growth period of the crop will have an influence. Likewise, the depth of rooting will determine the subsoil moisture available for absorption and eventually for transpiration. These interactions illustrate how the soil scientist and plant breeder must work together to determine the plant characteristics desired in improved varieties to enhance water conservation.

14.4

Magnitude of Evaporation Losses

There is a marked variation in the evapotranspiration (ET) of different climatic areas. During a growing season ET may vary from as little as 30 cm (12 in.), in cool mountain valleys where the growing seasons are short, to as much as 220 cm (87 in.) or more in irrigated desert areas. The ranges more commonly encountered are 35–75 cm (14–30 in.) in unirrigated, humid to semiarid areas and 50–125 cm (20–50 in.) in hot, dry regions where irrigation is used.

Daily ET figures are also high during the hot dry periods in the summer.

For example, daily ET rates for corn may be as high as 1.25 cm ($\frac{1}{2}$ in.) Even in deep soils having reasonably high capacities for storing available water, this rapid loss of moisture soon depletes the plant root zone of easily absorbed water. Sandy soils, of course, may lose most of their available moisture in a matter of a few days under these conditions. The importance of reducing vaporization losses is obvious.

SURFACE EVAPORATION VERSUS TRANSPIRATION

A number of factors determine the relative water losses from the soil surface and from transpiration: (a) plant cover in relation to soil surface (leaf area index); (b) efficiency of water use by different plants; (c) proportion of time the crop is on the land, especially during the summer months; and (d) climatic conditions.

Loss by evaporation from the soil is generally proportionately higher in drier regions than in humid areas. Such vapor loss is at least 60% of the total rainfall for dryland areas such as those in the central part of the United States. Losses by transpiration account for about 35%, leaving about 5% for runoff.

14.5

Efficiency of Water Use

The crop production obtained from the use of a given amount of water is an important figure, especially in areas where moisture is scarce. This efficiency may be expressed in terms of dry matter yield per unit of water transpired (T efficiency) or the dry matter yield per unit of water lost by evapotranspiration (ET efficiency).

T EFFICIENCY

The T efficiency for any given crop is markedly affected by climatic conditions. It is expressed in kilograms of water transpired to produce 1 kg of dry matter and generally ranges from 200 to 500 for crops in humid regions and almost twice as much for those of arid climates. The data in Table 14.3 illustrate the water transpired by different crops at different locations around the world.

There are differences in T efficiency of different plant species under the same climatic conditions. The data in Figure 14.5 illustrate this point. These data suggest that crops such as corn, sorghum, and millet have relatively high T efficiencies, that is, they require relatively small quantities of water to produce 1 kg of dry matter. In contrast, some legume forages such as alfalfa have low T efficiencies, requiring considerably higher quantities of water to produce 1 kg of dry matter. The cereal crops, such as wheat, oats, and barley, and vegetables such as potatoes are intermediate in their T efficiencies.

Note from Table 14.3 that the amount of water necessary to mature the average crop is very large. For example, a representative crop of wheat containing 5000 kg/ha of dry matter (about 4500 lb/acre) and having a transpiration ratio of 500 will withdraw water from the soil during the growing season equivalent to about 25 cm (10 in.) of rain. The corresponding figure for corn, assuming the dry matter at 10,000 kg/ha and the transpiration ratio at 350, would be 35 cm (14 in.). These amounts of water, *in addition* to that evaporated from the surface, must be supplied during the growing season. The possibility of moisture being the most critical factor in crop production is thus obvious.

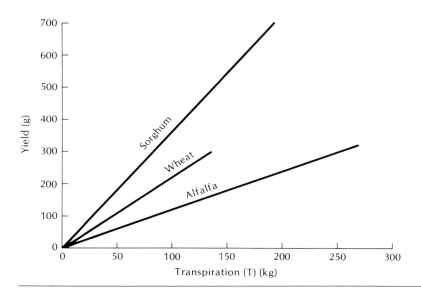

Figure 14.5
Relationship between the yield of three crops and the amount of water transpired. These data were obtained at several locations in the Great Plains of the United States. The plants were grown in containers, and evaporation from the soil surface was prevented by covering the soil. [Redrawn from graphs in Hanks (1983)]

ET EFFICIENCY

Since evapotranspiration (ET) includes both transpiration (T) and evaporation (E) from the soil surface, ET efficiency is somewhat more variable than T efficiency. Furthermore, certain soil and crop management practices that influence evaporation from the soil can affect ET and, in turn, ET efficiency.

As was the case with T efficiency, ET efficiency is affected by climate, being higher in moist areas than in dry areas. This is illustrated in Figure 14.6 showing higher ET efficiency in semiarid compared to arid conditions.

ET efficiency also responds to factors that drastically affect crop yields. Highest efficiency is attained where moisture and nutrient status are generally satisfactory for crop production, neither deficit nor excess. Figure 14.7 illustrates the effect on ET efficiency of adding phosphorus and nitrogen fertilizers to a nutrient-deficient soil. This figure confirms the advisability of maintaining optimum soil fertility to make most efficient use of available moisture.

TABLE 14.3
Transpiration Ratios of Plants as Determined by Different Investigators

Kilograms of water used in production of 1 kg of dry matter.

Crop	Harpenden, England	Dahme, Germany	Madison, Wisconsin	Pusa, India	Akron, Colorado
Barley	258	310	464	468	534
Beans	209	282	—	—	736
Buckwheat	—	363	—	—	578
Clover	269	310	576	—	797
Maize	—	—	271	337	368
Millett	—	—	—	—	310
Oats	—	376	503	469	597
Peas	259	273	477	563	788
Potatoes	—	—	385	—	636
Rape	—	—	—	—	441
Rye	—	353	—	—	685
Wheat	247	338	—	544	513

Data compiled by Lyon et al. (1952).

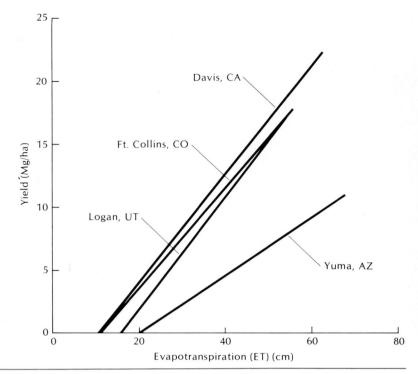

Figure 14.6
Evapotranspiration (ET) and yield of corn-fields at four locations in the western part of the United States. The site at Yuma AZ is very dry, with less than 10 cm annual precipitation, while the other three sites enjoy 40–45 cm annual precipitation. All locations received irrigation water. The effect of climate on ET efficiency is obvious. [Redrawn from Hanks (1983)]

14.6

Control of Water Infiltration

The efficient use of water is affected by two types of management practices: (a) those that increase the amount of water entering the soil and remaining there until taken up by plants and (b) those that increase crop production per unit of water taken up by plants.

The proportion of rain or irrigation water entering the soil is enhanced by practices that keep the soil surface open and receptive to water penetration

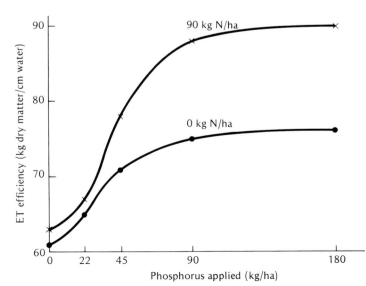

Figure 14.7
Water-use efficiency of wheat as affected by application of phorphorus fertilizer with and without nitrogen fertilizer (12-year average). ET efficiency increased with increased fertilizer rates. [Data from Black (1982)]

Figure 14.8
The effectiveness of small furrow dikes (right) in preventing water runoff and encouraging water infiltration. [Courtesy O. R. Jones, USDA Agricultural Research Service, Bushland, TX]

and that keep the soil surface covered to protect it from the beating action of rain drops. Included are selected tillage and engineering practices, especially those that leave considerable residues on the surface (see Section 15.12). Excessive tillage that destroys the surface roughness should be avoided in all cases. Tillage across the slope, leaving small ridges, encourages water infiltration. Likewise, terraces help control the amount of water lost by runoff (see Section 15.9).

In the Great Plains and Northwest areas of the United States, some use has been made of a system of small *furrow dikes* to capture rain or irrigation water (see Figure 14.8). Such systems increase water infiltration and subsequent crop yields.

In some areas water infiltration and root penetration are constrained by subsurface layers and pans. Special efforts are needed to break up or shatter these constraining layers. Moderately *deep plowing* on occasion can disrupt a plow pan that develops just under normal plowing depth, but special equipment is needed to *chisel* or break up dense layers called "hardpans" in some soils. Simultaneous attempts have been made to introduce chopped organic residues into the subsoil layers (termed *vertical mulching*) and to deep-place lime and fertilizers. While increased water infiltration and some yield increases have been achieved using these practices, their high costs have restricted their general use.

14.7
Control of Evapotranspiration

Preceding discussions remind us that transpiration (T) is a plant process subject to only minor control in a given climate if other plant processes are proceeding near normally. In contrast, evaporation (E) from the soil surface is not essential for plant growth and is subject to some degree of control by soil and crop management. Although some means of selecting and handling the crop may reduce T and thereby ET, major focus should be on those practices that reduce E, so as to maximize the water remaining to accommodate the plant process, T. Soil management considerations will follow those concerning crop management.

CROP SELECTION AND MANAGEMENT

The selection of a crop species and the time of year when it is grown will definitely influence both T and ET. For example, corn and sorghum have lower water requirements than alfalfa, and the requirement for wheat and other cereals is generally intermediate. Also, when possible, growing any crop during cool seasons when the vapor pressure gradients are apt to be low can decrease both T and ET.

Keeping a dense crop cover (high leaf area index) also can be effective in reducing E and ET. Figure 14.9 illustrates this point. Crop cover is sometimes maintained by the use of mixtures of legumes and nonlegumes in pastures. In dryland areas, however, there are some limitations to maintaining high leaf area indices. While dense cover can reduce ET, the high plant populations needed to produce the cover can deplete soil moisture prematurely and thereby drastically reduce crop yields.

FALLOW CROPPING

Farming systems that alternate summer fallow (noncropped) one year with traditional cropping the next are sometimes used to conserve soil moisture in semiarid and subhumid areas such as the Great Plains of the United States. The objective of fallow systems is to eliminate T every other year, thereby increasing the soil moisture storage for the subsequent crop. The previous year's stubble is commonly allowed to stand until the spring of the fallow year. The soil may then be disked, leaving most of the stubble near the surface. During the summer, weed growth is minimized by either light cultivation or the use of herbicides. For the fallow cropping system, vapor loss is a result primarily of evaporation at the soil surface in the year no crop is grown.

The effectiveness of summer fallow has varied greatly. However, moisture levels at planting time are commonly increased by the fallow cropping system, and yields have generally been increased through its use. Data in Table 14.4 illustrate this point. Fallow cropping certainly reduces the risk of crop failure from droughty conditions.

Figure 14.9
Relationship between the leaf area per unit of land area (leaf area index, LIA) of cotton and transpiration (T). Transpiration is expressed as a percentage of the maximum evapotranspiration (ET). Soil water was not limiting in these trials. Note that when the soil was effectively shaded (high LAI), most of the moisture vapor loss was by transpiration (T), evaporation from the soil surface (E) making up a small portion of the total, essentially none where the surface soil was dry. [Redrawn from Ritchie (1983)]

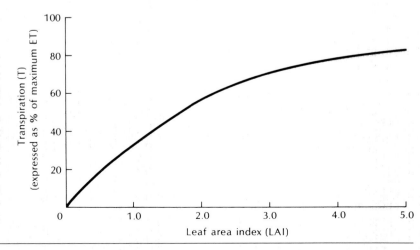

TABLE 14.4

Wheat Yields at Seven Locations in the Great Plains Area of the United States Where Continuous Cropping and Fallow Cropping were Compared

Location	Years of record	Yield (kh/ha) After fallow	After wheat
Havre, MT	35	2100	540
Dickinson, ND	44	1400	780
Newell, SD	40	1420	910
Akron, CO	60	1420	500
North Platte, NE	56	2140	830
Colby, KS	49	1320	620
Bushland, TX	29	1010	640

From USDA (1974).

14.8

Control of Surface Evaporation

More than half the precipitation in dryland areas is usually returned to the atmosphere by evaporation (E) directly from the soil surface. Evaporation losses are also high in arid-region irrigated agriculture. Even in humid-region rainfed areas, E losses are of significance in hot rainless periods. Such losses rob the plant of much of its crop production potential.

The most effective practices aimed at controlling E are those that provide some cover to the soil. This cover can best be provided by mulches and by selected conservation tillage practices.

MULCHES

Any material used at the surface of a soil primarily to reduce evaporation or to keep weeds down may be designated as a mulch. Examples are sawdust, manure, straw, leaves, crop residues, and other litter. Mulches are highly effective in checking evaporation and are most practical for home garden use and for high-valued crops, including strawberries, blackberries, fruit trees, and such other crops as require infrequent, if any, cultivation. Intensive gardening justifies the use of these moisture-saving materials.

Mulches comprised of crop residues are effective in reducing E and, in turn, in conserving soil moisture. Combining summer fallow with residue mulches saves soil moisture in dry areas, as shown in Table 14.5. Unfortunately, crop yields on these soils are not sufficiently high to leave adequate residues to provide the organic mulch level that is needed.

TABLE 14.5

Gains in Soil Water from Different Rates of Straw Mulch During Fallow Periods at Four Great Plains Locations

Location	Av. annual precipitation (cm)	Soil water gain (cm) at each mulch rate 0 Mg/ha	2.2 Mg/ha	4.4 Mg/ha	6.6 Mg/ha
Bushland, TX	50.8	7.1	9.9	9.9	10.7
Akron, CO	47.6	13.4	15.0	16.5	18.5
North Platte, NE	46.2	16.5	19.3	21.6	23.4
Sidney, MT	37.9	5.3	6.9	9.4	10.2
Average		10.7	12.7	14.5	15.7
Average gain by mulching			2.0	3.8	5.0

From Greb (1983).

Figure 14.10
For crops with high cash value, plastic mulches are being used. The plastic is installed by machine (left) and at the same time the plants are transplanted (right). Plastic mulches help control weeds, conserve moisture, encourage rapid early growth, and eliminate need for cultivation. The high cost of plastic makes it practical only with the highest value crops. [Courtesy K. Q. Stephenson, Pennsylvania State University]

PAPER AND PLASTIC MULCHES

Specially prepared paper and plastics are also used as mulches (see Figure 14.10). This cover is spread and fastened down either between the rows or over the rows. The plants in the latter case grow through suitable slits or other openings. Paper and plastic mulches can be used only with crops planted in rows or in hills. As long as the ground is covered, evaporation and weeds are checked, and, in some cases, remarkable crop increases have been reported. Unless the rainfall is very heavy, the paper or plastic does not seriously interfere with the infiltration of rain water.

Paper and plastic mulches have been employed with considerable success in the culture of pineapples in Hawaii and with vegetable crops in desert areas of the Middle East where water conservation is critical. This type of mulch is also being used to some extent in truck farming and vegetable gardening. The cost of the cover and the difficulty of keeping it in place limit the use of these materials to high-value crops.

CROP RESIDUE AND CONSERVATION TILLAGE

In recent years, conservation tillage practices, which leave a high percentage of the residues from the previous crop on or near the surface (Figure 14.11), have been widely adopted. These practices reduce markedly both vapor and liquid losses. In conservation tillage (see Section 15.12) the use of the traditional moldboard plow, which incorporates essentially all the residues into the soil, is replaced or minimized.

Among the most widely used conservation tillage practices is *stubble mulch* tillage. In this method, used mostly in subhumid and semiarid regions, the residues from the previous crop are uniformly spread on the soil surface. The land is then tilled with special tillage implements that permit much of the crop residue to remain on or near the surface. Wheat stubble, straw, cornstalks, and similar crop residues are among the residues used.

Other conservation tillage systems that leave residues on the soil surface

Figure 14.11
Planting soybeans in wheat stubble with no tillage (left). In another field corn has been planted in wheat stubble (right). In both cases the wheat residue will help reduce evaporation losses from the soil surface and will reduce erosion. [Courtesy USDA Soil Conservation Service]

include *no tillage* (Figure 14.11), where the new crop is planted directly in the sod or residues of the previous crop with no plowing or disking. Another, called *till planting*, is a strip tillage system where residues are swept away from the immediate area of planting, but most of the soil and surface area remain untilled. Like the other methods, till planting leaves most of the soil covered with crop residues to help reduce water vapor losses from soils. These systems will receive further attention in Section 15.12.

SOIL MULCH

One of the early contentions about the control of evaporation was that the formation of a *natural* or *soil mulch* was a desirable moisture-conserving practice. Years of experimentation showed, however, that a natural soil mulch does not necessarily conserve moisture, especially in humid regions. In fact, in some cases a soil mulch may encourage moisture loss. Only in regions with rather distinct wet and dry seasons, as in some tropical areas, can a soil mulch conserve moisture.

CULTIVATION AND WEED CONTROL

In addition to seedbed preparation, the most important reason for cultivation of the soil is to control weeds. Transpiration by these unwanted plants can extract soil moisture far in excess of that used by the crop itself, especially a row crop. The widespread use of chemical herbicides, however, provides weed control without cultivation of the soil. Herbicides thus have become one of the most important tools in controlling evaporation. Tillage is thereby limited to situations where this practice is needed to maintain the proper physical conditions for plant growth.

14.9

Climatic Zone Management Practices

There is variation in the need for practices that limit E in different climatic regions. Brief mention will be made of these regional differences.

Figure 14.12
Typical irrigation scene. The use of easily installed siphons or gated pipes reduces the labor of irrigation and makes it easier to control the rate of application of water. Note the upward capillary movement of water along the sides of the rows. [Courtesy USDA Soil Conservation Service]

HUMID REGIONS

For high-valued specialty crops and for home gardens and nurseries, paper and plastic mulches are in common use. Likewise, the use of conservation tillage practices that maximize residue coverage is spreading throughout humid areas of the United States.

Weed control is a practical means of reducing losses of soil moisture by ET in humid regions. Modern herbicides are tools that can be used to help remove this source of moisture loss.

SEMIARID AND SUBHUMID REGIONS

These regions sorely need means of managing water vapor loss. The two primary methods used are the *summer fallow* and *stubble mulch* systems. While evaporation is still high even with these water-conserving systems, years of field experience show that they take at least part of the risk out of dryland farming.

IRRIGATED FIELDS IN ARID REGIONS

Irrigation in dryland areas (Figure 14.12) presents a combination of the opportunities and problems of water management in both the humid and semiarid regions. Because of the high solar radiation and frequently high wind velocities, evaporation losses are very high. At the same time, during the crop-growing season, the soils may be at moisture levels fully as high as their counterparts in humid regions. In fact, one of the advantages of irrigation is the ability to control the timing, location, and quantity of the water application.

The only practical control of evaporation under dryland conditions is through irrigation practices. The surface of the soil should be kept only as moist as needed for good crop production, yet the irrigation schedule should be such as to keep more than just the surface soil wet. Deep penetration of roots should be encouraged.

One unique method of reducing evaporation from irrigated lands is to concentrate the added water in the immediate root zone of the crop plant. This is done by trickling water out of pipes or tubes alongside each plant, without wetting the soil between the rows. This system, termed "trickle" or "drip" irrigation, has the advantage of markedly reducing moisture loss by evaporation from the soil surface. In some situations, tubes are buried several centimeters below the soil surface, further reducing evaporation losses.

SALT ACCUMULATION

A phenomenon closely correlated with evaporation and often encountered on arid-region lands under irrigation is the concentration of soluble salts at or near the soil surface. The upper horizons of *saline* and *alkali* soils may contain sufficient quantities of soluble salts to inhibit the growth of many cultivated plants (see Figure 8.22). Irrigation practices on these soils can improve or impair their usefulness as crop soils. By flood irrigation, some of the soluble salts can be temporarily washed down from the immediate surface. Subsequent upward movement of the water and evaporation at the soil surface may result in salt accumulation and concentration, which are harmful to young seedlings (see Figure 14.13). Although little can be done to prevent evaporation and upward movement of the salt-containing water, timing and method of irrigation may prevent the salts from being concentrated in localized areas where plant roots are found.

14.10

Types of Liquid Losses of Soil Water

We now turn to the loss of water in the liquid form. Two types of liquid losses of water from soils are recognized: (a) the percolation or drainage water and (b) the runoff water (see Figure 14.1). Percolation results in the loss of soluble salts (leaching), thus depleting soils of certain nutrients. Runoff losses generally include not only water but also appreciable amounts of soil (erosion). Runoff and erosion will be considered in Chapter 15.

14.11

Percolation and Leaching—Methods of Study

Two general methods are used to study percolation and leaching losses: (a) underground pipe or *tile drains* specially installed for the purpose and (b) specially constructed *lysimeters*[2] (from the Greek word *lysis*, meaning loosening, and *meter*, to measure). For the first method, an area should be chosen where the tile drain receives only the water from the land under study and where the drainage is efficient. The advantage of the tile method is that water and nutrient losses can be determined from relatively large areas of soil under normal field conditions.

---

[2] For a review of lysimeter work, see Aboukhaled et al. (1982).

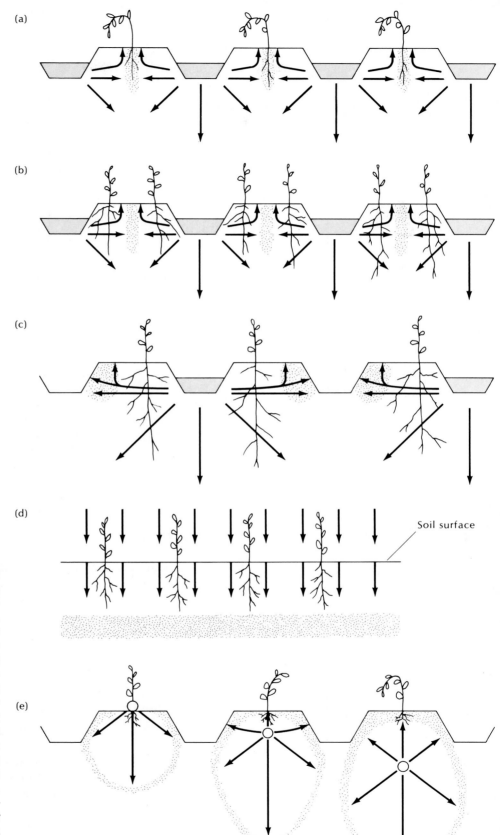

Figure 14.13
Effect of irrigation on salt movement and plant growth in saline soil. (a) Furrow irrigation on both sides of the row, the salts move to the center and damage plant root systems, (b) same as (a) except placing plants on the shoulders of each bed alleviate the salt toxicity, (c) water in every other furrow and plants near the watered furrow provides no damage. Sprinkle irrigation (d) or uniform flood irrigation moves salts downward and the problem is alleviated. (e) Drip or trickle irrigation results in salt removal or accumulation, depending on the placement of the trickle tube. [Courtesy Wesley M. Jarrell]

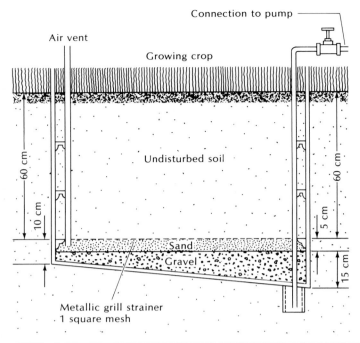

Figure 14.14
Field lysimeter used to collect percolation water from an undisturbed soil. The water moves down through the soil to the sand and gravel, then down the inclined slope to the container (lower right) from which it can be pumped and collected. Some more sophisticated systems are equiped with devises to weigh the entire lysimeter, thereby permitting the additional measurement of water intake and evapotranspiration. [Redrawn from UNDP/WMO (1974)]

The lysimeter method involves the measurement of percolation and nutrient losses under somewhat more controlled conditions. Soil is removed from the field and placed in concrete or metal tanks, or a small volume of field soil is isolated from surrounding areas by concrete or metal dividers (Figure 14.14). In either case, water percolating through the soil is collected and measured. The advantages of lysimeters over a tile-drain system are that the variations in a large field are avoided, the work of conducting the study is not so great, and the experiment is more easily controlled.

14.12
Percolation Losses of Water

When the amount of rainfall entering a soil becomes greater than the water-holding capacity of the soil, losses by percolation will occur. Percolation losses are influenced by the amount of rainfall and its distribution, by runoff from the soil, by evaporation, by the character of the soil, and by the crop.

PERCOLATION–EVAPORATION BALANCE

The relationship among precipitation, runoff, soil storage, and percolation for representative humid and semiarid regions and for an irrigated arid-region area is illustrated in Figure 14.15. In the humid temperate region, the rate of water infiltration into the soils (precipitation less runoff) is commonly greater, at least at some time of year, than the rate of evapotranspiration. As soon as the soil field capacity is reached, percolation into the substrata occurs.

In the example shown in Figure 14.15(a), maximum percolation occurs during the winter and early spring, when evaporation is lowest. During the summer little percolation occurs, and evapotranspiration that exceeds the infiltration results in a depletion of soil water. Normal plant growth is

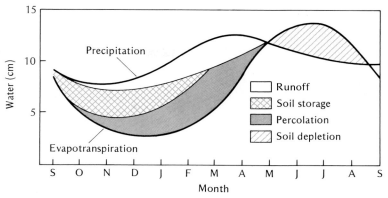

(a) Humid Region (Udic soil moisture regime)

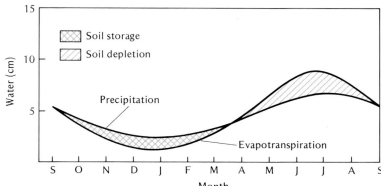

(b) Semiarid Region (Ustic soil moisture regime)

Figure 14.15
Generalized curves for precipitation and evapotranspiration for three temperate zone regions: (a) a humid region, (b) a semiarid region, and (c) an irrigated arid region. Note the absence of percolation through the soil in the semiarid region. In each case water is stored in the soil. This moisture is released later when evapotranspiration demands exceed the precipitation. In the semiarid region evapotranspiration would likely be much higher if ample soil moisture were available. In the irrigated arid region soil, the very high evapotranspiration needs are supplied by irrigation. Soil moisture stored in the spring is utilized by later summer growth and lost through evaporation during the late fall and winter.

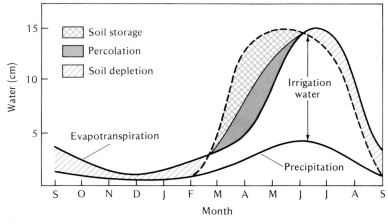

(c) Arid Region (Aridic soil moisture regime), irrigated

possible only because of moisture stored in the soil the previous winter and early spring.

The general trends in the temperate zone semiarid region are the same as for the humid region. Soil moisture is stored during the winter months and used to meet the moisture deficit in the summer. But because of the low rainfall, essentially no percolation out of the profile occurs. Water may move to the lower horizons, but it is absorbed by plant roots and ultimately lost by transpiration.

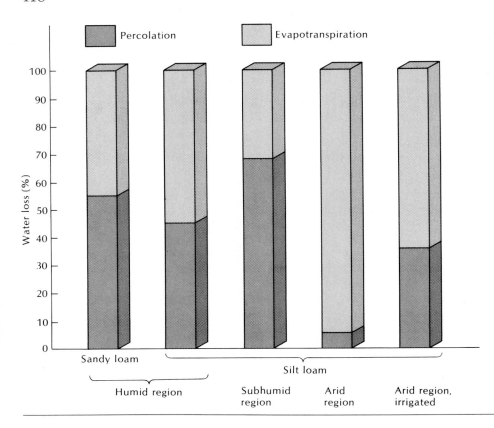

Figure 14.16
Percentage of the water entering the soil that is lost by downward percolation and by evapotranspiration. Representative figures are shown for different climatic regions.

The irrigated soil in the arid region shows a unique pattern. Irrigation in the early spring, along with a little rainfall, provides more water than is being lost by evapotranspiration. The soil is charged with water and some percolation may occur. During the summer, fall, and winter months, this stored water is depleted because the amount being added is less than the very high evapotranspiration that takes place in response to large vapor pressure gradients.

Figure 14.15 depicts situations common in temperate zones. In the tropics one would expect evapotranspiration to be somewhat more uniform throughout the year, although it will vary from month to month depending on the distribution of precipitation. In very high rainfall areas of the tropics, the percolation and especially the runoff would be higher than shown in Figure 14.15. In irrigated areas of the arid tropics, the relationships would likely not be too different from those in arid temperate regions.

The comparative losses of water by evapotranspiration and percolation through soils of different climatic regions are shown in Figure 14.16. These differences should be kept in mind as attention is given to nutrient losses through percolation.

14.13

Leaching Losses of Nutrients

The loss of nutrients through leaching is determined by climatic factors as well as by soil–nutrient interactions. In regions where water percolation is high, the potential for leaching is also high (see Figure 14.17). Such conditions exist in the United States in the humid east and in the heavily irrigated sections of the west. In these areas, percolation of excess water is the rule, providing opportunities for nutrient removal. In unirrigated semiarid areas, less nutrient leaching occurs because there is little percolation. Some nutrient leaching takes place in subhumid areas, although less than in those with humid climates. In all cases, growing crops reduce the loss of nutrients by leaching.

NUTRIENT–SOIL INTERACTION

Soil properties have a definite effect on nutrient-leaching losses. Sandy soils generally permit greater nutrient loss than do clays because of the higher rate of percolation and lower nutrient-adsorbing power of the sandy soils. For example, soluble phosphorus is quickly bound chemically in fine-textured soils with appreciable amounts of iron and aluminum oxides (see Section 12.7). Consequently, very little phosphorus is lost by leaching from these soils. Sulfates, and to a much lesser extent nitrates, react by anion exchange (Section 7.14) with iron and aluminum hydrous oxides. For this reason, sulfates and nitrates are less prone to leach from soils with red

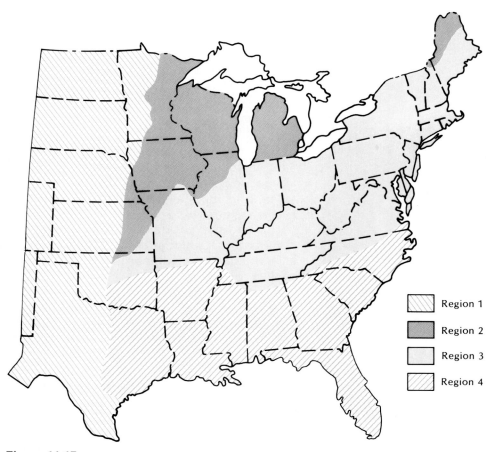

Region 1

Region 2

Region 3

Region 4

Figure 14.17
Eastern part of the United States, showing regions varying in their susceptibility to leaching during the winter months. In region 1 the losses are low, nutrients being carried down the profile but not out of it; in region 4 losses are high. Regions 2 and 3 are intermediate. [Modified from Nelson and Uhland (1955)]

TABLE 14.6

Average Annual Loss of Nutrients by Percolation Through Soils from Four Different Areas

The Monolith type of lysimeter was used, and runoff was allowed from Illinois and Wisconsin lysimeters only.

Condition	Loss per year (kg/ha)					
	N	P	K	Ca	Mg	S
Rotation on a Scottish soil (av. 6 yr)						
No treatment	8	Trace	10	56	17	—
Manure and fertilizers	7	Trace	9	63	18	—
Manure, fertilizers, and lime	9	Trace	9	89	21	—
Uncropped Illinois soils (av. $3\frac{1}{2}$ yr)						
Udoll (Muscatine, well-drained)	86	—	1	101	52	12
Aqualf (Cowden, poorly-drained)	7	—	1	12	4	2
Wisconsin soil—Udalf (Fayette, av. 3 yr)						
Fallow	—	—	1	42	21	3
Cropped to corn	—	—	<1	16	7	1
New York soils						
Udalf (Dunkirk, av. 10 yr)						
Bare	77	Trace	81	446	71	59
Rotation	9	Trace	64	258	49	48
Grass	3	Trace	69	291	56	49
Aquept (Volusia, av. 15 yr)						
Bare	48	Trace	72	362	46	39
Rotation	7	Trace	64	280	30	37
Canadian orchard soil (av. 5 yrs)						
Grass under orchard	—	—	19	51	19	—
Clear cultivated under orchard	—	—	45	374	104	—
Normal N fertilizer (162 kg/ha)	—	—	28	156	44	—
High N fertilizer (324 kg/ha)	—	—	36	267	80	—

Canadian data from Nielsen and Stevenson (1983) corrected for cation content of irrigation water; other data compiled by Buckman and Brady (1969).

subsoils (high in iron oxides) than from comparable soils where iron oxides are not so prominent.

The leaching of cations added in fertilizers is affected by the soil's cation exchange capacity; soils with a high capacity tend to hold the added nutrients and prevent their leaching. At the same time, because such soils are often naturally high in exchangeable cations, they provide a large reservoir of nutrients; a small portion of these exchangeable ions is continually subject to leaching.

Some of the factors affecting the loss of nutrients by leaching are illustrated in Table 14.6, which includes data from lysimeter experiments at five locations. The loss of phosphorus is negligible in all cases. The leaching of nitrogen depends on the crops grown, the fertilizer applied, and the amount of percolation occurring. The loss of cations is roughly proportional to their probable content in exchangeable form on the colloidal complex, calcium being lost in the greatest amounts and potassium in the least.

NUTRIENT LOSS AND FERTILIZATION

The rates of fertilizer additions in the lysimeter experiments reported in Table 14.6 are modest compared to those common today. For example, nitrogen applications at rates of 150–200 kg N/ha and greater are common for fields of crops such as corn, vegetables, and sugarcane. For this reason, nutrient losses today are generally considerably higher than those shown in Table 14.6. These lost nutrients either react with the soil or are removed by microbial action (for example, denitrification) or through leaching. Leaching losses of nitrogen as high as 100 kg/ha per year have been reported.

The practice of fall application of nitrogen for crops to be planted the next spring can increase leaching loss, especially in humid areas. The percolation losses are generally greatest in the spring prior to rapid plant growth when water loss by percolation is highest. Similarly, some irrigated soils of arid areas that receive heavy nitrogen application may suffer considerable loss of nitrates through leaching during the intensive irrigation periods.

SOIL AND WATER POLLUTION

There are two prime reasons for concern over the loss of essential elements by leaching. First is the obvious concern for keeping these nutrients in the soil so that essential elements are available to crop plants. A second and equally significant reason is to keep the nutrients out of streams, rivers, and lakes. Excessive growth of algae and other aquatic species takes place in water overly enriched with nitrogen, phosphorus and other nutrients. This process, called *entrophication*, ultimately depletes the water of its oxygen, with disastrous effects on fish and other aquatic animals. Also, in some areas underground sources of drinking water have become sufficiently high in nitrates to cause health concerns for humans. Likewise, surface runoff waters from heavily fertilized lands may contain levels of nitrates toxic to livestock.

CONTROL OF NUTRIENT LOSSES

Nutrient losses can be minimized by following a few simple rules. First, to the extent feasible, a crop should be kept on the land. Fall and winter cover crops can be grown following heavily fertilized cash crops such as corn, potatoes, and other vegetables. Second, fertilizer application rates should be no higher than can be clearly justified by scientific research trials. Third, in areas where water percolation from soils is common, the fertilizer should be added as close as feasible to the time of nutrient utilization by the crop plants. For example, fertilizer can be applied as a side dressing when the crop is growing vigorously rather than a broadcast application before planting. Even when following these suggestions, some leaching losses will occur. The aim should be to minimize these losses both for the sake of the farmer and for society generally.

14.14

Land Drainage

If a poorly drained soil is to be used effectively, the excess soil moisture must be removed by the installation of a *land-drainage system*.[3] The objective is to lower the moisture content of the upper layers of the soil so that oxygen can be available to the crop roots and carbon dioxide can diffuse from these roots into the atmosphere.

Land drainage is needed in select areas in almost every climatic region, but it is critical in areas such as the flat coastal plain sections of the eastern United States. Likewise, the fine-textured soils of river deltas, such as

---

[3] For excellent reviews of various aspects of this subject, see Van Schilfgaarde (1974).

(a)   (b)

Figure 14.18
Two common types of drainage ditches. (a) A small ditch with vertical sides such as is used to drain organic soil areas. Note the tile outlet from one side. (b) A larger sloping-sided ditch of the type commonly used to transport drainage water to a nearby stream. The source of drainage water may be either tile or open drain systems.

those of the Mississippi and the Nile, and lake-laid soils throughout the world usually require some form of artificial drainage. Even irrigated lands of arid regions often require extensive drainage systems to remove excess salts or to prevent their buildup in irrigated areas.

Two general types of land-drainage systems are used: (a) surface field drains and (b) subsurface drains. Each will be discussed briefly.

14.15

Surface Field Drains

The most extensive means of removing excess water from soils is through the use of surface drainage systems. Their purpose is to remove the water from the land before it infiltrates the soil. Surface drainage may involve deep and rather narrow field ditches such as those used to remove large quantities

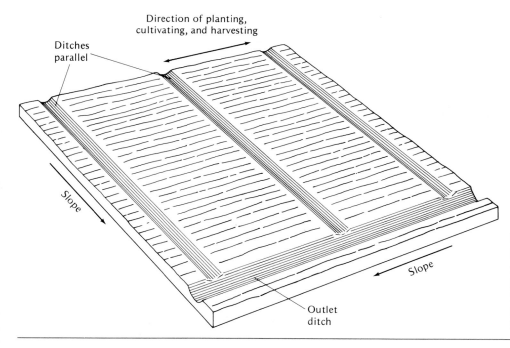

Figure 14.19
Example of an open ditch drainage system on afield with gentle slope. [From Hughes (1980); used with permission of Deere & Company, Moline, IL]

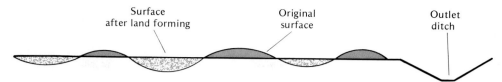

Figure 14.20
Land surface before and after land forming or smoothing. Note that soil from the ridges fills in the depressions. Land forming makes possible the controlled movement of surface water to the outlet ditch, which transports it to a nearby natural waterway.

of water from peat soil areas (Figure 14.18). More often, however, shallow ditches with gentle side slopes are used to help remove the water. Shallow ditches are commonly constructed with simple equipment and are not too costly. If there is some slope on the land, the shallow ditches are usually constructed across the slope and across the direction of planting and cultivating, thereby permitting the interception of runoff water as it moves down the slope (see Figure 14.19).

Open drainage ditches have the advantage of large carrying capacities. For this reason, they are an essential component of all drainage systems. In general, their cost per unit of water removed is relatively low. Disadvantages include cost of maintenance, and intereference with agricultural operations. Also, open ditches use valuable agricultural lands, which, if the area had underground drainage, would be available for cropping.

LAND FORMING

Surface drain ditches are combined with *land forming* or smoothing to rapidly remove water from soils. Depressions or ridges that prevent water movement to the drainage outlet are filled in with precision, using field-leveling equipment. The resulting land configuration permits excess water to move slowly over the soil surface to the outlet ditch and then on to a natural drainage channel (Figure 14.20).

Land smoothing is a common practice in irrigated areas. Surface irrigation is made possible as is the removal of excess water by the outlet ditch. In humid areas, the same methods are being put to use to remove excess surface water. In combination with selectively placed tile drains, smoothing the land permits orderly water removal.

14.16

Subsurface (Underground) Drains

Systems of underground channels to remove water from the zone of maximum water saturation provide the most effective means of draining land. The underground network can be created in several ways, including the following.

1. A *mole-drainage* system can be created by pulling through the soil at the desired depth a pointed cylindrical plug about 7–10 cm in diameter. The compressed wall channel thus formed provides a mechanism for the removal of excess water (Figure 14.21).

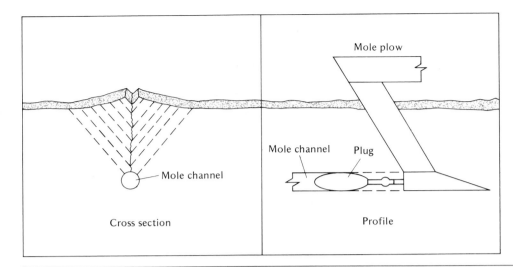

Figure 14.21
Diagrams showing how an underground mole drainage system is put in place. The plug is pulled through the soil, leaving a channel through which drainage water can move. [From Hughes (1980); used with permission of Deere & Company, Moline, IL]

2. A *perforated plastic pipe* can be laid underground using special equipment (Figure 14.22). Water moves into the pipe through the perforations and can be channeled to an outlet ditch. About 90% of the underground drain systems being installed today are of this type.

3. A *clay-tile system* made up of individual clay pipe units 30–40 cm long can be installed in an open ditch. The tiles are then covered with a thin layer of straw, manure, or gravel, and the ditch is then refilled with soil. Tile drains were most popular 20–30 years ago, but high installation costs make them less competitive than the perforated plastic systems.

The mole-drain system is the least expensive to install but is also more easily clogged than the other two systems, which involve a more stable channel for water flow.

OPERATION OF UNDERGROUND SYSTEMS

The same principles govern the function of all three types of underground drains. The tile, pipe, or mole channel is placed in the zone of maximum water accumulation. Water moves from the surrounding soil to the channel,

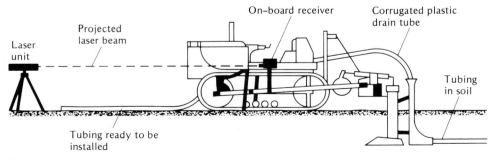

Figure 14.22
Drawing showing how corrugated plastic tubing may be placed in the soil. The flexible tile is fed through a pipe as the tractor moves forward. Appropriate grade level is assured by a laser beam controlled mechanism. [From Fouss (1974); used with permission of the American Society of Agronomy]

Figure 14.23
Demonstration of the saturated flow patterns of water toward a drainage tile. The water, containing a colored dye, was added to the surface of the saturated soil and drainage was allowed through the similation drainage tile shown on the extreme right. [Courtesy G. S. Taylor, The Ohio State University]

enters through perforations in a pipe or joints between the tiles, and is then transported through the channel to an outlet ditch. An illustration of the water flow pattern to the channel from the surrounding soil is shown in Figure 14.23. Different field layouts for underground systems are illustrated in Figure 14.24.

Care needs be taken to place the tile or other underground channels in the zone of maximum water accumulation. The channels should be at least

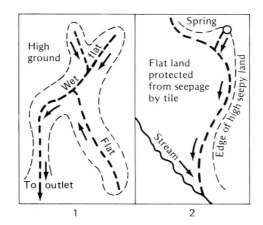

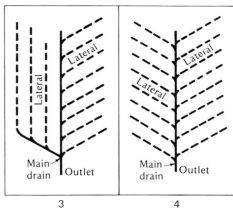

Figure 14.24
Tile drainage systems. (Top) Four typical systems for laying tile drain: (1) *natural*, which merely follows the natural drainage pattern; (2) *interception*, which cuts off water seeping into lower ground from higher lands above; (3) *gridiron* and (4) *fishbone* systems, which uniformly drain the entirety of an area. (Bottom) A freshly installed drainage system. Tile lines feed into main drains at the bottom, the direction of flow being shown by the arrows. [Photo courtesy U.S. Department of Agriculture]

TABLE 14.7
Suggested Spacing Between Tile Laterals for Different Soil and Permeability Conditions

Soil	Permeability	Spacing Meters	Feet
Clay and clay loam	Very slow	9–18	30–70
Silt and silty clay loam	Slow to moderately slow	18–30	60–100
Sandy loam	Moderately slow to rapid	30–90	100–300
Muck and peat	Slow to rapid	15–61	50–200

Modified from Beauchamp (1955).

75 cm below the soil surface because of the danger of damage from heavy machinery, and a depth of 1 m is usually recommended. The distance between the channel lines will vary with the soil conditions (Table 14.7). On heavy clay soils, the lines may be as close as 9–10 m, but a distance of 15–20 m is more common.

The grade or fall in the underground drain system is normally 12–25 cm/50 m (about 3–6 in./100 ft), although a fall as great 75 cm/50 m is used for rapid water removal.

Special care must be taken to protect the outlet of an underground drainage channel. If the outlet becomes clogged with sediment, the whole system is endangered (see Figure 14.25). It is well to embed the drainage outlet in a masonry concrete wall or block. The last 2.5–3 m of drain may even be replaced by a galvanized iron pipe or sewer tile, thus ensuring against damage by frost. The outlet may be covered by a gate or by wire in such a way as to allow the water to flow out freely and to prevent the entrance of rodents in dry weather.

DRAINAGE AROUND BUILDING FOUNDATIONS

Surplus water around foundations and underneath basement floors of houses and other buildings can cause serious and expensive damage. The removal of this excess water is commonly accomplished using underground drains similar in principle to those used for field drainage in agriculture.

Figure 14.25
This tile half-filled with sediment has lost much of its effectiveness. Such clogging can result from poorly protected outlets or inadequate slope of the tile line. In some areas of the western United States iron and manganese compounds may accumulate in the tile even when the system is properly designed. [Courtesy L. B. Grass, USDA Soil Conservation Service]

There must be free movement of the excess water to a tile drain placed alongside and slightly below the foundation or underneath the floor. The water must move rapidly through the tile drains to an outlet ditch or sewer. Care must be taken to ensure that outlet openings are kept free from sediment or debris.

14.17

Benefits of Land Drainage

GRANULATION, HEAVING, AND ROOT ZONE

Draining the land promotes many conditions favorable to higher plants and soil organisms and provides greater stability of building foundations and roadbeds. "Heaving"—the bad effects of alternate expansion and contraction due to freezing and thawing of soil water—is alleviated (see Section 6.7). Such heaving can break foundations and seriously disrupt roadbeds. Also, the heaving of small-grain crops and the disruption of such taprooted plants as alfalfa and sweet clover are especially feared (see Figure 6.11). By quickly lowering the water table at critical times, drainage maintains a sufficiently deep and effective root zone (see Figure 14.26). By this means, the volume of soil from which nutrients can be extracted is maintained at a higher level.

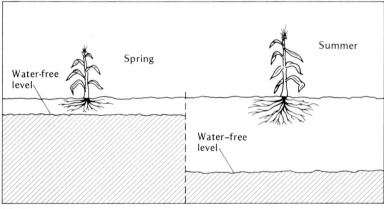

(a) Undrained Land

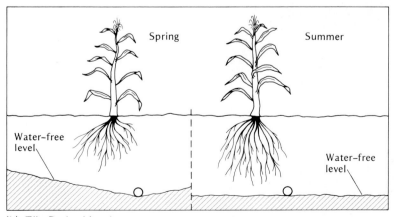

(b) Tile-Drained Land

Figure 14.26
Illustration of water levels of undrained and tile-drained land in the spring and summer. Benefits of the drainage are obvious. [Redrawn from Hughes (1980); used with permission of Deere & Company, Moline, IL]

SOIL TEMPERATURE

The removal of excess water also lowers the specific heat of soil, thus reducing the energy necessary to raise the temperature of the drained layers (see Section 6.9). At the same time, surface evaporation, which has a cooling effect, may be reduced. The two effects together tend to make the warming of the soil easier. The converse of the old saying that in the spring "a wet soil is a cold soil" applies. Good drainage is necessary if the land is to be satisfactorily cultivated and is imperative in the spring if the soil is to warm up rapidly. In the spring a wet soil may be 3–8 °C cooler than a moist one.

AERATION EFFECTS

Perhaps the greatest benefits of drainage, direct and indirect, relate to aeration. Good drainage promotes the ready diffusion of oxygen to, and carbon dioxide from, plant roots. The activity of the aerobic organisms depends on soil aeration, which in turn influences the availability of nutrients, such as nitrogen and sulfur. Likewise, the chance of toxicity from excess iron and manganese in acid soils is decreased if sufficient oxygen is present, because the oxidized states of these elements are quite insoluble. Removal of excess water from soils is at times just as important for plant growth as is the provision of water when soil moisture is low.

TIMELINESS OF FIELD OPERATIONS

A distinct advantage of tile-drain systems is the fact that drained land can be worked earlier in the spring. A well-functioning tile-drain system permits the farmer to start spring tillage operations one to two weeks earlier than where no drainage is installed. Efficient time use in the spring of the year is critical in modern farming.

BENEFITS IN IRRIGATED AREAS

Drainage plays a critical role in maintaining the productivity of irrigated soils of arid regions, especially if the irrigation water contains significant salts. The provision of good drainage systems on such soils permits the removal of excess salts if they are already present in the soil, and prevents their buildup in the upper soil horizons if they are present in the irrigation water. Unfortunately, the essential need for adequate drainage was not recognized when some irrigation systems were first constructed. As a result, over time the level of salts in the soils has increased to the extent that crop productivity has been seriously curtailed. Some areas in Pakistan, the Middle East, and India have been so affected.

This problem of salt buildup is accentuated by the practice of reusing water drained from one irrigated saline soil area to irrigate other fields further down the irrigation system. This water is likely to be high in salts, and may contain sufficient sodium to encourage the formation of a sodic soil, which is physically unsuited for good crop production. A combination of good soil drainage and careful monitoring of the salt content of irrigation water is essential for sustained agricultural production in irrigated arid region soils.

14.18

Conclusion

The behavior and movement of water in soils and plants are governed by one common set of principles. These principles can be used to manage water more effectively as it moves through the soil–plant–atmosphere continuum (SPAC). They are especially critical in the continuing effort to increase the efficiency of water use.

The major challenge to soil and crop scientists, as well as to farmers, is to maintain sufficient soil moisture to satisfy plant needs. This can be achieved by encouraging water to move into and penetrate a soil until the soil's retentive or field capacity is reached and then to remove the excess by controlled runoff or drainage.

Vapor loss of water from soils is through plant absorption followed by transpiration from the leaf surfaces (T) and through evaporation from the soil surface (E). Most soil water management practices are aimed at encouraging plant use (T) while minimizing surface soil evaporation losses (E).

Mulches and conservation tillage systems, which leave crop residues on or near the soil surface, provide the most effective means of reducing E. Increasing crop coverage to intercept solar energy and to reduce its impact on the soil surface evaporation is another significant option for lowering E. Crop management to permit fallow periods is also critical in some semiarid areas.

Study Questions

1. What is SPAC and why is it important?
2. Why is evapotranspiration (ET) so much higher in an irrigated area in an arid section of Colorado than in a humid section of North Carolina?
3. Which of the two major components of ET (E or T) is more subject to control by soil and crop management practices? Explain why this is the case.
4. What steps would you take to increase the amount of water that enters and penetrates a soil?
5. What is summer fallow, where is it practiced, and of what benefit is it?
6. How can conservation tillage practices help improve water-use efficiency?
7. Under what conditions would you expect to see plastic mulches used? Which will they most influence, E or T in ET?
8. What is trickle irrigation and what are the conditions under which it is used?
9. Contrast the disposition of the water added to soils annually in the following temperate zone regions: (a) humid (b) semiarid nonirrigated, and (c) semiarid irrigated.
10. What are the benefits of land drainage? Can you think of any disadvantages?
11. Over 90% of the new underground drainage systems being installed today use perforated plastic pipe. Why is this the case?
12. Why is the loss of nutrients by leaching more important today than it was 25 years ago?
13. Explain how a tile-drainage system works.

References

ABOUKHALED, A., A. ALFARO, and M. SMITH. 1982. *Lysimeters*, FAO Irrigation and Drainage Paper 39 (Rome, Italy: UN Food and Agriculture Organization).

BEAUCHAMP, K. H. 1955. "Tile drainage—its installation and upkeep," *The Year-book of Agriculture* (*Water*) (Washington, DC:U.S. Department of Agriculture), p. 513.

BERNSTEIN, L., et al. 1955. "The interaction of salinity and planting practice on the germination of irrigated row crops," *Soil Sci Soc. Amer. Proc.*, **19**:240–43.

BIERHUIZEN, J. F., and R. O. SLATYER, 1965. "Effect of atmospheric concentration of water vapor and CO_2 in determining transpiration-photosynthesis relationships of cotton leaves," *Agric. Meteor.*, **2**:259.

BLACK, A. L. 1982, "Long-term N-P fertilizer and climate influences on morphology and yield components of spring wheat," *Agron. J.*, **74**:651–57.

BUCKMAN, H. O., and N. C. BRADY. 1969. *The Nature and Properties of Soils*, 7th ed. (New York: Macmillan).

DREIBELBIS, F. R., and C. R. AMERMAN. 1965. "How much topsoil moisture is available to your crops?" *Crops and Soils*, **17**:8–9.

FOUSS, J. L. 1974. "Drain tube materials and installation," in J. Van Schilfgaarde (Ed.), *Drainage for Agriculture*, Agronomy Series No. 17 (Madison, WI: Amer. Soc. Agron.), pp. 147–77.

GREB, B. W. 1983. "Water conservation: central Great Plains," in H. E. Dregue and W. O. Willis (Eds.), *Dryland Agriculture*, Agronomy Series No. 23 (Madison, WI: Amer. Soc. of Agron).

HANKS, R. J. 1983. "Yield and water-use relationships: an overview," in H. M. TAYLOR et al. (Eds.), *Limitations to Efficient Water Use in Crop Production* (Madison, WI: Amer. Soc. Agron.).

HAYNES, J. L. 1954. "Ground rainfall under vegetative canopy of crops," *J. Amer. Soc. Agron.*, **46**:67–94.

HILLEL, D. 1980. *Applications of Soil Physics* (New York: Academic Press).

HUGHES, H. S. 1980. *Conservation Farming* (Moline, IL: John Deere and Company).

KELLY, O. J. 1954. "Requirement and availability of soil water," *Adv. Agron.*, **6**:67–94.

KITTRIDGE, J., 1948. *Forest Influences. The Effects of Woody Vegetation on Climate, Water and Soil* (New York: McGraw-Hill).

KOZLOWSKI, T. T. (Ed.). 1968–72. *Water Deficits and Plant Growth*, 3 vols. (New York: Academic Press).

KRAMER, P. J. 1983. *Water Relations of Plants* (New York: Academic Press).

LYON, T. L., H. O. BUCKMAN, and N. C. Brady. 1952. *The Nature and Properties of Soils*, 5th ed. (New York: Macmillan).

NELSON, L .B., and R. E. Uhland. 1955. "Factors that influence loss of fall-applied fertilizers and their probable importance in different sections of the United States," *Soil Sci. Soc. Amer. Proc.*, **19**:492–96.

NIELSEN, G. H., and D. S. STEVENSON. 1983. "Leaching of calcium, magnesium and potassium in irrigated orchard lysimeters," *Soil Sci. Soc. Amer. J.*, **47**:692–96.

PHILIP, J. R. 1966. "Plant water relations: some physical aspects," *Ann. Rev. Plant Physiol.*, **17**:245–68.

RITCHIE, J. T. 1983. "Efficient water use in crop production: Discussion on the generality of relations between biomass production and evapotranspiration," in H. M. Taylor et al. (Eds.), *Limitations to Efficient Water Use in Crop Production* (Madison, WI: Amer. Soc. Agron., Crop Sci. Soc. Amer., Soil Sci. Soc. Amer.).

Science of Food and Agriculture. 1983. "Current in CAST" Science of Food and Agriculture 1:1.

TALYOR, H. M., W. R. JORDAN, and T. R. SINCLAIR (Eds.). 1983. *Limitations to Efficient Water Use in Crop Production* (Madison, WI: Amer. Soc. Agron., Crop Sci. Soc. Amer., Soil Sci. Soc. Amer.).

UNDP/WMO. 1974. *Hydrometeorological Survey of the Catchments of Lakes Victoria, Kyoja and Albert*, Project RAF 66/025 (4 vols), pp. 498–509.

USDA 1974. *Summer Fallow in the Western United States*. Cons. Res. Rept. No. 17. (Washington, DC: USDA Agricultural Research Service).

VAN SCHILFGAARDE, J. (Ed.). 1974. *Drainage for Agriculture*. Agronomy Series No. 17 (Madison WI: Amer. Soc. Agron.).

Soil Erosion and Its Control

No soil phenomenon is more destructive worldwide than soil erosion. It involves losing not only water and plant nutrients but ultimately the soil itself. Furthermore, the soil that is removed finds its way into streams, rivers, and lakes and pollutes those resources and bodies. Erosion is a serious problem in all climates because wind as well as water can remove soil. Water erosion will be considered first.

15.1

Significance of Runoff and Soil Erosion[1]

EFFECTS OF RUNOFF

As discussed in Chapter 14, a primary principle of soil water management is to encourage water movement into rather than off the soil. If the water is allowed to penetrate, the soil can serve as a "reservoir" for future plant uptake. Surface runoff reduces or prevents penetration. Water that could be retained in the plant root zone for subsequent absorption is lost.

Runoff levels vary greatly from region to region and from soil to soil. In some humid regions, losses as high as 50–60% of the annual precipitation have occurred. While annual runoff losses are much lower in semiarid and arid regions, high rates of loss are not unusual during heavy storms, which are common in these regions. In any case, runoff losses seriously deter sustainable agricultural production, and attempts to prevent them must receive high priority.

SOIL EROSION

The loss of soil from both agricultural and nonagricultural lands is a serious problem throughout the world. In the United States about 5 billion metric

[1] For an interesting review of this subject, see Batie (1983) and Follett and Stewart (1985).

TABLE 15.1
Annual Sediment Loads of Eight Major Rivers from Other Continents Compared to That of the Mississippi River.

River	Countries	Annual sediment load (million Mg)	Erosion (Mg/ha drained)
Yellow	China	1600	479
Ganges	India, Nepal	1455	270
Amazon	Brazil, Peru, etc.	363	13
Irrawaddy	Burma	299	139
Kosi	India, Nepal	172	555
Mekong	Vietnam, Thailand, etc.	170	43
Red	China, Vietnam	130	217
Nile	Sudan, Egypt, etc.	111	8
Mississippi	United States	300	93

Data from different sources compiled by El-Swaify and Dangler (1982).

tons (Mg) of soil are moved annually by soil erosion, some two thirds being moved by water and one third by wind. More than half of the water erosion and about 60% of the wind erosion are on croplands that produce most of the country's food. The load of sediment carried by some of the world's major rivers to the ocean is enormous, as indicated by the data in Table 15.1. These data suggest that soil erosion problems are even more severe in some other areas of the world.

While soil erosion has serious implications for agriculture, its total cost to society may be even higher in nonagricultural areas. For example, the lifetime of water storage reservoirs is shortened dramatically by inputs of sediment from eroded lands. River channels are filled in and treatment plants for domestic water supplies are damaged by silt from the uplands. The annual costs of these off-farm damages in 1980 were estimated by the Conservation Foundation to be between $3.2–13 billion compared to about $2.2 billion for cropland damages. Soil erosion is most certainly a concern for all of society.

Soil erosion in the United States accelerated when the first European settlers chopped down the trees and began to farm the sloping to hilly lands of the humid eastern part of the country. General public recognition of this problem came only in 1930, however, when H. H. Bennett and associates recognized the damage being done and obtained federal support for erosion control efforts.

Although Bennett may have overemphasized the hazards of soil erosion, subsequent surveys have confirmed its seriousness. A 1967 national survey by the U.S. Department of Agriculture showed soil erosion to be a prominent problem on about 90 million ha (200 million acres) of croplands, about half the total of the cropland area. A second National Resources Inventory (NRI) made 10 years later identified areas of greatest soil loss by erosion (see Figure 15.1).

The 1982 NRI showed an average annual water and wind erosion loss from croplands of 16.4 Mg/ha (7.3 tons/acre). This study also showed that 44% of United States cropland has annual erosion losses for water or wind erosion in excess of 11 Mg/ha (5 tons/acre) per year, the maximum level that can be tolerated if soil productivity is to be maintained (see Section 15.15). About 23% of cropland lose 11–22 Mg/ha (5–10 tons/acre) per year. Furthermore, erosion losses from other than agricultural sources are very high. Some 60% of the sediment reaching streams in the United States is from stream roadbanks, pastures, and forests and from intensive land use for other than agriculture.

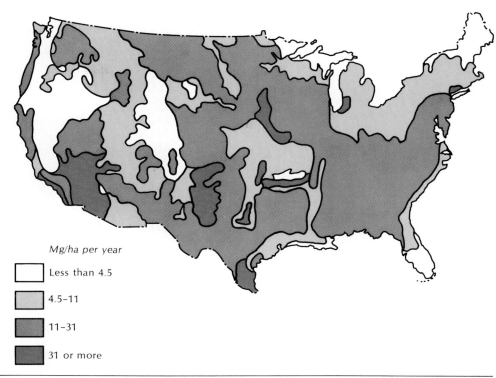

Figure 15.1
Extent of water (sheet and rill) and wind erosion on cropland in continental United States in 1983. Gully erosion is not included. For the average soil, erosion rates greater than 11 Mg/ha (5 tons/acre) per year will result in soil productivity decline. [From USDA (1984)]

Mg/ha per year

☐ Less than 4.5

▨ 4.5–11

▨ 11–31

■ 31 or more

In addition to the sediment carried in streams, another 1.5 billion Mg of sediment is deposited each year in the nation's reservoirs. Agriculture is affected directly because some of these reservoirs hold water for irrigation, but the reservoirs also provide water for domestic and industrial purposes. To the agricultural costs must be added those associated with the presence of clay and other chemicals in the runoff water and with the silting up of the reservoirs.

The effects of erosion are no less serious in countries other than the United States—China, USSR, and India to name but a few (Table 15.2). Nor is the problem a modern phenomenon. In ancient times, erosion evidently was a menace in Greece, Italy, Syria, North Africa, and Persia. And the fall of empires such as Rome was accelerated by the decrease in food production resulting from the washing away of fertile surface soils.

NUTRIENT LOSSES

The quantity of essential nutrients lost from the soil by erosion is quite high, as shown in Table 15.3. The total quantity of nitrogen, phosphorus, and

TABLE 15.2
Estimated Excessive World Soil Erosion Loss

Country	Total cropland (million ha)	Excessive soil loss (million mg)
United States	167	1,524
Soviet Union	251	2,268
India	140	4,716
China	99	3,628
Other	607	11,201
Total	1265	23,337

From Brown and Wolf (1984).

TABLE 15.3
Estimated Losses of Total and Available Nitrogen, Phosphorus, and Potassium in Eroded Sediments.

In thousands of metric tons.

Region	Nitrogen		Phosphorus		Potassium	
	Total	Available	Total	Available	Total	Available
Pacific	100	18	29	0.6	1,154	23
Mountain	176	32	64	1.3	2,550	51
Southern Plains	512	94	101	2.0	3,043	61
Northern Plains	2,068	380	293	5.9	11,711	234
Lake states	622	114	107	2.1	3,643	73
Corn Belt	4,360	802	624	12.5	24,959	499
Delta states	478	88	141	2.8	4,220	84
Southeastern states	202	37	101	2.0	1,007	20
Appalachian states	676	124	169	3.4	3,381	67
Northeastern states	300	55	75	1.5	2,252	45
Total	9,494	1,744	1,704	34.1	57,920	1,158

Data computed by Larson et al. (1983).

potassium removed by erosion from soils nationwide is estimated at more than 38 million Mg. Such losses can be counterbalanced only in part by adding fertilizers. Much of the nitrogen and phosphorus lost is in the soil organic matter, finer particles of which are included in the eroded sediments. Experiments have shown organic matter and nitrogen in the eroded material to be five times as high as in the original topsoil. Comparable figures for phosphorus and potassium are three and two, respectively.

Total potassium losses shown in Table 15.3 are higher than those for nitrogen or phosphorus because this element is commonly present in larger total quantities in soil solids. But the amount of available nitrogen lost through erosion is greater than for the other two elements. This may be because of increasing nitrogen fertilizer applications to the soil surface where the nitrogen may be more subject to loss in runoff and erosion.

15.2

Accelerated Erosion—Mechanics

Water erosion is one of the most common geologic phenomena. It accounts in large part for the leveling of mountains and the development of plains, plateaus, valleys, river flats, and deltas. The vast deposits that now appear as sedimentary rocks originated in this way. This normal erosion amounts annually to about 0.2–0.5 Mg/ha (0.1–0.2 ton/acre). It operates slowly, yet inexorably. Erosion that exceeds this normal rate becomes unusually destructive and is referred to as *accelerated erosion*. This is usually caused by water action and is of special concern to agriculture.

Two steps are recognized in accelerated erosion—the *detachment* or loosening influence, which is a preparatory action, and *transportation* by floating, rolling, dragging, and splashing. Freezing and thawing, flowing water, and rain impact are the major detaching agencies. Raindrop splash and especially running water facilitate carrying away of loosened soil. In channels cut into the soil surface, most of the loosening and cutting is due to

Figure 15.2
A raindrop (left) and the splash that results when the drop strikes a wet bare soil (right). Such rainfall impact not only tends to destroy soil granulation and encourage sheet and rill erosion but also effects considerable transportation by splashing. A ground cover, such as sod, will largely prevent this type of erosion. [Courtesy USDA Soil Conservation Service]

water flow; on comparatively smooth soil surfaces, the beating of raindrops causes most of the detachment.

INFLUENCE OF RAINDROPS

Raindrop impact has three important effects: (a) it detaches soil; (b) its beating tends to destroy granulation; and (c) its splash, under certain conditions, causes an appreciable transportation of soil (Figure 15.2). So great is the force exerted by raindrops that soil granules are not only loosened and detached but may even be beaten to pieces. Under such hammering, surface soil aggregates are broken up. The dispersed material may develop into a hard crust on drying that will prevent the emergence of seedlings of large seeded crops such as beans and will encourage runoff from subsequent precipitation.

TRANSPORTATION OF SOIL—SPLASH EFFECTS

In soil translocation, runoff water plays the major role. So familiar is the power of water to cut and carry that little more need be said regarding these capacities. In fact, so much publicity has been given to runoff that the public generally ascribes to it all of the damage done by heavy rainfall.

Under certain conditions, however, splash transportation is of considerable importance (see Figure 15.2). On a soil subject to easy detachment, a very heavy rain may splash as much as 225 Mg/ha of soil, some of the drops rising as high as 0.7 m and moving horizontally perhaps 1–2 m. On a slope or if the wind is blowing, splashing greatly aids and enhances runoff translocations of soil; the two together account for the total soil movement that finally occurs.

15.3

Types of Water Erosion

Three types of water erosion are generally recognized: *sheet, rill,* and *gully* (Figure 15.3). In *sheet erosion,* soil is removed more or less uniformly from every part of the slope. However, sheet erosion is often accompanied by tiny channels (rills) irregularly dispersed, especially on bare land newly planted or in fallow. This is *rill erosion.* The rills can be obliterated by tillage, but the damage is already done—the soil is lost.

Where the volume of runoff water is further concentrated, the formation of larger channels or gullies occurs by undermining and downward cutting. This is called *gully erosion.* The gullies are obstacles to tillage and cannot be removed by ordinary tillage practices. While all types may be serious, the losses due to sheet and rill erosion, although less noticeable, are responsible for most of the field soil deterioration. The cropland area in the United States classed by its erodibility by sheet and rill erosion is shown in Figure 15.4.

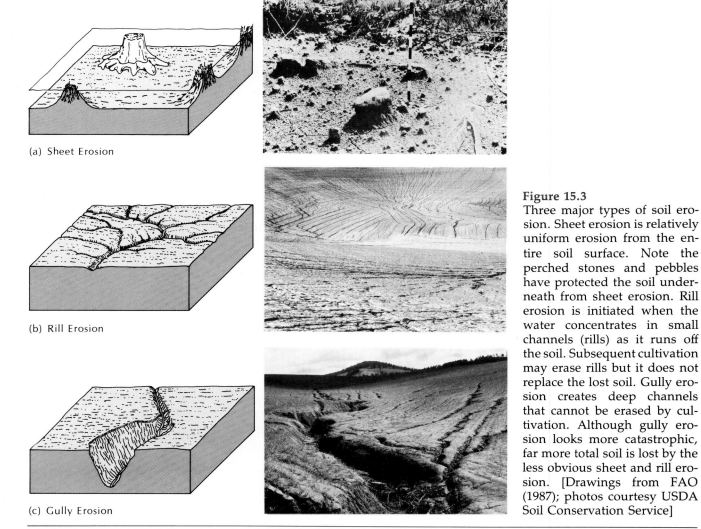

(a) Sheet Erosion

(b) Rill Erosion

(c) Gully Erosion

Figure 15.3
Three major types of soil erosion. Sheet erosion is relatively uniform erosion from the entire soil surface. Note the perched stones and pebbles have protected the soil underneath from sheet erosion. Rill erosion is initiated when the water concentrates in small channels (rills) as it runs off the soil. Subsequent cultivation may erase rills but it does not replace the lost soil. Gully erosion creates deep channels that cannot be erased by cultivation. Although gully erosion looks more catastrophic, far more total soil is lost by the less obvious sheet and rill erosion. [Drawings from FAO (1987); photos courtesy USDA Soil Conservation Service]

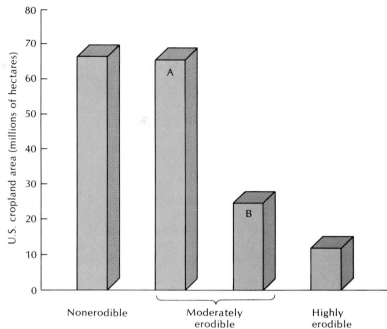

Figure 15.4
Hectares of United States cropland in 1982 classed by its erodibility by water. The nonerodible land will lose less than 11 Mg/ha (5 tons/acre) per year (the allowable limit) regardless of treatment. The moderately erodible land A will lose less than 11 Mg/ha per year if appropriately managed while land B probably will lose somewhat more than 11 Mg/ha per year even if well managed. Erosion from the highly erodible land is very high, and permanent vegetation should be established on this land. [Data from Heimlich and Bills (1986)]

15.4

Factors Affecting Accelerated Erosion—Universal Soil-Loss Equation

Decades of agricultural research coupled with centuries of farmers' experience have clearly identified the major factors affecting accelerated erosion. Thes factors are included in the universal soil-loss equation (USLE).[2]

$$A = RKLSCP$$

A, the predicted soil loss, is the product of

R = climatic erosivity (rainfall and runoff)

K = soil erodibility

L = slope length

S = slope gradient or steepness

C = cover and management

P = erosion control practice

Working together, these factors determine how much water enters the soil, how much runs off, and the manner and rate of its removal. A brief description of each factor and the extent of its influence on soil erosion indicate clearly how to control soil erosion.

[2] For an excellent discussion of this topic, see Wischmeier and Smith (1978). In this historic publication the factors used in the soil-loss equation are calculated in English units rather than SI units. Since most subsequent publications also use the same English-system-based factors, the practice will be continued in this text. The soil loss (A) calculated by the USLE is expressed in tons per acre, which can easily be converted to SI units by multiplying by 2.24.

15.5

Rainfall and Runoff Factor

The rainfall and runoff factor, R, measures the erosive force of rainfall and runoff. Of the two phases of rainfall, amount of *total rainfall* and its *intensity*, the latter is usually the more important. A high annual precipitation received in a number of gentle rains may cause little erosion, whereas a lower yearly rainfall descending in a few torrential downpours may result in severe damage. This latter type of rainfall accounts for the marked erosion often recorded in semiarid and even arid regions.

The *seasonal distribution* of the rainfall is also critical in determining soil erosion losses. For example, in the northern part of the United States, precipitation that runs off the land in the early spring when the soils are still frozen may bring about little erosion. The same amount of runoff a few months later, however, often carries considerable quantities of soil with it. In any climate, heavy precipitation occurring at a time of year when the soil is bare is likely to cause soil loss. The soil is particularly vulnerable when seedbeds are prepared with traditional tillage and after the harvesting of such crops as corn, cotton, beans, sugar beets, and potatoes.

The R factor, sometimes called the rainfall erosion index, takes into account the erosive effects of storms. The total kinetic energy of each storm (related to intensity and total rainfall) plus the average rainfall during the 30-minute period of greatest intensity is considered. The sum of the indexes for all storms occurring during a year provides an annual index. An average of such indexes for several years is used in the universal soil-loss equation.

Rainfall indexes computed for the United States are shown in Figure 15.5. Note that they vary from less than 20 in areas of the west to more than 550 along the coasts of the gulf states.

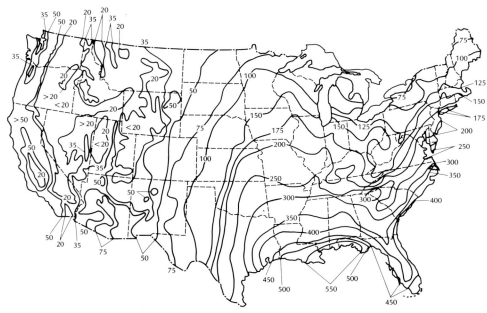

Figure 15.5
Average annual values of the rainfall erosion index in the United States. Note the high values in parts of the East and the generally low values in the West. [From Wischmeier and Smith (1978)]

15.6

Soil Erodibility Factor

The soil erodibility factor, K, indicates the inherent erodibility of a soil. K gives an indication of the soil loss from a unit plot 22 m (72 ft) long with a 9% slope and continuous fallow culture. The two most significant and closely related soil characteristics affecting erosion are (a) infiltration capacity and (b) structural stability. The infiltration capacity is influenced greatly by structural stability, especially in the upper soil horizons. In addition, organic matter content, soil texture, the kind and amount of swelling clays, soil depth, tendency to crust, and the presence of impervious soil layers all influence the infiltration capacity.

The stability of soil aggregates affects the extent of erosion damage in another way. Resistance of surface granules to the beating action of rain saves soil even though runoff does occur. The marked granule stability of certain tropical clay soils high in hydrous oxides of iron and aluminum accounts for the resistance of these soils to the action of torrential rains. Downpours of a similar magnitude on temperate region clays would be disastrous.

The soil erodibility, or K factor, normally varies from near zero to about 0.6. K is low for soils into which water readily infiltrates, such as well-drained sandy soils, friable tropical clays high in hydrous oxides of iron and aluminum, or kaolinite. Erodibility indexes of less than 0.2 are normal for these readily infiltrated soils (Table 15.4). Soils with intermediate infiltration capacities and moderate soil structural stability generally have a K factor of 0.2–0.3, and the more easily eroded soils with low infiltration capacities will have a K factor of 0.3 or higher.

TABLE 15.4
Computed K Values for Soils at Different Locations

Soil	Location	Computed K
Udalf (Dunkirk silt loam)	Geneva, NY	0.69
Udalt (Keene silt loam)	Zanesville, OH	0.48
Udult (Lodi loam)	Blacksburg, VA	0.39
Udult (Cecil sandy clay loam)	Watkinsville, GA	0.36
Udoll (Marshall silt loam)	Clarinda, IA	0.33
Udalf (Hagerstown silty clay loam)	State College, PA	0.31
Ustoll (Austin silt)	Temple, TX	0.29
Aqualf (Mexico silt loam)	McCredie, MO	0.28
Udult (Cecil sandy loam)	Clemson, SC	0.28
Udult (Cecil sandy loam)	Watkinsville, GA	0.23
Udult (Tifton loamy sand)	Tifton, GA	0.10
Ochrept (Bath flaggy silt loam)	Arnot, NY	0.05
Alfisols	Indonesia	0.14
Ultisols	Hawaii	0.09
Alfisols	Nigeria	0.06
Oxisols	Brazil	0.02
Andisols	Nigeria	0.02
Inceptisols	Puerto Rico	0.02

From Wischmeier and Smith (1978); data for tropical soils cited by Lal (1984).

TABLE 15.5

The Topographic Factor (*LS*) for Selected Combinations of Slope Length and Steepness

Note that the factor increases with both percent slope and length of slope.

Slope (%)	Approximate slope length (m)				
	15	30	45	60	90
2	0.16	0.20	0.23	0.25	0.28
4	0.30	0.40	0.47	0.53	0.62
6	0.48	0.67	0.82	0.95	1.17
8	0.70	0.99	1.21	1.41	1.72
10	0.97	1.37	1.68	1.94	2.37
12	1.28	1.80	2.21	2.55	3.13

From Wischmeier and Smith (1978).

15.7

Topographic Factor

The topographic factor, *LS*, reflects the influence of length and steepness of slope. Table 15.5 gives *LS* figures for selected slope characteristics. *LS* is the ratio of soil loss from the field in question to that of a unit plot with 9% slope, 22 m (72 ft) long and continuously fallowed. The greater the steepness of *slope*, other conditions being equal, the greater the erosion, partly because more water is likely to run off but also because of increased velocity of water flow. Theoretically, a doubling of the flow velocity enables water to move particles 64 times larger, allows it to carry 32 times more material in suspension, and makes the erosive power in total four times greater.

The *length* of the slope is important, because the greater the extension of the inclined area, the greater the concentration of the flooding water. For example, research in southwestern Iowa showed that doubling the length of a 9% slope increased the loss of soil by 2.6 times and the runoff water by 1.8 times. This influence of slope is, of course, greatly modified by the size and general *topography* of the drainage area.

15.8

Cover and Management Factor

The cover and management factor, *C*, indicates the influence of cropping systems and management variables on soil loss. *C* is the factor over which the farmer has the most control. Forests and grass provide the best natural protection known for soil and are about equal in their effectiveness, but forage crops, both legumes and grasses, are next in protective ability because of their relatively dense cover (Table 15.6). Small grains such as wheat and oats are intermediate and offer considerable obstruction to surface wash. Row crops such as corn, soybeans, and potatoes offer relatively little cover during the early growth stages and thereby encourage erosion. Most subject to erosion are fallowed areas where no crop is grown and all the residues have been incorporated into the soil. The effect on erosion of leaving wheat stubble as surface mulch is illustrated in Figure 15.6.

The marked differences among crops in their ability to maintain soil cover emphasize the value of appropriate crop rotation to reduce soil erosion.

TABLE 15.6

Area Under Different Land Use in Eastern United States and the Percent with Water Erosion Greater Than 11 Mg/ha per Year[a]

Note the high rates with row crops and especially soybeans

Land use	Land area (1000 ha)	Land with water erosion greater than 11 Mg/ha per year (%)
Cropland		
All	167,288	23.5
Corn	37,832	33.6
Soybean	24,020	44.3
Cotton	6,713	33.6
Wheat	28,995	12.9
Pasture	53,846	10.0
Rangeland	165,182	11.0
Forest		
Grazed	24,696	15.0
Ungrazed	124,696	2.0

From USDA (1980a).

[a] 11 Mg/ha (5 tons/acre) is the level above which the maintenance of reasonable soil productivity is very difficult.

The inclusion of a close-growing forage crop in rotation with row crops will help control both erosion and runoff. Likewise, the use of so-called "conservation tillage" systems, which leave most of the crop residues on the surface, greatly decreases erosion hazards.

The *C* value for a specific location depends on a number of factors including the crop or crops being grown, crop stage, tillage, and other management factors.

Technically, the *C* value is the ratio of soil loss under the conditions found in the field in question to that which would occur under clean tilled, continuous fallow conditions. This ratio (C) will be high (approaching 1.0) with little soil cover, such as bare soil in the spring before a crop canopy develops.

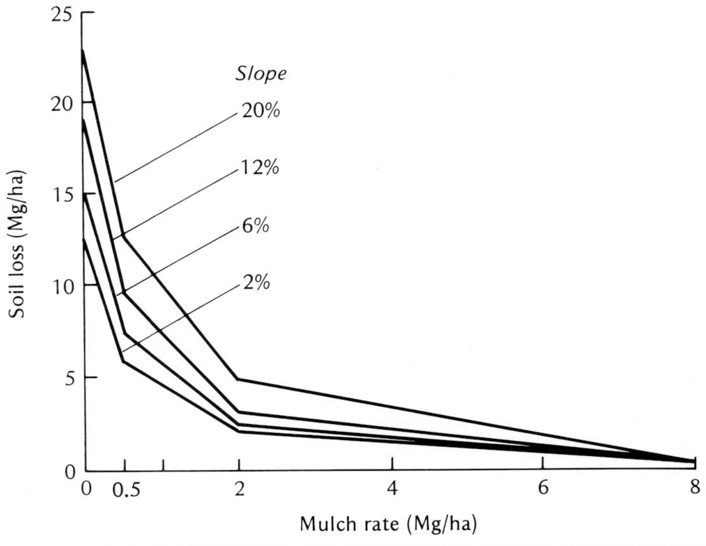

Figure 15.6

The effect of wheat straw mulch rate and soil slope on soil loss by erosion. Note the striking reduction in erosion with increased coverage of the soil surface (mulching) even on the steep slopes. A prime objective of conservation tillage is to keep the cover on the soil. [From Lattanzi et al. (1979); used with permission of the Soil Science Society of America]

TABLE 15.7

Crop Management or *C* Values for Different Crop Sequences in Northern Illinois

Note the dominating effect of tillage systems and of the maintenance of soil cover. Values would differ slightly in other areas but the principles illustrated would pertain.

Crop sequence[a]	Conventional tillage[b]		Minimum tillage Residue level (kg)		No tillage Residue level (kg)	
	Residue left	Residue removed	458–907	908–1816	458–907	908–1816
Continuous soybeans (Sb)	0.49	—	0.33	—	0.29	—
Continuous corn (C)	0.37	0.47	0.31	0.07	0.29	0.06
C–Sb	0.43	0.49	0.32	0.12	0.29	0.06
C–C–Sb	0.40	0.47	0.31	0.12	0.29	0.06
C–C–Sb–G–M	0.20	0.24	0.18	0.09	0.14	0.05
C–Sb–G–M	0.16	0.18	0.15	0.09	0.11	0.05
C–C–C–M	0.12	0.16	0.13	0.08	0.09	0.04

Selected data from Walker (1980).

[a] Crop abbreviations: C = corn; Sb = soybeans; G = small grain (wheat or oats); M = meadow.

[b] Spring plowed; assumes high crop yields.

It will be low (e.g., <0.10) where large amounts of crop residues are on the land or in areas of dense forests.

C values are usually computed by experienced scientists with knowledge of the effects of crop cover and management practices on soil erosion in a given area. Actual values are available through the state offices of the USDA Soil Conservation Service. Examples of selected *C* values are given in Table 15.7. Note the significant influence of tillage systems coupled with the soil cover that they provide. This influence is noted in Figure 15.7, which shows the expected increase in land eroding more than 11 Mg/ha per year as the *C* value is increased.

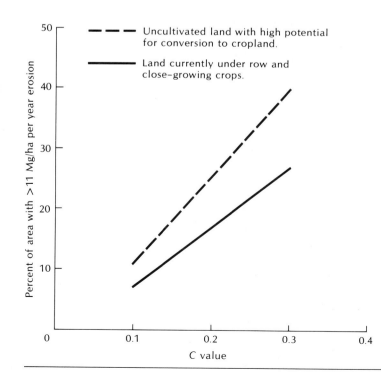

Figure 15.7

Relationship between the *C* value and the percentage of land in the United States that is expected to erode at a rate in excess of 11 Mg/ha per year. Note the percentage goes up as the *C* factor increases. [Redrawn from Pierce et al. (1986)]

Grassed waterway

Figure 15.8
Aerial photograph of fields in Kentucky where strip cropping is being practiced. [Courtesy USDA Soil Conservation Service]

15.9

Support Practice Factor

The support practice factor, P, reflects the benefits of contouring, strip cropping, and other supporting factors. P is the ratio of soil loss with a given support practice to the corresponding loss when crop culture is up and down the slope. If there are no support practices the P factor is 1.0. On sloping fields, the protection offered by surface coverage and crop management must be supported by other practices that help slow the runoff water and, in turn, reduce soil erosion. These support practices (factor P) include tillage on the contour, contour strip cropping, terrace systems, and grassed waterways, all of which will tend to reduce the P factor.

Except on fields with only modest slopes, it is important that cultivation be across the slope rather than with it. This is called *contour tillage*. On long slopes subject to sheet and rill erosion, the fields may be laid out in narrow strips across the incline, alternating the tilled crops, such as corn and potatoes, with hay and grain. Water cannot achieve an undue velocity on the narrow strips of cultivated land, and the hay and grain crops check the rate of runoff. Such a layout is called *strip cropping* and is the basis for much of the erosion control now advocated (see Figure 15.8).

When the cross strips are laid out rather definitely on the contours, the system is called *contour strip cropping*. The width of the strips will depend primarily on the degree of slope, the permeability of the land, and its erodibility. Widths of 30–125 m (100–400 ft) are common. Contour strip cropping often is guarded by diversion ditches and waterways between fields. When their grade is high, these waterways should be sown with grass (grassed waterways) to prevent underwashing.

There are a variety of terraces in use around the world today (see Figure 15.9). Bench terraces are used where rather complete control of the water runoff must be achieved, such as in rice paddies. Broad-base terraces are more common where large-scale farm machinery is used and where the entire land surface is to be cropped. The terrace channel has a grade of from

443

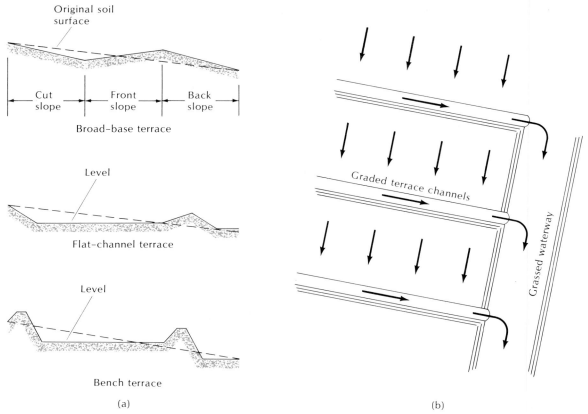

Figure 15.9

(a) Three types of terrace in use around the world. Broad-base terraces permit the entire surface to be cropped and are widely used in the United States. Flat-channel terraces allow large volumes of water to move off the soil without erosion. Bench terraces can keep water on the land and are used widely in rice production. (b) Diagram of controlled flow of runoff water from a field to terrace channels and on to a grassed waterway from which it can move to a canal or stream channel.

50 to 65 cm per 100 m (6–8 in./100 ft) to permit gentle but controlled runoff. The runoff water is commonly carried in the terrace channel to a broader based grassed waterway through which the water moves to a nearby canal, stream, or river.

Since some terraces are low and broad, they may be cropped without difficulty and offer no obstacle to cultivating and harvesting machinery. Such terraces waste little or no land and are quite effective if properly maintained. Where necessary, waterways can be sodded. Terracing of this kind is really a more or less elaborate type of contour strip cropping.

Examples of P values for contour tillage and strip cropping at different slope gradients are shown in Table 15.8. Note that P values increase with slope and that they are low for strip cropping, illustrating the importance to erosion control of this practice. Note that terracing also reduces the P values.

15.10

Calculating Expected Soil Losses

Estimates of the expected soil loss in any location can be calculated using the universal soil-loss equation (USLE). Assume for example a location in central

TABLE 15.8

P Factors for Contour and Strip Cropping at Different Slopes and the Terrace Subfactor at Different Terrace Intervals

The product of the contour or strip cropping factors and the terrace subfactor gives the P value for terraced fields.

Slope (%)	Contour P factor	Strip cropping P factor	Terrace interval (m)	Terrace subfactor Closed outlets	Terrace subfactor Open outlets
1–2	0.60	0.30	33	0.5	0.7
3–8	0.50	0.25	33–34	0.6	0.8
9–12	0.60	0.30	43–54	0.7	0.8
13–16	0.70	0.35	55–68	0.8	0.9
17–20	0.80	0.40	69–60	0.9	0.9
21–25	0.90	0.45	90	1.0	1.0

Contour and strip cropping factors from Wischmeier and Smith (1978); terrace subfactor from Foster and Highfill (1983).

Iowa on a Marshall silt loam with an average slope of 4% and an average slope length of 30 m (100 ft). Assume further that the land is clean tilled and fallowed.

Figure 15.5 shows that the R factor is about 150 for this location. The K factor for a Marshall silt loam in central Iowa is 0.33 (Table 15.4), and the topographic factor (LS) from Table 15.5 is 0.400. The C factor is 1.0, since there is no cover or other management practice to discourage erosion. If we assume the tillage is up and down the hill, the P value is also 1.0. Thus, the anticipated soil loss can be calculated by the USLE ($A = RKLSCP$)

$$A = (150)(0.33)(0.400)(1.0)(1.0) = 19.8 \text{ tons/acre or } 44.4 \text{ Mg/ha}$$

If the crop rotation involved wheat-hay-corn-corn and conservation tillage practices (e.g., minimum tillage) were used, a reasonable amount of residue would be left on the soil surface. Under these conditions, the C factor can be reduced to about 0.2. Likewise, if the tillage and planting were done on the contour, the P value would drop to about 0.5 (Table 15.8). P could be reduced further to 0.4 (0.5 × 0.8) if terraces with outlets were installed about 40 m apart. With these figures the soil loss becomes

$$A = (150)(0.33)(0.400)(0.2)(0.4) = 1.58 \text{ tons/acre or } 3.5 \text{ Mg/ha}$$

The benefits of good cover and management and support practices are obvious. The figures cited were chosen to provide an example of the utility of the universal soil-loss equation, but calculations can be made for any specific location. In the United States, pertinent factor values that can be used for erosion prediction in specific locations are generally available from state offices of the USDA Soil Conservation Service.

15.11

Sheet and Rill Erosion Control

From the preceding sections it is obvious that three objectives must be attained if sheet and rill erosion is to be held to a reasonable level. First, we must encourage as much of the precipitation as possible to run *into* the soil rather than *off* it. Second, the *sediment burden* (soil solids) of the runoff water should be kept low. And third, we must direct the concentration of the runoff water into channels, thereby minimizing its cutting action and the sediment burden it can carry.

The key to control of sheet and rill erosion is found in factors included in the universal soil-loss equation. The control measures needed will be determined by the characteristic of the rainfall, by fundamental properties of the soil, and by the slope gradient and length. Care must be exercised to keep vegetative cover on or near the soil surface. Close-growing crops such as grasses and small grains provide good cover, especially if adequate nutrients have been provided from fertilizer, lime, and manure. Furthermore, mechanical support practices such as *contour tillage* and *strip cropping* should continue to receive high priority on sloping fields. But perhaps the most significant and relatively inexpensive means of reducing sheet and rill erosion is the use of so-called "conservation tillage" practices. These will now be considered.

15.12

Conservation Tillage Practices [3]

Until fairly recently, conventional agricultural practice encouraged rather extensive soil tillage. Most of the crop residues were incorporated into the soil using the moldboard plow. The soil surface was then tilled further using a disc harrow to provide a seedbed devoid of clods. Once row crops were planted, a cultivator was used, often several times, to maintain weed control. Thus, the soil was tilled repeatedly at great cost in terms of time and energy. More important, however, the soil was usually left bare immediately after plowing until later in the year when crop growth was sufficient to provide some ground cover. This means that the soil was left unprotected during the spring of the year when runoff and erosion pressures were greatest.

Three developments of the past two decades have dictated drastic changes in tillage practices. First, herbicides capable of controlling most of the major weeds have become available at reasonable costs. This has reduced the need for cultivating and even for plowing in some cases. Second, dramatic increases in fuel costs forced tractor-dependent farmers to seek means of reducing their tillage operations. Third, public environmental concerns have forced a reevaluation of soil erosion as a source of water pollution. These three developments have stimulated farmers to adopt reduced-tillage methods, which allow less erosion than the conventional tillage systems. Reduced tillage practices are collectively known as *conservation tillage*, in contrast to the *conventional tillage* system described above.

CONSERVATION TILLAGE SYSTEMS

In Table 15.9 some of the modern conservation tillage systems are listed along with the specific operations involved. In studying them, keep in mind that conventional tillage involves plowing, from one to three passes with a harrow, crop planting, and sometimes subsequent tillage with a cultivator.

The conservation tillage systems vary from those that merely reduce excess tillage to the *no-tillage system*, which permits direct planting in the residue of the previous crop and uses only that localized tillage necessary to plant the seed. Figure 15.10 shows a moldboard plow in action and also

[3] For reviews of this topic, see SCSA (1983), D'Itri (1985), and Phillips and Phillips (1984).

TABLE 15.9

General Classification of Different Conservation Tillage Systems

All systems maintain at least 30% of the crop residues on the surface.

Tillage system	Operation involved
No till	Soil undisturbed prior to planting, which occurs in narrow seedbed, 2.5–7.5 cm wide. Weed control primarily by herbicides.
Ridge till (till, plant)	Soil undisturbed prior to planting, which is done on ridges 10–15 cm higher than row middles. Residues moved aside or incorporated on about one third of soil surface. Herbicides and cultivation to control weeds.
Strip till	Soil undisturbed prior to planting. Narrow and shallow tillage in row using rotary tiller, in-row chisel, etc. Up to one third of soil surface is tilled at planting time. Herbicides and cultivation to control weeds.
Mulch till	Soil surface disturbed by tillage prior to planting, but at least 30% of residues left on or near soil surface. Tools such as chisels, field cultivators, disks, and sweeps are used (e.g., stubble mulch). Herbicides and cultivation to control weeds.
Reduced till	Any other tillage and planting system that keeps at least 30% of residues on the surface.

System used by Conservation Technology Information Center, West Lafayette, IN.

illustrates a no-tillage system by which one crop is planted under the residues of another and organic cover is kept on the land.

(a)   (b)   (c)

Figure 15.10

Photographs showing action of the moldboard plow and two conservation tillage practices. The moldboard plow (a) essentially turns the top 15–20 cm of soil, covers the crop residues, and leaves bare soil on the surface. In contrast, the no-till systems leave all the residue on the surface, the new crop being planted in the previous crop's residue. (b) Corn is being planted in a rye mulch on a farm in Georgia; a disk (coulter) cuts through the rye and opens a slit in which the corn is planted. (c) On an Illinois field, corn was planted in wheat stubble and herbicides were used to control the weeds. [(a) Courtesy USDA National Tillage Laboratory; (b, c) courtesy USDA Soil Conservation Service]

TABLE 15.10

The Effect of Tillage Systems in Nebraska on the Percentage of Land Surface Covered by Crop Residues

Note that the conventional moldboard plow system provided essentially no cover while no till and planting on a ridge (ridge-till) provided best cover.

Tillage system	Number of fields	Percentage of fields with residue cover greater than			
		15%	20%	25%	30%
Moldboard	33	3	0	0	0
Chisel	20	40	15	5	0
Disk	165	40	20	9	4
Field cultivate	13	46	23	0	0
Ridge-till (till, plant)	2	100	50	50	0
No-till	3	100	100	100	100
All systems	236	36	18	9	4

From Dickey et al. (1987).

A prime objective of conservation tillage is to keep some plant residues on the soil surface. Extensive field research shows that the conventional moldboard plow systems leave only 1–5% of the soil covered with crop residues (Table 15.10). Reduced tillage systems (e.g., chiseling or disking) commonly leave 15–25% soil coverage, while with no-tillage systems we can expect 50–100% of the land to remain covered. These differences in land cover have marked effects on both soil erosion and runoff.

RUNOFF AND EROSION CONTROL

Since conservation tillage systems were initiated, hundreds of field trials have demonstrated that these tillage systems result in much less soil erosion than that experienced with conventional tillage methods. Surface runoff is also decreased, although the differences are not as pronounced as with soil erosion.

Typical of the results obtained are those presented in Figure 15.11. The minimum tillage (disk chisel) and no-till plots experienced far less erosion and significantly less runoff than was found on the conventional fall-plowed plot. The practical significance of conservation tillage is obvious.

The need for conservation tillage is determined in part by the erosion potential of the soil on which crops are to be grown. Estimates shown in Table 15.11 illustrate the point. Note that on soils with high and very high erosion potentials, estimated erosion losses from the conventional plow system are two to four times that considered tolerable if current soil productivity levels are to be maintained (11.2 Mg/ha). In contrast, when the no-till system is used, soil losses, even from soils with high erosion potential, are estimated to be far below the tolerable limit.

Table 15.11 indicates that conservation tillage systems also can significantly reduce losses of total soil nitrogen. The finer fractions of soil, which are among the first to be carried away through erosion, contain most of this nitrogen. Similar results were noted for phosphorus.

EFFECT ON CROP YIELDS

Conservation tillage systems generally provide yields equal to or greater than those from conventional tillage, provided the soil is not poorly drained and can be kept free of weeds through the use of chemicals. Typical of the results obtained from numerous field trials on well drained soils are those

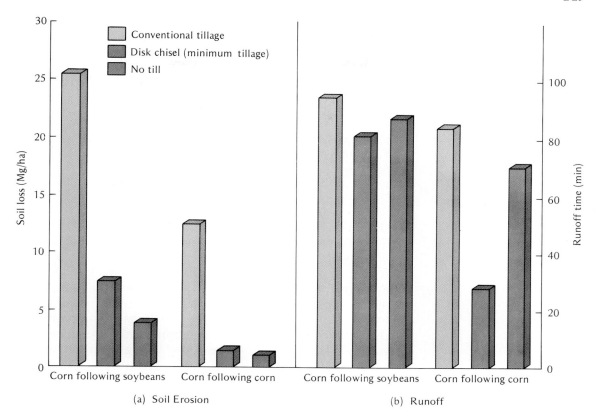

(a) Soil Erosion   (b) Runoff

Figure 15.11

Effect of tillage systems on soil erosion and runoff from corn plots in Illinois following corn and following soybeans. Soil loss by erosion was dramatically reduced by the conservation tillage practices. The period of runoff was reduced most by the disk chisel system where corn was grown after corn. The soil was a Typic Argiudoll (Catlin silt loam), 5% slope, up and down slope, tested in early spring. [Data from Oschwald and Siemens (1976)]

TABLE 15.11

Estimates of Annual Soil and Nitrogen Losses as Affected by Three Tillage Systems for Soybeans Following Corn and for Corn Following Soybeans, Each Grown on Soils with Different Erosion Potentials

Erosion potential[a]	Soil loss (Mg/ha)			Nitrogen loss (kg/ha)		
	Plow[b]	Chisel	No-till	Plow	Chisel	No-till
			Soybeans following corn			
	(4% cover)[c]	(26% cover)	(76% cover)	(4% cover)	(26% cover)	(76% cover)
Low ($\frac{1}{2}T$)	5.6	1.9	0.2	11.4	5.6	2.8
Med. (1T)	11.2	3.7	0.3	18.9	8.7	3.2
High (2T)	22.4	7.5	0.6	31.9	14.1	3.9
Very high (4T)	44.8	14.9	1.2	54.5	23.4	5.2
			Corn following soybeans			
	(2% cover)	(14% cover)	(42% cover)	(2% cover)	(14% cover)	(42% cover)
Low ($\frac{1}{2}T$)	6.2	3.4	0.8	4.4	2.9	1.5
Med. (1T)	12.4	6.8	1.7	7.5	4.9	2.1
High (2T)	24.8	13.6	3.4	13.0	8.2	3.2
Very high (4T)	49.5	27.2	6.7	22.6	14.1	5.1

Estimated by Baker and Laflen (1983).
[a] T = soil loss tolerance of 11.2 Mg/ha (5 tons/acre), thought to be the maximum allowable to maintain productivity of most soils.
[b] Plow = moldboard plow, disk, and plant; chisel = chisel, disk, and plant; no-till = plant in untilled land.
[c] Cover estimated assuming 90% residue cover for corn and 50% for soybeans, and that the plow system leaves 10% of the residue on the surface, the chisel 50%, and no-till 67%.

TABLE 15.12

Average Corn Grain Yields from Well-Drained Soils in Kentucky Under No-Tillage and Conventional Tillage Systems

| | | Grain yields (kg/ha) | |
		No tillage	Conventional tillage
Soil type	Years tested	No tillage	Conventional tillage
Maury silt loam	8	9,136	8,932
Crider silt loam	5	9,886	8,318
Tilsit silt loam	5	7,705	7,705
Allegheny silt loam	3	10,977	10,909

From Phillips et al. (1980).

shown in Table 15.12. Long-term trials show higher yields from conservation tillage systems than those where conventional tillage is used.

On soils with restricted drainage, yields with conservation tillage are sometimes inferior to those from conventionally tilled areas (Figure 15.12). Reasons for these lower yields include lower soil temperatures in the spring on the conservation tillage plots and the incidence of certain plant diseases, which may be higher under the somewhat higher moisture conditions that characterize these plots. Also, certain weeds tend to be more of a problem on wet soils, and rodent numbers may be higher where excess residues are left

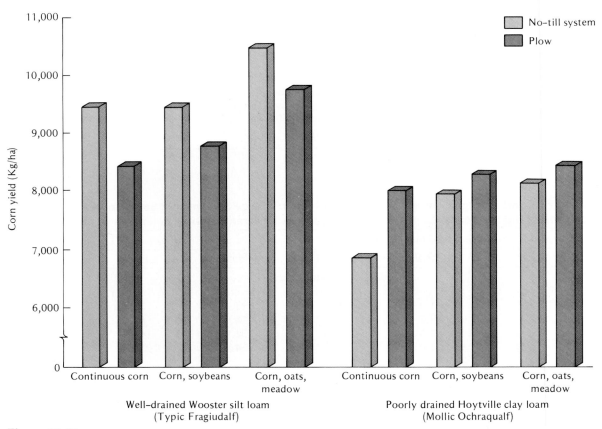

Figure 15.12

Effect of tillage systems on the yield of corn grown in different rotations on a well-drained and a poorly drained soil. The no-till system was superior on the well-drained soil but inferior on the soil with poor drainage. Data averages of 5 years. [Data from Van Doren et al. (1976)]

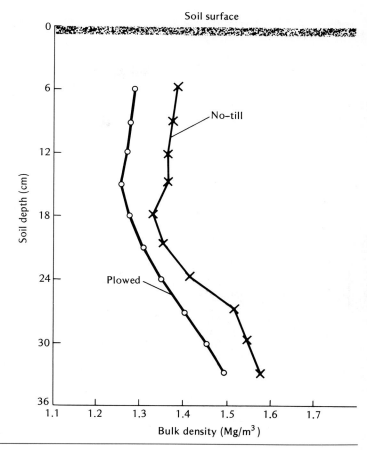

Figure 15.13
Effect of tillage on the bulk density of a loam soil that had been cropped for 6 years with barley. The no-till plot had higher bulk densities at all depths. Measurements made just before harvest. [Data selected from Soane and Pidgeon (1975)]

on the surface. These limitations reduce the effectiveness of conservation tillage on poorly drained soils and will likely limit the long-term adoption of these practices.

EFFECT ON SOIL PROPERTIES

Conservation tillage has variable effects on soil properties, depending on the particular system chosen. No-tillage systems, which maintain high surface soil coverage, have resulted in significant changes in soil properties, especially in the upper few centimeters.

Physical Properties. No-tillage (NT) plots are generally somewhat higher in soil water, especially in the upper portion of the profile, than the conventionally tilled (CT) plots. This is likely due to increased water infiltration from no-tillage systems and reduced evaporation losses, which are characteristic of any residue-covered plots.

No-tillage systems often leave the surface layers somewhat more compacted, with higher bulk density and lower total pore space (Figure 15.13). However, infiltration of water into the soil is often greater with NT since the organic residues protect the soil surface from the beating action of raindrops.

The higher moisture content in soil under no-tillage cultivation is often associated with reduced oxygen levels in soil pores. This bring about anaerobic conditions in local pockets of the soil. Soil organism numbers, however, are commonly higher in the upper few centimeters of no-tillage plots because of the higher organic matter levels at the surface. High soil moisture in

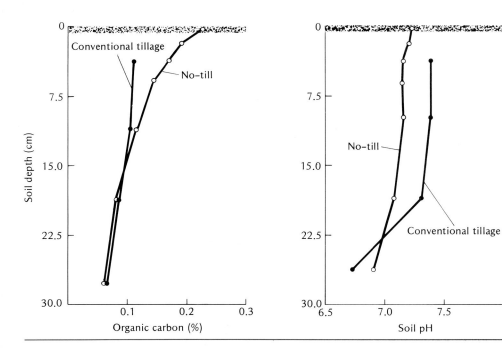

Figure 15.14
Comparative effects of 19 years of no tillage and conventional tillage on the organic carbon content and pH of a Typic Fragiudult (Wooster silt loam). Keeping organic residues near the surface (no tillage) increased the organic content of the upper 10 cm but reduced the soil pH in the upper 20 cm. The lower pH was likely due to increased nitrification where soil organic matter was higher. [From Dick (1982); used with permission of the Soil Science Society of America]

NT plots also affects soil temperature, especially in poorly drained soils in the spring. This is a constraint on adoption of no-tillage practices since the lower temperatures delay seed germination in cool temperate regions and reduce crop yields.

Soil cropped with no-tillage systems for years are generally considerably higher in organic matter in the upper 3–5 cm than comparable conventionally tilled soils (Figure 15.14). As a result, aggregate stability and resistance to crusting are increased when NT is followed.

Effects of conservation tillage on soil properties are related to the degree of surface cover or mulching. Consequently, conservation tillage practices such as stubble mulch farming, till-plant, and ridge planting, which leave part of the soil unprotected, would likely be intermediate between no tillage and conventional tillage in their effect on soil properties.

Chemical Properties.   As already mentioned, the no-tillage method significantly increases the organic matter content of the upper few centimeters of no-till soils. The higher soil microbe population needed to break down the surface residues also encourages the immobilization rather than mineralization of soil nutrients, especially nitrogen (Figure 15.15). High moisture and low oxygen levels are also known to stimulate denitrification at times. Coupled with greater leaching losses of nitrate ions sometimes associated with no-tilled areas, these gaseous losses can be of some concern.

Response to nitrogen fertilizer also may be affected by soil nitrogen reactions stimulated by the tillage system used. In some cases, yield increases from low nitrogen fertilizer applications are smaller under no-tillage systems than under conventional systems. This is likely due to the immobilization of nitrogen in the no-till surface layers as microbes decompose the organic residues. But when higher levels of nitrogen are applied, the response under no tillage is fully as great if not greater than where conventional "plow and cultivate" systems are used.

No tillage in humid areas has been found to significantly reduce soil pH, especially in the upper horizon (see Figure 15.14). Although some pH

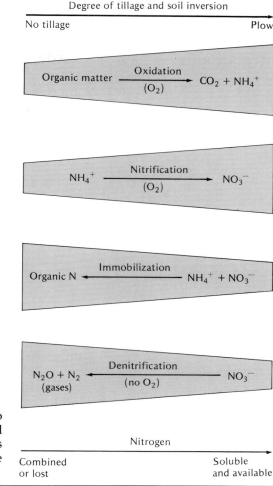

Figure 15.15

Diagram showing differences in nitrogen reaction in soils under no tillage compared with the conventional system of plowing. The no-till system tends to leave the soil somewhat less well aerated than does conventional tillage. [From Doran (1982); used with permission of the American Society of Agronomy]

reduction is apparently due to acids that form when crop residues are broken down, most of it results from the acidifying effect of surface-applied nitrogen fertilizers. This acidity must be countered by incorporation of liming materials into the soil.

The effects of no tillage on both physical and chemical properties of soils are considerably less marked in areas where crop residues do not cover a high percentage of the soil surface. Likewise, some effects are not as noticeable in subhumid and semiarid areas as in more humid regions. The effects of no tillage also vary somewhat depending on the cropping sequence. Crops with high residues, such as corn, produce more pronounced effects than crops with little residue, such as soybeans, cotton, potatoes, and sugar beets.

BIOLOGICAL EFFECTS

Conservation tillage practices, which leave most of the crop residues on or near the surface, tend to encourage higher microbial populations in the surface few centimeters. Also, numbers of soil fauna are commonly higher in minimum tillage plots as compared with conventionally tilled plots (Table 15.13). Higher numbers of earthworms on no-till plots may account for the ease with which water can infiltrate the soil even though minimum tillage has been used.

TABLE 15.13

Effect of Conventional and No-Tillage Systems on the Number and Weight of Soil Fauna

Except for nematodes, all organisms were more abundant in no-till plots.

Organism	Numbers (per m^2)		Dry weight (Mg/m^2)	
	Conv.	No-till	Conv.	No-till
Nematodes	3,008	2,473	344	252
Micro arthropods	49,430	95,488	135	343
Macro arthropods	14	78	7	44
Earthworms	149	967	3,129	20,307

From Hendrix et al. (1986).

LABOR AND ENERGY REQUIREMENTS

A primary reason for a farmer to adopt conservation tillage practices is their lower labor and energy requirements. Although the time and energy savings vary depending on the conservation tillage system chosen, a number of studies suggest labor requirements are cut significantly when some conservation tillage practices are used. Since the number of passes over a field is greatly reduced by adopting conservation practices, lowered labor costs are not surprising. Likewise, fewer machines are required in conservation tillage, which also tends to reduce costs.

Fuel consumption for conservation tillage is also markedly lower, comparative levels being only one third to one half those for the conventional systems. Even when the energy needed to manufacture the greater quantity of herbicides used to control weeds in the conservation tillage systems is taken into consideration, in most cases conservation tillage systems require less total energy than do their conventional counterparts.

PROBABLE FUTURE EXPANSION

Conservation tillage practices have spread rapidly in the United States in the last 20 years. Whereas in 1965 these practices were used on only 2–3% of the harvested cropland, by 1979 this figure had risen to about 20% and in 1986 to

TABLE 15.14

Land Area on Which Some Type of Conservation Tillage Was Used in 1986.

See Figure 15.23 for states in each region.

Region	Conservation tillage		
	Hectares (millions)	Acres (millions)	Percent of cropland
Corn Belt	13.6	33.7	43
Appalachian	2.4	6.0	40
Northeast	1.4	3.5	40
Mountain	3.3	8.2	36
Northern plains	9.6	23.6	36
Lake states	3.4	8.4	27
Southeast	0.9	2.3	21
Southern plains	2.7	6.6	20
Pacific	1.1	2.6	19
Delta	1.1	2.6	15
Total U.S.	39.5	97.5	33

Source: Conservation Technology Information Center, West Lafayette, IN.

33%. Regional use of conservation tillage systems in the United States is shown in Table 15.14.

Future projections suggest that this trend will continue, and that by the year 2000, about 75% of the country's cropland will be under conservation tillage. The expansion in area covered by conservation tillage practices is encouraging because of its probable constraint on soil erosion.

Up to this point, consideration has been given to sheet and rill water erosion. It is appropriate to have done so since most soil loss occurs from one of these processes. But in local areas, gully erosion is of greatest importance. This destructive process will now receive our attention.

15.13

Gully Erosion

Small channels (rills), though at first insignificant, may soon enlarge into large gullies that quickly eat into the land, exposing its subsoil and increasing the already detrimental sheet and rill erosion (see Figure 15.3). If small enough, such gullies can be plowed in and seeded to grass.

When the gully erosion is too active to be checked and the ditch is still small, dams of rotted manure or straw at intervals of 5–7 m are very effective. Such dams may be made more secure. Strips of wire netting below them can make such dams. After a time, the ditch may be plowed in and the site of the gully seeded and kept in sod permanently. The gully then becomes a *grassed waterway*, an important feature of most successful erosion control systems (see Figure 15.8).

With very large gullies, dams of earth, concrete, or stone are often used successfully. Most of the sediment is deposited above the dam, and the gully is slowly filled. The use of semipermanent check dams, flumes, and paved channels are also recommended on occasion. Unfortunately, if such engineering features are at all extensive, the cost may exceed the benefits derived or even the value of the land to be served.

15.14

Wind Erosion—Importance and Control

Wind erosion, although most common in arid and semiarid regions, occurs to some extent in humid climates as well. It is essentially a dry-weather phenomenon and, hence, is stimulated by moisture deficiency. All kinds of soils and soil materials are affected, and at times their more finely divided portions are carried to great heights and for hundreds of miles.

In the great dust storm of May 1934, which originated in western Kansas, Texas, Oklahoma, and contiguous portions of Colorado and New Mexico, clouds of powdery debris, silt, clay, and organic matter were carried eastward to the Atlantic seaboard and even hundreds of miles out over the ocean. Such activity is not a new phenomenon but has been common in all geologic ages. Strong winds gave rise to the wind-blown loess deposits now so important agriculturally in the United States (see Figure 2.18) and other countries.

The destructive effects of wind erosion are often very serious. Not only is the land robbed of its richest soils but crops can be blown away, left to die with roots exposed (Figure 15.16), or covered by the drifting debris. In 1982 some 37% of the soil moved by erosive forces in the United States was

Figure 15.16
The devastating effects of wind erosion. The roots of this desert grass in Niger were exposed when the soil was blown away. [From IUCN (1986)]

moved by wind erosion. Possibly 12% of the continental United States is somewhat affected by wind erosion—8% moderately so and perhaps 2–3% greatly. Figure 15.17 shows relative wind erosion forces at four locations in the United States.

In six of the Great Plains states, wind erosion exceeds water erosion. Wind erosion in these states is shown in Figure 15.18. Here the mismanagement of plowed lands and the lowered holding power of the range grasses because of overgrazing have greatly exacerbated wind action. In dry years, as experience has shown, the results have been most deplorable.

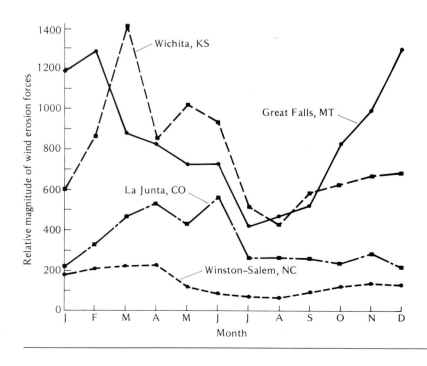

Figure 15.17
Monthly variation in the relative magnitude of wind erosion forces at four locations in the United States. One would expect little soil movement in Winston-Salem, NC, not only because of the low wind erosion forces but also because of the probable high soil moisture content. [From Skidmore and Woodruff (1968)]

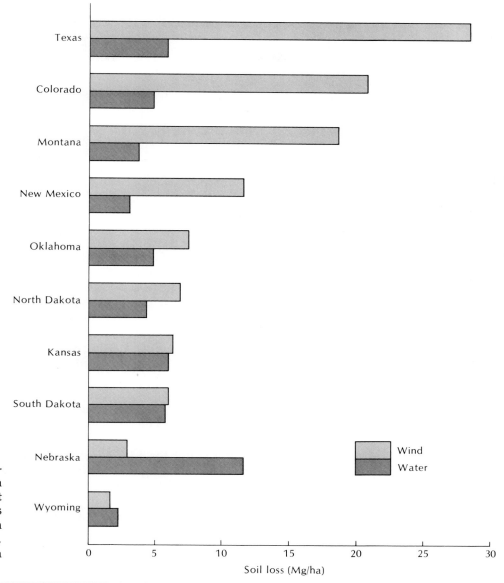

Figure 15.18
Average wind and water erosion rates on cropland in ten Great Plains states. In eight states wind erosion exceeds water erosion, especially in Texas, Colorado, Montana, and New Mexico. [From USDA (1987)]

Although most of the damage is confined to regions of low rainfall, some serious wind erosion occurs in humid sections. Sand dune movement is a good example. More important agriculturally, sandy soils and peat soils cultivated for many years are often affected by the wind. The drying out of the finely divided surface layers of these soils leaves them extremely susceptible to wind erosion.

Wind erosion is a concern not limited to the United States. It is a worldwide problem. Large areas of the Soviet Union have been subject to severe damage by wind erosion, and declines in soil productivity and crop production have resulted. Likewise, in recent years large areas of sub-Saharan Africa have been badly damaged by wind erosion. Overgrazing and other misuses of the fragile lands of arid and semiarid regions have brought starvation and human misery to millions of people in these areas. Water and wind erosion threaten humankind's capacity to sustain food and fiber production.

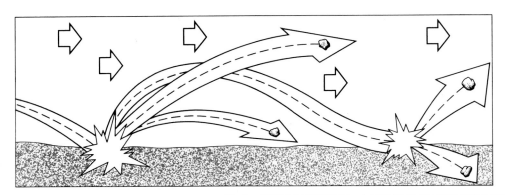

Figure 15.19
The process of saltation. Medium-sized particles (2.5–3.75 mm in diameter) bounce along the soil surface, striking and dislodging other particles as they move. They are too large to be carried long distances suspended in the air but small enough to be transported by the wind. [From Hughes (1980); used with permission of Deere & Company, Moline, IL]

MECHANICS OF WIND EROSION

As was the case for water erosion, the loss of soil by wind movement involves two processes: (a) detachment and (b) transportation. The lifting and abrasive action of the wind results in some detachment of tiny soil grains from the granules or clods of which they are a part. When the wind is laden with soil particles, however, its abrasive action is greatly increased. The impact of these rapidly moving grains dislodges other particles from soil clods and aggregates. The dislodged particles are now ready for movement.

The transportation of the particles once they are dislodged takes place in several ways. The first and most important is the process of *saltation*, or movement of soil by a short series of bounces along the surface of the ground (Figure 15.19). The particles remain fairly close to the ground as they bounce, seldom rising more than 30 cm or so. Depending on conditions, this process may account for 50–70% of the total movement of soil.

Saltation also encourages *soil creep*, or the rolling and sliding along the surface of larger particles. The bouncing particles carried by saltation strike the large aggregates and speed up their movement along the surface. Soil creep accounts for the movement of particles up to about 0.84 mm in diameter, which may amount to 5–25% of the total movement.

The most spectacular method of transporting soil particles is by movement in *suspension*. Here, dust particles of a fine-sand size and smaller are moved parallel to the ground surface and upward. Although some of the particles are carried at a height no greater than a few meters, the turbulent action of the wind results in others being carried kilometers upward into the atmosphere and many hundreds of kilometers horizontally. These particles return to the earth only when the wind subsides and/or when precipitation washes them down. Although it is the most striking manner of transportation, suspension seldom accounts for more than 40% of the total and is generally no more than about 15%.

FACTORS AFFECTING WIND EROSION

Susceptibility to wind erosion is related to the moisture content of soils. Wet soils do not blow. The moisture content is generally lowered by hot dry winds to the wilting point and lower before wind erosion takes place (Figure 15.20).

Other factors that influence wind erosion are (a) wind velocity and turbulence, (b) soil surface conditions, (c) soil characteristics, and (d) the nature and orientation of the vegetation. Obviously, the rate of wind movement,

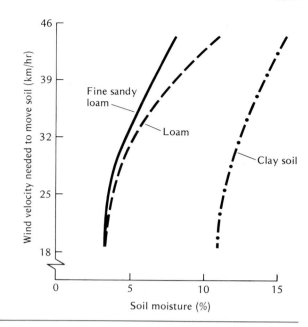

Figure 15.20
Effect of soil moisture and soil texture on the wind velocity required to move soils. Soil moisture and soil texture play a highly significant role in determining the susceptibility of soils to wind erosion. [Redrawn from Bisal and Hsiek (1966)]

especially gusts having greater than average velocity, will influence erosion. Tests have shown that wind speeds of about 20 km/hr (12 mph) are required to initiate soil movement. At higher wind speeds, soil movement is proportional to the cube of the wind velocity. Thus, the quantity of soil carried by wind goes up very rapidly as speeds above 30 km/hr are reached.

Wind turbulence also influences the capacity of the atmosphere to transport matter. Although the wind itself has some direct influence in picking up fine soil, the impact of wind-carried particles on those not yet loose is probably more important.

Wind erosion is less severe where the soil surface is rough. This roughness can be obtained by proper tillage methods, which leave large clods or ridges on the soil surface. Leaving stubble mulch is probably an even more effective way of reducing wind-borne soil losses.

In addition to moisture content, several other soil characteristics affecting wind erosion are (a) mechanical stability of dry soil clods and aggregates, (b) stability of soil crust, and (c) bulk density and size of erodible soil fractions. The clods must be resistant to the abrasive action of wind-carried particles. If a soil crust resulting from a previous rain is present, it too must be able to withstand the wind's erosive power without deteriorating. The importance of clay, organic matter, and other cementing agents is quite apparent. Sandy soils, which do not have such agents, are easily eroded.

Soil particles or aggregates about 0.1 mm in diameter are most erodible, those larger or smaller in size being less susceptible to movement. Thus, fine sandy soils are quite susceptible to wind erosion. Particles or aggregates about 0.1 mm in size are also responsible to a degree for the movement of larger or smaller particles. By saltation, most of the erodible particles bounce against larger particles, causing surface movement (creep), and against smaller soil particles, resulting in movement in suspension.

Vegetation or a stubble mulch will reduce wind erosion hazards, especially if the rows run perpendicular to the prevailing wind direction. This effectively presents a barrier to wind movement. In addition, plant roots help bind the soil and make it less susceptible to wind damage.

WIND EROSION EQUATION

As was the case for sheet and rill erosion, a wind erosion equation (WEE) has been developed.

$$E = f(ICKLV)$$

The potential quantity of erosion per unit area (E) is a function (f) of soil erodibility (I), a local wind erosion climate factor (C), the soil surface roughness (K), width of the field (L), and quantity of vegetative cover (V). The influence of several of these factors is illustrated in Figure 15.21. Although wind and soil characteristics are generally beyond the farmer's control, the other factors are subject to control through the choice of cultural practices.

CONTROL OF WIND EROSION

The factors of the wind erosion equation give clues to methods of reducing wind erosion. Obviously, if the soil can be kept moist, there is little danger of wind erosion. A vegetative cover also discourages soil blowing, especially if the plant roots are well-established. In dry-farming areas, however, sound

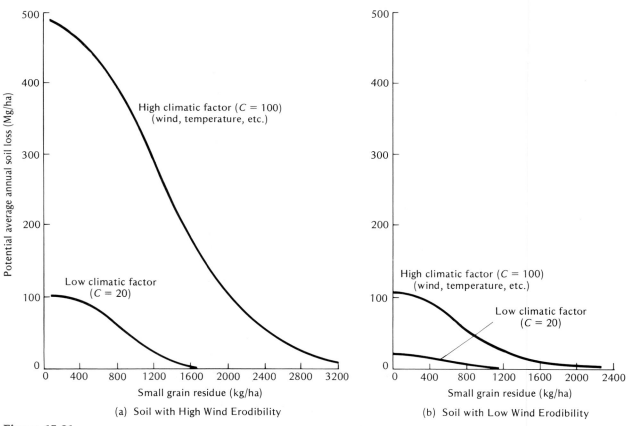

Figure 15.21

Effect of small grain residues on the potential wind erosion of soils with high (a) and low (b) erodibility. Strong dry winds and high temperatures encourage erosion, especially on the soil with high wind-erodibility characteristic. Surface residues can be used to help control this wind erosion. [From Skidmore and Siddoway (1978); used with permission of the American Society of Agronomy]

(a)

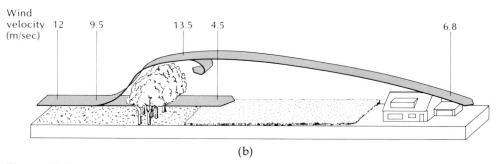

Wind velocity (m/sec) 12   9.5   13.5   4.5   6.8

(b)

Figure 15.22
(a) Shrubs and trees make good windbreaks and add beauty to a North Dakota farm homestead. (b) Effect of a windbreak on wind velocity. The wind is deflected upward by the trees and is slowed down even before reaching them. On the leeward side further reduction occurs, the effect being felt as far as 20 times the height of the trees. [(a) Courtesy USDA Soil Conservation Service; (b) from FAO (1987)]

moisture-conserving practices require summer fallow on some of the land, and hot, dry winds reduce the moisture in the soil surface. Consequently, other means must be employed on cultivated lands in these areas.

Conservation tillage practices described in Section 15.12 were used for wind erosion control long before they became popular as water erosion control practices. Keeping the soil surface rough and maintaining some vegetative cover is accomplished by using appropriate tillage practices. However, the vegetation should be well-lodged into the soil to prevent it from blowing away. Stubble mulch has proved to be effective for this purpose (see Section 14.8).

Tillage that provides a cloddy surface condition should be at right angles to the prevailing winds. Likewise, strip cropping and alternate strips of cropped and fallowed land should be perpendicular to the wind. Barriers such as shelter belts (Figure 15.22) are effective in reducing wind velocities for short distances and for trapping drifting soil.

461

Various devices are used to control blowing of sands, sandy loams, and cultivated peat soils (even in humid regions). Windbreaks and tenacious grasses and shrubs are especially effective. Picket fences and burlap screens, though less efficient as windbreaks than such trees as willows, are often preferred because they can be moved from place to place as crops and cropping practices are varied. Rye, planted in narrow strips across the field, is sometimes used on peat lands. All of these devices for wind erosion control, whether applied in arid or humid regions and whether vegetative or purely mechanical, are closely associated with the broad problem of soil moisture control.

15.15

Soil Loss Tolerance

The loss of any amount of soil by erosion generally is not considered beneficial, but years of field experience as well as scientific research indicate that some loss can be tolerated.[4] Scientists of the USDA Soil Conservation Service working in cooperation with field personnel throughout the country have developed tentative soil loss tolerance limits, known as *T-values*, for most cultivated soils of the country.

A tolerable soil loss (*T*-value) is considered as the maximum combined water and wind erosion that can take place on a given soil without degrading that soil's long-term productivity. As might be expected, there is insufficient research data to ascertain accurately the *T*-values for all soils. However, the soil-loss equations coupled with years of practical experience and the judgment of knowledgeable soil scientists have made it possible to set these values for the more widely used soils.

The *T*-values for soils in the United States commonly range from 5 to 11 Mg/ha. They depend on a number of soil quality and management factors, including soil depth, organic matter content, and the use of water control practices. At the present time, 11 Mg/ha is the maximum *T*-value assigned for most soils in this country; many have lower values.

The soil tolerance values set in the United States should be used with caution in other regions and particularly in the tropics. In some tropical soils, the nutrient-supplying power of the subsoil horizons is low. Since the subsoil becomes part of the plow layer in eroded soils, the *T*-value in some deep soils of the tropics may well be higher than that in the United States.

There is considerable controversy as to whether the current *T*-values should be increased or lowered. For soils with deep favorable rooting zones, the current 11 Mg/ha *T*-value may be too low. Some scientists contend, for example, that for a soil with a rooting depth of 200 cm, a *T*-value of 15 or even 20 Mg/ha would be appropriate. Others are concerned, however, not only about soil productivity but about the sediment from eroded fields as it affects environmental quality, including the quality of water for downstream users. These scientists contend that the *T*-values currently in use may be too high. In any case, the concept of *T*-values is useful if for no other reason than to focus attention on practical means of reducing soil erosion.

MAINTENANCE OF SOIL PRODUCTIVITY

The soil-loss tolerance (*T*) concept suggests that the productive capacity of a soil is reduced only when the erosion loss exceeds the tolerable level. The

---

[4] For a discussion of this subject, see Schertz (1983).

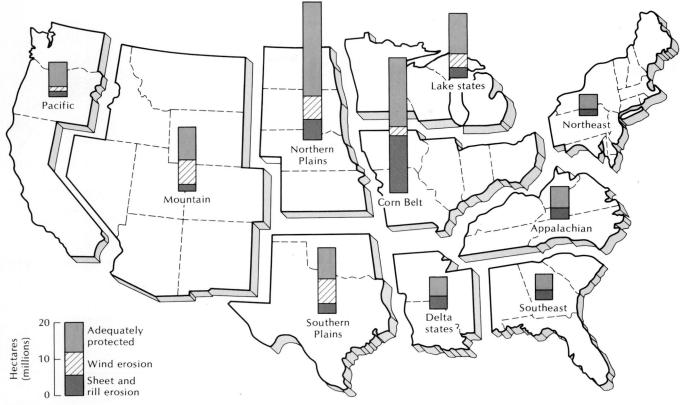

Figure 15.23
Millions of hectares of cropland that are suffering wind and water erosion by more than the allowable tolerance (T) in different farming regions of the mainland United States. [From USDA (1987)]

1982 U.S. National Resource Inventory suggests that under current land use and management, soil productive capacity is going down on about 116 million hectares (286 million acres) of nonfederal land in the United States. Forty percent of all cropland is eroding at rates in excess of T, and about 23% at twice the tolerable rate ($2T$). The cropland eroding at rates in excess of T in different regions of the United States is shown in Figure 15.23. It is obvious that progress must still be made to bring these excessive losses under control.

The ultimate influence of soil loss by erosion on soil productivity is determined by such soil properties as soil depth and permeability. As shown in Figure 15.24, a deep well-managed and well-drained soil may not lose its crop productivity even though it suffers some erosion. In contrast, erosion from a shallow poorly drained soil may bring about rapid decline in soil productivity.

15.16

Land Capability Classification

As discussed in Section 3.13, the U.S. Department of Agriculture has developed a land capability classification system. While the system is useful in ascertaining the appropriate use of soils for many purposes, it is especially helpful in identifying practices that can minimize soil erosion.

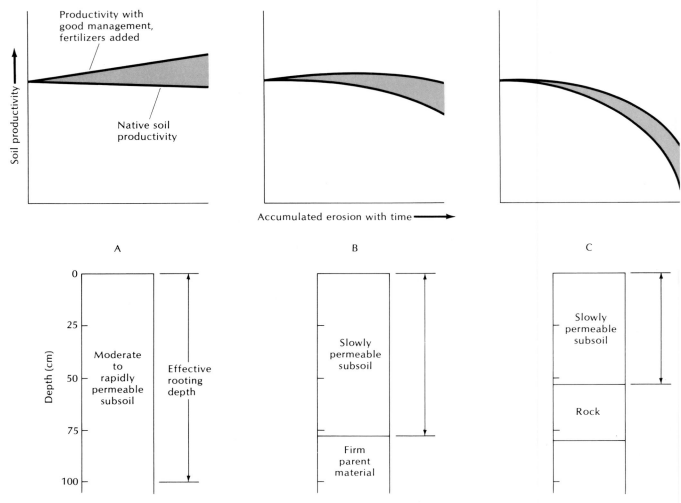

Figure 15.24
Effect of erosion over time on the productivity of three soils differing in depth and permeability. Productivity on soil A actually increases with time because of good management practices and fertilizer additions, even though the native soil productivity declines as a result of erosion. Because soil C is shallow and has restricted permeability, its productivity declines rapidly as a result of erosion, a decline that good management and fertilizers cannot prevent. Soil B, which is intermediate in both depth and permeability, suffers only a slight decline in productivity due to erosion. Obviously, the effect of erosion on soil productivity is influenced by soil characteristics.

Eight land capability classes are recognized in this classification system. Class I is least susceptible to erosion and Class VIII the most susceptible. Figure 15.25 shows the appropriate intensity of use allowable for each of these land capability classes if damage from erosion is to be avoided. A brief description of the characteristics and safe use of each land capability class follows.

CLASS I

Soils in this land class have few limitations on their use. They can be cropped intensively, used for pasture, range, woodlands, or wildlife preserves. The soils are deep and well drained, and the land is nearly level

Increasing intensity of land use →

Land capability class	Wildlife	Forestry	Grazing			Cultivation			
			Limited	Moderate	Intense	Limited	Moderate	Intense	Very Intense
I	▓	▓	▓	▓	▓	▓	▓	▓	▓
II	▓	▓	▓	▓	▓	▓	▓	▓	
III	▓	▓	▓	▓	▓	▓	▓		
IV	▓	▓	▓	▓	▓	▓			
V	▓	▓	▓	▓	▓				
VI	▓	▓	▓	▓					
VII	▓	▓	▓						
VIII	▓								

Increasing limitations and hazards →
← Decreasing adaptability and freedom of choice of uses

Shaded portion shows uses for which classes are suitable

Figure 15.25
Intensity with which each land capability class can be used with safety. Note the increasing limitations on the safe uses of the land as one moves from Class I to Class VIII. [Modified from Hockensmith and Steele (1949)]

(Figure 15.26). Class I lands are either naturally fertile or have characteristics that encourage good response of crops to applications of fertilizer.

The soils in Class I need only prudent crop management practices to maintain their productivity. Good management entails the use of fertilizer and lime and the return of manure and crop residues, including green manures.

Figure 15.26
Several land capability classes in San Mateo County, California. A range is shown from the nearly level land in the foreground (Class I), which can be cropped intensively, to the badly eroded hillsides (Classes VII and VIII). Although topography and erosion hazards are emphasized here, it should be remembered that other factors—drainage, stoniness, droughtiness—also limit soil usage and help determine the land capability class. [Courtesy USDA Soil Conservation Service]

Class II

Soils in Class II have some limitations that reduce the choice of crops or require moderate conservation practices. Less intensive cropping systems must be used on Class II than Class I land, or, if the same cropping systems are used, Class II lands require some conservation practices.

The use of soils in Class II may be limited by one or more factors, such as (a) gentle slopes, (b) moderate erosion hazards, (c) inadequate soil depth, (d) less than ideal soil structure and workability, (e) slight to moderate alkaline or saline conditions, and (f) somewhat restricted drainage.

The management practices that may be required for soils in Class II include terracing, strip cropping, contour tillage, rotations involving grasses and legumes, and grassed waterways (see Section 15.8). In addition, those prudent management practices that were suggested for Class I land are also generally required for soils in Class II.

Class III

Soils in Class III have severe limitations that reduce the choice of plants or require special conservation practices or both. The same crops may be grown on Class III land as on Classes I and II land, but crops that provide soil cover, such as grasses and legumes, must be more prominent in the rotations used. The amount of land used for row crops is severly restricted.

Limitations in the use of soils in Class III result from factors such as (a) moderately steep slopes, (b) high erosion hazards, (c) very slow water permeability, (d) shallow depth and restricted root zone, (e) low water-holding capacity, (f) low fertility, (g) moderate alkalinity or salinity, and (h) unstable soil structure.

Soils in Class III often require special conservation practices. The methods mentioned for Class II land must be employed, frequently in combination with restrictions in kinds of crops. Tile or other drainage systems also may be needed in poorly drained areas.

Class IV

Soils in Class IV can be used for cultivation, but there are severe limitations on the choice of crops. Also, careful management may be required. The alternative uses of Class IV soils are more limited than for those of Class III. Close-growing crops must be used extensively, and row crops cannot be grown safely in most cases. The choice of crops may be limited by excess moisture as well as by erosion hazards.

Limiting factors on Class IV soils may include (a) steep slopes, (b) severe erosion susceptibility, (c) severe past erosion, (d) shallow soils, (e) low water-holding capacity, (f) poor drainage, and (g) severe alkalinity or salinity. Soil conservation practices must be used more frequently than for soils in Class III and must usually be combined with strict limitations in choice of crop.

Class V

Soils in Classes V–VIII are generally not suited to cultivation. Those in Class V are limited in their safe use by factors other than erosion hazards. Such limitations include (a) frequent stream overflow, (b) growing season too short for crop plants, (c) stony or rocky soils, and (d) ponded areas where drainage is not feasible. Often, pastures can be improved on this class of land.

Class VI

Soils in Class VI have extreme limitations that restrict use largely to pasture or range, woodland, or wildlife. The limitations are the same as those for Class V land, but they are more rigid.

Class VII

Soils in Class VII have severe limitations that restrict their use to grazing, woodland, or wildlife. The physical limitations are the same as for Class VI except they are so strict that pasture improvement is impractical.

Class VIII

Class VIII land comprises soils that should not be used for any kind of commercial plant production. Land use is restricted to recreation, wildlife, water supply, or aesthetic purposes. Examples of kinds of soils or landforms included in Class VIII are sandy beaches, river wash, and rock outcrop.

Subclasses

In each of the land capability classes are *subclasses* that have the same kind of dominant limitations for agricultural uses. The four kinds of limitations recognized in these subclasses are risks of erosion (e); wetness, drainage, or overflow (w); root-zone limitations (s); and climatic limitations (c). Thus, a soil may be found in Class III(e), indicating that it is in Class III because of risks of erosion.

Land Use Capability in the United States

In 1977, the U.S. Department of Agriculture made a national inventory of soil and water conservation needs for the United States (USDA, 1981). This inventory included information on land capability classification. Figure 15.27 presents summary information from this inventory.

About 44% of the land in the United States (248 million ha) is suitable for regular cultivation (Classes I, II, and III). Another 13% (76 million ha) is marginal for growing cultivated crops. The remainder is suited primarily for grasslands and forests and is used mostly for this purpose.

The land classification scheme illustrates the practical use that can be made of soil surveys. The many soils delineated on a map by the soil surveyor are viewed in terms of their safest and best longtime use. The eight land capability classes have become the starting point in the development of farm plans so useful to thousands of American farmers.

15.17

Conservation Treatment in the United States

Cost-Sharing Assistance

For years, farmers in the United States have received cost-sharing assistance in the establishment and use of conservation practices. The adoption of these practices has helped reduce soil erosion loss, as can be seen from the data in Table 15.15. The adoption of each of the erosion control practices resulted in significant decreases in erosion loss, in most cases to a level lower

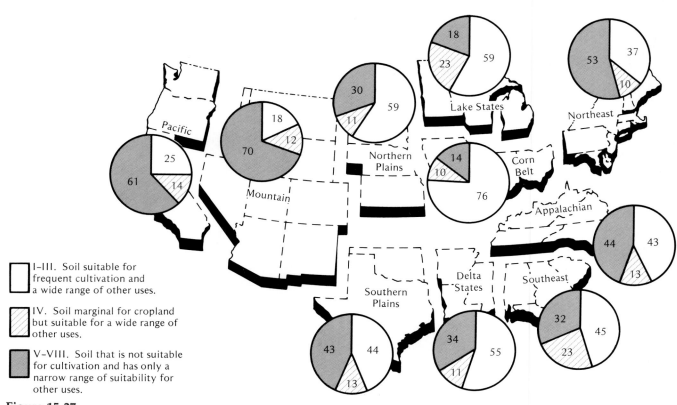

Figure 15.27

Percentages of land in capability Classes I–VIII by crop production regions of mainland United States. [From USDA (1981)]

than 11 Ma/ha per year, the maximum soil loss tolerance (*T*-value) for most soils.

Data in Table 15.16 are disturbing, however, because they show comparatively little cost-sharing assistance in the United States has been provided for the most highly erodible lands. Areas that lose annually more than 30 Mg/ha of soil, and from which 84% of the total soil loss occurs, receive only 21% of the cost sharing. These findings may well dictate changes in policies to ensure better erosion control in areas suffering very high erosion losses. They also confirm findings that conservation treatments are adequate on only a small portion of cropland in the United States. Unfortunately the inadequacy applies to pastures, rangelands, and forests as well as to cropland.

TABLE 15.15

Impact of the U.S. Agricultural Conservation Programs on Sheet and Rill Erosion by Type of Practice from 171 Sample Counties 1975–1978

		Average erosion rates (Mg/ha)[a]	
Erosion control practice	No. of cases	Before assistance	After assistance
Diversions	429	49 (22)	25 (11)
Terraces	1,754	32 (14)	11 (5)
Establish permanent cover	10,315	27 (12)	8 (3)
Minimum tillage	119	22 (10)	9 (4)
Improve permanent cover	6,978	18 (8)	8 (3)
Strip cropping	172	18 (8)	11 (5)

From USDA (1980b).

[a] Losses in tons per acre in parentheses.

TABLE 15.16

The Percentage of Total Erosion Loss from Land with Differing Rates of Erosion per Hectare and the Proportion of Cost-Shared Erosion Practices on That Land

Unfortunately, most of the cost sharing is on land with the least erosion hazards.

Erosion loss category (Mg/ha per yr)	Percent of total erosion loss	Percent of cost-shared erosion practices
0–11	2	52
11–31	14	27
>31	84	21

From USDA (1980b).

15.18

Conclusion

Runoff and erosion losses, particularly on prime farmland, have profound significance for crop production and for long-term soil productivity. Erosion losses occur worldwide and not only damage the soils in upstream areas but also cause equal if not greater damage to downstream reservoirs, waterways, and domestic water plants.

Soil loss by sheet and rill water erosion is most critical in subhumid and humid areas, whereas wind erosion exacts a higher toll in semiarid and arid areas. For both types of erosion the maintenance of soil cover at or near the soil surface offers the most effective means of controlling soil and water loss. Such practices, known collectively as conservation tillage, have become increasingly popular in the United States and are used on about one third of the harvested cropland. Conservation tillage systems are undoubtedly the most significant soil and water conservation practices developed in modern times.

Study Questions

1. A stream remains relatively free of soil sediment even after a moderately heavy rainfall. What does this tell you about the condition of the watershed feeding into this stream?

2. If you wanted to take a photograph of sheet and rill erosion, where would you go? For wind erosion?

3. What erosive forces are drastically reduced if a soil has significant residues on its surface?

4. A field located in central Missouri has a Mexico silt loam soil, a slope of about 6%, and a slope length of 45 meters. If the soil receives conventional tillage and the crop is planted up and down the row, what would be the anticipated rate of soil loss by erosion in tons per acre? In Mg/ha?

5. If conservation (minimum) tillage was used on the field described in question 4, corn was grown with residues (1000 kg/ha) left on the soil surface, and the land was strip cropped, what would be the expected soil erosion loss (tons/acre and Mg/ha)?

6. What characteristics of minimum tillage (and related conservation tillage practices) are responsible for relatively low water runoff and soil erosion?

7. Corn yields in some Minnesota soils are not as high with conservation tillage as with conventional tillage. What are the likely characteristics of these soils and why do you think the corn yields are lower?

8. Minimum tillage was practiced on a given soil for 15 years. Contrast the probable soil properties of this soil after the 15 years with those of a nearby field (same type of soil) where conventional tillage was practiced.

9. Wind erosion removes more soil sediment than water erosion in what area of the United States? How can you best control this wind erosion?

10. Small applications of nitrogen fertilizers to the surface of a no-till plot planted to corn produced lower yield increases than did an equal amount of nitrogen added to a nearby conventionally tilled field (same soil). How do you account for this difference?

11. The erosion losses from a certain soil were about 0.5 Mg/ha per year for about 20 years and yet the crop productivity on this soil was somewhat higher after 20 years than it was when the production started. What are the likely characteristics of this soil, and how do you account for the increased soil productivity even though there were erosion losses?

References

BAKER, J. L., and J. M. LAFLEN. 1983. "Water quality consequences of conservation tillage," *J. Soil Water Cons.*, **38**:186–94.

BATIE, S. S. 1983. *Soil Erosion—Crisis in America's Cropland* (Washington, DC: The Conservation Foundation).

BISAL, F., and J. HSIEK. 1966. "Influence of moisture on erodibility of soil by wind," *Soil Sci.*, **102**:143–46.

BROWN, L. R., and E. C. WOLF. 1984. *Soil Erosion: Quite Crisis in the World Economy*, Worldwatch Paper 60 (Washington, DC: Worldwatch Institute).

DICK, W. A. 1982. "Organic carbon, nitrogen and phosphorus concentrations and pH in soil profiles as affected by tillage intensity," *Soil Sci. Soc. Amer. J.*, **47**:102–107.

DICKEY, E. C., et al. 1987. "Conservation tillage: Perceived and actual use," *J. Soil Water Cons.*, **42**:431–34.

D'ITRI, F. M. (Ed.). 1985. *A Systems Approach to Conservation Tillage* (Chelsea, MI: Lewis Publishers, Inc.).

DORAN, J. W. 1982. "Tilling changes soil," *Crops and Soils*, **34**:10–12.

EL-SWAIFY, S. A., and E. W. DANGLER. 1982. "Rainfall erosion in the tropics: A state-of-the-art," *Soil Erosion and Conservation in the Topics*, ASA Special Publication No. 43 (Madison, WI: Amer. Soc. Agron.).

FAO. 1987. *Protect and Produce* (Rome, Italy: UN Food and Agriculture Organization).

FOLLETT, R. F., and B. F. STEWART (Eds.). 1985. *Soil Erosion and Crop Productivity* (Madison, WI: Amer. Soc. Agron.).

FOSTER, G. R., and R. E. HIGHFILL. 1983. "Effect of terraces on soil loss: USLE *P* factor values for Terraces," *J. Soil Water Cons.*, **38**:48–51.

HEIMLICH, R. E., and N. L. BILLS. 1986. "An improved soil erosion classification: Update, comparison and extension," *Soil Conservation, Assessing the National Resources Inventory*, Vol 2 (Washington DC: National Academy Press).

HENDRIX, P. F., et al. 1986. "Detritus food webs in conventional and no-tillage agroecosystems," *BioScience*, **36**:374–79.

HOCKENSMITH, R. D., and J. G. STEEL. 1949 "Recent trends in the use of the land-capability classification," *Soil Sci. Soc. Amer. Proc.*, **14**:383–88.

HUGHES, H. A. 1980. *Conservation Farming* (Moline, IL: Deere & Company).

IUCN. 1986. *The IUCN Sabel Report* (Gland, Switzerland: Intl. Union Cons. Nature).

LAL, R. 1984. "Soil erosion from tropical arable lands and its control," *Adv. Agron.*, **37**:187–248.

LARSON, W. E., F. J. PIERCE, and R. H. DOWDY. 1983. "The threat of soil erosion to long-term crop productivity," *Science*, **219**:458–65.

LATTANZI, A. R., L. D. MEYER, and M. F. BAUMGARDNER. 1979. "Influence of mulch rate and slope steepness in interrill erosion," *Soil Sci. Soc. Amer. Proc.*, **38**:946–50.

OSCHWALD, W. R., and J. C. SIEMENS. 1976. "Conservation tillage: a perspective," SM-30, *Agron. Facts* (Urbana, IL: Univ. of Illinois).

PHILLIPS, R. E., et al. 1980. "No tillage agriculture," *Science*, **208**:1108–13.

PHILLIPS, R. E., and S. H. PHILLIPS (Eds.). 1984. *No-tillage Agriculture: Principles and Practices* (New York: Van Nostrand Reinhold Co.).

PIERCE, F. J., W. E. LARSON, and R. H. DOWDY. 1986. "Field estimates of *C* factors: How good are they and how do they affect calculations of erosions?" in *Soil Conservation, Assessing the National Resources Inventory*, Vol 2, (Washington, DC: National Academy Press).

SCHERTZ, D. L. 1983. "The basis for soil loss tolerance," *J. Soil Water Cons.*, **38**:10–14.

SCSA. 1983. Special issue on conservation tillage, *J. Soil Water Cons.*, **38**:134–319 (Akeny, IA: Soil Cons. Soc. Amer.).

SKIDMORE, E. L., and F. H. SIDDOWAY. 1978. "Crop residue requirements to control wind erosion," in W. R. Oschwald (Ed.), *Crop Residue Management Systems*, ASA Special Publication No. 31 (Madison, WI: Amer. Soc. Agron., Crop Sci. Soc. Amer., and Soil Sci. Soc. Amer.).

SKIDMORE, E. L., and N. P. WOODRUFF. 1968. *Wind Erosion Forces in the United States and Their Use in Predicting Soil Loss*, Agr. Handbook 346 (Washington, DC: U.S. Department of Agriculture).

SOANE, B. D., and J. C. PIGEON. 1975. "Tillage requirement in relation to soil physical properties," *Soil Sci.*, **119**:376–84.

USDA. 1975. *Minimum Tillage: A Preliminary Technology Assessment*, Part II of a report for the Committee on Agriculture and Forestry, U.S. Senate, Publ. No. 57–398. (Washington, DC: Govt. Printing Office).

USDA. 1980a. *Appraisal 1980 Soil and Water Resources*, Conservation Act Review Draft, Part I (Washington, DC: U.S. Department of Agriculture).

USDA. 1980b. *National Summary: Evaluation of the Agricultural Conservation Program* (Washington, DC: U.S. Department of Agriculture).

USDA. 1987. *The Second RCA Appraisal, Analysis of Conditions and Trends*, Review Draft (Washington, DC: U.S. Department of Agriculture).

VAN DOREN, D. M., Jr., G. B. TRIPETT, Jr., and J. E. HENRY. 1976. "Influence of long-term tillage, crop rotation and soil type combinations on corn yields," *Soil Sci. Soc. Amer. J.*, **40**:100–105.

WALKER, R. D. 1980. "USLE, a quick way to estimate your erosion losses," *Crops and Soils*, **33**:10–13.

WISCHMEIER, W. J., and D. D. SMITH. 1978. *Predicting Rainfall Erosion Loss—A Guide to Conservation Planning*, Agr. Handbook No. 537 (Washington, DC: U.S. Department of Agriculture).

16

Fertilizers and Fertilizer Management

Although the use of animal manure was common as far back as agricultural records can be traced, chemical fertilizers have been extensively employed for little more than 100 years. They are now an economic necessity on many soils. Any inorganic salt, such as ammonium nitrate, or an organic substance, such as urea, used to promote crop production by supplying plant nutrients is considered to be a "commercial" fertilizer.

16.1

The Fertilizer Elements

There are at least fourteen essential elements, including six macronutrients, that plants obtain from the soil. Two of the macronutrients, *calcium* and *magnesium*, are applied as lime in regions deficient in these elements. Although not usually rated as a fertilizer, lime does exert a profound nutritive effect. *Sulfur* is applied in commercial fertilizers, especially in areas where little sulfur is returned to the soil from the atmosphere. Three other macro elements—*nitrogen*, *phosphorus*, and *potassium*—are commonly applied in commercial fertilizers and are often referred to as the *fertilizer elements*.[1]

The quantities of nitrogen, phosphorus, and potassium consumed in the United States in the past 20 years or so are shown in Figure 16.1. Note that the use of nitrogen increased markedly from 1965 to 1981 but that the consumption of all fertilizers has declined since then because of temporary surplus food production.

[1] For more specific information on fertilizers, see Englestad et al. (1985).

471

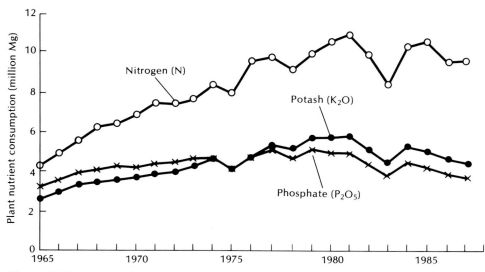

Figure 16.1

Primary nutrient content of fertilizers consumed in the United States, 1965–1987. Nitrogen use has increased very rapidly since the 1960s, although the consumption of all fertilizers declined in the mid-1980s because of surplus food production. Note that by fertilizer convention the phosphorus and potassium are expressed in the oxide form, although they are actually present in various fertilizer compounds. [From Hargett et al. (1987)]

The fertilizer carriers of each of these three elements will be discussed in a general way, and the outstanding characteristics of the more important ones will be noted. Inorganic nitrogen sources will be considered first.

16.2

Inorganic Nitrogen Carriers

GENERAL CONSIDERATIONS

Many inorganic carriers are used to supply fertilizer nitrogen, as shown in Table 16.1. There is a wide range in the nitrogen contents of these carriers— from 8% in ammonium polyphosphates to 82% in anhydrous ammonia.

The materials listed in Table 16.1 have one thing in common—they can all be produced synthetically starting with atmospheric nitrogen. Thus the quantity of nitrogen available to produce these compounds is in any practical sense unlimited. Unfortunately, the energy required to synthesize nitrogen fertilizers is very high and generally far exceeds that required to produce all other fertilizers. This high energy requirement places constraints on the production and use of nitrogen fertilizers and is a strong argument for their efficient use.

AMMONIA AND ITS SOLUTIONS

Ammonia gas is formed from the elements hydrogen and nitrogen under very high temperatures and pressures and in the presence of an appropriate catalyst.

$$N_2 + 3 H_2 \rightarrow 2 NH_3$$

TABLE 16.1
Nitrogen Carriers

Fertilizer	Chemical form	Source	Nitrogen (approx. %)
Sodium nitrate	$NaNO_3$	Chile saltpeter and synthetic	16
Potassium nitrate	KNO_3	Synthetic	13
Ammonium sulfate	$(NH_4)_2SO_4$	Synthetic; by-product from coke and gas	21
Ammonium nitrate	NH_4NO_3	Synthetic	33
Calcium nitrate	$Ca(NO_3)_2$	Synthetic	15
Cal-nitro and A.N.L.	NH_4NO_3 and dolomite	Synthetic	20
Urea	$CO(NH_2)_2$	Synthetic	45–46
Calcium cyanamid	$CaCN_2$	Synthetic	22
Anhydrous ammonia	Liquid NH_3	Synthetic	82
Aqua ammonia	Dilute NH_4OH	Synthetic	20–25
Nitrogen solutions	NH_4NO_3 + urea in water	Synthetic	28–32
Monoammonium phosphate	$NH_4H_2PO_4$ (mostly)	Synthetic	11 (48% P_2O_5)
Diammonium phosphate	$(NH_4)_2HPO_4$ (mostly)	Synthetic	21 (53% P_2O_5)
Ammonium polyphosphates	$(NH_4)_3HP_2O_7$; $NH_4H_2PO_4$; $(NH_4)_3H_2P_3O_{10}$	Synthetic	12–15 (60–62% P_2O_5)

   Natural gas is the source of hydrogen and the atmosphere the source of nitrogen. The ammonia (NH_3) formed by this reaction is the least expensive compound per unit of nitrogen of any listed in Table 16.1. Furthermore, the quantity of nitrogen supplied by NH_3 in the United States far exceeds that of any other nitrogen carrier (Table 16.2).

   The NH_3 gas is used in at least three ways. It may be liquefied under pressure; large quantities of the anhydrous ammonia thus formed are used as a separate material for direct application. Some NH_3 gas may be dissolved in water, yielding NH_4OH (aqua ammonia), which is used in the production of some liquid fertilizers or is applied directly to the soil. The NH_3 gas is also used to manufacture other inorganic nitrogen fertilizers. A careful study of

TABLE 16.2
**Total Nutrients Contained in Selected Fertilizers
Consumed in the United States in 1987**

In thousands of metric tons.

Nitrogen materials	Nutrient N	Phosphorus carriers	P_2O_5	P
Ammonia[a]	3498	Diammonium phosphate	1346	588
Nitrogen solutions	1778	Concentrated superphosphate	272	119
Urea	1393	Phosphoric solutions	26	11
Diammonium phosphate	527	Potassium carriers	K_2O	K
Ammonium nitrate	517			
Ammonium sulfate	121	Muriate of potash	2715	2253
		Potassium sulfate	31	26

Calculated from Hargett et al. (1987).

[a] Primarily anhydrous ammonia.

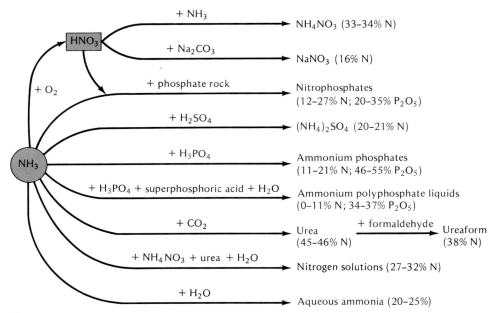

Figure 16.2
How various fertilizer materials may be synthesized from ammonia. This gas is obtained as a by-product of coke manufacture and, in even larger quantities, by direct synthesis from elemental nitrogen and hydrogen. In recent years, ammonia and other synthetic materials shown have supplied most of our fertilizer nitrogen.

Figure 16.2 will help to establish the relationship between ammonia and the materials derived synthetically from it. These substances will be considered briefly.

NITROGEN SOLUTIONS

Next to anhydrous ammonia, nonpressure nitrogen solutions are the most widely used nitrogenous fertilizers in the United States. They are comprised primarily of urea and ammonium nitrate dissolved in water and contain 28–32% nitrogen. Since they are not under pressure, these nitrogen solutions are safe to handle and easy to apply directly to the soil. They are also used in the formulation of mixed soil and liquid fertilizers.

UREA

Another important synthetic is urea [$CO(NH_2)_2$], a fertilizer containing about 45% nitrogen. The use of this material is increasing markedly. World-wide it is the most extensively used nitrogen carrier. Urea readily undergoes hydrolysis in the soil, producing ammonium carbonate.

$$CO(NH_2)_2 + 2\,H_2O \xrightarrow{\text{urease}} (NH_4)_2CO_3$$

Since the ammonium carbonate produced can be nitrified, urea applications ultimately provide both NH_4^+ and NO_3^- ions for plant absorption. Unfortunately, ammonium carbonate is unstable at pH values above 7, releasing gaseous NH_3 to the atmosphere. Consequently, it is best to incorporate urea into the soil rather than to leave it on the soil surface, especially if the soil is alkaline. In addition, the NH_3 from hydrolyzed urea has some toxicity to seeds and seedlings. Consequently, it must be properly placed to prevent damage.

SULFATE OF AMMONIA

Sulfate of ammonia, $(NH_4)_2SO_4$, is produced synthetically, as shown in Figure 16.2, and is also a by-product of the manufacture of coke and synthetic fibers. This material is used in the manufacture of mixed fertilizers. Its nitrogen is somewhat more expensive than that of the liquid forms and of urea, and it has a strong residual acidifying effect (see Section 16.9). It is a useful fertilizer for acid-loving plants.

AMMONIUM, SODIUM, POTASSIUM, AND CALCIUM NITRATES

Nitric acid (HNO_3) can be manufactured by oxidizing ammonia (see Figure 16.2) and is used in making ammonium salts, and nitrates of sodium, potassium, and calcium. Ammonium nitrate has the advantage of supplying both NH_4^+ and NO_3^- ions. The percentage of nitrogen in ammonium nitrate carriers ranges from 20% for the lime-containing Cal-nitro and ammonium nitrate–lime (A.N.L.) to about 33% for the higher grades of ammonium nitrate (Table 16.1).

Sodium nitrate is also obtained as a natural product, saltpeter, from salt beds in Chile. Because of its high cost per unit of nitrogen, however, saltpeter is only a minor supplier of nitrogen.

Calcium nitrate is manufactured by combining nitric acid and calcium carbonate. It is also a by-product of the manufacture of nitrophosphate fertilizers, which are popular in Europe. Calcium nitrate is very hygroscopic (water absorbing), so it must be stored in water-tight containers. Potassium nitrate is made by combining nitric acid and potassium chloride. It supplies both K^+ and NO_3^- ions and is used for high-value vegetables and fruit crops.

AMMONIUM PHOSPHATES

The ammonium phosphates are important fertilizers because they carry both phosphorus and nitrogen. These compounds are made from phosphoric acid and ammonia (see Figure 16.2). Since both their phosphorus and their nitrogen are water soluble, ammonium phosphates are in demand where a high degree of water solubility is required.

Ammonium polyphosphates (Table 16.1) are used especially in liquid (solution and suspension) fertilizers. In solid form these phosphates are very high in phosphorus (58–61% P_2O_5; 25–27% P) and, in addition, contain 12–15% nitrogen. Solutions and suspensions of these materials contain 34–37% P_2O_5 and 10–11% nitrogen.

OTHER SYNTHETIC NITROGEN CARRIERS

By acidulating rock phosphate with nitric acid, fertilizers called "nitrophosphates" are formed. They are apparently as effective as other materials in supplying nitrogen and are used extensively in the manufacture of complete fertilizers in Europe.

SLOW-RELEASE NITROGEN CARRIERS

For some purposes, the nitrogen supplied by most fertilizers is too readily available. For example, the homeowner wants a fertilizer that, when applied to the lawn in the spring, will maintain a grass cover throughout the summer. Unfortunately, most nitrogen fertilizers are quickly absorbed by the lawn grasses in the spring, and little is left for midsummer growth.

Some slow-release nitrogen fertilizers have been developed. An example is urea–formaldehyde complexes that contain about 38% slowly available nitrogen. Unfortunately, this material and other slow-release fertilizers are relatively expensive, a fact that limits their use to specialty crops, lawns, and turf.

Another way to reduce nitrogen release is to coat conventional fertilizers with substances that slow down their rate of solution and microbial attack. Waxes, paraffin, acrylic resins, and elemental sulfur are among the materials that have been used with some success. These substances slow down the rate of moisture penetration of the granule and the outward movement of the soluble nitrogen.

NITRIFICATION INHIBITORS

In recent years, a number of compounds have been developed to prevent nitrification (see Section 11.11), thereby slowing down the rate of leaching and possible loss of nitrogen by denitrification. These compounds are mixed with the nitrogen fertilizers or applied as a surface coating on individual pellets. The compound most thoroughly researched in the United States is nitrapyrin [2-chloro-6-(trichloromethyl) pyridine], sold under the trade name N-serve. The effect of these compounds on crop yields varies with soil and climate but is usually most beneficial in humid regions.

16.3

Phosphatic Fertilizer Materials

The primary source of phosphorus fertilizers is rock phosphate, the essential component of which is the mineral apatite, $[3Ca_3 (PO_4)_2] \cdot CaX_2$, where the X may be F, OH, Cl, and so on. Since the phosphorus in apatite is at best slowly available, this mineral must be treated with phosphoric, sulfuric, or nitric acid to change the phosphorus into more readily available forms such as $CaHPO_4$ and $Ca(H_2PO_4)_2$. The overriding importance of phosphorus availability must be kept in mind.

CLASSIFICATION OF PHOSPHATE FERTILIZERS

The various phosphorus compounds present in phosphatic fertilizers are classified in an arbitrary yet rather satisfactory way as follows.

1. Water soluble: $Ca(H_2PO_4)_2 \cdot H_2O$, $NH_4H_2PO_4$, $(NH_4)_2HPO_4$, K phosphates
2. Citrate soluble (in 15% neutral ammonium citrate): $CaHPO_4$ — available
3. Insoluble: phosphate rock $[3(Ca_3(PO_4)_2] \cdot CaX$ (X = F, OH, or Cl) — unavailable

This classification has some practical limitations because *available* phosphates may be rendered quite insoluble when they interact with the soil. Likewise, over a period of time some immediately *unavailable* phosphates may be absorbed by plant roots, especially if the soil is quite acid (see Figure 16.3).

Most state laws require that the phosphorus content of fertilizers be expressed as if it were in the oxide form, as phosphorus pentoxide (P_2O_5)

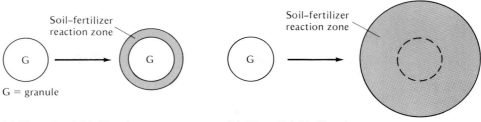

(a) Water–Insoluble Phosphate   (b) Water–Soluble Phosphate

Figure 16.3

Reaction of water-soluble and water-insoluble phosphates with soil. The water-insoluble granules react with only the small volume of soil in their immediate vicinity. Soluble phosphates move into the soil from water-soluble granules, reacting with iron, aluminum, and manganese compounds in acid soils and with calcium and magnesium compounds in alkaline soils. [From Engelstad (1965)]

rather than as the element P. However, it should be stressed that fertilizers do not contain phosphorus pentoxide as such, the oxide term merely being used as a means of expressing the phosphorus content. The most important phosphorus carriers are shown in Table 16.3.

SUPERPHOSPHATE

When rock phosphate is treated with sulfuric acid, *ordinary* superphosphate (16–20% available P_2O_5) is formed (see Figure 16.4). The reaction may be expressed as follows [representing the rock phosphate as simple $Ca_3(PO_4)_2$].

$$2\ Ca_3(PO_4)_2 + 3\ H_2SO_4 + 3\ H_2O \rightarrow Ca(H_2PO_4)_2 \cdot H_2O + 2\ CaHPO_4 + 3\ CaSO_4 \cdot 2H_2O + \text{impurities}$$
(insoluble)   (water soluble)   (citrate soluble)

Note that ordinary superphosphate consists of about 31% phosphates, 50% gypsum, and 19% impurities of various kinds. The total *phosphorus* (16–20% P_2O_5 or 7–8% P) is rather low, although there are significant quantities of sulfur and calcium present.

   Triple superphosphate contains 40–48% available P_2O_5 (17–21% P) and no

TABLE 16.3

Phosphorus Carriers

Fertilizer	Chemical form	Available P_2O_5 (%)	Phosphorus (%)
Superphosphates	$Ca(H_2PO_4)_2$; $CaHPO_4$	16–50	7–22
Ammoniated superphosphate	$NH_4H_2PO_4$; $CaHPO_4$; $Ca_3(PO_4)_2$; $(NH_4)_2SO_4$	16–18 (3–4% N)	7–8
Monoammonium phosphate	$NH_4H_2PO_4$ (mostly)	48–55 (11–12% N)	21–24
Diammonium phosphate	$(NH_4)_2HPO_4$ (mostly)	46–53 (16–18% N)	20–23
Ammonium polyphosphates	$(NH_4)_3HP_2O_7$; $(NH_4)_3H_2P_3O_{10}$; $NH_4H_2PO_4$	58–60 (12–15% N)	25–26
Basic slag	$(CaO)_5 \cdot P_2O_5 \cdot SiO_2$	15–25	7–11
Steamed bone meal	$Ca_3(PO_4)_2$	23–30	10–13
Rock Phosphate	Fluoro-, chloro-, and hydroxyapatites	25–40	11–17
Calcium metaphosphate	$Ca(PO_3)_2$	62–63	27–28
Phosphoric acid	H_3PO_4	52–54	22–24
Superphosphoric acid	H_3PO_4; $H_4P_2O_7$; $H_5P_3O_{10}$	68–76	29–33

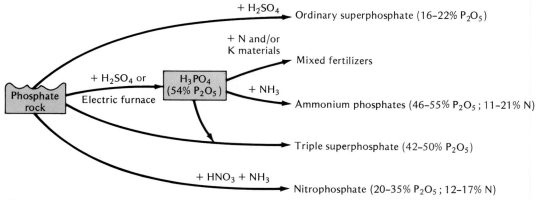

Figure 16.4

How several important phosphate fertilizers are manufactured. [Concept from Travis Hignett, Tennessee Valley Authority]

gypsum. It is synthesized by treating a high-grade phosphate rock with phosphoric acid, as shown in a simplified equation.

$$Ca_3(PO_4)_2 + 4\ H_3PO_4 + 3\ H_2O \rightarrow 3\ Ca(H_2PO_4)_2 \cdot H_2O + \text{impurities}$$

Triple superphosphate is used in higher analysis fertilizers, although in recent years it has been replaced by diammonium phosphate for many formulations.

AMMONIUM PHOSPHATES

Ammonium phosphates are the most widely used phosphorus-containing fertilizers in the United States (Table 16.2). *Diammonium phosphate (DAP)*, commonly containing 16–18% nitrogen and 46–48% P_2O_5 (20–21% P), provides more fertilizer phosphorus than any other material. DAP is popular as one of the materials in bulk blending (see Section 16.8) and as a constituent of high-analysis fertilizers.

Diammonium phosphate is manufactured by producing phosphoric acid from rock phosphate and reacting the acid with ammonia (see Figure 16.4).

$$Ca_3(PO_4)_2 + 3\ H_2SO_4 + 2\ H_2O \rightarrow 2\ H_3PO_4 + 3\ CaSO_4 \cdot 2H_2O$$

$$H_3PO_4 + 2\ NH_3 \rightarrow (NH_4)_2HPO_4$$

Other ammonium phosphate–containing materials include monoammonium phosphate (MAP) (11% N, 48–55% P_2O_5) and ammoniated superphosphate, which contains 3–4% nitrogen and 16–18% P_2O_5.

Ammonium polyphosphates, made by ammoniating superphosphoric acid, are used extensively in the manufacture of liquid fertilizers (see Section 16.8). They provide reasonably high analyses, prevent the precipitation of iron and other impurities, and keep most of the micronutrients in solution. In addition to orthophosphates (salts of H_3PO_4), the polyphosphates contain pyrophosphates and tripolyphosphates, which are salts of $H_4P_2O_7$ and $H_5P_3O_{10}$, respectively (Table 16.3).

BASIC SLAG AND BONE MEAL

Basic slag is a by-product of open-hearth steel production. It is commonly used in Europe but is on the market only to a limited extent in the United

States. Because of its alkalinity, basic slag seems to be especially effective on acid soils. Bone meal is an expensive form of phosphorus, and it is rather slowly available in the soil; its use is limited mostly to organic gardening.

ROCK PHOSPHATE

Raw rock phosphate, which is comprised of complex apatites, is the least soluble of the phosphatic fertilizers mentioned. When finely ground, it is beneficial on some acid soils and on mineral soils with considerable organic matter. Because of its low solubility, however, it will continue to be used primarily as a source for the manufacture of other much more soluble forms.

HIGH-ANALYSIS PHOSPHATES

Superphosphoric acid, the highest analysis phosphorus-containing material used in fertilizer manufacture today (68–76% P_2O_5; 29–33% P), is a mixture of orthophosphoric, pyrophosphoric, and other polyphosphoric acids. Its high phosphate concentration makes superphosphoric acid attractive when transportation distance is a factor. This liquid can be used in the manufacture of liquid-mixed fertilizers or to make a high-analysis superphosphate containing 54% P_2O_5 (24% P). In the formulation of liquid fertilizers, the polyphosphates help keep iron, aluminum, and micronutrients in solution.

NITROPHOSPHATES

An important fertilizer-manufacturing process uses nitric acid instead of sulfuric or phosphoric acid to increase the solubility of the phosphorus in rock phosphate (see Figure 16.4). The products of this process are called nitrophosphates. Reactions such as the following occur (again, simple $Ca_3(PO_4)_2$ is used to represent rock phosphate).

$$2\ Ca_3(PO_4)_2 + 6\ HNO_3 + H_2O \rightarrow$$
$$2\ CaHPO_4 + Ca(H_2PO_4)_2 \cdot H_2O + 3\ Ca(NO_3)_2$$

Nitrophosphate processes are used extensively in Europe, where sulfuric acid sources are not abundant. Also, ammonium nitrate, one of the products of the nitrophosphate process, is popular in Europe, as are mixed fertilizers in contrast to single-element materials.

16.4
Potassium Fertilizer Materials

Potassium is obtained primarily by mining underground salt (KCl) beds. Brine from salt lakes is also an important source. Kainit and manure salts (Table 16.4) are the most common of the crude potash sources. The high-grade chloride and sulfate of potash are their refined equivalents.

All potash salts used as fertilizers are water soluble and are therefore rated as readily available. Unlike nitrogen salts, most potassium fertilizers, even if employed in large amounts, have little or no effect on the soil pH. Potassium chloride (muriate) is considered to lower crop quality of potatoes and especially tobacco when used in large dosages. Hence, when large amounts of potash are to be applied for tobacco, the sulfate form is preferred.

Potassium magnesium sulfate, although rather low in potassium, is used in areas likely to be magnesium deficient. Because of magnesium's ready

TABLE 16.4
Common Potassium Fertilizer Materials

The oxide means of expression (K_2O) is commonly used to identify the nutrient level, although the elemental expression (K) is technically more correct.

Fertilizer[a]	Chemical form	K_2O (%)	K (%)
Potassium chloride	KCl	48–60	40–50
Potassium sulfate	K_2SO_4	48–50	40–42
Potassium magnesium sulfate[b]	Double salt of K and Mg	25–30	19–25
Manure salts	KCl mostly	20–30	17–25
Kainit	KCl mostly	12–16	10–13
Potassium nitrate	KNO_3	44 (13% N)	37

[a] All contain other potash salts than those listed.
[b] Contains 25% $MgSO_4$ and some chlorine.

availability in this material, it is a more desirable source of the element than either dolomitic limestone or dolomite.

16.5
Sulfur in Fertilizers

Until recently, there was little reason to be concerned with sulfur deficiency in most soils of the United States. Ample sulfur was added in rain and snow and through mixed fertilizers that contained two sulfur-containing ingredients,—ordinary superphosphate, which contains gypsum or calcium sulfate, and ammonium sulfate.

Public concern over atmospheric pollution has resulted in some reductions in sulfur additions from the atmosphere to agricultural lands. Also, modern high-analysis fertilizers contain little superphosphate and ammonium sulfate. As a result, in areas far removed from atmospheric sources of sulfur, such as eastern Washington and Oregon and parts of the Southeast and the Midwest, sulfur deficiencies have become common. The deficiencies are being met by adjusting the fertilizer constituents to include sulfur-containing materials. However, the automatic supply of sulfur through fertilizers can no longer be taken for granted. This element must be consciously added where deficiencies are apt to occur. For example, thiosulfates are used as sulfur sources for NPK solution fertilizers.

16.6
Micronutrients

The amount of micronutrients in fertilizers must be much more carefully controlled than the macronutrients. The difference between the deficiency and toxicity levels of a given micronutrient is extremely small. Consequently, micronutrients should be added only when their need is certain and when the amount required is known.

When a trace element deficiency is to be corrected, a salt of the lacking nutrient is usually added separately to the soil (Table 16.5). Copper, manganese, iron, and zinc are commonly supplied as sulfate salts, and boron as borax. Molybdenum is added as sodium molybdate. Iron, manganese, and

TABLE 16.5
Salts of Micronutrients Commonly Used in Fertilizers

Compound	Formula	Nutrient content (%)
Sodium borate (borax)	$Na_2B_4O_7 \cdot 10H_2O$	11
Cupric oxide	CuO	75
Copper sulfate	$CuSO_4 \cdot 5H_2O$	35
Ferric sulfate	$Fe_2(SO_4)_3 \cdot 4H_2O$	23
Ferrous sulfate	$FeSO_4 \cdot 7H_2O$	19
Manganous oxide	MnO	41–68
Manganous sulfate	$MnSO_4 \cdot 4H_2O$	26–28
Ammonium molybdate	$(NH_4)_6Mo_7O_{24} \cdot 2H_2O$	54
Sodium molybdate	$Na_2MoO_4 \cdot 2H_2O$	39
Zinc oxide	ZnO	80
Zinc sulfate	$ZnSO_4 \cdot 7H_2O$	23

Selected data from Tisdale et al. (1985).

zinc are often sprayed in small quantities as chelates or sulfates on the leaves (foliar application) rather than being applied directly to the soil. Porous, glassy (fritted) silicates prepared by fusing compounds of boron, manganese, iron, and zinc with silica may also be used to supply these nutrients.

Chelates are used as suppliers of iron, zinc, manganese, and copper (see Section 13.6), especially on soils of high pH. Because of their high cost, however, these materials are often used as foliar sprays, which permit much lower application rates.

Micronutrients are also being applied in liquid fertilizers in areas where the quantities of these nutrients required has been carefully ascertained. The presence of polyphosphates in many of these fluid materials protects the micronutrients from precipitation as insoluble iron and other compounds. By including micronutrients in the liquid mixes, application costs for these materials are kept low.

The total quantity of micronutrient fertilizer used is small compared to that of the macronutrients. In 1987 about 360,000 tons of fertilizer materials of all kinds were used in the United States to supply specific nutrients other than nitrogen, phosphorus, and potassium.

16.7

Organic Sources of Nutrients

In recent years, some individuals and groups have expressed preferences for foodstuffs grown on soils to which only natural organic materials have been added. The demand for these foods is being met by farmers and gardeners who practice what some have termed "organic gardening." Although most of the natural organic materials are from manure or from crop residues grown on these farms, some commercially available organic materials are also used. Since Chapter 18 deals with farm manure and other organic wastes, brief reference will be made here only to commercially available organic fertilizers.

In 1987, about 200,000 tons of so-called "natural organic" fertilizers were sold in the United States. These commercial organic materials include dried poultry and cow manures, peats of various kinds, mixtures of peat and manure, composted organic residues, and dried domestic wastes. Ground bone meal, dried blood, oil seed meals, fish tankage, and other food-processing wastes may be included. The readily available nutrients in these materials are only a fraction of those in commercial inorganic fertilizers, and the cost of the nutrients they supply is frequently high. However these materials do provide slowly available essential elements and generally have beneficial effects on the physical condition of soils. If sufficient quantities of the organic materials are applied, good crop yields can be obtained. Furthermore, because the prices for organically grown foods are generally higher than for those produced using inorganic chemicals, farmers and gardeners can justify the higher fertilizer costs.

16.8

Mixed Fertilizers

For years, farmers have used mixtures of materials that contain at least two of the "fertilizer elements" and usually all three. For example, an ammonia solution, triple superphosphate, muriate of potash, and a very small quantity of an organic might be used if a mixed fertilizer is desired. Such complete (NPK) fertilizers supply about one fourth of the total fertilizer nutrient consumption in the United States, the remainder coming from separate materials (Table 16.6). The nutrient use per hectare for various regions of the United States is shown in Figure 16.5.

BULK BLENDING

As of the late 1950s, essentially all mixed fertilizers were bagged at the manufacturing plant and shipped to distribution points for sale to farmers. Today, bulk handling and blending of fertilizers are the common procedures, with more than half of the total fertilizers applied in the United States being sold this way. Free-flowing granular materials are shipped in bulk to a small blending plant (Figure 16.6), the fertilizers are mixed to the customer's desired analysis and then are sent directly to the farm where they are spread on the field.

TABLE 16.6

Consumption of Single- and Multiple-Nutrient Fertilizers in the United States During Year Ending June 30, 1987

Materials	Consumption Thousand Mg	Percent of total
Single-nutrient fertilizers		
Nitrogen	16,474	43.6
Phosphorus	982	2.6
Potassium	4,771	12.6
Total	(22,227)	(58.9)
Multiple-nutrient fertilizers		
NP mixtures	5,273	14.0
NK mixtures	785	2.1
PK mixtures	631	1.7
NPK mixtures	8,846	23.4
Total	(15,534)	(41.1)
Grand total	37,761	100

From Hargett et al. (1987).

Figure 16.5
Nutrients added in fertilizers applied to cropland in different regions of the United States (represented as N, P_2O_5, and K_2O in kg/ha). While the South Atlantic uses more nutrients per hectare than the other regions, total fertilizer usage is highest in the central part of the country. [Courtesy U.S. Department of Agriculture]

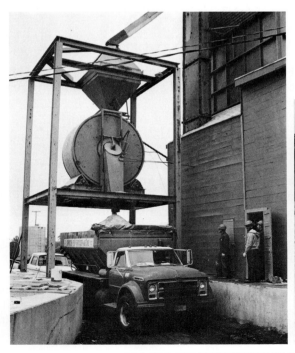

Figure 16.6
Bulk blending plant and bulk spreading trucks. Different materials are blended in a small plant and loaded directly on a truck (left). Truck in the field spreading the fertilizer (right). [Courtesy Tennessee Valley Authority]

Bulk blending and handling reduce the costs of labor, storage, production, transportation, and spreading the fertilizer. Also, the bulk-handled fertilizers are generally high in analysis and may include micronutrients as needed in specific field situations. In some cases, pesticides are mixed with the bulk fertilizers, either at the blending plant or in the field, and are then applied to the soil.

Fertilizer materials commonly used in the bulk-blending process are urea, ammonium nitrate, ammonium sulfate, ammonium phosphates, triple superphosphate, and potassium chloride. These compounds are also often applied as separate materials rather than in mixtures.

LIQUID FERTILIZERS

Another innovation in the formulation and handling of mixed fertilizers is the use of liquid fertilizers, sometimes termed *fluid* fertilizers. In the United States in 1987 nearly 40% of the total tonnage of fertilizers and more than half of the single-nutrient materials were applied in the liquid form. As with bulk blending, liquid fertilizers have the advantage of low labor costs because the materials are stored in tanks and pumped out for application. However, the analyses of the multinutrient liquid fertilizers are somewhat lower than for solid fertilizers. Slightly higher analyses are possible by using superphosphoric acid or ammonium polyphosphates as a source of at least part of the phosphorus. Also, "suspension fertilizers," wherein a small amount of solids is suspended in the liquid fertilizer and 1–2% clay is added to keep the solids in suspension, permit higher analyses, such as 13–13–13. Suspension fertilizers make up about 40% of the fluid market in the United States.

16.9

Effect of Mixed Fertilizers on Soil pH

ACID-FORMING FERTILIZERS

Fertilizers that supply ammonia or produce ammonia when added to the soil tend to increase soil acidity. The following reaction accounts for this acidifying effect.

$$NH_4^+ + 2\ O_2 \rightarrow 2\ H^+ + NO_3^- + H_2O$$

In addition to ammonium compounds, materials such as urea, which upon hydrolysis yield ammonium ions, are potential sources of acidity. The phosphorus and potash fertilizers commonly used have little effect on soil pH unless they also contain nitrogen. The approximate acid-forming capacities of some fertilizer materials, expressed in kilograms of calcium carbonate needed to neutralize the acidity produced by 20 kg of nitrogen supplied, are as follows.

Ammonium sulfate	107	Ammonium nitrate	36
Ammo-phos	100	Cottonseed meal	29
Anhydrous ammonia	36	Castor pomace	18
Urea	36	High-grade tankage	15

The base-forming capacity, expressed in the same terms, for certain fertilizers is

Nitrate of soda	36	Potassium nitrate	86
Calcium nitrate	27	Cocoa meal	12
Calcium cyanamid	57		

It should be recognized that some nutrient-containing materials are added specifically to increase soil acidity. For example, elemental sulfur, ammonium sulfate, and compounds such as iron or aluminum sulfates can reduce the soil pH and thereby enhance the growth of plants such as azaleas and rhododendron (see Section 8.9).

NON-ACID-FORMING FERTILIZERS

In some instances, the acid-forming tendency of nitrogen fertilizers is completely counteracted by adding dolomitic limestone to the mixture. Such fertilizers are termed *neutral* or *non-acid-forming* and exert little residual effect on soil pH. However, it is economically preferable in commercial agriculture to use acid-forming fertilizers and subsequently neutralize the soil acidity with separate bulk applications of lime.

16.10

The Fertilizer Guarantee

Laws generally require that every fertilizer material carry a guarantee as to its content of nitrogen, phosphorus, and potassium (and other elements such as sulfur, if guaranteed). The *total nitrogen* is usually expressed in its *elemental form* (N). The phosphorus is quoted in terms of *available phosphorus* (P) or *available phosphoric acid* (P_2O_5); the potassium as water-soluble potassium (K) or *water-soluble potash* (K_2O); and the sulfur as *elemental sulfur* (S). It should be emphasized that most state laws require that the P and K contents be expressed in terms of the oxides. In any case, it is the phosphorus and potassium supplied by fertilizers that are of concern to us.

Thus, an 8-16-16 fertilizer contains 8% *total nitrogen*, 16% *available* P_2O_5 (7% available P), and 16% *water-soluble* K_2O (13% water-soluble K). In some states, the percentages of *water-soluble nitrogen*, *water-insoluble nitrogen*, and *available insoluble nitrogen* are required by law to be shown on the label, particularly for lawn fertilizers.

If any of the other essential elements is identified as being present in the fertilizer, its guaranteed content must be listed. Sulfur, magnesium, and the micronutrients are among those elements whose content is most commonly guaranteed.

Commercial fertilizers are sometimes grouped according to their nutrient *ratio*. For instance, a 5-10-10, a 10-20-20, and a 12-24-24 all have a 1:2:2 ratio. These fertilizers should give essentially the same results when applied in equivalent amounts. Thus, 1000 kg of 10-20-20 furnishes the same amounts of N, P, and K as does 2000 kg of 5-10-10.

This grouping according to ratios is valuable when several analyses are offered and the comparative costs become the deciding factor as to which should be purchased.

16.11

Fertilizer Inspection and Control

To ensure product quality, laws controlling the sale of all fertilizers are necessary. Such regulations protect the public as well as the reliable fertilizer companies by keeping products of unknown value off the market. The manufacturer is commonly required to print the following data on the bag or on an authorized tag in the case of bulk shipments.

1. Number of net pounds or kilograms of fertilizer to a package.
2. Name, brand, or trademark.
3. Chemical composition guaranteed.
4. Potential acidity in terms of pounds of calcium carbonate per ton.
5. Name and address of manufacturer.

An example of how this information often appears on the fertilizer bag is shown in Figure 16.7.

16.12

Fertilizer Economy

HIGH VERSUS LOW ANALYSES

In general, the higher the analysis of a mixed fertilizer, the lower is the cost per kilogram of nutrient. This is due to lower per unit handling costs of the high analysis fertilizers (Figure 16.8). Obviously, from the standpoint of economy, and 8-16-16 fertilizer furnishes more nutrients *per dollar* than a 4-8-8 or a 5-10-10.

RELATIVE COSTS OF NITROGEN, PHOSPHORUS, AND POTASSIUM

Another price factor to remember in purchasing fertilizers is the relative costs of nitrogen, phosphorus, and potassium. Phosphorus is generally more costly per unit of P than a similar unit of nitrogen while potassium carriers are usually least expensive. However, there is considerable variation in the cost of different carriers for a given element. Consequently, the manufactured cost of nitrogen in some solid carriers (e.g., ammonium nitrate) is about double that in anhydrous ammonia. Likewise, at the manufactur-

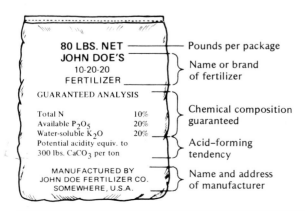

Figure 16.7

Fertilizer bag on which is printed the information commonly required by law in most states. Note the acid-forming tendency of this particular fertilizer.

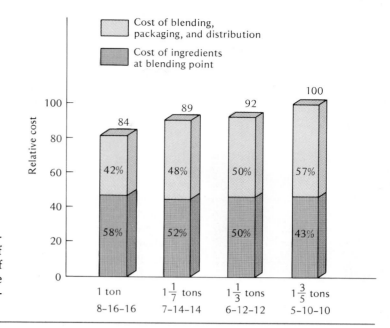

Figure 16.8
Relative farm costs of nutrients in fertilizer of different concentrations. Note that the same amount of *nutrients* is supplied by the indicated quantities of each fertilizer. These nutrients are furnished more cheaply by the higher analysis goods, mainly because of the lower handling costs.

ing plant, potassium sulfate (sulfate of potash) is about twice as expensive as an equivalent amount of potassium chloride (muriate of potash).

16.13

Movement of Fertilizer Salts in the Soil

The movement of fertilizer salts in the soil is of practical importance. As previously indicated, phosphorus compounds tend to move very little, except in the more sandy soils and some organic soils (see Section 14.13). Consequently, for maximum effectiveness this element should be placed in the root zone. Surface applications, unless worked into the soil, do not supply the deeper roots with phosphorus. Also, because phosphorus is immobile, its loss by leaching is minimal in mineral soils.

Potassium, on the other hand, and nitrate nitrogen, to an even greater extent, tend to move from their zones of placement. This movement is largely *vertical*, the salts moving up or down depending on the direction of water movement. These translocations greatly influence the time and method of applying nitrogen and potassium. For example, it is undesirable to supply nitrogen in one annual application because of the leaching hazard. Yet this tendency of nitrates to move downward is sometimes an advantage when fertilizer is applied on the soil surface. Water movement in the latter case carries the dissolved nitrate salts down to the plant roots. However, top dressings of nitrogen solutions and urea present problems in some cases because of the danger of volatilization (see Section 11.7).

The movement of nitrogen and, to a lesser degree, potassium must be considered in the *placement* of the fertilizer with respect to the seed. If the fertilizer salts are located in a band directly under the seeds, the upward movement of nitrates and some of the potassium salts by capillary water often results in injury to the plants. Rain immediately after planting followed by a long dry spell encourages such damage to seedlings. Placing the fertilizer immediately above the seed or on the soil surface also may result in injury, especially to row crops. The possibility of such injuries should be kept in mind while reading the following section.

16.14

Application of Solid Fertilizers

A fertilizer should be placed in the soil in such a position that it will serve the plant to the best advantage. This involves not only different zones of placement but also the time of year the fertilizer is to be applied.

ROW CROPS

Cultivated crops such as corn, cotton, and potatoes are usually fertilized in the *row*, with part or all of the fertilizer being applied at the time of planting. The fertilizer usually is placed in a narrow *band* on one or both sides of the row, 5–6 cm away and a little below the seed level (Figure 16.9).

When the amount of fertilizer is large, as is often the case with vegetable crops, it is wise to spread at least part of it over the soil surface (broadcast it) and thoroughly work it into the soil before the planting is done. In some cases an additional amount of fertilizer is applied on the surface alongside the plants (side dressed) later in the season.

SMALL GRAINS

With small grains such as wheat, the planting drill is equipped with a fertilizer distributor, so that the fertilizer enters the soil more or less in contact with the seed. Since there are many rows per hectare for small grains, the quantity of fertilizer applied in each row is small. Consequently, as long as the fertilizer is low in analysis and the amount applied does not exceed 300 or 400 kg/ha, germination injury is not serious. Higher rates, especially of high-analysis fertilizers containing nitrogen and potassium, may result in serious injury if the seed and fertilizer are placed together. Modern grain drills are equipped to place the fertilizer alongside the seed rather than in contact with it.

PASTURES AND MEADOWS

With meadows, pastures, and lawns, fertilizer may be applied with the seed or, better, broadcasted and worked thoroughly into the soil as the seedbed is prepared. During succeeding years it may be necessary to top-dress such crops with a suitable fertilizer mixture, being careful to avoid injury to the foliage.

TREES

Orchard trees usually are treated individually, the fertilizer being applied around each tree within the spread of the branches but beginning several feet from the trunk. The fertilizer is worked into the soil as much as possible. When the orchard cover crop needs fertilization, it is treated separately, the fertilizer being drilled in at the time of seeding or broadcast later. In untilled orchards, fertilizers should be placed where roots are most active and where rainfall or irrigation water will wash nitrogen and potassium into the root zone.

Ornamental trees may be fertilized by what is called the *perforation* method. Numerous small holes are dug around each tree within the outer half of the branch-spread zone and extending well into the upper subsoil. A suitable amount of an appropriate fertilizer is placed into each of these holes,

Figure 16.9
Best fertilizer placement for row crops is to the side and slightly below the seed. This placement eliminates danger of fertilizer "burn" and concentrates the nutrients near the seed. [Courtesy National Plant Food Institute, Washington, DC]

which are afterward filled up. This method of application places the nutrients within the root zone and avoids an undesirable stimulation of the grass that may be growing around ornamental trees.

BULK SPREADING

For economic reasons, much of the fertilizer used in the United States is spread by truck directly on the soil surface (see Figure 16.6). The fertilizer should be spread on the land as close to planting time as possible because losses from the soil surface to the atmosphere of some nitrogen compounds such as ammonia can occur. Unfortunately, efficiency of use of bulk-spread fertilizer for row crops is not as high as is the row placement. Consequently, row placement of part of the fertilizer at planting time can supplement the bulk spreading before planting.

16.15

Application of Liquid Fertilizers

Three primary methods of applying liquid fertilizers have been used: (a) direct application to the soil, (b) application in irrigation water, and (c) spraying plants with suitable fertilizer solutions.

APPLIED DIRECTLY TO SOIL

The practice of making direct applications of anhydrous ammonia, nitrogen solutions, and mixed fertilizers to soils is rapidly increasing throughout the United States (see Sections 16.2 and 16.8). Anhydrous ammonia and pressure solutions must be injected into the soil to prevent losses by volatilization. Depths of 15 cm and 5 cm, respectively, are considered adequate for these two materials.

APPLIED IN IRRIGATION WATER

In the western part of the United States, particularly in California and Arizona, liquid fertilizers are sometimes applied in irrigation waters. Liquid ammonia, nitrogen solutions, phosphoric acid, and even complete fertilizers are dissolved in the irrigation stream or the overhead sprinkler system. The nutrients are thus carried into the soil in solution. Application costs are reduced and relatively inexpensive nitrogen carriers can be used. Some care must be taken, however, to prevent ammonia loss by evaporation.

The use of drip irrigation systems has greatly facilitated the application of nutrients in irrigation water. Because such water applications are made quite frequently, the nutrient-containing water is delivered readily to the root zone and the efficiency of nutrient use is high.

APPLIED AS SPRAY ON LEAVES

Diluted NPK fertilizers, micronutrients, or small quantities of urea can be sprayed directly onto plants, although care must be taken to avoid significant concentrations of Cl^- or NO_3^-, which can be toxic to some plants. This type of fertilization is unique. It does not involve extra procedures or machinery because the fertilizer is often applied simultaneously with insecticides. Pineapples, citrus, and apple trees respond especially well to urea because much of the nitrogen is absorbed by the leaves. Moreover, what drips or is washed off is not lost, for it falls on the soil and may later be absorbed by the plants.

16.16

Factors Affecting the Kind and Amount of Fertilizers to Apply

Many factors influence decisions as to the kind and amount of fertilizers to be applied. There must be some assurance that factors other than nutrient supply are not seriously limiting crop growth. Deficiencies or excesses of soil moisture will limit fertilizer effectiveness, and fertilizer response may be meager on very acid soils containing toxic quantities of aluminum.

KIND AND ECONOMIC VALUE OF FERTILIZED CROP

The kind and amount of fertilizer to be applied is influenced by the responsiveness of the crop to the added nutrients and by the economic value of the crop. Since cereal crops such as corn are quite responsive to nitrogen, high rates of this element are economically justified for these crops. Likewise, high-value crops such as some vegetables will provide economic returns from relatively high rates of fertilizer, applications as high as 1500 kg/ha of analyses such as 15-15-15 being common.

With a given crop, the profitable rate of fertilization will depend on the ratio of the value of the crop being produced to fertilizer costs. Data in Table 16.7 show that if this ratio is high, higher rates of fertilizer will be profitable. If the ratio is low, high fertilizer rates will be uneconomical.

It must always be remembered that the very highest yields obtainable under fertilizer stimulation are not always the ones that give the best return on the money invested (Figure 16.10). In other words, the law of diminishing returns is a factor in fertilizer practice regardless of the crop being grown.

NUTRIENT CONTENT OF SOILS

To determine the kind and quantity of fertilizer to add to soil, it is necessary to know what nutrient elements (or element) are deficient. The total quanti-

TABLE 16.7

Most Profitable Nitrogen Rate for Corn Based on a Computer Model for Predicting Yield and Corn/Nitrogen Price Ratios

Corn/nitrogen price ratio	Most profitable nitrogen rate (lb/acre) to reach each of four yield potentials[a]			
	100 bu/acre (6.3)	130 bu/acre (8.2)	160 bu/acre (10.0)	190 bu/acre (12.0)
5:1	90 (101)	110 (123)	140 (157)	170 (191)
10:1	110 (123)	140 (157)	180 (202)	210 (235)
15:1	120 (135)	160 (179)	200 (224)	240 (269)
20:1	130 (146)	170 (191)	210 (235)	250 (280)
25:1	140 (157)	180 (202)	220 (247)	260 (291)

Calculated from Michigan State University data quoted by Spies (1976).

[a] Metric units for yield potential (Mg/ha) and for nitrogen rate needed (kg/ha) are given in parentheses.

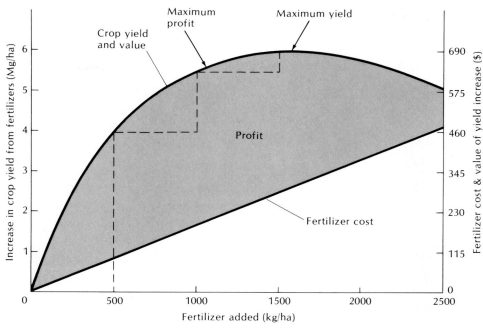

Figure 16.10
Relationship among rate of fertilizer addition, crop yield increase, fertilizer costs, and profit from adding fertilizers. Note that the yield increase (and profit) from the first 500 kg of fertilizer was much greater than from the second and third 500 kg. Also note that the maximum profit is obtained at a lower fertilizer rate than that needed to give maximum yield.

ties of the elements present in soils give little if any information on the availability of the essential elements to plants. Consequently, analyses suitable for measuring only the available portion of a given nutrient constituent must be employed for agricultural purposes. Such analyses may be termed *partial* because only a portion of the total quantity of a soil constituent is determined. Partial analyses will now receive attention.

16.17

Quick Tests for Available Soil Nutrients

In general, *partial analyses* attempt to extract from the soil amounts of essential elements that are correlated with the nutrients taken up by plants. The extraction solutions employed have varied from strong acids such as sulfuric (H_2SO_4) to weak solutions of carbon dioxide in water. Buffered salt solutions such as sodium or ammonium acetate are the extracting agents now most commonly used.

GENERAL CONSIDERATIONS

The group of tests most extensively used for nutrient availability are the *rapid* or *quick tests.* As the name implies, the individual determinations are quickly made, a properly equipped laboratory being able to handle thousands of samples a month.

A weak extracting solution such as buffered ammonium acetate is generally employed in rapid tests, and only a few minutes are allowed for the extraction of calcium, potassium, and magnesium. Most of the nutrients removed are those rather loosely held by the colloidal complex. Available phosphorus is estimated using acidic extracting solutions for humid region soils and alkaline solutions for soils of dry areas. The test for a given constituent may either be reported in general terms such as *low, medium,* or *high* or be specified in kilograms per hectare. Tests are most commonly conducted for those constituents held largely in inorganic combination— phosphorus, potassium, calcium, and magnesium. Aluminum, iron, and manganese are sometimes included. Although nitrogen is sometimes included, the expected release of this element through action of microorganisms is difficult to predict.

LIMITATIONS OF QUICK TESTS

Perhaps the first limitation is the difficulty of proper *field sampling of soils.* Generally, only a very small sample is taken, perhaps a pint or so from one or more hectares of land. The chance of error, especially if only a few borings are made, is quite high. Therefore, a composite sample of at least 15–20

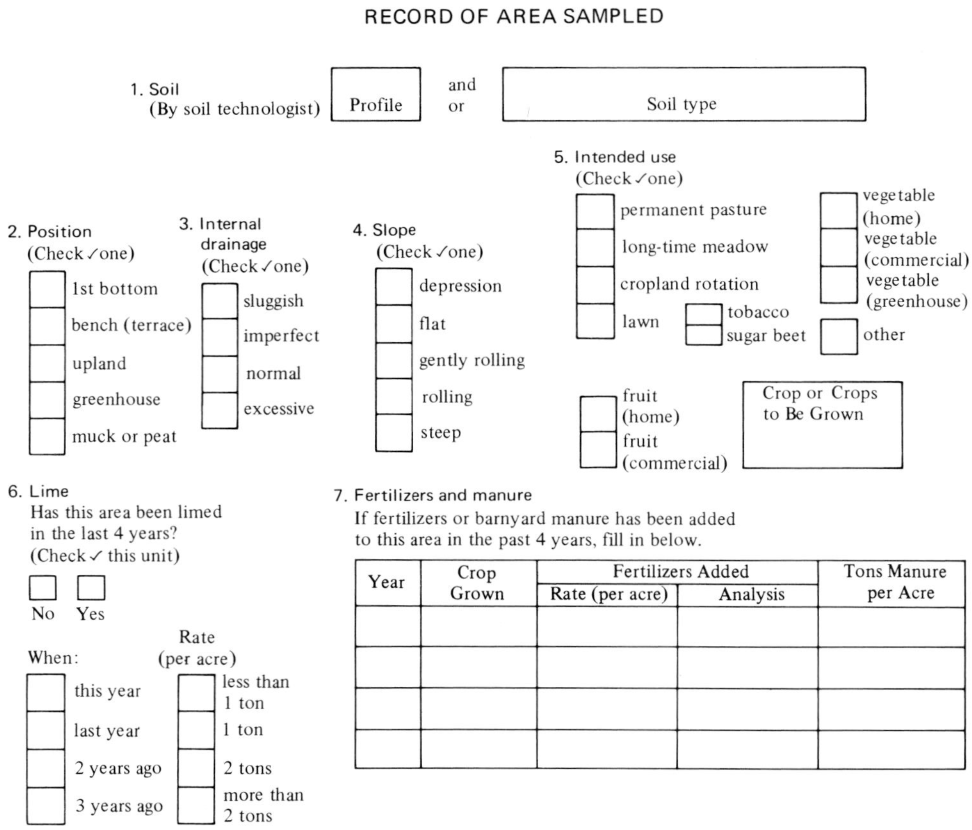

RECORD OF AREA SAMPLED

Figure 16.11

Satisfactory interpretation of a quick test analysis of a soil cannot be made without certain pertinent field data. This record sheet suggests the items that are most helpful to the expert in making a fertility recommendation. [Modified from a form used by the Ohio Agricultural Experiment Station]

REPORT ON SOIL TESTS
AUBURN UNIVERSITY
SOIL TESTING LABORATORY
Auburn University, AL 36849

NAME   ALABAMA RESIDENT
ADDRESS   118 MAIN STREET
CITY   HOMETOWN, AL 36830

COUNTY   LEE
DISTRICT   2
DATE   1/04/80

LAB. NO.	SENDER'S SAMPLE DESIGNATION	AND	CROP TO BE GROWN	SOIL* GROUP	pH**	PHOSPHORUS P***		POTASSIUM K***		MAGNESIUM Mg***		LIME-STONE TONS/ACRE	N	P₂O₅	K₂O
													POUNDS PER ACRE		
23887	1		SOYBEANS	2	5.3	L	70	M	70	H	160	2.0	0	80	40

COMMENT 224–SOIL ACIDITY (LOW pH) CAN BE CORRECTED WITH EITHER DOLOMITIC OR CALCITIC LIME.

| 23888 | 2 | | CORN | 1 | 5.6 | L | 70 | M | 70 | H | 160 | 1.0 | 120 | 80 | 40 |

SEE COMMENT 224 ABOVE.

COMMENT 15–CORN ON SANDY SOILS MAY RESPOND TO NITROGEN RATES UP TO 150 LBS. PER ACRE. ON SANDY SOILS APPLY 3 LBS. ZINC (Zn) PER ACRE IN FERTILIZER AFTER LIMING OR WHERE pH IS ABOVE 6.0.

| 23889 | 3 | | BAHIA | 1 | 6.0 | M | 100 | H | 140 | H | 240 | 0.0 | 60 | 40 | 0 |
| 23890 | GARDEN | | VEGETABLES | 3 | 5.2 | M | 90 | M | 70 | H | 160 | 3.0 | 120 | 120 | 120 |

SEE COMMENT 224 ABOVE.

COMMENT 82–PER 100 FT. OF ROW APPLY 6 LBS. 8-8-8 (3 QUARTS) AT PLANTING AND SIDEDRESS WITH 4 LBS. 8-8-8 (2 QUARTS).

***ON SUMMERGRASS PASTURES APPLY P AND K AS RECOMMENDED AND 60 LBS. OF N BEFORE GROWTH STARTS. UP TO SEPTEMBER 1 REPEAT THE N APPLICATIONS WHEN MORE GROWTH IS DESIRED.

***1.0 TON LIMESTONE PER ACRE IS APPROXIMATELY EQUIVALENT TO 50 LBS. PER 1,000 SQ. FT.

***FOR CAULIFLOWER, BROCCOLI AND ROOT CROPS, APPLY 1.0 LB. OF BORON (B) PER ACRE. (FOR HOME GARDENS, 1 TABLESPOON BORAX PER 100 FT. OF ROW.)

THE NUMBER OF SAMPLES PROCESSED IN THIS REPORT IS 4.

*1. Sandy soils
2. Loams & light clays
3. Heavy clays (excluding Blackbelt)
4. Heavy clays of the Blackbelt

**7.4 or higher   Alkaline   6.5 or lower   Acid
6.6-7.3   Neutral   5.5 or lower   Very acid

***Rating & fertility (percent sufficiency)

APPROVED

SOIL TESTING FORM B

Figure 16.12
Example of a soil test report giving the soil test levels and the recommendations for lime and fertilizer application. [From Cope et al. (1981)]

borings from the upper 15 cm is recommended to increase the probability that a representative portion of the soil has been obtained.

A second limitation of these rapid tests is the difficulty of extracting from a soil sample in a few minutes the amount of a nutrient, or even a constant proportion thereof, that a plant will absorb from that soil in the field during the entire growing season. The test results must be correlated with crop responses before reliable fertilizer recommendations can be made. The recommendations are made with regard to practical knowledge of the crop to be grown, the characteristics of the soil under study, and other environmental conditions. Supplemental information that can help relate soil test data to fertilizer use is shown in Figure 16.11.

NEED FOR TRAINED PERSONNEL TO INTERPRET TESTS

The interpretation of quick test data is best accomplished by experienced and technically trained personnel, who fully understand the scientific principles underlying the common field procedures. In modern laboratories, the factors to be considered in making fertilizer recommendations are programmed into a computer, and the interpretation is efficiently printed out for the farmer's or gardener's use (see Figure 16.12).

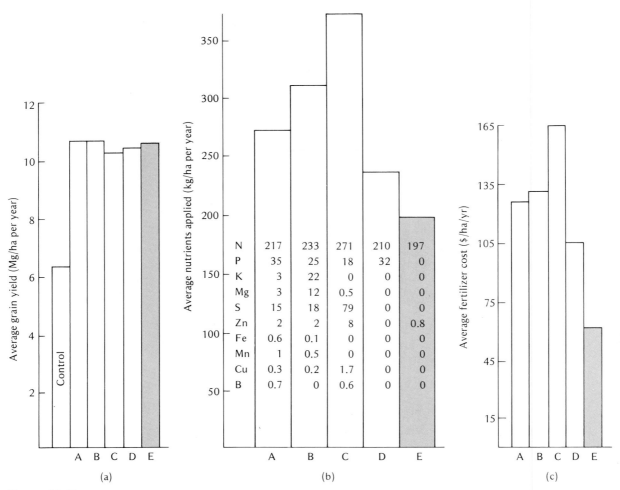

The table within figure (b):

	A	B	C	D	E
N	217	233	271	210	197
P	35	25	18	32	0
K	3	22	0	0	0
Mg	3	12	0.5	0	0
S	15	18	79	0	0
Zn	2	2	8	0	0.8
Fe	0.6	0.1	0	0	0
Mn	1	0.5	0	0	0
Cu	0.3	0.2	1.7	0	0
B	0.7	0	0.6	0	0

Figure 16.13

(a) The yield of corn from plots near North Platte, NE, receiving five different rates of fertilizer as recommended by five soil-testing laboratories (A–E). (b) The nutrient levels recommended by these laboratories and (c) the cost of the fertilizer applied each year. Note that the yields from all fertilized plots were about the same, even though rates of fertilizer application differed markedly. The fertilizer rates recommended by laboratory E utilized a "sufficiency level" concept that was based on the calibration of soil tests with field yield responses. Those recommended by laboratories A–D were based either on a "maintenance" concept that required replacement of all nutrients removed by a crop or on the supposed need to maintain set cation ratios (Ca/Mg, Ca/K, and Mg/K). Obviously, the "sufficiency level" concept provided more economical results in this trial. Data averaged over the period 1974–1980. The soil was a Cozad silt loam (Fluventic Haplustoll). [From Olson et al. (1982); used with permission of the American Society of Agronomy]

MERITS OF RAPID TESTS

It must not be inferred from the preceding discussion that the limitations of rapid tests outweigh their advantages. When the precautions already described are observed, rapid tests are an invaluable tool in making fertilizer recommendations. These tests are most useful when they are correlated with the results of field fertilizer experiments (see Figure 16.13).

16.18

Broader Aspects of Fertilizer Practice

Fertilizer practice involves many intricate details regarding soils, crops, and fertilizers. Because of the great variability from place to place in each of these three factors, it is difficult to arrive at generalizations for fertilizer use. However, the prominence in most cropping systems of nitrogen-responsive crops (e.g., corn and vegetables) leads to initial focus on nitrogen in most fertilizer schemes. Applications of phosphorus and potassium are made to balance and supplement the nitrogen supply whether it be from the soil, crop residues (especially legumes), or added fertilizers.

A second aspect of fertilizer practice relates to economics. Farmers do not use fertilizers just to grow big crops or to increase the nutrient content of their soils. They do so to make a living. As a result, any fertilizer practice, no matter how sound it may be technically, that does not give a fair economic return will not long stand the test of time.

A third aspect relates to the effects of fertilizer use on other than the production of a crop. Wise fertilizer practices can result in increased soil fertility over a period of time, especially if the rates of application exceed the removal rates of nutrients by the crop. On the other hand, excessive fertilizer use is not only wasteful of the nutrients applied but can adversely affect the quality of the environment. Undesired levels of phosphates and nitrates in drainage water from heavily fertilized areas can be harmful to humans and wildlife. Also, nitrogen oxides resulting from denitrification can have adverse effects on global climate as they move into the upper atmosphere. For these reasons, environmental quality considerations must play a role in determining fertilizer practices.

16.19

Conclusion

Chemical fertilizers play a major role in supplementing the soil's ability to provide both macro- and micronutrients for crops. There are a large variety of fertilizers on the market today to satisfy the demands of farmers, gardeners, and home owners.

In the United States, bulk spreading of solid fertilizers, both mixed- and single-nutrient carriers, is increasingly common in large-scale commercial agriculture. Also, liquid fertilizers have become popular because of the low labor costs required for their use.

The choice of the fertilizer to use in a given situation will be determined by a number of factors, including the soil's ability to provide its nutrients and the crop's need for them. In an increasingly competitive world, the costs of the fertilizers applied to the land and the value of increased yields that the fertilizers help produce play a major role in determining the kind and amounts of fertilizers to be applied. Also, the effects of the fertilizers on soil properties (such as pH) and on crop quality must be considered.

It is important that excessive use of fertilizer be avoided, not only because of the potential for the waste of the plant nutrients but also because of the potential to damage environmental quality. The challenge to the soil and crop scientists is to determine the kinds and amounts of fertilizers needed to meet the world's demands for food and other products while helping to assure environmental quality.

Study Questions

1. Anhydrous ammonia is said to be the most important inorganic nitrogen fertilizer. Why is this the case?

2. What fertilizers would you *not* use if you were concerned about the problem of soil acidification? How do these fertilizers bring about soil acidity?

3. What methods are used to try to reduce the change of nitrogen fertilizers to the leachable nitrate state?

4. The addition of phosphorus in fertilizers in the United States exceeds the removal of this element in crops. What is the reason for this?

5. Changes in the kinds of fertilizers used in the United States during the past 30 years have resulted in sulfur deficiencies. Explain.

6. The use of diammonium phosphates has increased rapidly in the past 10 years. What are the probable reasons for the wide use of this material?

7. What is the most widely used potassium fertilizer material and how do you account for its popularity?

8. Assume you have 300 kg of triple superphosphate (48% P_2O_5) and you want to make 1 Mg (1000 kg) of a 10-20-20 fertilizer. How much of each of the following materials would you need to formulate this fertilizer. Diammonium phosphate (18-46-0), muriate of potash (0-0-60), and a nitrogen solution (32-0-0)?

9. Assume you apply 150 kg of a 10-10-10 fertilizer on your lawn. The next year you learn from soil tests that the phosphorus and potassium levels are high and therefore you want to apply nitrogen only. How much urea (46-0-0) would you need to provide the same amount of nitrogen you applied the previous year?

10. What major changes have occurred in the last 30 years in the types, analyses, and means of applying fertilizers used in the United States?

11. What are the advantages and limitations of soil tests as guides to determine the kinds and rates of fertilizers to be used?

12. What harm can result from excessive rates of fertilizer application?

References

ASA. 1980. *Nitrification Inhibitors—Potentials and Limitations*, ASA Publication No. 38 (Madison, WI: Amer. Soc. Agron. and Soil Sci. Soc. Amer.).

COPE, J. T., C. E. EVANS, and H. C. WILLIAMS. 1981. *Soil Test Fertilizer Recommendations for Alabama Crops*, Circular 251, Agr. Exp. Sta., Auburn Univ.

ENGLESTAD, O. P. (Ed.). 1985. *Fertilizer Technology and Use*, 3rd ed. (Madison, WI: Soil Sci. Soc. Amer.).

HARGETT, N. L., J. T. BERRY, and S. L. McKINNEY. 1987. *Commercial Fertilizers—1987*, Bull. Y199 (Muscle Shoals, AL: Tennessee Valley Authority).

OLSON, R. A., et al. 1982. "Economic and agronomic impacts of varied philosophies of soil testing," *Agron. J.*, **74**:492–99.

PARR, J. F. 1972. "Chemical and biochemical considerations for maximizing the efficiency of fertilizer nitrogen," in *Effects of Intensive Fertilizer Use on the Human Environment*, Soils Bull. 16 (Rome, Italy: UN Food and Agriculture Organization; published by Swedish International Development Authority).

SPIES, Cliff D. 1976. "Toward a more scientific basis," *Proceedings of TVA Fertilizer Conference, July 27–28*, pp. 68–72.

TISDALE, S. L., W. L. NELSON, and J. D. BEATON. 1985. *Soil Fertility and Fertilizers*, 4th ed. (New York: Macmillan).

Recycling Nutrients Through Animal Manures and Other Organic Wastes

Nutrient recycling is becoming an increasingly important element of environmentally sound sustainable agriculture. This involves return to the soil of essential elements that are taken up by plants and then find their way into animal, domestic, and industrial products. Such recycling not only reduces the need for additional fertilizer elements but simultaneously provides organic matter and soil cover that are essential for sustainable agriculture.

Three major sources of nutrients to be returned to the soil are (a) crop residues, (b) animal manures, and (c) domestic and industrial wastes.[1] Some of the organic residues and wastes available in the United States are listed in Table 17.1 along with the percentage of each that is used on the land. The largest quantity is crop residues, about which much has already been said. Returning these residues to the land, rather than burning them or otherwise preventing their being recycled, is a wise conservation practice. The next largest source of organic residues is animal manures.

17.1

Farm Manure Significance and Quantity

For centuries the use of farm manure has been synonymous with a successful and stable agriculture. Not only does manure supply organic matter and plant nutrients to the soil but it also is associated with animal agriculture and

[1] For discussions of waste management and utilization, see Elliot and Stevenson (1977) and Curi (1985).

TABLE 17.1
Quantity of Major Organic Wastes in the United States and Percentages of These Materials That Are Used on the Land

Organic waste	Annual production (thousand dry metric tons)	Used on land (%)[a]
Crop residues	431,087	68
Animal manure	175,000	90
Municipal refuse	145,000	(1)
Logging and wood manufacture	35,714	(5)
Industrial organics	8,216	3
Sewage sludge and septage	4,369	23
Food processing	3,200	(13)

From USDA (1980).

[a] Numbers in parentheses are estimates.

with forage crops, both of which protect and conserve soil. A high proportion of the solar energy captured by growing plants ultimately is embodied in farm manure (Figure 17.1). Crop and animal production and soil conservation are enhanced by its use.

Huge quantities of farm manure are available each year for possible return to the land. For each 1000 kg live weight of farm animals about 4 Mg of manure is generated (Table 17.2). In the United States the total quantity of

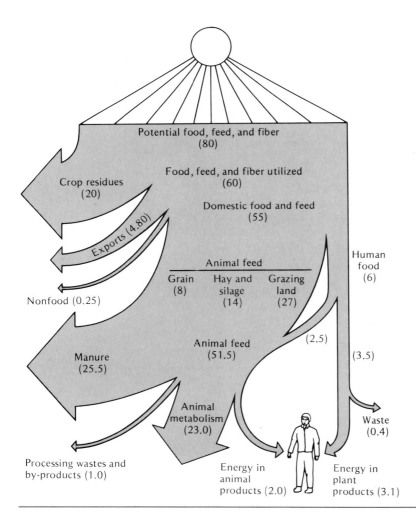

Figure 17.1
Diagram of the approximate energy flow for the United States food chain (in billions of joules) showing the high proportion of the energy ultimately found in animal manures. [From Stickler et al. (1975); courtesy Deere & Company, Moline, IL]

Figure 17.2
Aerial view of a 320-acre feed-lot in Colorado where 100,000 cattle are on feed at one time. Problems of manure disposal and utilization are difficult to cope with under these conditions of concentrated animal feeding. Farm animals produce at least ten times the biological waste produced by people in the United States. The challenge is to use this waste effectively for crop production and at the same time to minimize its potentially adverse effects on environmental quality. [Courtesy Monfort Feed Lots, Greeley, CO]

manure voided by these animals is more than 2 billion Mg annually. This is 10 times the amount produced by the human population in this country.

More than 1 billion Mg of this manure are produced in feedlots or giant poultry or swine complexes where manure *disposal* as a problem tends to overshadow manure *utilization* as an opportunity (Figure 17.2). A 50,000-head beef feedlot operation, for example, produces about 90,000 Mg of manure annually, after some decomposition and considerable moisture loss. At a conventional application rate of 22 Mg/ha (10 tons/acre), 3600 ha (9000 acres) of land would be required to utilize the manure. The enormity of the disposal problem is obvious even for operations one half or one third this size.

17.2

Chemical Composition of Animal Manures

Since manure as it is applied in the field is a combination of feces and urine along with bedding (litter) and feed wastage, its composition is quite variable (see Table 17.3).

TABLE 17.2
Representative Annual Rates of Manure Production from Different Animals

Note the relative constancy of the annual rate of dry matter.

	Annual production (Mg/Mg live weight)	
Animal	Fresh excrement	Dry matter
Cattle	25.2	3.78
Poultry	11.2	4.28
Swine	26.4	3.96
Sheep	11.8	4.00
Horse	11.6	3.96

MOISTURE AND ORGANIC CONSTITUENTS

The moisture content of fresh manure is high, commonly varying from 60 to 85%. This excess water is a nuisance if the fresh manure is spread directly on the land. But if the manure is handled and digested in a liquid form or slurry and applied to the land as such, the water is necessary.

Manures are essentially partially degraded plant materials, since the animals utilize only about half of the organic matter they eat. Thus the bulk of the solid matter in manures is composed of organic compounds very similar to those found in the feed the animals consumed. Much of the cellulose, starches, and sugars are decomposed, but hemicellulose and lignin have been only modified as have the lignoprotein complexes. There are plenty of plant materials ready for further degradation, as evidenced by the ready decomposition of at least the soluble components when they are added to soil or a digestion tank.

Manures, especially those from ruminant animals (e.g., cattle and sheep) are teeming with bacteria and other microorganisms. Up to one half and more of the fecal matter of ruminants may consist of microorganism tissue and of compounds synthesized by these microorganisms. Some of these organisms continue to break down constituents in the voided feces and participate in decomposition of the manure in storage.

NITROGEN AND MINERAL ELEMENT CONTENTS

Portions of the nutrient elements consumed in animal feeds are found in the voided excrement. As a generalization, three fourths of the nitrogen, four fifths of the phosphorus, and nine tenths of the potassium ingested in feeds are voided by the animals, and appear in the manures. For this reason, animal manures are valuable sources of both macro- and micronutrients.

The quantities of nitrogen, phosphorus, and potassium that might be expected in different manures are given in Table 17.3. The range of other nutrients commonly found, in kilograms per metric ton (Mg), is as follows (Benne et al., 1961).

Calcium	1.2–37.0	Boron	0.01–0.06
Magnesium	0.8–2.9	Manganese	0.005–0.09
Sulfur	0.5–3.1	Copper	0.005–0.015
Iron	0.04–0.93	Molybdenum	0.00051–0.0055
Zinc	0.015–0.09		

As shown in Table 17.3, the ratio of feces to urine in farm manure (except for poultry manure, which has no urine) varies from 2:1 to 4:1. On the average, a little more than *one half of the nitrogen*, almost *all of the phosphorus*,

TABLE 17.3
Moisture and Nutrient Content of Manure from Farm Animals

Animal	Feces/urine ratio	H_2O (%)	Nutrients (kg/Mg)				
			N	P_2O_5	P	K_2O	K
Dairy cattle	80:20	85	5.0	1.4	0.6	3.8	3.1
Feeder cattle	80:20	85	6.0	2.4	1.0	3.6	3.0
Poultry	100:0	62	15.0	7.2	3.1	3.5	2.9
Swine	60:40	85	6.5	3.6	1.6	5.5	4.5
Sheep	67:33	66	11.5	3.5	1.6	10.4	8.6
Horse	80:20	66	7.5	2.3	1.0	6.6	5.5

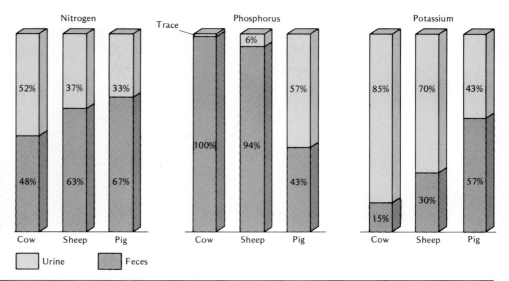

Figure 17.3
Distribution of nitrogen, phosphorus, and potassium between the solid and liquid portions of farm manure from cattle, sheep, and pigs. [Calculated from data of Van Slyke (1932)]

and about *two fifths of the potassium* are found in the solid manure (Figure 17.3). Nevertheless, this higher nutrient content of the solid manure is offset by the ready availability of the constituents carried by the urine. Care must be taken in handling and storing the manure to minimize the loss of the liquid portion.

The data in Table 17.3 suggest a relatively low nutrient content in comparison with commercial fertilizer, and a nutrient ratio that is considerably lower in phosphorus than in nitrogen and potassium. Modern commercial fertilizers usually carry 20–40 times the nutrient content of manure, and most mixed fertilizers have higher ratios of phosphorus to nitrogen and potassium than are found in manures.

THREE GENERALIZATIONS

First, animal manures are relatively low in nitrogen, phosphorus, and potassium. It takes about 2 Mg of an average farm manure to supply the same amount of N, P_2O_5, and K_2O as only 100 kg of 10-5-10 fertilizer (based on an average manure composition of 0.5% N, 0.25% P_2O_5, and 0.5% K_2O). However, since field rates of manure application are commonly 20–40 Mg/ha (9–18 tons/acre), the total nutrients supplied under practical conditions are substantial and sometimes more than are needed for the current year's crop. In the United States, a significant proportion of primary nutrients, particularly nitrogen, added to soils comes from farm manures (Table 17.4).

TABLE 17.4
Percentages of the Total N, P, and K Added to Soils of the United States Through Fertilizers, the Return of Crop Residues, and Manure

Nutrient	Source of nutrient (%)		
	Fertilizers	Crop residues	Manure
Nitrogen	70	24	24
Phosphorus	69	13	18
Potassium	51	36	13

From Follett et al. (1987).

TABLE 17.5
Effect of Handling Cow Manure on Yield and on Recovery of Nitrogen, Phosphorus, and Potassium by Corn Grown on Miami Silt Loam in a Greenhouse Experiment

Type of manure	Yield (g/pot)	Recovery by crop (%)		
		N	P	K
No manure	11.0	—	—	—
Cow manure, 15 tons/acre				
Fresh	19.5	44.0	19.5	40.5
Fermented	19.5	42.0	22.5	49.5
Aerobic liquid	17.0	18.5	19.5	38.0
Anaerobic liquid	22.5	52.5	29.0	48.0

From Hensler (1970).

A *second* generalization relates to the slow nutrient availability of manures. In general, nearly one half of the nitrogen, one fifth of the phosphorus, and one half of the potassium may be recovered by the first crop grown after manure is applied (Table 17.5). Thus, on the basis of readily available nutrients, 2 Mg of average manure supplies 5 kg of N, 1 kg of P_2O_5, and 5 kg of K_2O.

The *third* practical generalization is the need to balance the N/P/K ratio by supplying phosphorus in addition to that contained in manures. This may not be a problem for soils to which significant fertilizer has been applied, however, because the application of phosphorus in fertilizers in the United States commonly exceeds the amount of this element removed by harvested crops.

17.3

Storage, Treatment, and Management of Animal Manures

A generation ago, manure storage and application was a simple matter. Farmers spread manure daily from their barns or allowed it to pile up until time and soil conditions permitted it to be spread. Animal feeding operations were small enough to permit the spreading in the spring of manure that had accumulated during the fall and winter.

The coming of confined and concentrated animal management has drastically changed this situation. The problem of disposal of manure has taken precedence over its utilization, bringing about a revolution in the management of these manures. Offensive odors, along with the possibilities of pollution of streams and drinking water with organic materials, nitrates, and phosphates from decaying manures, have helped stimulate the revolution.

Today four general management systems are being used to handle farm manures: (a) collection and spreading of the fresh manure daily; (b) storage and packing in piles and allowing the manure to ferment before spreading; (c) aerobic liquid storage and treatment of the manure prior to application; and (d) anaerobic liquid storage and treatment prior to application. Since these different methods of handling manure affect the biological value of manure, each will be discussed briefly.

APPLYING FRESH MANURE DAILY

This system is commonly used where small to medium-sized stanchion dairy barns are employed. The manure is scraped or otherwise moved mechanically into spreaders, sometimes reinforced with superphosphate, and spread daily on the land. Daily spreading prevents nutrient loss by decomposition or volatilization, but subsequent loss of nutrients may occur if the manure is spread on frozen ground. Spring thaws carry off much potassium and nitrogen, thereby reducing crop response and increasing the probability of pollution of streams, ponds, and lakes. As much as 25% of the applied potassium can be lost in runoff and leaching from frozen ground.

STORING OR PACKING IN PILES (FERMENTATION)

Manure from dairy barns and cattle feedlots may be allowed to accumulate under the animals over a period of time or may be removed to a pile nearby. Depending on the moisture content and degree of compaction, both anaerobic and aerobic breakdown takes place. The most abundant products of decay are carbon dioxide, water, and heat, although reactions involving elements such as nitrogen, phosphorus, and sulfur are of great practical significance. For example, hydrolysis of urea during the decomposition process yields ammonia, which may be released to the atmosphere as shown.

$$CO(NH_2)_2 + 2\,H_2O \rightarrow (NH_4)_2CO_3$$

$$(NH_4)_2CO_3 \rightarrow 2\,NH_3\uparrow + CO_2\uparrow + H_2O$$

If conditions are favorable for nitrification, nitrate ions will appear in abundance. Either form of concentrated inorganic nitrogen can be lost and become a pollution hazard. The ammonia lost to the atmosphere may be captured by rain and snow and returned to the surface. The nitrate ions, being soluble and unadsorbed by the soil or the manure, are subject to leaching and to movement in runoff water. Also they may be denitrified and lost through volatilization.

AEROBIC LIQUID TREATMENT

One of the more sophisticated animal-waste treatment procedures is that of aerobic digestion of a liquid slurry containing the manure. Used primarily in swine operations, the manure may fall through slotted floors into an oxidation ditch or may be transported to a nearby lagoon. By vigorous stirring of the ditch or of the lagoon, oxygen is incorporated into the system, bringing about continuous oxidation. Offensive odors are kept to a minimum, although some nitrogen is lost, probably as ammonia. The primary products of decomposition are carbon dioxide, water, and inorganic compounds. Periodically the solids can be removed along with the resistant organic residues and applied to the land.

A diagram showing an oxidation ditch and lagoon in use for swine is shown in Figure 17.4. This method of treatment minimizes pungent odors, but its costs of construction and operation are high and it permits some nutrient loss.

ANAEROBIC LIQUID TREATMENT

This method is similar to its aerobic counterpart except that no gaseous oxygen is added to encourage aerobiosis of the liquid slurry. The reactions are similar to those taking place in a septic tank. The gaseous product is 60–80% methane, the remainder being mostly carbon dioxide. The methane

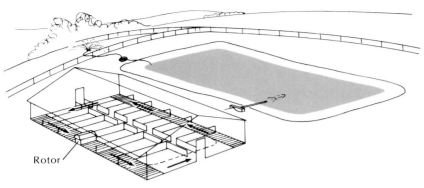

Rotor

Figure 17.4
Use of an oxidation ditch and lagoon to handle wastes in a confinement swine building. A slotted floor around the outside of the building permits the wastes to drop into a continuous channel or ditch filled with water. A rotor on the left center of the building stirs oxygen from the air into the slurry and drives the mixture counter-clockwise around the ditch. Aerobic organisms oxidize most of the organic wastes, and some of the nitrogen is volatilized. The partially purified residue is pumped or allowed to flow into a lagoon, from which it can later be applied to the land by irrigation for final purification. [Courtesy A. J. Muehling, University of Illinois]

gas produced in anaerobic digestion is used as a fuel, particularly in Asia and Central America. So-called "biogas" units make use of farm manures and other agricultural wastes as well as municipal wastes. Methane-producing digestion tanks serving individual families or even small communities are in operation. Such units are viable sources of energy in some rural areas, particularly in low-income countries where energy costs are high and where electrical power is not widely available.

This method of treatment, along with the others, results in considerable loss of organic carbon. However, crop response to the treated product generally is as good as where fresh manure is used. Data in Table 17.5 illustrate this point. Only for the aerobic processed liquid was there evidence that nitrogen had been lost during the animal waste treatment.

Scientific developments during the past few years have markedly improved animal-waste management technology. These technologies continue to provide products that are utilized in commercial agriculture.

17.4
Agricultural Utilization of Animal Manures

Biologically, manure has many attributes. It supplies a wide variety of nutrients along with organic matter that improves the physical characteristics of soils. Its beneficial effects on plant growth are sometimes difficult to duplicate with other materials. In spite of its high labor and handling costs and low analysis, manure remains a most valuable soil organic resource.

RESPONSIVE CROPS

Manure is an effective source of nutrients for most crops, especially those with relatively high nitrogen requirements. Crops such as corn, sorghum, small grains, and grasses respond well to manure as do vegetable and ornamental crops.

The rate of application of manure will depend upon the specific needs of the crop. However, rates of 20–40 Mg/ha (9–18 tons/acre) are commonly employed. In general, rates of application heavier than these give lower response per metric ton of manure and are justified only where manure disposal does not result in water pollution.

Special Uses

There are a number of special uses of manure. Applications to soil areas denuded by erosion or land leveling for irrigation are good examples. Initial applications of 75–100 Mg/ha (33–45 tons/acre) may be worked into the soil in the affected areas. These rates may be justified by the need to supply organic matter as well as nutrients.

Special cases of micronutrient deficiency can be ameliorated with manure application. Such treatments are sometimes used when there is some uncertainty as to the specific nutrient deficiency. Manure applications can be made with little concern about adding toxic quantities of the micronutrients.

The water-holding capacities of very sandy soils are increased with heavy manure applications as are structural stability and tilth of heavy-textured plastic clays. In both cases, the physical effects of manure justify its use.

Home gardeners commonly use manures at rates far in excess of those employed in commercial agriculture. Their aim is to provide a friable, easily tilled soil, and the cost is a secondary matter. Further, in planting trees and shrubs, one-time applications are common, therefore high initial rates of application are justified.

17.5

Long-Term Effects of Manures

Only one fifth to one half of the nutrients supplied by animal manures are recovered by the first crop following the application. Much of the remainder is held in humus-like compounds subject to very slow decomposition. In these forms, the elements are released only very slowly, rates of 2–4% per year being common. Thus, the humus-like compounds in manure will have continuing effects on soils years after their application.

An example of the long-term effects of manure is shown in Figure 17.5. At the Rothamsted Experiment Station in England, starting in 1852, 34 Mg of farm manure was applied each year for 19 years to a plot on which barley was grown continuously. Then in 1872 no more manure was applied, but the nitrogen content of the soil was monitored. For a period of more than 100 years the beneficial effect of the 19 years' manuring has been noted. Although such remarkable long-term effects may not be expected at all locations and on all soils, Figure 17.5 reminds us of the residual and highly beneficial effects of animal manures.

17.6

Urban and Industrial Wastes

Public concern about environmental pollution has focused attention on the disposal of urban and industrial wastes. Land application is used extensively for such disposal. Although the primary reason for adding these wastes to land is to safely dispose of them, under some circumstances they can simultaneously enhance crop production.

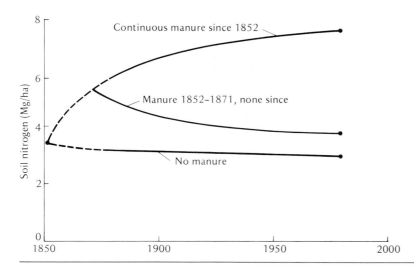

Figure 17.5
Effect of farm manure treatments on the nitrogen content of the top 23 cm of soil at the famous Rothamsted Experiment Station in England. Barley was grown continuously on all plots. Adding 35 Mg/ha of manure annually since 1852 greatly increased the soil nitrogen content. Note that the plots to which manure was applied from 1852 to 1871, and none since, lost nitrogen slowly over the next 100 years. The long-lasting effect of manure is obvious. [From Jenkinson and Johnston (1977); used with permission of the Rothamsted Experiment Station, Harpenden, England]

TYPES OF WASTE

Four major types of organic wastes are of significance in land application: (a) municipal garbage, (b) sewage effluents and sludges, (c) food processing wastes, and (d) wastes of the lumber industry. Because of their uncertain content of toxic chemicals, other industrial wastes may or may not be acceptable sources of organic matter and plant nutrients for land application.

GARBAGE

Municipal garbage is used extensively to enhance crop production in China and other Asian countries and in a few locations in Europe; the municipal organic wastes are composted (partially decomposed in piles) and later applied to the land. In the United States, most of the garbage is disposed of in other ways, and little finds its way to the farmer's fields for reasons that are mostly economic. The composted material contains only about 0.5% N, 0.4% P, and 0.2% K. Because of its low nutrient content and great bulk, the cost of transporting and handling municipal garbage compost generally makes it uneconomical for land application to supply nutrients in comparison with commercial inorganic fertilizers. However, the garbage can be a good soil conditioner.

FOOD-PROCESSING WASTES

Land application of food-processing wastes is being practiced in selected locations, but the practice is focused almost entirely on pollution abatement and not on crop production. Liquid wastes are commonly applied through sprinkle irrigation to permanent grass (Figure 17.6). Plant processing schedules dictate the timing and rate of application, and these might not be suitable for optimum crop production.

SAWDUST

Sawdust and other organic wastes from the lumber industry have long been sources of soil conditioners, especially for the home gardener. These wastes are high in lignin and related materials and have very high C/N ratios; therefore they decompose very slowly. As a consequence, lumber-mill wastes are not ready suppliers of plant nutrients. In fact, if an alternate

Figure 17.6
Agricultural processing wastes being sprayed on the land. A line of sprinklers from underground pipes applies the wastes coming from a nearby processing plant. [Courtesy L. C. Gilde, Campbell Soup Co.]

source of nitrogen is not applied with the sawdust, plants may show nitrogen deficiencies even when significant quantities of nitrogen have been added in the sawdust.

The primary beneficial effect of the sawdust is on the physical properties of soil. Productive flower and vegetable home gardens are possible, even on heavy-textured soils, if liberal and regular applications of sawdust are made.

17.7

Sewage Effluent and Sludge

Concerted efforts to prevent stream and ocean pollution by city sewage effluent have encouraged land application of either the sewage effluent or the solids emanating from sewage systems, sewage *sludge* (Figure 17.7). No other nonfarm source of organic waste for land application is more important for home use and farm crop production than is domestic sewage. About one third of the sludge produced in the United States is applied to the land.

LIQUID SEWAGE SLUDGE

Asian countries, especially China, use human wastes extensively and have done so for years. The wastes, termed "night soil," are collected and applied

Figure 17.7
Domestic sludge being removed from the Blue Plains wastewater treatment near Washington, DC, using vacuum filters. This material is available for direct application to the land. [U.S. Department of Agriculture photo by R. C. Bjork]

to the soil with little or no processing, even for crops intended for human consumption. The Chinese have a remarkable record of recycling wastes, particularly for agricultural purposes.

Sewage effluents also have been applied to land for decades in Europe and in selected sites in the United States. The city of Chicago has utilized land application for disposal of treated sewage. The effluent is transported by barge to the disposal site, held temporarily in lagoons, and then injected into the soil or applied by sprinkle irrigation. Using these disposal techniques, nearly 7 million m^3 of wastes each day are applied to the land. Corn, reed canary grass, alfalfa, and soybeans are among the crops that benefit from this treatment. Sewage effluent may become more important in the future as a source of organic matter and nutrients for crop production.

SEWAGE SLUDGE

Sewage sludge is the solid by-product of domestic and/or industrial waste-water-treatment plants (Figure 17.8). It also has been spread on the land for decades, and its use will likely increase in the future. The product "Milorganite," a dried sludge sold by the Milwaukee Sewerage Commission, has been used widely in North America since 1927. Philadelphia has developed a system of composting and of land application of sewage sludge as a basis for its sewage disposal.

As might be expected, the composition of the sludge varies from one sewage treatment plant to another, depending on the nature of the treatment and the degree to which the organic material is allowed to digest. Levels of the heavy metals, such as zinc, lead, copper, iron, manganese, and cadmium, are determined largely by the degree to which industrial wastes have been mixed in with domestic wastes. If the content of these elements is too high, the sludge may be of little value or even harmful for agricultural purposes (see Section 18.7.)

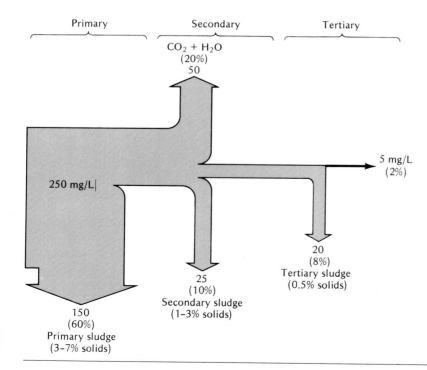

Figure 17.8
Diagram showing the removal of suspended solids from wastes. Primary treatment permits the separation of most of the solids from raw sewage. Secondary treatment encourages oxidation of much of the organic matter and separation of more solids. Tertiary treatment usually involves the use of calcium, aluminum, or iron compounds to remove phosphorus from the sewage water. [Redrawn from Loehr et al. (1979); used with permission of Van Nostrand Reinhold Company, New York]

TABLE 17.6
Median Level of Organic Carbon and Total N, P, and K in Sewage Sludge from Cities in Seven States

State	No. of sewage plants	Median levels of			
		Organic C (%)	Total N (%)	Total P (%)	Total K (%)
Wisconsin	38	35.8	5.4	2.7	1.13
Michigan	47	31.7	1.6	1.6	0.14
Indiana	14	22.7	3.1	1.9	0.33
Minnesota	19	29.9	5.3	3.5	0.30
New Jersey	13	—	2.6	1.7	0.16
New Hampshire	28	38.0	2.5	0.9	0.30
Ohio	15	—	3.8	—	0.15
All seven states		30.4	2.5	1.8	0.24

Data from NC118 North Central Regional Project, published by McCalla et al. (1977).

Data in Table 17.6 illustrate the variability in composition of sewage sludge from plants in seven midwestern and eastern states. In comparison with inorganic fertilizers, the sludges are generally low in nutrients, especially phosphorus and potassium. Representative levels of N, P, and K are 3%, 2.0%, and 0.4%, respectively. Obviously, rates of application of sludge far exceeding those for commercial fertilizers would be needed to provide comparable levels of essential elements, and especially for potassium. Supplemental applications of either manure or chemical fertilizers are necessary to provide the nutrient balance for good crop production.

PRACTICAL EFFECTS OF SLUDGE APPLICATION

Typical responses of corn to sludge applications at different soil temperatures are shown in Figure 17.9. Increased response at higher temperatures suggests the importance of microbial action in releasing plant nutrients. Note, however, that the responses obtained from these high rates of sludge application (up to 112 Mg/ha or 50 tons/acre) are generally lower than would be expected from animal manures applied at only 20–40 Mg/ha. This is likely due to the fact that the sewage sludge had already undergone significant degradation in the sewage plant before it was applied to the soil. Other studies verify that sludges decompose in the soil much more slowly than plant residues and even more slowly than animal manures.

The heavy sewage sludge applications have some favorable effects on the physical properties of soil. Sludge can serve as an organic mulch, thereby protecting the soil and conserving soil moisture. Unfortunately, considerable nitrogen can be lost from sludge by volatilization either as ammonia gas or through denitrification (Figure 17.10). While these losses may be of no consequence to those who merely want to dispose of the wastes, they are of concern to the farmer, whose crops could effectively use the nitrogen if it were kept in the soil. For that reason some farmers inject the sludge into the soil or plow in at least part of the applied sludge rather than leaving it on the soil surface.

Figure 17.10 also suggests that heavy annual land applications of sewage sludge can increase the organic matter and nitrogen contents of the soil. These increases are reflected most in complex organic compounds that are only slowly decomposed, but they do add to the large pool of organic matter and associated nitrogen, sulfur, and phosphoric compounds that, through the years, can be recycled for plant and animal use.

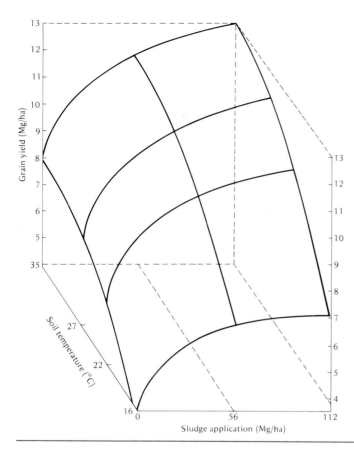

Figure 17.9
Effect of the application of sewage sludge on the yield of corn for grain at different soil temperatures. Response was obtained from only the first 56 tons at 16 °C soil temperature, but at higher temperatures further response was obtained to the 112-ton rate. [Redrawn from Sheaffer et al. (1979); used with permission of the *Journal of Environmental Quality*]

17.8

Composts

Composting is the time-honored practice of encouraging partial rotting by microorganisms and other soil organisms of organic materials of either plant or animal origin. The aerobic decomposition takes place in piles and in bins kept sufficiently moist to support the decay. Sewage sludge solids can be

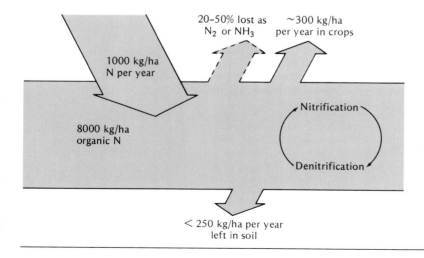

Figure 17.10
Fate of nitrogen applied to land through sewage sludge. About 1000 kg of nitrogen is shown as being added each year. Between 20 and 50% is lost through volatilization as N_2 or NH_3, and approximately 300 kg/ha is removed in crops. Some residual nitrogen is left in the soil and adds to the approximately 8000 kg/ha of soil organic nitrogen. [From Loehr et al. (1979); used with permission of Van Nostrand Reinhold Company, New York]

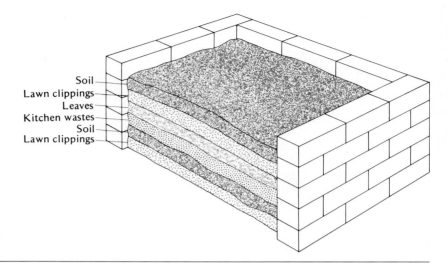

Figure 17.11
Example of a backyard compost heap. Various wastes from the home and yard along with a little manure and soil are added in layers, kept moist, and partially decomposed by soil organisms.

composted along with other municipal wastes such as leaves and prunings and, after drying, can be sold to gardeners.

Home production of compost is more widespread and has increased in recent years due to restrictions on the burning of leaves and other domestic wastes in residential areas. This is a blessing in disguise since composting ensures the conservation of some nutrients that are lost when residues are burned.

Almost any plant material can be composted. Leaves, weeds, lawn cuttings, small prunings, and garden wastes are the primary sources of the organic materials. These can be supplemented with materials such as sawdust, vacuum cleaner dust, and fireplace or furnace ashes. To hasten decay, small quantities of fertilizer may be added along with a little soil to assure the presence of decay organisms. The materials can be kept in a pile, in a wooden bin, or in an area surrounded by wire netting. The pile must be kept moist (50–70% water) but not too wet because the breakdown must be aerobic in nature. The materials are best packed down to help keep the pile from drying out. Figure 17.11 illustrates how a compost pile can be constructed.

The heat of combustion of organic material increases the temperature of the compost pile, temperatures of 50–72 °C (122–161 °F) being reached when decay is occurring rapidly. At these temperatures most weed seeds are killed along with most plant and animal disease organisms.

China has probably made more extensive use of composts than any other country. Crop residues, animal manures, and human wastes are mixed with soil and sludge from ponds and stream beds in elongated compost piles. The piles are then plastered on the outside with wet soil to prevent rapid loss of moisture (Figure 17.12). Bamboo rods are used to make holes in the pile, thus assuring a supply of air throughout the decomposing mass. The large piles are used to provide compost not only for home gardens but for fields as well.

CROP RESPONSE TO COMPOSTS

Properly prepared composts, if supplemented by small quantities of commercial fertilizers and lime when needed, are as effective as animal manures in enhancing crop production. Rates of application of 20–40 Mg/ha (9–18 tons/acre) are necessary since the nutrient element content is low. Fur-

Figure 17.12
Compost piles in China. Crop wastes, manure, and soil are mixed together alongside fields and after partial decomposition of the organic matter added to the nearby fields.

thermore, as with animal manures, the nutrients are not all readily available for plant uptake. The favorable response of different crops to compost in China is shown in Table 17.7.

Composted materials are sometimes used as mulches in vegetable or flower gardens (Figure 17.13). This practice provides not only plant nutrients but soil cover and moisture conservation as well. Indoor potted plants also thrive on mixtures high in composted materials.

TABLE 17.7
Effects of Added Compost on the Yield of Several Important Food and Feed Crops in China

Crop	Amount added (Mg/ha)	Yield (Mg/ha)	
		No compost	With compost
Corn	30.4	4,408	5,700
Potato	38.0	7,737	14,630
Sugarbeet	15.2	26,741	33,600
Wheat	38.0	2,336	3,230
Millet	38.0	2,257	3,341
Sorghum	38.0	1,664	3,078
Soybean	30.4	1,877	2,310

From FAO (1977)

Figure 17.13
Composts are widely used as mulches on flower beds and vegetable gardens. [Courtesy USDA Soil Conservation Service]

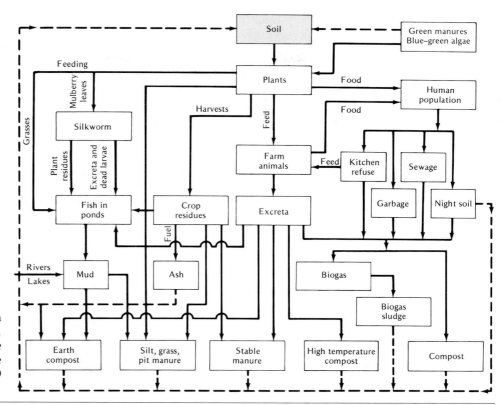

Figure 17.14
Recycling of organic wastes in the Peoples Republic of China. Note the degree to which the soil is used as a recipient of the wastes. [Modified from FAO (1977)]

17.9

Integrated Recycling of Wastes

For most of the industrialized countries, widespread recycling of organic wastes other than animal manures is a relatively recent phenomenon. In heavily populated areas of Asia, however, and particularly in China and Japan, such recycling has long been practiced. Figure 17.14 illustrates the many ways in which organic wastes are used in China. Agriculture is only one of the recipients of these wastes. Much is used for biogas production, as food for fish, and as a source of heat from compost piles to help warm homes, greenhouses, and household water. Most importantly, the plant nutrients and organic matter are recycled and returned to the soil for future plant utilization. Such conservation is likely to be practiced more widely in the future by other countries, including the United States.

17.10

Conclusions

Organic wastes are important sources of nutrients that can be recycled through the soil for plant use. In the United States, more than 2 billion Mg of farm animal manures is available for such recycling, as are large quantities of organic industrial and municipal wastes.

The application of these materials to soils can serve two purposes. First, they provide beneficial chemical and physical effects on soils and they stimulate crop production. These organic wastes help maintain or even increase soil organic matter levels and provide significant quantities of plant nutrients. Second, the soils can serve as important disposal sites for these organic waste materials. Although the second of these purposes is increasingly pertinent because of the huge quantities of organic wastes to be disposed of, the beneficial effects of organic materials on stable agricultural production cannot be disregarded.

The judicious cycling of agricultural wastes is a sound practice that has characterized successful agricultural systems through the centuries. It is no less important today in view of the high nutrient demands of the crops needed to feed an ever-increasing human population. Farm animal wastes should be recycled to the extent that sound management will permit.

Municipal wastes and some industrial wastes are also of some importance in nutrient recycling. However, some of these materials contain toxic levels of organic or inorganic chemicals that are not beneficial to plants or animals. These materials are among the subjects of the next chapter.

Study Questions

1. For centuries farm manures have been known as useful for sustainable agriculture production, yet in the United States they are generally considered today as much a disposal problem as a utilization opportunity. Why is this the case?

2. What are the advantage of farm manures compared to chemical fertilizers when they are added to soil?

3. Assume you have been applying 50 kg of a 10-10-10 fertilizer to a garden plot and you want to use farmyard manure in place of the chemical fertilizer. How much of a common manure would you use to provide the same amount of total nitrogen, and how much superphosphate (20% P_2O_5) would you need to supplement the manure to provide the same nutrient balance as existed in the 10-10-10?

4. When you go to purchase the manure, the salesman points out that you need about three times as much manure as you had calculated in the answer to question 3 to provide the same available nutrients for this year's crop as was provided by the 10-10-10 fertilizer purchased last year. Is there any truth to his advice or is the salesman merely taking advantage of you? Explain.

5. Contrast aerobic and anaerobic systems of fermentation of animal wastes and indicate the advantages and disadvantages of each. Which is used in "biogas" systems to provide methane?

6. Compare the chemical and biological properties of crop residues and animal manures.

7. What are the properties of sewage sludge that make it (a) a desirable and (b) an undesirable addition to soils?

8. In what countries are domestic wastes and sewage sludge used widely in agriculture?

9. How many tons of sewage sludge with 3.0% N would be needed to provide the same total nitrogen as is supplied by 800 kg of a 20-20-20 chemical fertilizer?

10. What materials are commonly used for backyard composts?

11. How does the chemical composition of a compost pile compare with that of the original plant materials?

References

BENNE, E. J., et al. 1961. *Animal Manures*, Circ. 291, Michigan Agr. Exp. Sta.

CURI, K. 1985. *Appropriate Waste Management for Developing Countries* (New York: Plenum Press).

ELLIOTT, L. F., and F. J. STEVENSON (Eds.). 1977. *Soils for Management of Organic Wastes and Waste Waters* (Madison, WI: Soil. Sci. Soc. Amer., Amer. Soc. Agron., Crop Sci. Soc. Amer.).

FAO. 1977. *China: Recycling of Organic Wastes in Agriculture*, FAO Soils Bull. 40 (Rome: UN Food and Agriculture Organization).

FOLLETT, R. F., S. C. GUPTA, and P. G. HUNT. 1987. "Conservation practices: Relation to the management of plant nutrients for crop production," pp. 19–51 in *Soil Fertility and Organic Matter as Critical Components of Production Systems*, SSSA Spec. Pub. No. 19 (Madison, WI: Soil Sci. Soc. Amer.).

HENSLER, R. F. 1970. "Cattle manure: I. Effect on crops and soils; II. Retention properties for Cu, Mn, and Zn," Ph.D. Thesis, University of Wisconsin.

JENKINSON, D. S., and A. E. JOHNSTON. 1977. "Soil organic matter in the Hoosfield barley experiment,"

Report of Rothamsted Experiment Station for 1976, Pt 2:87–102.

LOEHR, R. C., et al. 1979. *Land Application of Wastes*, Vol. I (New York: Van Nostrand Reinhold Co.).

McCALLA, T. M., J. R. PETERSON, and C. LUE-HING. 1977. "Properties of agricultural and municipal wastes," in L. F. Elliott and F. J. Stevenson (Eds.), *Soils for Management of Organic Wastes and Waste Waters* (Madison, WI: Soil Sci. Soc. Amer., Amer. Soc. Agron., Crop Sci. Soc Amer.).

SHEAFFER, C. C., et al. 1979. "Soil temperature and sewage sludge effects on corn yield and macronutrient content," *J. Environ. Qual.*, 8:450–54.

STICKLER, F. C., et al. 1975. *Energy from Sun to Plant to Man* (Moline, IL: Deere & Company).

USDA. 1980. *Appraisal, 1980 Soil and Water Resources Conservation Act, Review Draft*, Part I (Washington, DC: U.S. Department of Agriculture).

VAN SLYKE, L. L. 1932. *Fertilizers and Crop Production* (New York: Orange Judd), pp. 216–26.

Soils and Chemical Pollution

The soil is a primary recipient, intended or otherwise, of many of the waste products and chemicals used in modern society. Once these materials enter the soil, they become part of a cycle that affects all forms of life. At least a general understanding of the pollutants themselves, their reactions in soils, and available means of managing, destroying, or inactivating them is essential.

Six general kinds of pollutants commonly reach the soil. First are the thousands of *pesticide* formulations, most of which are used for agricultural purposes. Second is a group of *inorganic pollutants*, such as mercury, cadmium, and lead. Third are the *organic wastes*, such as those from concentrated feedlots and food-processing plants as well as municipal and industrial wastes, some of which may be added to soils. *Salts, radionuclides*, and *acid rain* are the remaining contaminants to be discussed.

18.1

Chemical Pesticides—Background

More than 10,000 species of insects, 600 weed species, 1500 plant diseases, and 1500 species of nematodes are known to be injurious to humans, animals, and plants. Since the early Greek civilization, chemicals have been used to control these pests. The use of chemicals expanded in the nineteenth century when Pasteur discovered that microbes caused plant and animal diseases and that chemicals could control these diseases. But it was not until synthetic organic pesticides came into being in the middle of the twentieth century that using chemicals became widespread.

SYNTHETIC PESTICIDES

When the insecticidal properties of DDT were discovered in 1939 along with the herbicidal effects of 2,4-D in 1941, the chemical revolution in agriculture

517

TABLE 18.1

Expenditures for Different Pesticides in the United States in 1970 and 1985 and Projected Expenditures for 1990

	Expenditures ($ millions)		
Pesticides	*1970*	*1985*	*1990 (est.)*
Herbicides	413	2675	3475
Insecticides	286	1360	1700
Fungicides	120	365	450
Others	83	300	380
Nematocides	(21)	(85)	(110)
Rodenticides	(13)	(50)	(64)
Defoliants/desiccants	(10)	(45)	(55)
Other	(17)	(80)	(104)
Total	902	4700	6000

Data quoted by Storck (1987).

truly began. These chemicals would kill pests and they were not too expensive. Since their discovery, tens of thousands of such chemicals and multichemical formulations have been developed, tested, and put to use. In 1985, nearly $4.7 billion was spent in the United States for about 500 million kg of these pesticides, more than three-fourths of which was used in agriculture (Table 18.1). Some 600 chemicals in about 50,000 formulations are used extensively worldwide to control pests, as shown in Figure 18.1.

Pesticides have provided many benefits to society. They have helped control mosquitos and other vectors of human diseases such as yellow fever and malaria. They have protected crops and livestock against insects and diseases. Without chemical weed control, minimum tillage practices would likely be uneconomical. Also, pesticides reduce the spoilage of food as it moves from the farm to the dinner table.

As these chemicals move into the soil, however, some have become troublesome. A few are not readily biodegradable and persist in the soil or in water for many years. Others are detrimental to nontarget organisms such as beneficial insects and some soil organisms. Moreover, as plant residues containing these chemicals are consumed by such soil organisms as earthworms, the chemicals tend to concentrate in the earthworm bodies. When birds and fish eat the earthworms, the pesticides can build up further to lethal levels. Damage to these creatures sounded the warning that pesticides can have devastating environmental complications.

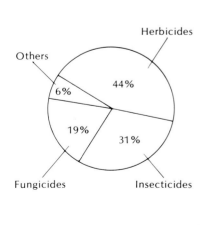

(a)

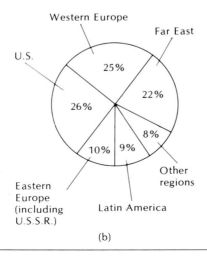

(b)

Figure 18.1

Worldwide usage of pesticides in 1986. (a) The porportions of different classes. (b) Major areas of usage. [Courtesy British Agricultural Association]

TABLE 18.2

Classes of Pesticides Commonly Used in the United States and Examples of Each Group

Chemical group	Examples
Insecticides	
Chlorinated hydrocarbons	Aldrin,[a] chlordane, heptachlor,[a] DBCP,
Organophosphates	Diazinon, disulfoton, parathion,[a] malathion, methylparathion[a]
Carbamates	Carbaryl, carbofuran, methomyl, aldicarb, dithiocarb, oxamyl
Pyrethrin	Permethrin
Other	EDB
Fungicides	
Benzimidazoles	Benomyl, thiabenzazole
Thiocarbamates	Ferbam, maneb
Triazoles	Triadimefon, bitertanol
Others	Copper sulfate, chlorothalonil
Herbicides	
Phenoxyalkyl acids	2,4-D, 2,4-DB, 2,4,5-T, MCPA, MCPB
Triazines	Atrazine, simazine, propazine, cyanazine, metribuzin
Phenylureas	Diuron, linuron, fluometuron, bromacil
Aliphatic acids	Dalapon
Carbamates	Butylate, vernolate, thiobencarb, EPTC
Nitrophenols	Dinoseb
Dinitroanilines	Trifluralin, benefin
Dipyridyls	Paraquat, diquat
Amides	Alachlor, propanil, alanap, metolachlor
Benzoics, phthalates	Dicamba, dacthal

[a] Restricted-use chemical.

18.2

Kinds of Pesticides

Pesticides are commonly classified according to the target group of pest organisms: (a) insecticides, (b) fungicides, (c) herbicides (weedicides), (d) rodenticides, and (e) nematocides. In practice, all find their way into soils. Since the first three are used in largest quantities and are therefore more likely to contaminate soils, they will be given primary consideration. Table 18.2 lists the names of some commonly used pesticides, and Figure 18.2 shows the great variability in their structures.

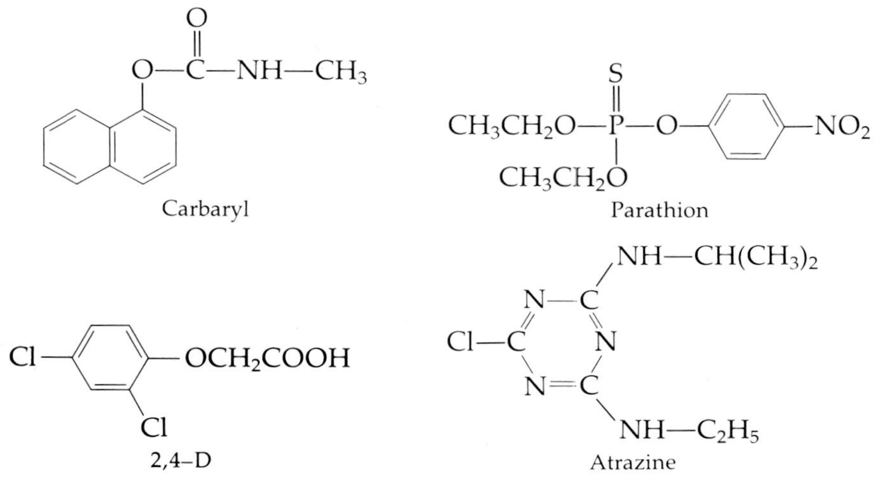

Figure 18.2
Structural formulas of some widely used pesticides. Carbaryl and parathion are insectides; 2,4-D and atrazine are herbicides. This variety of structures dictates great variability in properties and reactivity in the soil.

INSECTICIDES

Most insectides belong to three general groups. The *chlorinated hydrocarbons*, such as DDT, were the most extensively used until the early 1970s. However, their low biodegradability and persistence as well as toxicity to birds and fish made it necessary to restrict or eliminate the use of chlorinated hydrocarbons.

The *organophosphate* pesticides are generally biodegradable and thus less likely to build up in soils and water. Unfortunately they are relatively much more toxic to humans than are the chlorinated hydrocarbons, so great care must be used in handling and applying them. The *carbamates* are also popular among most environmentalists because of their ready biodegradability and relatively low mammal toxicity.

FUNGICIDES

Fungicides, which are used far less in the United States than either herbicides or insecticides, are applied mostly to control the field diseases of fruits and vegetables. Some are also used to protect harvested fruits and vegetables from decay and rot, to prevent wood decay, and to protect clothing from mildew. In the past significant quantities of fungicides containing mercury, arsenic, and copper were used. These have been largely replaced because of the toxicities of these elements to humans and animals. Organic materials such as the dithiocarbamates and organophosphates are currently in use.

HERBICIDES

The quantity of herbicides used in the United States exceeds that of the other two types of pesticides (Figure 18.3). Starting with 2,4-D (chlorinated phenoxyacetic acid), dozens of chemicals in literally hundreds of formulations have been placed on the market. These include the *triazines* (rather specific for weed control in corn), *phenylureas, aliphatic acids, carbamates, dinitroanilines, and dipyridyls* (Table 18.2). As one might expect, this wide variation in chemical makeup provides an equally wide variation in properties. However, herbicides are generally biodegradable, and most of them are relatively low in mammal toxicity. Some are quite toxic to fish and perhaps to other

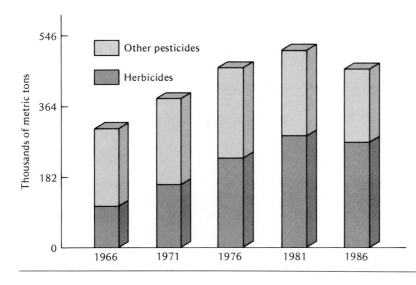

Figure 18.3
Increase in the use of pesticides in the United States since 1966. Note that essentially all the increase has been in herbicides. [U.S. Department of Agriculture data]

wildlife, however, emphasizing once again the need to consider the indirect effects of any pesticide used.

NEMATOCIDES

Although nematocides are not widely used, some of them are known to contaminate soils and the water draining from treated soils. For example, some carbamates used as nematocides are quite soluble in water, are not adsorbed by the soil, and consequently leach downward and into the groundwater.

18.3

Reactions of Pesticides in Soils[1]

Whether pesticides are applied to plant foliage, to the soil surface, or are incorporated into the soil, a high proportion of the chemicals eventually moves into the soil. These chemicals then move in one or more of six major directions.

1. They may vaporize into the atmosphere without chemical change.
2. They may be adsorbed by humus and clay particles.
3. They may move downward through the soil in liquid or solution form and be lost from the soil by leaching.
4. They may undergo chemical reactions within or on the surface of the soil.
5. They may be broken down by soil microorganisms.
6. They may be absorbed by plants and detoxified within the plants.

The specific fate of these chemicals will be determined at least in part by their chemical structures, which as shown in Figure 18.2 are quite variable. Various processes affecting organic chemicals in the soil are represented in Figure 18.4.

VOLATILITY

Pesticides vary greatly in their volatility and subsequent susceptibility to atmospheric loss. Some soil fumigants, such as methyl bromide, are selected because of their very high vapor pressure, which permits them to penetrate soil pores to contact the target organisms. This same characteristic encourages rapid loss to the atmosphere after treatment unless the soil is covered or sealed. A few herbicides (e.g., trifluralin) and fungicides (e.g., PCNB) are sufficiently volatile to make vaporization a primary means of their loss from soil.

Earlier assumptions that disappearance of pesticides from soils was evidence of their breakdown are now known to be questionable. That some chemicals are lost to the atmosphere only to return to the soil or surface waters in rain is now known.

ADSORPTION

The adsorption of pesticides by soil is determined largely by the characteristics of the pesticides and of the soils to which they are added. The presence of certain functional groups, such as —OH, —NH$_2$, —NHR, —CONH$_2$,

[1] For reviews on this subject, see Krueger and Seiber (1984) and Garner et al. (1986).

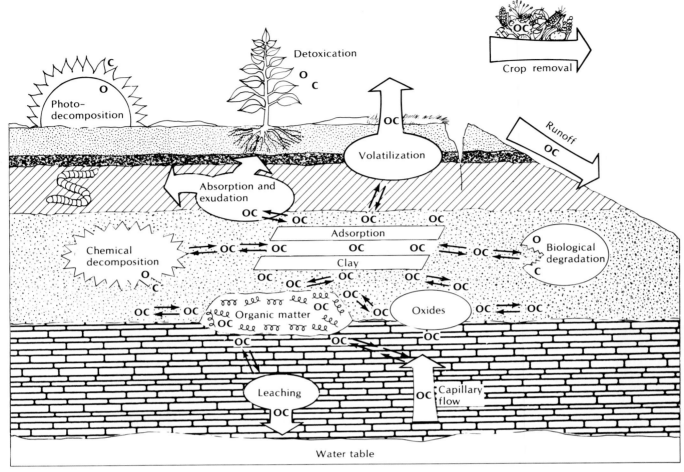

Figure 18.4

Processes affecting the dissipation of organic chemicals (OC) such as pesticides in soils. Note the degradation processes split the organic chemical (C and O), whereas in transfer processes the OC remains intact. [From Weber and Miller (1989)]

—COOR, and —$^+NR_3$, in the chemical structure encourages adsorption, especially on the soil humus. Hydrogen bonding (see Section 5.1) and protonation [adding H^+ to a group such as an —NH_2 (amino) group] probably promotes some of the adsorption. In general, the larger the size of the pesticide molecule, everything else being equal, the greater is its adsorption.

Some pesticides with positively charged groups, such as the herbicides diquat and paraquat, are also adsorbed by silicate clays. Adsorption by clays of some pesticides tends to be pH dependent (Figure 18.5), maximum adsorption occurring at low pH levels where protonation occurs. Addition of H^+ ion to functional groups (e.g., —NH_2) yields a positive charge on the herbicide, resulting in an attraction by the negatively charged soil colloids.

LEACHING

The tendency of pesticides to leach from soils is closely related to their solubility and their potential for adsorption. Strongly adsorbed molecules are not likely to move down the profile (Table 18.3). Likewise, conditions that encourage adsorption will discourage leaching. Leaching is apt to be favored by water movement, taking place most readily in permeable sandy

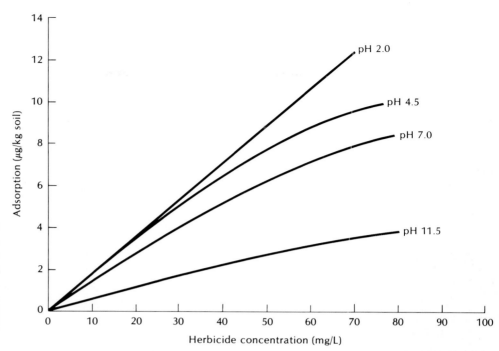

Figure 18.5
The effect of pH of kaolinite on the adsorption of glyphosate, a widely used herbicide. [Reprinted with permission from J. S. McConnell and L. R. Hossner, *J. Agric. Food Chem.* **33**:1075–78 (1985). Copyright 1985 American Chemical Society.]

soils that are low in clay and organic matter. In general, herbicides seem to be more mobile than either fungicides or insecticides, and their extensive use has resulted in groundwater pollution in some areas (Table 18.4).

CHEMICAL REACTIONS

Some pesticides undergo chemical modification independent of soil organisms. For example, DDT, diaquat, and the triazines are subject to slow *photodecomposition* activated by solar radiation (see Figure 18.4). The triazine herbicides (e.g., atrazine) and organophosphate insecticides (e.g.,

TABLE 18.3
The Degree of Adsorption of Selected Herbicides

Weakly adsorbed herbicides are more susceptible to movement in the soil than those that are more tightly adsorbed.

Common name or designation	Trade name	Adsorptivity to soil colloids
Dalapon	Dowpon	None
Chloramben	Amiben	Weak
Bentazon	Basagran	Weak
2,4-D	Several	Moderate
Propachlor	Ramrod	Moderate
Atrazine	AAtrex	Strong
Alachor	Lasso	Strong
EPTC	Eptam	Strong
Diuron	Karmex	Strong
Paraquat	Paraquat	Very strong
Trifluralin	Treflan	Very strong
DCPA	Dacthal	Very strong

Selected data from DMI (1981).

TABLE 18.4

Pesticides Found as Pollutants in the Groundwaters and the Number of States in Which the Pollution Has Been Reported

Pesticide	No. of states with known pollution
Aldicarb	15
EDB	8
Atrazine	5
DBCP	5
Alachlor	4
1,2-Dichloropropane	4
Carbofuran	3
Simazine	3
Cyanazine	2
Metolachlor	2
Metribuzin	2
TCP	2
Oxamyl	2

Selected Data from Cohen et al. (1986).

malathion) are subject to hydrolysis and subsequent degradation in the soil. While the complexities of molecular structure of the pesticides suggest different mechanisms of breakdown, it is important to realize that degradation independent of soil organisms does in fact occur.

MICROBIAL METABOLISM

Biochemical degradation by soil organisms is the single most important method by which pesticides are removed from soils. Certain polar groups such as —OH, —COO$^-$, and —NH$_2$ on the pesticide molecules provide points of attack for the organisms.

DDT and other chlorinated hydrocarbons such as aldrin, dieldrin, and heptachlor are very slowly broken down in most soils. In contrast, the organophosphate insecticides such as parathion are degraded quite rapidly, apparently by a variety of organisms (Figure 18.6). Likewise, the most widely used herbicides, such as 2,4-D, the phenylureas, the aliphatic acids, and the carbamates, are readily attacked by a host of organisms. Exceptions are the triazines (e.g., atrazine), which are degraded slowly, primarily by chemical action. Most organic fungicides are also subject to microbial decomposition, although the rate of breakdown of some is slow, causing troublesome residue problems.

PLANT ABSORPTION

Pesticides are commonly absorbed by higher plants, a necessary process for the effectiveness of most herbicides and of some insecticides. The absorbed chemicals may remain intact inside the plant or they may be degraded. Some degradation products are quite harmless to humans and other creatures, but others are toxic, in some cases even more toxic than the original pesticide. Pesticide residues, especially those found in the edible portion of plants (fruits, nuts, etc.), are of critical concern to humans. Such residue levels are strictly regulated by law to assure human safety.

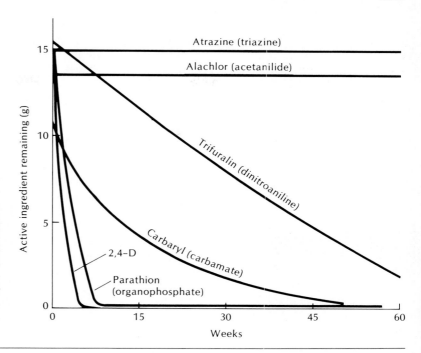

Figure 18.6
Degradation of four herbicides (alachlor, atrazine, 2,4-D, and trifuralin) and two insecticides (parathion and carbaryl), all of which are used extensively in the Midwest of the United States. Note that atrazine and alachlor are quite slowly degraded, whereas parathion and 2,4-D are quickly broken down. [Reprinted with permission from R. G. Krueger and J. N. Seiber, *Treatment and Disposal of Pesticide Wastes*, Symposium Series 259. Copyright 1984 American Chemical Society.]

PERSISTENCE IN SOILS

The persistence of pesticides in soils is a summation of all the reactions, movements, and degradations affecting these chemicals. Marked differences in persistence are the rule (Figure 18.6). For example, organophosphate insecticides may last only a few days in soils. The most widely used herbicide, 2,4-D, persists in soils for only 2–4 weeks, whereas DDT and other chlorinated hydrocarbons may persist for from 3 to 15 years or longer (Table 18.5). The persistence times of other herbicides, fungicides, and insecticides generally are between the extremes cited. The majority of pesticides degrade

TABLE 18.5
Common Range of Persistence of a Number of Pesticides

Risks of environmental pollution are highest with those chemicals with greatest persistence.

Pesticide	Persistence
Arsenic	Indefinite
Chlorinated hydrocarbon insecticides (e.g., DDT, chlordane, dieldrin)	2–5 yr
Triazine herbicides (e.g., atrazine, simazine)	1–2 yr
Benzoic acid herbicides (e.g., amiben, dicamba)	2–12 mo
Urea herbicides (e.g., monuron, diuron)	2–10 mo
Phenoxy herbicides (2,4-D, 2,4,5-T)	1–5 mo
Organophosphate insecticides (e.g., malathion, diazinon)	1–12 wk
Carbamate insecticides	1–8 wk
Carbamate herbicides (e.g., barban, CIPC)	2–8 wk

rapidly enough to prevent buildup in soils. Those that resist degradation have potential for environmental damage.

Continued use of the same pesticide on the same land can increase the rate of microbial breakdown of the chemical. This is an advantage in relation to environmental quality, but the breakdown may become sufficiently rapid to reduce the herbicide's effectiveness.

REDUCING SOIL PESTICIDE LEVELS

Among the practices suggested to reduce pesticide levels in soils is the addition of easily decomposed organic matter. The growth of high-nitrogen cover crops or the additions of large quantities of animal manures also should be helpful. Apparently degradation of even the most resistant pesticides is encouraged by conditions favoring overall microbial action. Other practices suggested to reduce pesticide levels are cropping to plants that accumulate the pesticide and leaching the soil. Unfortunately, some of these procedures merely transfer the chemical from the soil to some other part of the environment, a process of dubious value.

This brief review of the behavior of pesticides in soils reemphasizes the complexity of the changes that take place when new and exotic chemicals are added to our environment. Clearly the ecological effects of new chemicals must be evaluated as thoroughly as possible before their extensive use is permitted.

18.4

Effects of Pesticides on Soil Organisms

Since the purpose of pesticides is to kill organisms, it is not surprising that some of them are toxic to specific soil organisms. At the same time, the diversity of the soil organism population is so great that, except for a few fumigants, most pesticides do not kill a broad spectrum of soil organisms.

FUMIGANTS

Fumigants are compounds used to free a soil of a given pest, such as nematodes, and they have a more drastic effect on both the soil fauna and flora than do other pesticides. For example, 99% of the microarthropod population is usually killed by the fumigants DD and Vapam, and it may take as long as 2 years for the population to recover fully. Fortunately, the recovery time for the microflora is generally much less.

Fumigation reduces the number of species of both flora and fauna, especially if the treatment is repeated, as is often the case where nematode control is attempted. At the same time, the total number of bacteria is frequently much greater following fumigation than before. This is probably due to the relative absence of competitors and predators following fumigation.

EFFECTS ON SOIL FAUNA

The effects of pesticides on soil animals vary greatly from chemical to chemical and organism to organism. Nematodes are not generally affected except by specific fumigants. Mites are generally sensitive to most organophosphates and to the chlorinated hydrocarbons, except for aldrin. Springtails vary in their sensitivity to both chlorinated hydrocarbons and organophosphates, some chemicals being quite toxic to these organisms.

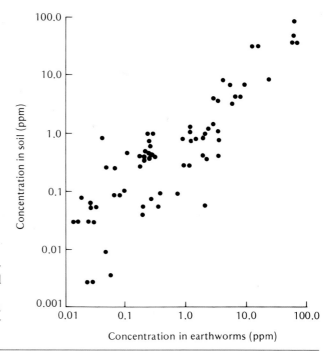

Figure 18.7

Effect of concentration of pesticides in soil on their concentration in earthworms. Birds eating the earthworms at any level of concentration would further concentrate the pesticides. [Data from several sources gathered by Thompson and Edwards (1974); used with permission of Soil Science Society of America]

Fortunately, many pesticides have only mildly depressing effects on earthworm numbers. Exceptions are most of the carbamates and some nematocides, which are quite toxic to earthworms. The concentrations of pesticides in the bodies of the earthworms are closely related to the levels found in the soil (Figure 18.7).

Pesticides have significant effects on the numbers of certain predators and, in turn, on the numbers of prey organisms. For example, an insecticide that reduces the numbers of predatory mites may stimulate numbers of springtails that serve as prey for the mites (Figure 18.8). Such organism interaction is normal in most soils.

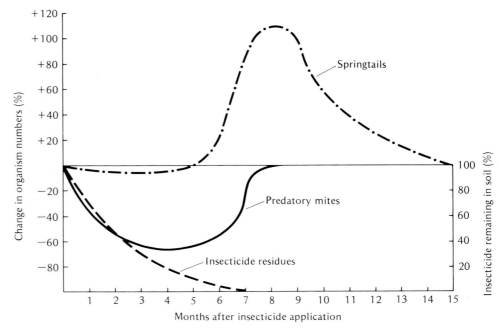

Figure 18.8

The direct effect of insecticide on predatory mites in a soil and the indirect effect of reducing mite numbers on the population of springtails (insect) that serve as prey for the mites. [Replotted from Edwards (1978); used with permission of Academic Press, Inc., London]

EFFECTS ON SOIL MICROORGANISMS

The overall levels of bacteria in the soil are generally not too seriously affected by pesticides. However, the organisms responsible for nitrification and nitrogen fixation are sometimes adversely affected. Insecticides and fungicides affect both processes more than do most herbicides, although some of the latter can reduce the numbers of organisms carrying out these two reactions. Recent evidence suggests that some pesticides can enhance biological nitrogen fixation by reducing the activity of protozoa and other organisms that are competitors or predators of the nitrogen fixing bacteria. These findings illustrate the complexity of the life of the soil.

Fungicides, especially those used as fumigants, can have marked adverse effects on soil fungi and actinomycetes, thereby slowing down the degradation of soil organic matter. Interestingly enough, however, the process of ammonification is often benefited by pesticide use.

The negative effects of most pesticides on soil microorganisms are temporary, and after a few days or a few weeks, organism numbers generally recover. But the exceptions are common enough to dictate caution in the use of the chemicals. Care should be taken to apply them only at recommended levels.

18.5

Contamination with Toxic Inorganic Compounds

Soils also are contaminated by a number of inorganic compounds that, to a greater or less degree, are toxic to humans and other animals. Cadmium, arsenic, chromium, and mercury are extremely poisonous; lead, nickel, molybdenum, and fluorine are moderately so; and boron, copper, manganese, and zinc are relatively low in toxicity. Although some of these metallic elements (which exclude fluorine and boron) are not normally included among the heavy metals, for simplicity this term is often used in referring to them.

SOURCES AND ACCUMULATION

There are many sources for the inorganic chemical contaminants that can accumulate in soils (see Table 18.6). The burning of fossil fuels, smelting, and other processing techniques release into the atmosphere tons of these elements, which can be carried for miles and later deposited on the vegetation and soil. Lead, nickel, and boron are gasoline additives that are released into the atmosphere and carried to the soil through rain and snow.

Borax is used as a detergent and in fertilizer, both of which commonly reach the soil. Superphosphate and limestone usually contain small quantities of cadmium, copper, manganese, nickel, and zinc. Cadmium and chromium are used in plating metals, and cadmium in the manufacture of batteries. Arsenic, for many years used as an insecticide on cotton, tobacco, and fruit crops, is still being used as a defoliant or vine killer and on lawns. Some of these elements are found as constituents in specific organic pesticides and in domestic and industrial sewage sludge.

The quantities of most of the products in which these inorganic contaminants are used have increased notably in recent years, enhancing the opportunity for contamination. They are present in the environment in increasing amounts and are daily ingested by people either through the air or through food and water.

TABLE 18.6
Sources of Selected Inorganic Soil Pollutants

Chemical	Major uses and sources of soil contamination
Arsenic	Pesticides, plant desiccants, animal feed additives, coal, and petroleum; mine tailings and detergents
Cadmium	Electroplating, pigments for plastics and paints, plastic stabilizers, and batteries
Chromium	Stainless steel, chrome-plated metals, pigments, and refractory brick manufacture
Copper	Mine tailings, fly ash, fertilizers, wind blown copper-containing dust
Lead	Combustion of oil, gasoline, and coal; iron and steel production
Mercury	Pesticides, catalysts for synthetic polymers, metallurgy, thermometers
Nickel	Combustion of coal, gasoline, and oil; alloy manufacture, electroplating, batteries
Zinc	Galvanized iron and steel, alloys, batteries, brass, rubber manufacture

Data selected from Moore and Ramamoorthy (1984).

CONCENTRATION IN ORGANISM TISSUE

Whatever their sources, toxic elements can and do reach the soil, where they become part of the life cycle of soil → plant → animal → human (see Figure 18.9). Unfortunately, once the elements become part of this cycle they may accumulate in animal and human body tissue to toxic levels. This situation is especially critical for fish and other wildlife and for humans at the end of the food chain. It has already resulted in restrictions on the use for human food of certain fish and wildlife. Also, it has become necessary to curtail the release of these toxic elements in the form of industrial wastes.

18.6

Potential Hazards of Chemicals in Sewage Sludge

The domestic and industrial sewage sludges considered in Chapter 17 are major sources of potentially toxic chemicals, and at least one third of these wastes in the United States is being applied to the soil. Sewage sludges commonly carry significant quantities of inorganic as well as organic chemicals that can have harmful environmental effects.

Figure 18.9
Sources of heavy metals and their cycling in the soil–water–air–organism ecosystem. It should be noted that the content of metals in tissue generally builds up from left to right, indicating the vulnerability of humans to heavy metal toxicity.

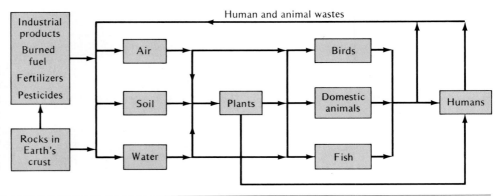

TABLE 18.7
Comparative Concentrations of Several Potentially Toxic Elements in Municipal Sewage Sludge and Cow Manure

Element	In municipal sewage sludge (mg/kg)		In cow manure (mg/kg)
	Small village	Range from 15 larger cities[a]	
Antimony	3	4–44	0.5
Arsenic	3	4–30	4
Cadmium	7	9–444	1
Chromium	169	207–14,000	56
Copper	821	458–2,890	62
Mercury	11	4–18	0.2
Manganese	128	32–527	286
Molybdenum	1	2–33	14
Nickel	36	51–562	29
Lead	136	329–7,627	16
Zinc	560	601–6,890	71

Selected data from Furr et al. (1976).
[a] The 15 larger cities had significant industrial inputs.

In Table 18.7 ranges in levels of several chemical inorganic elements are given for sewage sludges from 15 municipal waste disposal plants in large cities in the United States. The plants all received effluent from industrial as well as domestic sources. Concentrations of most of these elements were much higher in sewage sludge from large cities than in animal manure or in domestic sewage sludge from a small village. For example, the cadmium concentration of the sludge from one city was more than 400 times that of cow manure and 60 times that of domestic sludge from the village plant. Industrial wastes are the primary sources for these chemicals (Table 18.7).

When combined with the uncertainties as to the nature of many of the organic chemicals found in the sludge, these inorganic element levels dictate caution in the unmonitored application of sludge to soils. The effect of such applications on heavy metal content of soils and of earthworms growing in the soil is illustrated in Table 18.8. The sludge-treated soil areas as well as the earthworms growing in the soils were higher in some of these elements than where sludge had not been applied. One would expect further concentration to take place in the tissues of birds and fish, many of

TABLE 18.8
The Effect of Sewage Sludge Treatment on the Content of Heavy Metals in Soil and in Earthworms Growing in the Soil

Note the high concentration of cadmium and zinc in the earthworms.

Metal	Concentration of metal (mg/kg)			
	Soil		Earthworms	
	Control	Sludge-treated	Control	Sludge-treated
Cd	0.1	2.7	4.8	57
Zn	56	132	228	452
Cu	12	39	13	31
Ni	14	19	14	14
Pb	22	31	17	20

From Beyer et al. (1982).

Figure 18.10
Collecting soybean leaf samples from a field fertilized with sewage sludge. The samples are to be analyzed for their heavy metal content, thereby helping to establish the levels of these metals in sludge that can be tolerated. [U.S. Department of Agriculture photo by R. C. Bjork]

which consume the earthworms. The farmer must be assured that the levels of inorganic chemicals in sludge are not sufficiently high to be toxic to plants or to humans and other animals who consume the plants. Scientists measure the effects of sludge applications on heavy metal levels in plants (see Figure 18.10).

18.7

Reactions of Inorganic Contaminants in Soils [2]

HEAVY METALS IN SEWAGE SLUDGE

Concern over the possible buildup of heavy metals in soils, from large land applications of sewage sludges, has prompted research on the fate of these chemicals in soils. Most attention has been given to zinc, copper, nickel, cadmium, and lead, which are commonly present in significant levels in these sludges. Studies have shown that these elements are generally bound by soil constituents, that they do not easily leach from the soil, and that they

[2] For a review of this subject see Lake et al. 1984.

TABLE 18.9

Forms of Six Heavy Metals Found in a Greenfield Sandy Loam (Coarse Loamy, Mixed, Thermic Typic Haloxeralf) That Had Received 45 Mg/ha Sewage Sludge Annually for 5 Years

Forms	Percentage of elements in each form					
	Cd	Cr	Cu	Ni	Pb	Zn
Exchangeable/adsorbed	1	1	2	5	1	2
Organically bound	20	5	34	24	3	28
Carbonate/iron oxides	64	19	36	33	85	39
Residual[a]	16	77	29	40	12	31

From Chang et al. (1984).

[a] Sulfides and other very insoluble forms.

are not readily available to plants. Only in moderately to strongly acid soils is there significant movement down the profile from the layer of application of the sludge. Monitoring soil acidity and judicious lime applications can prevent leaching of these elements into ground waters.

By using chemical extractants, researchers have found these metals to be associated with soil solids in four major ways. (Table 18.9). *First*, a very small proportion is held in adsorbed or exchangeable forms that are available for plant uptake. *Second*, the elements are bound by the soil organic matter and by the organic materials in the sludge. A high proportion of the copper is commonly found in this form; lead is not so highly attracted. Organically bound elements are not readily available to plants but can be released over a period of time.

The *third* association of heavy metals in soils is with carbonates and with oxides of iron and manganese. These forms are less available to plants than either the exchangeable or organically bound forms, especially if the soils are not allowed to become too acid. The *fourth* association is commonly known as the residual form, which consists of sulfides and other very insoluble compounds that are less available to plants than any of the other forms.

It is fortunate that soil-applied heavy metals are not readily absorbed by plants and that they are not easily leached from the soil. Care must be taken, however, not to add such large quantities that the soil's capacity to react with a given element is exceeded. Annual additions of sewage sludge high in these chemicals may eventually tax the soil's ability to tie up these chemicals.

CHEMICALS FROM OTHER SOURCES

Arsenic has accumulated in orchard soils following years of application of arsenic-containing pesticides. Being present in an anionic form (e.g., $H_2AsO_4^-$), this element is absorbed (as are phosphates) by hydrous iron and aluminum oxides. In spite of the capacity of most soils to tie up arsenates, long-term additions of arsenical sprays can lead to toxicities for sensitive plants. The arsenic toxicity can be reduced by applications of sulfates of zinc, iron, and aluminum that tie up the arsenic in insoluble forms.

Contamination of soils with *lead* comes primarily from airborne lead from automobile exhausts, a source that has decreased in quantity during the past decade or so. Some lead is deposited on the vegetation and some reaches the soil directly. In any case, most of the lead is tied up in the soil as insoluble carbonates and sulfides and in combination with iron, aluminum, and manganese oxides (Table 18.9). Consequently the lead is largely unavailable to plants.

Soil contamination by *boron* can occur from irrigation water high in this element or by excessive fertilizer application. The boron can be adsorbed by organic matter and clays but is still available to plants, except at high soil pH. Boron is relatively soluble in soils and toxic quantities are leachable, especially from acid sandy soils. Boron toxicity is usually considered a localized problem and is probably much less important than a deficiency of the element.

Fluorine toxicity is also generally localized. Fluorine appears in drinking water for animals and in fluoride fumes from industrial processes. The fumes can be ingested directly by the animals or deposited on nearby plants. If the fluorides are adsorbed by the soil, their uptake by plants is restricted. The fluorides formed in soils are highly insoluble, the solubility being least if the soil is well supplied with lime.

Mercury contamination of lake beds and of swampy areas has resulted in toxic levels of mercury in certain species of fish. Insoluble forms of mercury in soils, not normally available to plants and in turn to animals, are converted by microorganisms to an organic form (methylmercury); methylmercury is quite soluble and available for plant and animal absorption. It is concentrated in fatty tissue as it moves up the food chain until it accumulates in some fish to levels that may be toxic to humans. This illustrates how reactions in soil can influence human toxicities.

18.8

Prevention and Elimination of Inorganic Chemical Contamination

Two primary methods of alleviating soil contamination by toxic inorganic compounds are (a) to eliminate or drastically reduce the soil application of the toxins, and (b) to so manage the soil and crop as to prevent further cycling of the toxins.

REDUCING SOIL APPLICATION

First, action is required to reduce unintentional aerial contamination from industrial operations and from automobile, truck, and bus exhausts. Decision makers must recognize the soil as an important natural resource that can be seriously damaged by contamination from accidental addition of inorganic toxins. Such contamination must be curtailed. Also, there must be judicial reductions in intended applications to soil of the toxins through pesticides, fertilizers, irrigation water, and solid wastes.

REDUCING RECYCLING

Soil and crop management can help reduce the continued cycling of toxic inorganic chemicals. This is done primarily by keeping the chemicals in the soil and reducing their uptake by plants. The soil becomes a "sink" for the toxins, and thereby breaks the cycle of soil–plant–animal (humans) through which the toxin exerts its effect. The cycle is broken by immobilizing the toxins in the soil. For example, most of these elements are rendered less mobile and less available if the soil pH is kept near neutral or above (Figure 18.11). Liming of highly acid soils should expedite the immobility of toxic elements. The draining of wet soils should also be beneficial, since the oxidized forms of the several toxic elements are generally less soluble and less available for plant uptake than are the reduced forms (see Section 6.3).

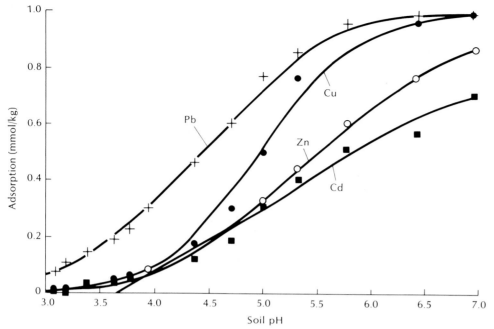

Figure 18.11
The effect of soil pH on the adsorption of four heavy metals. Maintaining the soil near neutral provides the highest adsorption of each of these metals and especially of lead and copper. The soil was a Typic Paleudult (Christiana silty clay loam). [From Elliot et al. (1986)]

Heavy phosphate applications reduce the availability of toxic cations but may have the opposite effect on arsenic, which is found in the anionic form. Leaching may be effective in removing excess boron, although moving the toxin from the soil to nearby waterways may not be of any real benefit.

Advantage can be taken of differences in the abilities of crop species or varieties to extract the toxins (Table 18.10). Even though none of the crops tested take up a high proportion of the elements, some absorb more than others. "Accumulators" should be avoided if the harvests are to be fed to humans or domestic animals. Moreover, forage crops should be harvested at the stage of maturity when the concentration of the toxin is lowest. It is obvious that soil and crop management offers some potential for alleviating contamination by inorganic elements.

TABLE 18.10

Quantities of Several Metals Added in 15 Mg of Sewage Sludge and the Uptake of These Metals by Three Crops Grown on a Udoll (Plano Silt Loam)

Element	Amount applied (kg/ha)	Concentration in crop (%)		
		Rye	Sorghum–Sudan grass	Corn
Cu	22	0.16	0.17	0.05
Zn	45	0.35	1.14	3.39
Cd	1	0.13	0.37	0.08
Ni	11	0.05	0.20	0.06
Cr	28	0.01	0.04	0.01

From Kelling et al. (1977).

18.9

Soils as Organic Waste Disposal Sites

In the United States nearly 250 million Mg of domestic wastes are generated each year. To this must be added the nearly 2 billion Mg of farm animal wastes, and millions of metric tons (Mg) of organic wastes from food- and fiber-processing plants and industrial operations.

Environmental considerations have led to restrictions on the disposal of organic wastes, both urban and rural, into waterways and into the atmosphere by burning. The soil offers an alternative disposal sink that is being used more and more widely. For example, in 1980 about 25% of the sewage sludge in the United States was applied to cropland. Many organic wastes can improve soil physical and chemical properties and can provide nutrients for increased crop yields. The positive effects will likely encourage continued land application of these wastes. However, when excess quantities are applied, yields may be depressed (Figure 18.12) and soil and water pollution can occur.

The effects of excessive concentration of organic wastes on nitrate accumulation in soils is illustrated in Figure 18.13. Nitrate ions are subject to leaching into the groundwater, which may later be used for drinking purposes by livestock and/or humans. While nitrate-toxic waters are not too common, care must be taken to prevent their occurrence.

LANDFILLS

Soils have long been used as disposal "sinks" for municipal refuse. "Sanitary landfills" are widely employed to dispose of a variety of wastes from our towns and cities. These wastes include paper products, garbage, and nonbiodegradable materials such as glass and metals. The sites are often located in swampy lowland areas that eventually become built up by the dumping to create upland areas for such uses as city parks and other facilities.

Unfortunately, sanitary landfills are sometimes not so sanitary. Leaching and runoff from these sites can contaminate both surface water and groundwater. Contaminants include heavy metals as well as soluble and biodegrad-

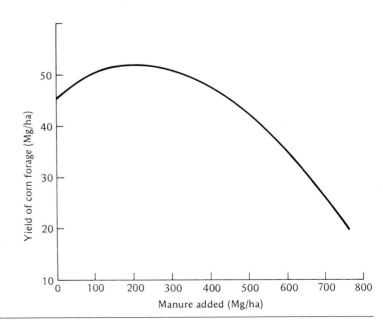

Figure 18.12
Effects of applying exceedingly large amounts of manure in 1969 on the forage yield of corn in 1970. Depressed yields at rates above 200 Mg/ha are probably due to high salt content in the manure. [From Murphy et al. (1972)]

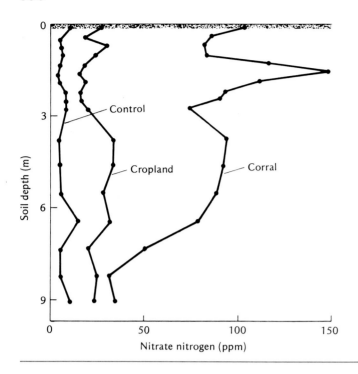

Figure 18.13
Nitrate nitrogen concentrations under a dairy corral as compared to those of a nearby cropped area and a control area to which no manure or irrigation water had been added. [From Adriano et al. (1971)]

able organic materials. The environmental hazards associated with landfills require marked restrictions on their continued use.

DISADVANTAGES OF LAND APPLICATIONS

The primary disadvantages of disposing of organic wastes in soils stem from exceeding the soil's capacity to accommodate the wastes. Heavy metal contamination and excessive nitrate leaching into groundwater have already been discussed. It is difficult to regulate the chemicals, organic or inorganic, that are dumped in landfills. While contamination of soils with these substances is serious, their movement from soils into streams leads to the greatest human problems. Constant monitoring is necessary to guard against such contamination.

18.10

Soil Salinity

Contamination of soils with salts is one form of soil pollution primarily agricultural in origin. Furthermore, it is not a new problem. Ancient civilizations in both the New and Old Worlds crumbled because salts built up in their irrigated soils. The same principles govern the management of irrigated soils today and the same dangers exist of salt buildup and concomitant soil deterioration (see Section 8.21).

Salts accumulate in soils because more salts move into the plant rooting zone than move out. This may be due to application of salt-laden irrigation waters or it may be caused by irrigating poorly drained soils. Salts move up from the lower horizons and concentrate in the surface soil layers.

EXAMPLES

The problem of salt buildup in soils of river valleys is common throughout the world. Noted examples are the soils of the lower basins of the Indus River in Pakistan, the Nile River in Egypt, and the Euphrates River in Iraq. Areas in the western part of the United States with major salinity problems are shown in Figure 18.14; some 4 million ha (10 million acres) are seriously salt-affected. The Colorado River Basin has the most extensive salinity problems. The lower Rio Grande Basin and the San Joaquin Valley of Central California are other relatively large areas seriously affected by salt buildup.

Smaller but significant areas are found in most irrigated valleys of the western states. In all cases, salt concentration in the water increases from the river head to the mouth of the river. Much of this increase is from return flow to the rivers of salt-laden drainage from irrigated areas.

SALTS IN HUMID REGIONS

Some salt buildup occurs in rivers whose watersheds are in humid regions. The salts are found in heavily populated and industrialized areas where water is returned to streams following domestic or industrial use. Sewage plant treatments commonly remove soluble inorganic salts only if they are known toxicants. Some sewage sludges have sufficiently high levels of salt to cause crop plant damage when the sludge is applied. When salts added from waste are combined with those leached from the watershed soils, the salt level may approach that found in rivers flowing through more arid areas.

The control of salinity depends almost entirely on the quality and management of water. In some areas, removal of excess water by the installation of

Figure 18.14
Areas in the western United States with major salinity problems. Shaded areas have most serious problems. [From El-Ashry et al. (1985)]

good drainage systems is required. In local areas sulfur or gypsum applications can be used to help eliminate toxic sodium bicarbonate (see Section 8.21). Even where sulfur or gypsum can be used, however, water management following chemical treatment is most vital. Water quality is determined by appropriate public policies and by individual farm management such as assure good soil drainage and optimum irrigation practices.

18.11

Acid Rain[3]

Scientists in Europe and North America have called attention to marked increases in the acidity of precipitation over recent decades, especially in relation to forest decline (Figure 18.15). In the absence of contaminating atmosphere gases, rainwater has a pH of about 5.6 owing to the presence of carbonic acid formed from the carbon dioxide in the atmosphere. But in or near areas with large-scale combustion of fossil fuel or with smelting of sulfide ores, the pH of precipitation may be as low as 4.0. In extreme cases of dense fog, the pH may drop to nearly 2.0, which is a serious potential hazard to society.

Acid precipitation, popularly called *acid rain*, is apparently due to the oxidation of nitrogen- and sulfur-containing gases that dissolve in the water vapor of the atmosphere to form nitric and sulfuric acids. Reactions such as the following are thought to occur.

$$2\ NO + O_2 \rightarrow 2\ NO_2 \xrightarrow{H_2O} HNO_3 + HNO_2$$

| Nitric oxide | Nitrogen dioxide | Nitric acid | Nitrous acid |

[3] For a review of this subject, see NRC (1983).

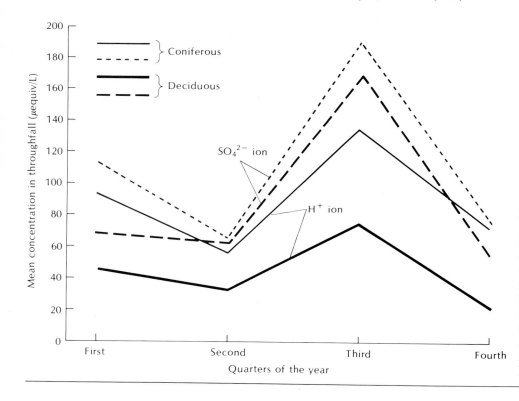

Figure 18.15
Effect of forest canopies on mean concentrations of SO_4^{2-} and H^+ ions that fall through the trees onto the soil (throughfall). Note the correlation between the two ions' concentrations. Also note that concentrations are higher in coniferous than in deciduous forests. [Drawn from data in Puckett (1987)]

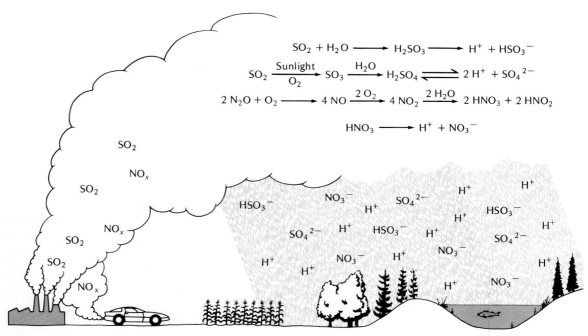

Figure 18.16

Illustration of the formation of nitrogen and sulfur oxides from the combustion of fuel in sulfide ore processing and from motor vehicles. The further oxidation of these gases and their reaction with water to form sulfuric acid and nitric acid are shown. These help acidify rainwater, which falls on the soil as "acid rain." NO_x indicates a mixture of nitrogen oxides, primarily N_2O and NO. [Modified from NRC (1983)]

$$2\,SO_2 + O_2 \rightarrow 2\,SO_3 \xrightarrow{2\,H_2O} 2\,H_2SO_4$$

Sulfur   Sulfur   Sulfuric
dioxide   trioxide   acid

Figure 18.16 shows how these nitrogen and sulfur oxides can move into the atmosphere, be converted to inorganic acids, and return to the land in rain and snow. Such cycling is responsible for lowering the pH of precipitation in the northeastern part of the United States and in eastern Canada. As illustrated in Figure 18.17, there are significant areas in the eastern United States and Canada where the pH of precipitation is 5 or less.

EFFECTS ON SOIL pH

Acid rain is assumed to be responsible for increased acidity in some lakes of the Adirondacks and in other regions of the Northeast. Most fish will not tolerate pH levels below about 4.5. As a consequence the increased acidity of lake water is thought to have essentially eliminated most fish in some Adirondack lakes.

Effects of acid rain are more pronounced on the acidity of water than on soil acidity. Soils generally are sufficiently buffered to accommodate acid rain with little or no increase in soil acidity on an annual basis. But continued inputs of acid rain at pHs of 4.0–4.5 would have significant effects on the pH of soils, especially those that are weakly buffered. This is also serious for soils that are already quite acid, since increased acidity could well make them even less fertile.

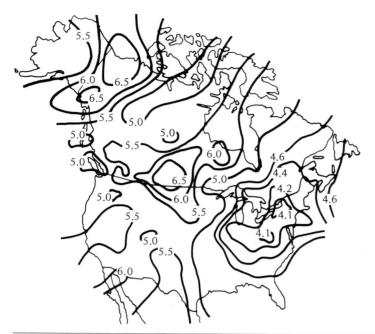

Figure 18.17
Map showing concentrations (pH) of acid rain in North America. Note the heaviest acid concentrations (lowest pH values) are found in the industrial northeastern areas of the United States. [Redrawn from De Young (1982); data from Canadian and U. S. government agencies; reprinted with permission, *High Technology Illustrated Magazine*, Sept./Oct. 1982, copyright © 1982 by High Technology/Technology Publishing Corporation, 38 Commercial Wharf, Boston, MA 02110]

ALLEVIATION OF ADVERSE EFFECTS

There are two obvious ways to alleviate the effects of acid rain on soils. First, the emission of sulfur and nitrogen oxides must be reduced drastically. This is a remedy already suggested, but economic and political decisions may delay its implementation. Second, the effects of the acid rain on soil pH can be overcome by adding lime. This solution may be less costly than the first, but just as difficult to achieve, since it will require commitment of resources by farmers, small and large alike. Unfortunately many adverse effects of acid rain are being felt on forestlands and on steep mountainous areas where the soils may be shallow and where application of lime is difficult if not impossible. Obviously, both potential solutions must receive attention.

18.12

Radionuclides in Soil

Nuclear fission in connection with atomic weapons testing provides another source of soil contamination. To the naturally occurring radionuclides in soil (e.g., ^{40}K, ^{87}Rb, and ^{14}C), a number of fission products have been added. Only two of these are sufficiently long-lived to be of significance in soils: strontium-90 (half-life = 28 yr) and cesium-137 (half-life = 30 yr). The average levels of these nuclides in soil in the United States are about 388 millicuries (mCi)/km^2 (150 mCi/mi^2) for ^{90}Sr and 620 mCi/km^2 (240 mCi/mi^2) for ^{37}Cs. A comparable figure for the naturally occurring ^{40}K is 51,80 mCi/km^2 (20,000 mCi/mi^2). These normal soil levels of the fission radionuclides are not high enough to be hazardous. Even during periods of weapons testing, the soil did not contribute significantly to the level of these nuclides in plants. Direct fallout from the atmosphere on the vegetation was the primary source of contamination. Consequently, only in the event of a catastrophic supply of fission products could toxic soil levels of ^{90}Sr and ^{137}Cs be expected. Fortunately, considerable research has been done on the soil behavior and nutrient uptake characteristics of these two nuclides.

Strontium-90

Strontium-90 behaves in soil much the same as does calcium, to which it is closely related chemically. It enters soil from the atmosphere in soluble forms and is quickly adsorbed by the colloidal fraction, both organic and inorganic. It undergoes cation exchange and is available to plants much as is calcium. The possibility that strontium is involved in the same plant reactions as calcium probably accounts for the fact that high soil calcium tends to decrease the uptake of ^{90}Sr.

Cesium-137

Although chemically related to potassium, cesium tends to be less readily available in many soils. This is apparently because ^{137}Cs is firmly fixed by vermiculite and related interstratified minerals. The fixed nuclide is nonexchangeable, much as is fixed potassium in some interlayers of clay. Plant uptake of ^{137}Cs from such soils is very limited. Where vermiculite and related clays are absent, as in some tropical soils, ^{137}Cs uptake is more rapid. In any case, the soil tends to dampen the movement of ^{137}Cs into the food chain of animals and man.

Radioactive Wastes[4]

Aside from radionuclides added to soils as a result of weapons testing, low-level radioactive waste materials are sometimes handled by burying them in soils. Even though the materials may be in the solid form when placed in shallow-land burial pits, some dissolution and subsequent movement in the soil are possible. Plutonium, uranium, americium, neptunium, curium, and cesium are among the elements whose nuclides occur in wastes.

Nuclides in wastes vary greatly in water solubility; uranium compounds are quite soluble, compounds of plutonium and americium are relatively insoluble, and cesium compounds are intermediate in solubility. Cesium, a positively charged ion, is adsorbed by soil colloids. Uranium is thought to occur as a UO_2^{2+} ion that is also adsorbed by soil. The charge on plutonium and americium appears to vary depending on the nature of the complexes these elements form in the soil.

There is considerable variability in the actual uptake by plants of these nuclides from soils, depending on soil properties such as pH and organic matter content. The uptake from soils by plants is generally lowest for plutonium, highest for neptunium, and intermediate for americium and curium. Crop fruits and seeds are generally much lower in these nuclides than are leaves, suggesting that human foods may be less contaminated by nuclides than forage crops.

Since soils are being used as burial sites for low-level radioactive wastes, care should be taken to be certain the soil properties will discourage leaching or significant plant uptake of the chemicals. Data in Table 18.11 illustrate differences in the ability of different soils to hold breakdown products of two radionuclides. It is evident that monitoring of selected sites will likely be needed to assure minimum transfer of the nuclides to other parts of the environment.

---

[4] This summary is based largely on papers on this subject in *Soil Science*, **132** (July 1981).

TABLE 18.11

Concentrations of Several Breakdown Products of Uranium-238 and Thorium-232 (Nucleotides) in Six Different Soil Suborders in Louisiana

Note marked differences among levels in the different soils.

Soil suborder	No. of samples	^{238}U breakdown products			^{232}Th breakdown products		
		^{226}Ra	^{214}Pb	^{214}Bi	^{212}Pb	^{137}Cs	^{40}K
Udults	22	37.3	27.7	28.9	27.4	16.7	136
Aquults	24	30.4	36.7	38.1	50.0	10.9	100
Aqualfs	37	51.1	38.3	36.6	59.7	13.5	263
Aquepts	93	92.2	47.6	45.2	63.8	16.1	636
Aquolls	57	90.4	45.8	44.7	59.5	8.7	608
Hemists	18	136.3	49.4	49.0	74.9	19.4	783

From Meriwether et al. (1988).

RADON GAS

The soil is the primary source of *radon*, a colorless and odorless radioactive gas that can cause lung cancer. Radon is formed from the radioactive decay of radium, a breakdown product of uranium found in minute quantities in most soils (Figure 18.17). The radon enters homes and other buildings from the surrounding soil. Since modern airtight buildings permit little exchange of air with the outside, radon can accumulate to undesirable levels. Radon movement into the buildings is usually through cracks in the basement walls and floors; these must be sealed to reduce the penetration. Likewise, basements must be ventilated with outside air to prevent radon buildup.

Since radon is an inert gas, it does not react with the soil, which merely serves as a channel through which the gas moves. The health hazard from this gas stems from its radioactive decay. One product is alpha particles, which can penetrate the lung tissue and cause cancer (see Figure 18.18).

The air in closed spaces, especially in basements, should be tested to ascertain the radon level. Where dangerously high levels exist, steps must be taken to seal off the floor and walls that have contact with soil.

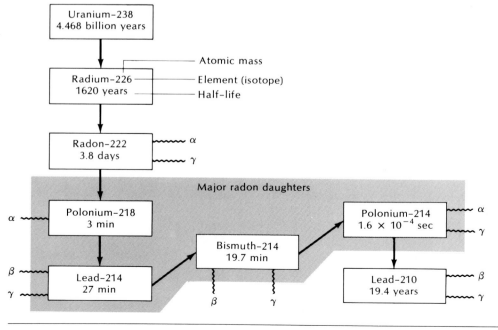

Figure 18.18
Radioactive decay of uranium-238 in soils that results in the formation of inert but radioactive radon. This gas emits alpha (α) particles and gamma (γ) rays and forms "radon daughters" that are capable of emitting alpha (α) and beta (β) particles and gamma (γ) rays. The alpha particles damage lung tissue and cause cancer. Radon gas accounts for about 10,000 deaths annually in the United States. [Modified from Boyle (1988)]

18.13

Soils and the Greenhouse Effect[5]

Widespread concern has been expressed that the Earth is warming up. Furthermore, it is predicted that this warming trend will continue and even accelerate in the future. The cause of this warming is thought to be the so-called greenhouse effect. When certain gases are emitted from the earth, they move into the upper atmosphere, capture, and return to the earth radiant heat that would ordinarily escape into space. In so doing the gases serve the same purpose as the glass in a greenhouse.

With the advent of modern industrialization the content of these greenhouse gases in the atmosphere has increased markedly. For example the carbon dioxide content of the upper atmosphere is thought to have been about 280 parts per million (ppm) in preindustrial times. It has increased to about 350 ppm in the past 30 years. Significant increases have also occurred in nitrogen- and sulfur-bearing compounds and in organic synthetics such as the chlorofluorocarbons (CFCs) used as aerosol propellants, refrigerants, and solvents.

The soil is a source of several of the gases involved in the greenhouse phenomenon including carbon dioxide (CO_2), nitrous oxide (N_2O), and methane (CH_4). The breakdown of soil organic matter as land is cleared and put under cultivation has been a major source of released CO_2. This loss is small however, in comparison with the much larger releases from the combustion of fossil fuels (petroleum and coal) and the burning of tropical forests.

Nitrous oxide (N_2O) content of the atmosphere has increased about 25% in this century. About two thirds of this increase is thought to be due to the combustion of coal and oil, the remainder to agricultural practices. The N_2O is released during the process of denitrification, which is fueled by the presence of nitrates in the soil and in the organic matter cover. Although heavy nitrogen fertilization may be a source of part of the nitrates that undergo denitrification, much nitrate reduction occurs in unfertilized areas. For example, tropical forests have been found to be a significant year-round source of N_2O.

The methane (CH_4) content of the upper atmosphere has essentially doubled in the past hundred years, and there is no consensus as to the reasons for this increase. However, the soil is a known source of methane. It is released under anaerobic conditions such as are common in swamps or in a rice paddy. Also, termites can and do produce methane, and some investigators have suggested that 20–40% of the methane reaching the atmosphere may come from this source.

POSSIBLE CONTROLS

Several management practices considered as essential for the maintenance of soil productivity are relevant to the release of "greenhouse" gases. For example, practices such as conservation tillage, which slow down the destruction of soil organic matter, should be followed. Likewise, applications of excessive rates of nitrogen fertilization should be avoided. Furthermore, the fertilizer should be applied at a time when the plants can readily absorb the nitrogen and when the soil is not excessively wet.

Further research is needed to better understand reasons for the buildup of methane. It is possible that management of rice paddies can be altered in such a way as to minimize the release of methane from this source.

[5] For further information, see Patrusky (1988).

TABLE 18.12

Nitrate Nitrogen Losses in Ground and Surface Waters from the Big Spring Basin in Iowa, 1982–1984

Total shown along with losses as percent of previous year's nitrogen fertilizer application.

Nitrate losses	Water year		
	1982	1983	1984
Total discharged in water (Mg)	821	1300	817
As percent of fertilizer application	33	55	33

From Hallberg et al. (1985).

18.14

Fertilizer Contamination of Water[6]

Fertilizer applications that supply nutrients in quantities far in excess of those taken up by plants can result in contamination of both surface and drainage waters (see Table 18.12). Nitrates and phosphates are the chemicals most often involved. Nitrate contamination can occur in both surface runoff and drainage waters, while excessive levels of phosphates generally occur only in surface runoff.

The loss of nitrogen and phosphorus from the soil has adverse effects on soil fertility, but the effect on water quality is even more serious. Nitrate levels in drinking water above about 10 mg per liter are considered a human health hazard. In some heavily fertilized areas, the drainage waters are sufficiently high in nitrates to be a problem (Figure 18.19). Some rural wells have been found to contain nitrates significantly above this safe limit.

A second problem stemming from high nutrient-bearing waters coming from soils is the "overfertilization" of lakes. Nitrogen and phosphorus in lake waters stimulate the growth of algae and other water-loving plants in the lakes. Algal growth depletes the water of oxygen, which is essential for fish. Other aquatic plants (weeds) are stimulated and produce heavy mats near the shoreline interfering with recreational uses of the lakes.

Applications of fertilizer far in excess of plant uptake should be discouraged, and the timing of fertilizer applications should coincide with plant needs. The fertilizer should be mixed with at least some soil, especially where conservation tillage practices are employed, to reduce surface runoff of the fertilizer compounds.

18.15

Three Conclusions

Three major conclusions may be drawn about soils in relation to environmental quality. First, since soils are valuable resources, they should be protected from environmental contamination, especially that which does permanent damage. Second, because of their vastness and remarkable capacity to absorb, bind, and break down added materials, soils offer promising mechanisms for disposal and utilization of many wastes that otherwise might contaminate other parts of the environment. Third, products of soil reactions can be toxic to humans and other animals if they are absorbed by plants or if they move from the soil into the air and particularly into water.

---

[6] For discussions of groundwater contamination see D'Itri and Wolfson (1987).

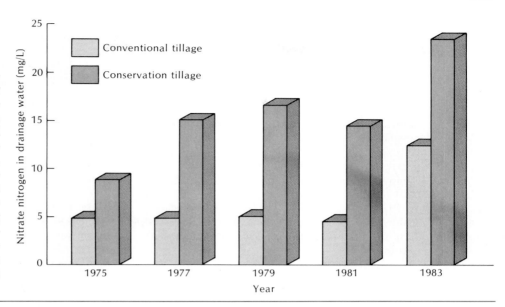

Figure 18.19
Concentration of nitrate nitrogen in the subsurface water from two well-fertilized corn-producing watersheds in Iowa, one with conventional tillage and the other with conservation tillage (till, plant). The limit for safe drinking water is 10 mg of nitrate nitrogen per liter. Apparently, the conservation tillage, by reducing runoff, encouraged more percolation of water, which carried nitrates downward in the soil. [Redrawn from Alberts and Spomer (1985)]

To gain a better understanding of how soils might be used and yet protected in waste management efforts, soil scientists must devote a fair share of their research effort to environmental quality problems.

Study Questions

1. What are the six major kinds of soil pollutants?
2. Conservation tillage, which minimizes plowing of the land, is said to be responsible for some increases in pollution of soil and water. Explain.
3. Chlorinated hydrocarbons were among the first and most effective insecticides. Why has their use been drastically reduced if not eliminated?
4. Which class of pesticides is most widely used in the United States?
5. What is the fate of pesticides after they enter the soil?
6. The land application of sewage sludge, which is largely of organic composition, is of concern primarily because of inorganic pollution of soils. Why?
7. Even though large amounts of the so-called heavy metals are applied to soils, relatively small quantities

of these metals are taken up by plants. why?
8. What management practices are important in preventing environmental degradation from the application of sewage sludge to soils?
9. Explain why modern beef production practices can result in nitrate build up in ground water.
10. What soil and water management practices must be used to prevent soil salinity?
11. What are the causes of acid rain?
12. Acid rain has not brought about marked reduction in soil pH over wide areas. Why?
13. Once products of thermonuclear explosions have reached the soil, their uptake by plants is quite limited. Why?

References

ALBERTS, E. E., and R. G. SPOMER. 1985. "Dissolved nitrogen and phosphorus in runoff from watersheds in conservation and conventional tillage," *J. Soil Water Cons.*, **40**:153–57.

ADRIANO, D. C., et al. 1971. "Nitrate and salt in soil and ground waters from land disposal of dairy manures," *Soil Sci. Soc. Amer. Proc.*, **35**:759–62.

ALLAWAY, W. H. 1970. "Agronomic controls over the environmental cycling of trace elements," *Adv. Agron.*, **20**:235–74.

BEYER, W. N., R. L. CHANEY, and B. M. MULHERN. 1982. "Heavy metal concentration in earthworms from soil amended with sewage sludge," *J. Environ. Qual.*, **11**:381–85.

BOYLE, M. 1988. "Radon testing of soils," *Environ. Sci. Tech.*, **22**:1397–99.

CHANG, A. C., et al. 1984. "Sequential extraction of soil heavy metals following a sludge application," *J. Environ. Qual.*, **13**:33–38.

COHEN, S. Z. et al. 1986. "Monitoring ground water for pesticides," pp. 170–96 in W. Y. Garner et al. (Eds.), *Evaluation of Pesticides in Ground Water*, ACS Symposium Series 315 (Washington DC: Amer. Chem. Soc.).

DE YOUNG, H. G. 1982. "Acid rain regulators shift into low gear," *High Technology*, 2:82–86.

D'ITRI, F. M., and L. G. WOLFSON. 1987. *Rural Groundwater Contamination* (Chelsea, MI: Lewis Publishers Inc.).

DMI. 1981. *Farming in the Profit Zone Through Plant Nutrition and Conservation Tillage* (Goodfield, IL: DMI Inc.).

EDWARDS, C. A. 1978. "Pesticides and the micro-fauna of soil and water," pp. 603–22 in I. R. Hill and S. J. Wright (Eds.), *Pesticide Microbiology* (London: Academic Press).

EL-ASHRY, N. T., J. van Schilfgaarde, and S. Schiffman. 1985. "Salinity pollution from irrigated agriculture," *J. Soil Water Cons.*, 40:48–52.

ELLIOTT, H. A., M. R. LIBERATI, and C. P. HUANG. 1986. "Competitive adsorption of heavy metals in soils," *J. Environ. Qual.*, 15:214–19.

FURR, A. K., et al. 1976. "Multielement and chlorinated hydrocarbon analysis of municipal sewage sludges of American cities," *Environ. Sci. Tech.*, 10:683–87.

GARNER, W. Y., R. C. HONEYCUTT, and H. N. NIGG. 1986. *Evaluation of Pesticides in Ground Water* (Washington, D C: Amer. Chem. Soc).

HALLBERG, G. R., R. D. LIBRA, and B. E. HOYER. 1985. "Nonpoint source contamination of groundwater in Karst–Carbonate aquifers in Iowa," in *Perspectives on Nonpoint Source Pollution*, EPA 440/5 85001 (Washington, DC: USEPA).

JUNK, G. A., et al., 1984. "Degradation of pesticides in controlled water–soil systems" in R. F. Krueger and J. N. Seiber (Eds.), *Treatment and Disposal of Pesticide Wastes*, ACS Symposium Series 259 (Washington, DC: Amer. Chem. Soc.).

KELLING, K. A., et al. 1977. "A field study of the agricultural use of sewage sludge: III. Effect on uptake and extractability of sludge-borne metals," *J. Environ. Qual.*, 6:352–58.

KRUEGER, R. F., and J. N. SEIBER (Eds.). 1984. *Treatment and Disposal of Pesticide Wastes* (Washington, DC: Amer. Chem. Soc.).

LAGERWERFF, J. V., and A. W. SPECHT. 1970. "Contamination of roadside soil and vegetation with cadmium, nickel, lead and zinc," *Environ. Sci. Tech.*, 4:583–86.

LAKE, D. L., P. W. W. KIRK, J. N. LESTER. 1984. "Fractionation, characterization and speciation of heavy metals in sewage sludge and sludge-amended soils: A review," *J. Environ. Qual.*, 13:175–83.

McCONNELL, J. S., and L. R. HOSSNER. 1985. "pH-dependent adsorption isotherm of glyphosate," *J. Agr. Food Chem.*, 33:1075–78.

MERIWETHER, J. R., et al. 1988. "Radionuclides in Louisiana soils," *J. Environ. Qual.*, 17:562–68.

MOORE, J. W., and S. RAMAMOORTHY. 1984. *Heavy Metals in Natural Waters* (New York: Springer–Verlag).

MURPHY, L. S., et al. 1972. "Effects of solid beef feedlot wastes on soil conditions and plant growth," in *Waste Management Research Proceedings*, Cornell Agricultural Waste Management Conference, Ithaca, NY.

NRC. 1983. *Acid Deposition: Atmospheric Processes in Eastern North America*, a National Research Council Report (Washington, DC: National Academy Press).

PATRUSKY, B. 1988. "Dirtying the infrared windows," *Mosaic*, 19:25–37.

PUCKETT, L. J. 1987. "The influence of forest canopies on the chemical quality of water and the hydrologic cycle," pp. 3–22 in R. C. Averett and D. M. McKnight (Eds.), *Chemical Quality of Water and the Hydrologic Cycle* (Chelsea, MI: Lewis Publishers, Inc.)

SANDERS, H. J. 1981. "Herbicides," *Chem. Eng. News*, Aug. 3, pp. 20–35.

STORK, W. J. 1987. "Pesticide growth slows," *Chem. Eng. News*, Nov. 16, pp. 35–37.

THOMPSON, A. R., and C. A. EDWARDS. 1974. "Effects of pesticides on nontarget invertebrates in freshwater and soil," pp. 341–86 in W. D. Guenzi (Ed.), *Pesticides in Soil and Water* (Madison, WI: Soil Sci. Soc. Amer.).

USDA. 1980. *Agricultural Statistics 1980* (Washington, DC: U.S. Government Printing Office).

WEBER, J. B., and C. T. MILLER. 1989. "Organic chemical movement over and through soil," pp. 305–34 in B. L. Sawhney and K. Prown (Eds.), *Reactions and Movement of Organic Chemicals in Soils*, SSSA Spec. Pub. 22 (Madison, WI: Soil Sci. Soc. Amer.).

Soils and the World's Food Supply

Hunger is not new to the world. It has always been a threat to human survival. Through the centuries at some place on earth scarcity of food has brought human misery, disease, and death. Never in recorded history has the threat of widespread hunger been greater than it is today. This threat is not due to the reduced capacity of the world to supply food. Indeed, this capacity is greater today than it has ever been and is continuing to grow at a reasonable rate. The problem lies in the even more rapid rate at which world population is increasing. World food production per person is barely holding its own. In selected areas, it is declining.

19.1

Expansion of World Population[1]

Science is in part responsible for the marked expansion in world population growth, especially that which has occurred in the developing nations. Until about the midpoint of the 20th century, high birth rates in most of South America, Africa, and Asia were largely negated by equally high death rates. High infant mortality, poor health facilities, inadequate medical personnel, and disease-spreading insects took their toll. Population expansion was held in check.

During the past few decades, advances in medical science and their application throughout the world have changed this situation (Figure 19.1). Death rates have been drastically reduced, especially among the young. Pesticides have helped check mosquitoes and other disease-carrying pests. Medical services have become available in remote areas of the developing nations. The result is unprecedented population growth. The population is

---

[1] A stimulating discussion of the relationship between population growth and food production is given in FAO (1984).

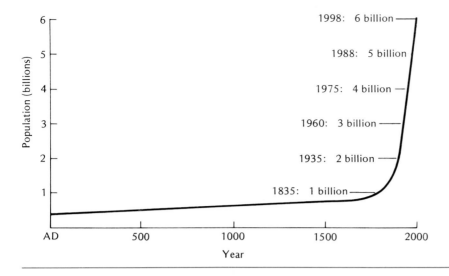

Figure 19.1
From the beginning of the human race until 1960, the world's population increased to 3 billion. Population estimates indicate that less than 40 years will be needed to provide the second 3 billion. Furthermore, most of the population growth will occur in countries already struggling to feed themselves. Projections for 1987 and 1997 from U.S. Bureau of the Census.

doubling every 18–30 years in many developing areas, where two thirds of the world's population now live. Experts predict the world's population will be between 6 and 7 billion by the year 2000—nearly 25% higher than it is today. Furthermore, *about 90% of this increase will occur in developing nations* where food supplies are already critical and where the technology for increased food production is wholly inadequate. The world food supply continues to be among humanity's most serious problems.

19.2

Factors Affecting World Food Supplies

The ability of a nation to produce food is determined by many social, economic, and political factors that affect the farmer's incentive to produce. Food production is also affected by physical and biological factors, such as the following.

1. The natural resources available, especially soil and water.
2. Available technology, including the knowledge of proper management of plants, animals, and soils.
3. Improved plant varieties and animal breeds that respond to proper management.
4. Supplies of production inputs such as fertilizers, insecticides, and irrigation water.

How each of these factors apply to food production depends on the area and quality of soils—their natural productivity and response to management. There is good reason to place satisfactory soil properties high on the list of requisites for an adequate world food supply.

19.3

The World's Land Resources

There is a total of more than 13 billion ha of land on the major continents (Table 19.1), but most of it is not suited for cultivation. About half the

TABLE 19.1
World Land Area in Different Climatic Zones

	Area (million ha)			
Climatic zone	Potentially arable	Grazing	Nonarable	Total
Polar and subpolar	0	0	560	560
Cold temperate boreal	50	190	1730	1,970
Cool temperate	910	1000	1000	2,910
Warm temperate subtropical	550	840	1370	2,760
Tropical	1670	1630	1650	4,950
Total	3180	3650	6310	13,150

From The President's Science Advisory Panel on World Food Supply (1967), Vol. II, p. 23.

land is completely nonarable. It is mountainous, too cold, or too steep for tillage; it is swampland; or it is desert that is too dry for any but the sparsest vegetation.

About one fourth of the land area supports enough vegetation to provide grazing for animals, but for various reasons cannot be cultivated. This leaves only about 25% of the land with the physical potential for cultivation (Table 19.2), and less than half of this potentially arable land was actually cropped in 1987. It is obvious that the kinds of soils on this arable 25% and their response to management may eventually hold the key to adequate food production in many areas of the world.

CONTINENTAL DIFFERENCES

Data in Table 19.2 suggest the role that soils may play in helping to meet the world's food requirements. While the total potentially arable land is more than double that being cultivated today, there is great variation from continent to continent. In Asia and Europe, where population pressures have been strong for years, most of the potentially arable land is under cultivation. In contrast, only 21% of the potentially arable land is cultivated in South America, 25% in Africa, and 31% in Oceania, including Australia and New Zealand. In these last three areas the physical potential for greater utilization of arable land is high.

TABLE 19.2
Population and Cropped Land on Each Continent, Along With Cropland Per Person and Percent of Potentially Arable Land That Was Cropped in 1987

	Population in 1987 (millions)	Area (million ha)			Cropland per person (ha)	Arable land cropped (%)
Area		Total	Potentially arable	1987 cropland		
Africa	589	2966	733	183	0.31	25
Asia	2913	2679	627	455	0.16	73
Europe	495	473	174	140	0.28	80
North America	412	2139	465	274	0.66	59
South America	279	1753	680	139	0.50	20
U.S.S.R.	284	2272	356	232	0.82	65
Oceania	25	843	154	48	1.92	31
Total	4998	13,081	3189	1472	0.29	46

Data for potentially arable land from President's Science Advisory Committee Panel on World Food Supply (1967); all other from *World Resources 1987*.

It is unfortunate that arable land is not better distributed in relation to population densities. The area of cultivated land *per person* is high in North America, the U.S.S.R., and Oceania. It is low in Asia, Europe, and Africa. This does not present a serious problem in Europe or the more economically developed parts of Asia. They can readily purchase food from the countries with excess supply. Only transportation, trade, and marketing problems must be overcome.

In the developing countries of South and Southeast Asia, Latin America, and especially Africa the situation is much more critical. Their populations are increasing very rapidly, straining their capacity to produce enough food. Africa's per capita food production has actually declined in the past 20 years (Figure 19.2). Countries that formerly exported food crops now must import them. In addition the national economic growth rate of these countries is too slow to provide the resources to pay for the needed food. They must either be provided with food aid by their more fortunate neighbors or must increase dramatically their own capacity to produce food.

CHOICES OF ACTION

There are three routes that nations may follow to increase their food production: (a) clear and cultivate arable land that has heretofore not been tilled, (b) increase cropping intensity (number of crops per year), or (c) intensify production on lands already under cultivation. Some nations, notably those in Europe and Asia, have only the latter choice. They have little opportunity to expand land under cultivation since most of their arable land is already in use and they are growing as many crops annually as feasible. Only by increasing annual yields per hectare can they produce more food.

In areas outside Asia and Europe, the physical potential for increasing land under cultivation is great. For example, in Africa and South America more than 1 billion ha of arable land are not now being cultivated. However, much of this land is inaccessible to modern transportation. Furthermore the

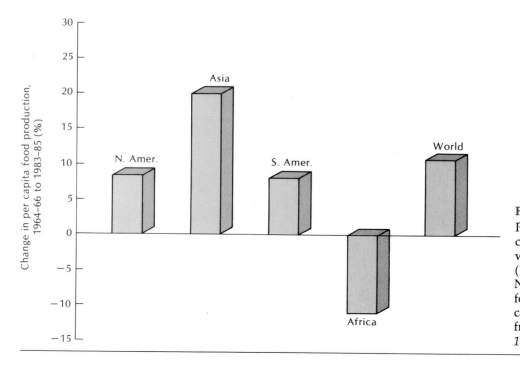

Figure 19.2
Percentage change in per capita food production in the world and in several regions (1964–1966 to 1983–1985). Note that Africa's per capita food production fell dramatically over this period. [Drawn from data in *World Resources 1987*]

cost of clearing the land, transporting the fertilizer and other needed inputs, and distributing the food produced from it is high. Also large areas on those continents have humid, tropical climates and tropical soils—for which the optimum management is yet to be determined.

ENVIRONMENTAL QUALITY CONSTRAINTS

There are other reasons for caution in expanding land under cultivation. To do so requires that the native vegetation be removed and replaced with agricultural crops. Except in relatively level areas, this invariably leads to increased runoff and erosion. Aside from the resultant destruction of the soils, downstream floods and the pollution of domestic water supplies are the consequences. In large areas of the tropics, the clearing of sloping and hilly lands has already resulted in serious erosion damage. In these areas, lands now used for agricultural crops should be allowed to return to forests and other natural vegetation, a sequence that has been followed in much of Europe and the Eastern part of the United States.

Other environmental considerations would dictate caution in clearing forested lands for agricultural use. First, the destruction of forests, particularly those of the tropics, results in a loss of biological diversity. Literally millions of species of plant and animal life that exist in tropical rain forests will become extinct and be lost forever if these forests are destroyed. Aside from their own intrinsic value, these species have economic potential for human use. For example, many medicinal plants have already been discovered in these forested areas.

Second, the destruction of tropical forests releases vast quantities of carbon dioxide into the atmosphere, and thereby contributes to the so-called greenhouse effect which can markedly influence world climates (see Section 18.13). Since there are viable alternatives to agricultural land expansion as means of providing future world food needs, the clearing of tropical forested areas for agriculture should be an alternative of last resort after very careful scrutiny.

In most areas of the world, intensification of cropping of land already under cultivation is the preferred immediate method of increasing food production. In time, however, as economic development, transportation, and knowledge of soil management progress, selective expansion of land under cultivation is almost certain to occur. In Table 19.3 are shown estimates of future increases in crop production from growth in arable land compared to cropping intensity and increased yields per hectare.

19.4

Potential of Different Soil Orders

In Table 19.4 is shown the percentage distribution of the ten soil orders worldwide and in the humid and semiarid tropics. Note the high proportion of the humid tropical areas of Africa and Latin America that are dominated by Oxisols or Ultisols. These are highly leached acid soils relatively low in available plant nutrients. With proper management, including the application of lime and fertilizers, these soils can be made quite productive. However, large areas of Oxisols and Ultisols are located far from supplies of either lime or fertilizer, and crop production is thereby constrained. The huge Amazon Basin of South America and the Congo River Basin of Central Africa have a high proportion of these highly weathered soils. While in time these soil areas may be more fully utilized for agricul-

TABLE 19.3

Percentage Contribution to Increased Food Production from Increases in the Area of Arable Land, Cropping Intensity, and Yields per Hectare in Selected Developing Countries from 1982–84 to 2000.

Note that, except for Latin America, the majority of the expected production increase is from increased yields per hectare.

Region	Contribution to crop output growth (%)		
	Increased area of arable land	Increased cropping intensity	Increased yield per hectare
Sub-Saharan Africa	26	17	57
Near East/North Africa	0	22	77
Asia (excl. China)	11	20	69
Latin America	39	12	49

From FAO (1987).

ture, the cost of making them suitable for crops and the environmental consequences of deforestation make this choice of action extremely questionable.

Soils developed on alluvial parent materials commonly have high crop production potential. Many of these soils are classified as Entisols or Inceptisols and are found in the river valleys of Asia, Africa, and Latin America. They are commonly used for paddy rice production for which they are well suited. Although rice yields are high in some areas, average yields of this crop in the humid tropics are commonly no more than half those of countries in temperate areas such as Japan, Korea, Italy, and the United States. Obviously these soils are not being utilized to their full potential.

TABLE 19.4

Percentages of the Land Area of the World and of the Humid and Semiarid Tropics of Three Continents Classified as the Different Soil Orders of *Soil Taxonomy*

Soil order	In the world	In humid tropics			In semiarid tropics		
		Latin America	Africa	Asia	Latin America	Africa	Asia
Oxisols	9	50	40	4	—	—	—
Ultisols	6	32	15	35	3	2	6
Inceptisols	9	9	17	24	—	3	9
Entisols	8	5	20	24	5	17	—
Alfisols	13	3	5	4	34	32	38
Histisols	1	—	1	6	—	—	—
Spodosols	4	2	1	2	—	—	—
Mollisols	9	—	—	2	—	—	—
Vertisols	2	<1	<1	<1			
Aridisols	19	—	<1	<1	11	30	15
Others[a]	20	—	—	—	47	16	32
Total area (million ha)	10,504	666	445	379	313	1462	319

World distribution from U.S. Department of Agriculture (excludes miscellaneous lands); other humid tropics data calculated from NAS (1982); semiarid tropics data calculated from USDA (1986).

[a] For semiarid tropics, others category includes areas of other soil orders that are not specified.

In the semiarid tropics about one third of the soil areas are classified as Alfisols. This suggests they have not been as highly leached as have the Oxisols and Ultisols of humid tropical areas. With proper soil and water management these soils can be made quite productive. Unfortunately, the semiarid tropics are subject to considerable variability in annual precipitation rates. This means that periods of drought are quite common. The widespread droughts in sub-Saharan Africa in recent years indicate the inherent difficulty of maintaining sustainable crop productivity in these semiarid areas. The larger hectarage of these areas in Africa (Table 19.4) is a reminder of the water constraint on crop production on that continent.

The highly productive, base-rich Mollisols so common in the midwestern part of the United States are not found extensively in the tropics.

The crop production potential of Aridisols is largely dependent on their access to irrigation water. Such soils are very important agriculturally in irrigated arid areas of India, Pakistan, the Middle East, and the United States. They can produce bumper crops when their nutrients are supplemented by chemical fertilizers. The high solar radiation that typifies these areas helps stimulate high crop productivity.

Europe and Asia are already utilizing most of their arable land. North America and Australia are areas of excess food production. For these reasons, attention is focused appropriately on Africa, where food shortages are occurring and where the soil resources are not being effectively utilized or managed.

MAJOR PRODUCTION CONSTRAINTS

Table 19.5 illustrates the limitations of soils for agriculture in major geographic areas of the world. A deficiency of soil moisture is probably the most significant worldwide constraint. This constraint is most notable in Australia, Africa, and South Asia (India, Pakistan).

Mineral deficiencies provide serious limitations in all areas but are most severe in Southeast Asia and South America. Shallow soils are problems in North, Central, and South Asia. Excess water is a more serious problem in Southeast Asia than in any other region. The heavy monsoon rains in this area account for this constraint.

TABLE 19.5
World Soil Resources and Their Major Limitations for Agriculture

Percent of total land area in each category.

			Limitation			
Region	Drought	Mineral stress[a]	Shallow depth	Water excess	Permafrost	No serious limitation
North America	20	22	10	10	16	22
Central America	32	16	17	10	—	25
South America	17	47	11	10	—	15
Europe	8	33	12	8	3	36
Africa	44	18	13	9	—	16
South Asia	43	5	23	11	—	18
North and Central Asia	17	9	38	13	13	10
Southeast Asia	2	59	6	19	—	14
Australia	55	6	8	16	—	15
World	28	23	22	10	6	11

Data compiled from FAO/UNESCO soil map of the world by Dent (1980).

[a] Nutritional deficiencies or toxicities related to chemical composition or mode of origin.

Europe has the highest proportion of soils without serious limitation (30%). The comparable figure for Central and North America is slightly more than 20%.

19.5

Problems and Opportunities in the Tropics

One cannot help but wonder why soils of the humid tropics have not been more effectively utilized. They seemingly have many advantages over their temperate zone counterparts. The crop growing season is usually year round. In the humid areas ample moisture is available throughout the growing season. And some of the soils have physical characteristics far superior to the soils of the temperate zones. The hydrous oxide and kaolinitic types of clays that dominate in these areas permit cultivation under very high rainfall conditions.

A comparison of the distribution of some major soil groupings in the humid tropics and in other nondryland areas of the world is shown in Table 19.6. The high proportion of the soils that are acid and infertile (Oxisols and Ultisols), especially in South America and Africa, helps explain why soils of these continents have not been more extensively utilized. These soils are low in nutrients and often are far from sources of limestone or fertilizers, making it difficult to increase their fertility.

LIMITING FACTORS

The primary factor limiting more effective utilization of tropical soils is their infertility. In Table 19.7 are shown the major soil constraints on crop productivity in the Amazon region of South America. Note that 90% of the soils are deficient in phosphorus and more than two thirds have toxic levels of

TABLE 19.6

Distribution of General Soil Groups Found in the Humid Tropics Compared to Other Non-dryland Areas

Note the high proportion of acid, infertile soils, especially in South America and Africa.

General soil grouping	Humid tropical soil areas (%)				Other dryland soil areas (%)
	Central and South America	Africa	Asia	World	
Acid, infertile soils (Oxisols and Ultisols)	82	56	38	63	10
Moderately fertile, well-drained soils (Alfisols, Vertisols, Mollisols, Andepts, Tropepts, Fluvents)	7	12	33	15	32
Poorly drained soils (Aquepts)	6	12	6	8	10
Very infertile sandy soils (Psamments, Spodosols)	2	16	6	7	13
Others	3	4	16	17	35[a]
Total	100	100	100	100	100

From National Research Council (1982) and U.S. Department of Agriculture data.
[a] Mostly mountainous areas.

TABLE 19.7
Main Soil Constraints in the Amazon Basin Under Native Vegetation

Soil constraint	Millions of hectares	Percent of Amazon Basin
Phosphorus deficiency	436	90
Aluminum toxicity	315	73
Low potassium reserves	242	56
Poor drainage and flooding	116	24
High phosphorus fixation	77	16
Low cation exchange capacity	64	15
High erodibility	39	8
No major limitations	32	6
Steep slopes (>30%)	30	6
Laterite[a] hazards	21	4

After Cochrane and Sanchez (1981).

[a] Plinthite; see Figure 19.3.

aluminum due to their strong acidity. The small reserves of essential elements in these soils are depleted quickly when agricultural crops replace the natural vegetation.

Special problems relate to tropical rain forest areas, which are characterized by high rainfall and by dense forest. Canopies made up of trees and other woody species vary in height from 40 m to essentially ground level. This vegetative cover is extremely rich in organic materials (biomass) and in mineral nutrients. The quantities of these constituents in the biomass are often greater than can be found in the soil profile. Note in Table 19.8 the comparative distribution of chemical nutrients and organic matter in the biomass and soil in a rain forest area in Brazil. Obviously, the biomass is the dominant component of a nutrient cycle that is largely responsible for life in this area. If the forests are harvested or cut down and burned and the land used for agriculture, the small quantities of nutrients and organic matter remaining in the soil will soon be depleted or leached from the soil.

As a continent, Africa has a problem of low water availability. As shown in Table 19.9 inadequate water is a serious limiting factor for much of Africa. Water conserving efforts must be paramount for crop production on this continent. At the same time, vast areas of arable soils in Africa receive adequate rainfall but are low in productivity. Soil and crop scientists are challenged to devise systems that will change this situation.

TABLE 19.8
Distribution of Major Nutrient Elements and Organic Materials in the Soil and in the Living and Dead Plant Species (Biomass) in a Tropical Forested Area in Brazil

Element	Percentage distribution in Biomass	Soil
Nitrogen	26.9	73.1
Phosphorus	31.9	68.1
Potassium	89.6	10.3
Calcium	100.0	0.0[a]
Magnesium	92.3	7.7
Organic matter	68.4	31.6

From Salati and Vose (1984).

[a] This sampling showed essentially all the calcium in the biomass.

TABLE 19.9

Water Characteristics of Africa Compared to Those of North and South America and the World

Note that Africa is dramatically drier than the other areas.

Water characteristics	Africa	North America	South America	World
Precipitation (cm/yr)	70.7	64.7	151.6	70.2
Evaporation (cm/yr)	58.0	38.9	86.8	45.3
Precip. minus evap. (cm/yr)	12.7	25.8	64.8	24.9

Calculated from Mather (1984).

RESEARCH ON TROPICAL SOILS

Unfortunately, all too little is known about tropical soils and their management. Research on these soils, their properties, and their potential for crop production is insignificant compared to that on temperate region soils. Knowledge of their characteristics makes possible identifying only the broadest categories of classification. More intensive study will undoubtedly show many different kinds of soils where now only a few can be identified.

The little we have learned about some tropical soils is encouraging. For example, some of the soils (Oxisols) of Hawaii and the Philippines are excellent for pineapple and sugar cane production. They respond well to modern management and mechanization. In contrast, modern mechanized farming was a dramatic failure for the "Groundnut Scheme" carried out by the British in Tanganyika (Tanzania) following World War II. In that case, apparently, exposing the cleared soil to tropical rains resulted in catastrophic erosion. Knowledge of the nature of the soil and the adoption of management practices to keep vegetative cover on the soil might well have prevented much of the project's $100 million loss.

There is a good likelihood that more intensive study will identify complexities among tropical soils similar to those known for temperate regions. We already know of much variability among soils of tropical areas. Some are deep and friable, easily manipulated and tilled. At the other extreme are soils that, when denuded of their upper horizon, expose layers that harden into a surface resembling a pavement (Figure 19.3). Such soils when eroded are essentially worthless from an agricultural point of view. Fortunately, these soil areas are not extensive.

Figure 19.3
Area in central India with a pavement-like surface where little crop growth occurs. The cementation results from removal of the upper layers by erosion and the exposure at the surface of iron-rich materials called *plinthite* (Gr. *plinthis*, brick) that harden irreversibly when repeatedly wetted and dried. Sizeable areas with such soils are found in India, Africa, South America, and Australia.

The chemical characteristics of Oxisols differ drastically from those of soils of temperate regions. The high hydrous oxide content dictates enormous phosphate-fixing capacities. The low cation exchange capacities and heavy rainfall result in removal of not only macronutrients but micronutrients as well. Indeed, the level of technology needed to manage Oxisols is fully as high as that required for temperate zone soils.

PLANTATION MANAGEMENT SYSTEMS IN THE TROPICS

The plantation system of agriculture has been successful in raising crops such as bananas, sugar cane, pineapples, rubber, coffee, and cacao. These crops are usually grown for export and require considerable skill and financial inputs for their production. The plantation system generally imports the best available technology from developed countries and sometimes has associated with it sizable research staffs to gain new knowledge for improved technology. Although it has been generally successful in producing and marketing crops and animals, it provides little benefit for the indigenous small-scale producer. Consequently social and political problems have plagued this system.

19.6

Shifting Cultivation

At the opposite extreme from the plantation approach are indigenous systems that require little from the outside and have evolved mostly by trial and error by the native cultivators. One of the most widespread of these systems—that of *shifting cultivation*—will be described briefly to illustrate what the natives have learned from centuries of experience (Figure 19.4).

While there are variations in the practice of shifting cultivation, in general it involves three major steps.

Figure 19.4
Nigerian farmer clearing land of natural vegetation in preparation for the planting of crops. The cleared underbrush and trees will be burned, and the ash derived therefrom will support crop production for a few years. The farmer will then move to another site, where the steps will be repeated. [Courtesy International Institute of Tropical Agriculture, Ibadan, Nigeria]

1. The cutting and burning of trees or other native plants with their ashes left on the soil. Sometimes only the vegetation in the immediate area is burned. In other cases this is supplemented with plants brought in from nearby areas.
2. Growing crops on the cleared area for a period of 1–5 yr, thereby utilizing nutrients left from the clearing and burning of native plants. Vegetation from outside the area may be brought in and burned between plantings.
3. Fallowing the area for a period of 5–12 yr, thereby permitting regrowth of the native trees and other plants that "rejuvenate" the soil. Nutrients are accumulated in the native plants and some, such as nitrogen, are released to the soil. The cutting and burning are repeated, and the cycle starts again.

ADVANTAGES OF THE SYSTEM

Shifting cultivation is primarily a system of nutrient conservation, accumulation, and recycling. The native plants absorb available nutrients from the soil and from the atmosphere. Some are nitrogen fixers. Others are deep rooted and bring nutrients from lower horizons to the surface. All help protect the soil from the devastating effects of rain and sunshine. Also, the cropped area is generally small in size and is surrounded by native vegetation. This reduces the chances of gully formation or severe sheet erosion from runoff water.

There are benefits of the system other than those relating to nutrients and erosion control. The burning is likely to destroy some weed seeds and even some unwanted insects and disease organisms. The short period of actual cropping (1–5 yr out of 10–20 yr) discourages the buildup of weeds, insects, and diseases harmful to the cultivated crops. Although shifting cultivation seems primitive, it deserves careful study. In some tropical areas it is more successful than the seemingly more efficient temperate zone systems.

SOIL DETERIORATION

Unparalleled increases in human populations have placed great strain on the traditional shifting cultivation systems in some areas of the tropics. The more than 300 million people who depend upon shifting cultivation for most of their food represent a dramatic increase in numbers during the past two decades. To provide food for their growing families, farmers have been forced to shorten the period of fallow—to recrop a given area after only 3–6 yr fallow compared to 10 yr or longer in the past. As a consequence insufficient time is allowed for soil rejuvenation between cropping periods. This leads to lower crop yields and greater exposure to erosion. The mechanism by which shortened fallow periods have reduced soil fertility is illustrated in Figures 19.5 and 19.6. Today's shorter fallow period is one of the reasons for a decline in per capita food production in Africa during the past decade.

PROMISING ALTERNATIVES

Some soils in the tropics must have continuous vegetative cover to remain productive. If they dry out, and especially if erosion removes the surface layers, the doughy subsoil layers harden irreversibly, making plant growth impossible. This turn of events may be prevented by planting the crop desired among native plants or other crop plants without completely removing the latter. Since the plants involved are in most cases trees, this system

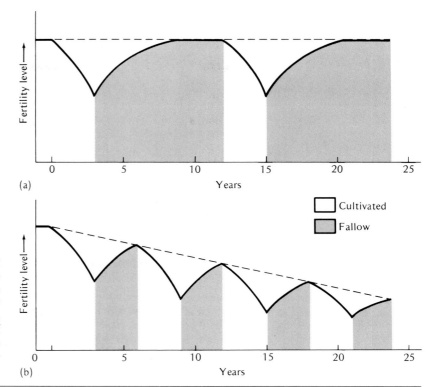

Figure 19.5

Changes in fertility levels under two shifting-cultivation systems. (a) In the system with a long fallow period, the natural vegetation is able to help "regenerate" the fertility level after cropping. (b) With a shorter fallow period so common in areas of high population pressure, insufficient regeneration time is available and the fertility level declines rapidly.

has been termed the *mixed tree crop* system. The desired crop or crops are introduced by removing some of the existing plants and replacing them with crop plants. In time, a given area may be planted entirely to a number of crop plants. The essential feature of the system, however, is that the soil will at no time be free of vegetation. As primitive as the mixed tree crop system may appear, up to now science has not been able to develop more successful alternatives for the management of some of these tropical soils.

Attempts are being made to develop viable alternatives to the shifting-cultivation system—alternatives that would permit crop production without slashing and burning. Some research has shown that modest inputs of fertilizer and lime, the selection of acid-tolerant crop cultivars, and the inclu-

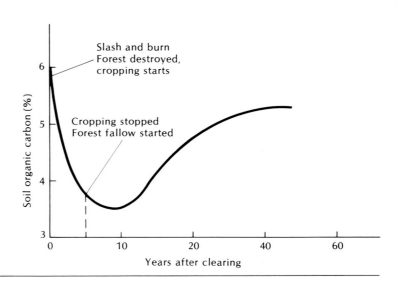

Figure 19.6

Effect of slash and burn, cropping for 5 years to pineapple, and then returning the land to forest (fallow) on the organic carbon content of Paleudults soils of Mexico. Reduction in organic matter continues even after cropping is stopped until forest species residues are sufficient to begin to replenish the soil organic matter. [Redrawn from Wadsworth et al. (1988)]

TABLE 19.10

Average Yields of Four Crops with and Without Modest Fertilizer Applications Following Burning in a Slash and Burn System in Peru

Crop	No. of harvests	Average harvest yield (Mg/ha)	
		No fertilization or lime	Fertilizer and lime added
Rice	37	0.99	2.71
Corn	17	0.21	2.81
Soybeans	24	0.24	2.30
Peanut	10	0.69	3.46

From Sanchez et al. 1983).

sion of legumes in crop rotations help sustain crop yields. Data in Table 19.10 from research in Peru illustrate this point. Small quantities of fertilizer and lime along with better means of weed control were effective in stabilizing yields and maintaining reasonable soil cover. Small-scale farmers nearby the research plots have adopted the scheme and are pleased with its greater stability.

A second alternative closely related to the mixed-free-crop system already discussed is that of *alley cropping*. This is an agroforestry system that involves growing food crops in alleys, the borders of which are formed by fast-growing trees or shrubs. These woody species are usually leguminous and can provide fixed nitrogen to the system. The hedgerows are pruned to prevent shading of the food crop. The cuttings obtained by pruning are used as mulch and as a source of nutrients for the food crop. The mulch helps control weeds, prevents runoff and erosion, and reduces evaporation from the soil surface.

Figure 19.7 provides an example of the functioning of an alley cropping system. Note that during the dry season the border trees are permitted to grow and then are cut back at crop-planting time. In some cases the hedgerow woody plant leaves are fed to livestock and the manure is then applied to the land.

When carefully monitored, this system appears to function effectively. Data in Table 19.11 illustrate this point. The yield of corn was sustained at a reasonable level over a period of five years when the prunings of the woody legume were used as a mulch. In contrast, yields dropped off sharply where no mulch material was returned to the soil. This system is being tried on farmers' fields; should it prove to be acceptable, a major step will have been taken toward a viable alternative to the slash and burn system.

Alley cropping systems not only maintain crop yields but also maintain soil organic matter and nutrient element levels (Table 19.12). Apparently the deep-rooted woody species used as hedgerows cycle mineral nutrients from the lower soil horizons to leaves and stems and eventually to the soil surface. These residues are incorporated into the soil by soil organisms, thereby enhancing organic matter and mineral element contents while simultaneously maintaining crop yields. Also, the leguminous woody species fix nitrogen annually in quantities that compare favorably with annual amounts fixed by some forage legumes in temperate regions.

POTENTIAL OF TROPICAL AGRICULTURE

One cannot help but be optimistic about the future of agriculture in the tropics. The basic requirements for maximum year-round production appear to be higher in the humid tropics than anywhere else. Total annual solar radiation and warm to hot climates provide unmatched photosynthetic

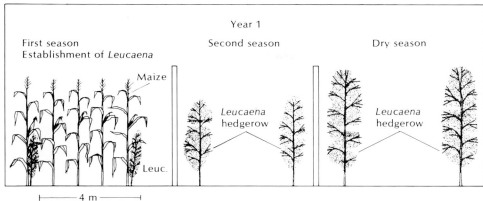

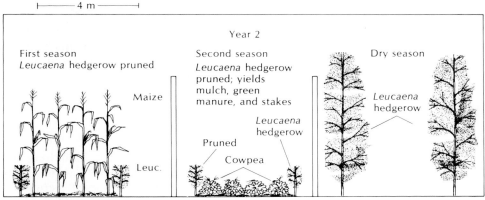

Figure 19.7
Cropping sequence diagram for establishing *Leucaena leucocephala* hedgerows (spaced 4 m) for alley cropping with sequentially cropped maize and cowpeas. Note that the hedgerow is never allowed to shade the food crop and that the clippings from the *Leucaena* are placed as a mulch on the food crop rows. [From Kang et al. (1984)]

TABLE 19.11
Yield of Corn Over a Period of 6 Years in an Alley Cropping Experiment Using *Leucaena leucocephala* as the Hedgerow Component on a Psammentic Ustorthent Soil in Nigeria

Nitrogen rate[a] (kg/ha)	Leucaena prunings	Yield of corn[a] (Mg/ha)						
		1979	1980	1981	1982	1983	1984	1986
0	Removed	—	1.04	0.48	0.61	0.26	0.69	0.66
0	Used as mulch	2.15	1,91	1.21	2.10	1.92	1.99	2.10
80	Used as mulch	3.40	3.26	1.89	2.91	3.16	3.67	3.00

From Kang et al. (1989).

[a] All plots received P, K, Mg, and Zn. Crop affected by drought in 1983; land fallowed in 1985.

TABLE 19.12
Effect of 6 Years of Alley Cropping Using *Leucaena* in the Hedgerows on Selected Chemical Properties of the Upper 15 cm of a Psammentic Ustorthent

Nitrogen rate (kg/ha)	Leucaena prunings	pH	Organic matter (g/kg)	Exchangeable cations (cmol/kg)		
				K	Ca	Mg
0	Removed	6.0	6.5	0.19	2.90	0.35
0	Retained	6.0	10.7	0.28	3.45	0.50
80	Retained	5.8	11.9	0.26	2.80	0.45

From Kang et al. (1989).

potential. Several crops a year can be grown on a given field in the tropics. Already researchers have developed new crop varieties especially adapted to warm climates. Means of controlling pests and fertilizer usage are becoming more common in tropical areas. The potential for food production there is enormous.

19.7

Requisites for the Future

The world's ability to feed itself depends upon many factors, not the least of which is improved agricultural technology in the developing nations of the world. This technology depends largely on research and education, which must have direct relevance to the developing countries and not be a mere transplant of what is available in the more developed nations. We cannot assume that the technology of western Europe or the United States can be transferred directly to the underdeveloped countries. Disastrous failures have shown the fallacy of this approach.

TWO REQUISITES

Past accomplishments suggest that there are two major requirements for increased agricultural production in the tropics. *First*, improved technologies focused specifically on tropical agriculture must be developed and put into place. *Second*, the political, social, and economic climate in the developing countries must be such as to make it profitable for the cultivator to adopt and use the new technologies. Assuming the second of these requirements can be achieved, we will focus on some of the technologies needed.

CROP VARIETIES

New crop varieties adapted to conditions in the developing areas along with methods of pest control are among the first requirements. For example, the dwarf wheats and rices that are highly responsive to fertilizer applications were the twin engines of the "green revolution" that helped feed Asia and Latin America during the past quarter century. The average yield of wheat in Mexico doubled as a result of introducing these new varieties, especially in irrigated areas. The production of wheat in India has tripled since the mid-1960s. Similarly, rice production in Indonesia more than doubled in the past 20 years. Luckily the rice and wheat varieties are adapted to these and other food-deficit countries. Disease and insect problems are being countered through the development of resistant varieties.

IRRIGATION AND DRAINAGE

Remarkable increases in the amount of irrigated land have taken place in the past 35 years, especially in Asia. Worldwide, more than 275 million ha of land are now being irrigated, 185 million of which are in Asia. Irrigation coupled with high yielding varieties and fertilizers have stimulated unparalleled increases in food production in Asia and will likely continue to do so.

Unfortunately, in Africa, which is the major "dry" continent of the world, only 13 million acres are irrigated. This is likely a major reason for Africa's slow agricultural growth. While there is considerable potential for increased irrigation in Africa, a number of economic and managerial factors have limited achievements.

There is growing recognition that irrigation without proper soil drainage can be disastrous. Salt buildup in some irrigated arid region soils results from poor drainage. Salinization seriously affects crop production from more than 20 million ha or about 7% of the world's irrigated land. Drainage simply must be considered as an integral part of any irrigation scheme.

FERTILIZER

To move from yields common in subsistence agriculture to those dictated by today's food requirements, dramatic increases in supplies of fertilizer nutrients have occurred. From the 1960s to 1979 fertilizer use rose eightfold in the developing countries. The total $N–P_2O_5–K_2O$ consumption in these countries was about 30 million Mg in 1986. This has resulted in a demand for information on soil characteristics to determine the kinds of fertilizers that are needed. In highly leached areas evaluations are being made of micronutrient deficiencies as well as those of nitrogen, phosphorus, potassium, and lime. Considerable attention is being given to fertilizer that resists rapid reaction with the soil or volatilization by microbial action.

Africa lags far behind the other areas in rates of fertilizer application. In 1984 an average of only 18.5 kg fertilizer per hectare was consumed in Africa, compared to about 32 kg/ha in Latin America and more than 80 kg/ha in Asia. Fertilizer trials have demonstrated the response that can be obtained from fertilizers in Africa, but economic, political, and supply issues have prevented more extensive use.

While the use of commercial fertilizers is being encouraged, practical considerations dictate the search for alternative sources, especially of nitrogen. Native and improved legumes can and should be used. And animal manures will help supplement the manufactured fertilizers.

SOIL MANAGEMENT

Research is demonstrating that proper soil management is a critical factor in tropic agriculture, not only in underpinning good crop yields but also in sustaining long-term productivity. Even though soils of the tropics differ from those of temperate regions, some temperate-zone management principles are pertinent for the tropics. For example, conservation tillage coupled with alley cropping helps maintain crop yields and simultaneously minimize runoff and soil erosion. Data in Table 19.13 from experiments in Africa illustrate this point. Likewise, the beneficial effects of organic residues on crop yields are obvious from data shown in Figure 19.8.

Agricultural development activities have emphasized the critical need for the characterization of soils. For example, the salinity and alkalinity status can well determine the probable success of an irrigation project. Nutrient

TABLE 19.13
The Influence of Conservation Tillage and Alley Cropping (with Woody Species, *Leucaena* and *Gliricidia*) on the Yields of Corn and Cowpeas, and the Losses of Runoff Water and Soil by Erosion

Treatment	Crop yield (Mg/ha)		Runoff (mm)	Erosion (Mg/ha)
	Corn	Cowpea		
Plowed	4.2	0.5	232	15
No till	4.3	1.1	6	0.03
Leucaena	3.9	0.6	10	0.2
Gliricidia	4.0	0.6	20	1.7

From Lal (1987).

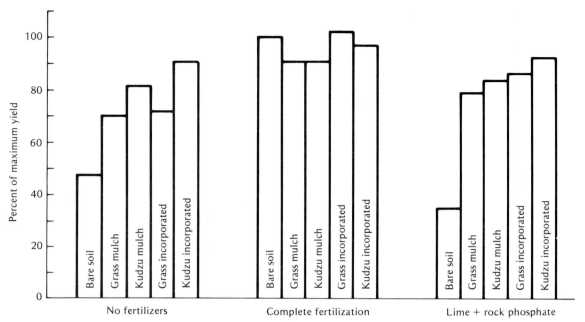

Figure 19.8
Effect of crop residues with and without fertilizers and lime on the yield of five consecutive crops grown in the Amazon Basin. Yields are expressed as percent of that obtained from complete fertilization on bare soil. [From Wade and Sanchez (1983); used with permission of the American Society of Agronomy]

deficiencies can be identified, as can the potential for erosion and drainage problems. The time and effort being devoted to soil characterization are insignificant compared to the need for this kind of information.

SOIL SURVEY

Reliable soil survey information is inadequate or unavailable in most of the developing areas. Sometimes this is due to lack of information about the soil characteristics upon which a classification scheme can be based. More frequently it is due to ignorance of the significance and value of the soil survey. Planners sometimes look upon soils as soils, without regard to the vast differences that exist among them—differences that could affect markedly the success of plans that are made to use the soils.

Soil surveys are of special significance in two ways. First, they make possible the extrapolation of research results from a given area to other areas where the same kinds of soil are found. Second, they provide one of the criteria to determine the economic feasibility of clearing and preparing for tillage lands that have as yet not been used for agriculture.

HUMAN RESOURCES

A final requisite for increased food production is trained personnel. The range needed goes from basic scientists (from whose test tubes and field plots new technologies and food products are to come) to the cultivators and their families. Researchers whose interest relates directly to the solution of the world food problem are needed. Likewise, technicians, farm managers, field service personnel, and individuals trained in processing and marketing trades are essential. And we must have educators to teach not only the students but the farmers as well.

The fight to feed the world is not yet lost. But to win it will require technological and scientific inputs of a magnitude not yet realized. And among the most important of these inputs are those relating to soils and soil science.

19.8

Conclusion

The world population explosion of this century presents humankind with one of its most complex challenges. This challenge is greatest in the low income countries of the world where the bulk of the 90 million annual population increase is occurring. To feed an increase of population equivalent every three years to the entire population of the United States will tax the world's ingenuity and determination.

It is in the developing countries and especially Africa where the challenge is most severe. The conservation and improvement of soils is one of the keys to meeting this challenge. Soil and crop management systems must be developed and adapted to increase food production and simultaneously prevent soil deterioration. Soil scientists will play a key role in developing these systems.

Soil scientists must also gain greater knowledge of the characteristics and potential of arable but uncultivated lands. This knowledge will be used in determining the extent to which new lands should be brought into agricultural production to help meet the world's food demands.

There is potential for the world to produce adequate food for current and future generations. Whether this potential will be reached will depend in part on the wisdom that is used in managing and conserving soils. This wisdom will be based largely on the service of dedicated soil scientists around the world.

Study Questions

1. Why are food availability problems so much more severe in India than in the United States?
2. What are the two major alternatives for increasing the world's food production? Which will be the more viable in the future and why?
3. Considering only the nature and properties of soils, how would you compare the food-producing potential of Africa with that of Europe or of the United States? Explain.
4. Asia has been able to increase its per capita food production slightly over the past 20 years, in spite of its enormous increases in population, while Africa has not been able to do so. What are the soil-related reasons for this difference?
5. Vast areas of potentially arable but uncultivated land exist in tropical South America and Africa. What are the arguments for and against the conversion of these lands to agricultural use?
6. Oxisols are being used successfully in Hawaii and the Philippines for large-scale pineapple production, and yet the use of these soils by small-scale farmers is quite limited elsewhere. What are the limitations on the productivity of these soils?
7. The shifting-cultivation or slash and burn system has been used for centuries without causing serious problems of soil deterioration. Why have these problems become more serious today?
8. What is meant by alley cropping, and what are its advantages and disadvantages over the shifting cultivation systems?
9. What are the major improvements that will be needed in the future if the developing countries are to feed their people successfully?
10. What is the role of soil scientists in helping the World achieve its food production goals?

References

COCHRANE, T. T., and P. A. SANCHEZ. 1981. "Land resources, and soil properties and their management in the Amazon region: A state of knowledge report," in *Proceedings of Conference on Amazon Land Use Research.* (Cali, Colombia: Centro Internacional de Agricultura Tropical [CIAT]).

DENT, F. J. 1980. "Major production systems and soil-related constraints in Southeast Asia," pp. 78–106 in *Priorities for Alleviating Soil-related Constraints to Food Production in the Tropics* (Manila: International Rice Research Institute and Ithaca, NY: Cornell University).

FAO. 1978. *Production Yearbook,* Vol. 31 (Rome, Italy: UN Food and Agriculture Organization).

FAO. 1984. *Land, Food and People* (Rome, Italy: UN Food and Agriculture Organization).

FAO. 1987. *Agriculture: Toward 2000* (Rome, Italy: UN Food and Agriculture Organization).

KANG B. T., G. F. WILSON, and T. L. LAWSON. 1984. *Alley-Cropping: A Stable Alternative to Shifting Cultivation* (Ibadan, Nigeria: Int. Inst. of Tropical Agric.).

KANG, B. T., L. REYNOLDS, and A. N. ATTA-KRAH. 1989. "Alley Cropping", *Adv. Agron.,* **43** (In press).

MATHER, J. R. 1984. *Water Resources: Distribution, Use and Management* (New York: Wiley).

NAS. 1982. *Ecological Aspects of Development in the Humid Tropics* (Washington, DC: National Academy Press).

NATIONAL RESEARCH COUNCIL. 1982. *Ecological Aspects of Development in the Humid Tropics,* Report of the Committee on Selected Biological Problems in the Humid Tropics (Washington, DC: National Academy of Sciences Press).

SALATI, E., and P. B. VOSE. 1984. "Amazon basin: A system in equilibrium," *Science,* **225**:129–38.

SANCHEZ, P. A., J. H. VILLACHICA, and D. E. BANDY. 1983. "Soil fertility dynamics after clearing a tropical rainforest in Peru," *Soil Sci. Soc. Amer. J.,* **47**:1171–78.

The President's Science Advisory Committee Panel on World Food Supply. 1967. *The World Food Problem,* Vols. I–III (Washington, DC: The White House).

USDA. 1986. *Symposium on Low Activity Clay (LAC) Soils.* SMSS Technical Monograph No. 14, U. S. Department of Agriculture and U. S. Agency for International Development.

WADE, M. K., and P. A. SANCHEZ. 1983. "Mulching and green manure applications for continuous crop production in the Amazon basin," *Agron. J.,* **75**:39–45.

WADSWORTH, G., R.J. SOUTHHARD, and M.J. SINGER. 1988. "Effects of fallow length on organic carbon and soil fabric of some tropical Udults," *Soil Sci Soc. Amer. J.,* **52**:1424–30.

World Resources 1987, A report of International Institute for Environment and Development, and World Resources Institute, Washington, DC.

Soil Taxonomy Maps and Simplified Key

Soil surveys have been prepared for most counties of the continental United States. Likewise, scientists have made soil surveys and have otherwise characterized soils in other areas around the world. The Soil Survey Staff of the U.S. Soil Conservation Service, in cooperation with scientists from other countries, have used information from these surveys and analyses to develop generalized soils maps based on *Soil Taxonomy*. Maps for the United States and the world are shown in Figures A.1 and A.2. Areas dominated by specific soil orders and suborders are delineated on the maps using map symbols (A.3a, D.1s, etc.). The soil orders and suborders to which these symbols refer are shown on the page facing the map.

Keys have been developed to help scientists determine which soil order, suborder, great group, etc., is present in a given field site. A simplified key is shown in Figure A.3. This key is based strictly on soil properties that can be measured by physical and chemical means. Use of this and more complex keys assures that two or more independent soil surveyors will arrive at the same classification of a given soil area.

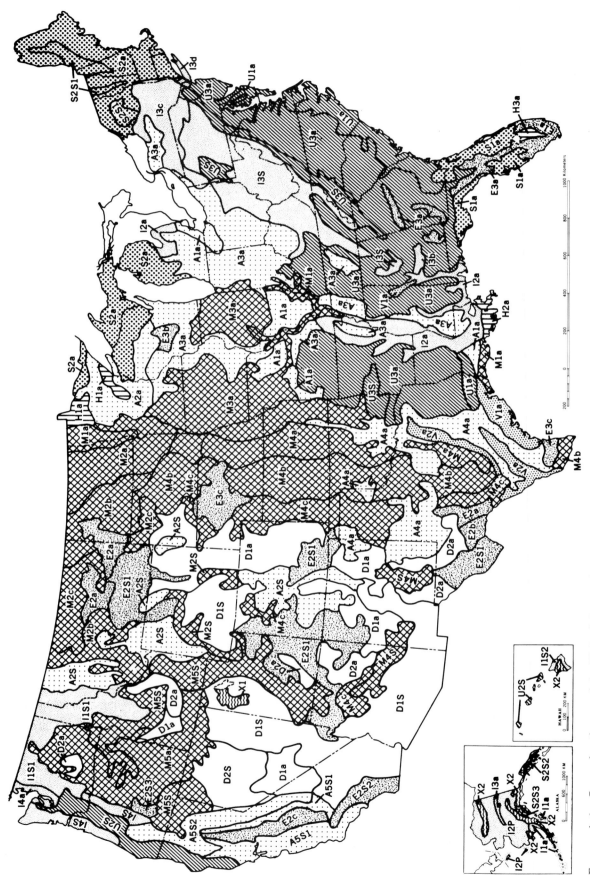

FIGURE A.1 General soil map of the United States showing patterns of soil orders and suborders based on *Soil Taxonomy*. Explanations of symbols follow. [Courtesy USDA Soil Conservation Service, Soil Survey Staff]

568

ALFISOLS

AQUALFS
A1a Aqualfs with Udalfs, Haplaquepts, Udolls; gently sloping.

BORALFS
A2a Boralfs with Udipsamments and Histosols; gently and moderately sloping.

A2s Cryoboralfs with Borolls, Cryochrepts, Cryorthods, and Rock outcrops; steep.

UDALFS
A3a Udalfs with Aqualfs, Aquolls, Rendolls, Udolls, and Udults; gently or moderately sloping.

USTALFS
A4a Ustalfs with Ustochrepts, Ustolls, Usterts, Ustipsamments, and Ustorthents; gently or moderately sloping.

XERALFS
A5S1 Xeralfs with Xerolls, Xerorthents, and Xererts; moderately sloping to steep.

A5S2 Ultic and lithic subgroups of Haploxeralfs with Andepts, Xerults, Xerolls, and Xerochrepts; steep.

ARIDISOLS

ARGIDS
D1a Argids with Orthids, Orthents, Psamments, and Ustolls; gently and moderately sloping.

D1S Argids with Orthids, gently sloping; and torriorthents, gently sloping to steep.

ORTHIDS
D2a Orthids with Argids, Orthents, and Xerolls; gently or moderately sloping.

D2S Orthids, gently sloping to steep, with Argids, gently sloping; lithic subgroups of Torriorthents and Xerorthents, both steep.

ENTISOLS

AQUENTS
E1a Aquents with Quartzipsamments, Aquepts, Aquolls, and Aquods; gently sloping.

ORTHENTS
E2a Torriorthents, steep, with borollic subgroups of Aridisols; Usterts and aridic and vertic subgroups of Borolls; gently or moderately sloping.

E2b Torriorthents with torrerts; gently or moderately sloping.

E2c Xerorthents with Xeralfs, Orthids, and Argids; gently sloping.

E2S1 Torriorthents; steep, and Argids, Torrifluvents, Ustolls, and Borolls; gently sloping.

E2S2 Xerorthents with Xeralfs and Xerolls; steep.

E2S3 Cryorthents with Cryopsamments and Cryandepts; gently sloping to steep.

PSAMMENTS
E3a Quartzipsamments with Aquults and Udults; gently or moderately sloping.

E3b Udipsamments with Aquolls and Udalfs; gently or moderately sloping.

E3c Ustipsamments with Ustalfs and Aquolls; gently or moderately sloping.

HISTOSOLS
H1a Hemists with Psammaquents and Udipsamments; gently sloping.

H2a Hemists and Saprists with Fluvaquents and Haplaquepts; gently sloping.

H3a Fibrists, Hemists, and Saprists with Psammaquents; gently sloping.

INCEPTISOLS[1]

ANDEPTS[1]
I1a Cryandepts with Cryaquepts, Histosols, and Rock land; gently or moderately sloping.

I1S1 Cryandepts with Cryochrepts, Cryumbrepts, and Cryorthods; steep.

I1S2 Andepts with Tropepts, Ustolls, and Tropofolists; moderately sloping to steep.

AQUEPTS
I2a Haplaquepts with Aqualfs, Aquolls, Udalfs, and Fluvaquents; gently sloping.

I2P Cryaquepts with cryic great groups of Orthents, Histosols, and Ochrepts; gently sloping to steep.

OCHREPTS
I3a Cryochrepts with cryic great groups of Aquepts, Histosols, and Orthods; gently or moderately sloping.

I3b Eutrochrepts with Uderts; gently sloping.

I3c Fragiochrepts with Fragiaquepts, gently or moderately sloping; and Dystrochrepts, steep.

I3d Dystrochrepts with Udipsamments and Haplorthods; gently sloping.

I3S Dystrochrepts, steep, with Udalfs and Udults; gently or moderately sloping.

UMBREPTS
I4a Haplumbrepts with Aquepts and Orthods; gently or moderately sloping.

I4S Haplumbrepts and Orthods; steep, with Xerolls and Andepts; gently sloping.

MOLLISOLS

AQUOLLS
M1a Aquolls with Udalfs, Fluvents, Udipsamments, Ustipsamments, Aquepts, Eutrochrepts, and Borolls; gently sloping.

BOROLLS
M2a Udic subgroups of Borolls with Aquolls and Ustorthents; gently sloping.

M2b Typic subgroups of Borolls with Ustipsamments, Ustorthents, and Boralfs; gently sloping.

M2c Aridic subgroups of Borolls with Borollic subgroups of Argids and Orthids, and Torriorthents; gently sloping.

M2S Borolls with Boralfs, Argids, Torriorthents, and Ustolls; moderately sloping or steep.

UDOLLS
M3a Udolls, with Aquolls, Udalfs, Aqualfs, Fluvents, Psamments, Ustorthents, Aquepts, and Albolls; gently or moderately sloping.

USTOLLS
M4a Udic subgroups of Ustolls with Orthents, Ustochrepts, Usterts, Aquents, Fluvents, and Udolls; gently or moderately sloping.

M4b Typic subgroups of Ustolls with Ustalfs, Ustipsamments, Ustorthents, Ustochrepts, Aquolls, and Usterts; gently or moderately sloping.

[1]Many Andepts would now be included in the Andisols order.

M4c Aridic subgroups of Ustolls with Ustalfs, Orthids, Ustipsamments, Ustorthents, Ustochrepts, Torriorthents, Borolls, Ustolls, and Usterts, gently or moderately sloping.

M4S Ustolls with Argids and Torriorthents; moderately sloping or steep.

XEROLLS

M5a Xerolls with Argids, Orthids, Fluvents, Cryoboralfs, Cryoborolls, and Xerorthents; gently or moderately sloping.

M5S Xerolls with Cryoboralfs, Xeralfs, Xerorthents, and Xererts; moderately sloping or steep.

SPODOSOLS

AQUODS

S1a Aquods with Psammaquents, Aquolls, Humods, and Aquults; gently sloping.

ORTHODS

S2a Orthods with Boralfs, Aquents, Orthents, Psamments, Histosols, Aquepts, Fragiochrepts, and Dystrochrepts; gently or moderately sloping.

S2S1 Orthods with Histosols, Aquents, and Aquepts; moderately sloping or steep.

S2S2 Cryorthods with Histosols; moderately sloping or steep.

S2S3 Cryorthods with Histosols, Andepts and Aquepts; gently sloping to steep.

ULTISOLS

AQUULTS

U1a Aquults with Aquents, Histosols, Quartzipsamments, and Udults; gently sloping.

HUMULTS

U2S Humults with Andepts, Tropepts, Xerolls, Ustolls, Orthox, Torrox, and Rock land; gently sloping to steep.

UDULTS

U3a Udults with Udalfs, Fluvents, Aquents, Quartzipsamments, Aquepts, Dystrochrepts, and Aquults; gently or moderately sloping.

U3S Udults with Dystrochrepts; moderately sloping or steep.

VERTISOLS

UDERTS

V1a Uderts with Aqualfs, Eutrochrepts, Aquolls, and Ustolls, gently sloping.

USTERTS

V2a Usterts with Aqualfs, Orthids, Udifluvents, Aquolls, Ustolls, and Torrerts; gently sloping.

Areas with Little Soil

X1 Salt flats.

X2 Rock land (plus permanent snow fields and glaciers).

Slope Classes

Gently sloping—Slopes mainly less than 10%, including nearly level.

Moderately sloping—Slopes mainly between 10 and 25%.

Steep—Slopes mainly steeper than 25%.

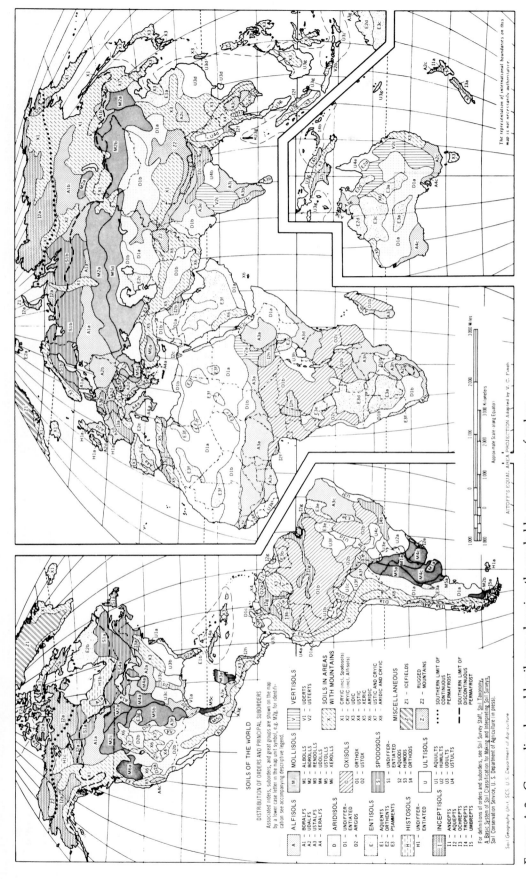

SOILS OF THE WORLD

DISTRIBUTION OF ORDERS AND PRINCIPAL SUBORDERS

Associated orders, suborders, and great groups are shown on the map
by a lower case letter in the map unit symbol, e.g. M2a, for identification see accompanying descriptive legend.

A	ALFISOLS	M	MOLLISOLS	V	VERTISOLS
	A1 — BORALFS		M1 — ALBOLLS		V1 — UDERTS
	A2 — UDALFS		M2 — BOROLLS		V2 — USTERTS
	A3 — USTALFS		M3 — RENDOLLS		
	A4 — XERALFS		M4 — UDOLLS		SOILS IN AREAS
			M5 — USTOLLS		WITH MOUNTAINS
D	ARIDISOLS		M6 — XEROLLS	X	X1 — CRYIC (incl. Spodosols)
	D1 — UNDIFFER-	O	OXISOLS		X2 — CRYIC (incl. Altisols)
	ENTIATED		O1 — ORTHOX		X3 — UDIC
	D2 — ARGIDS		O2 — USTOX		X4 — USTIC
					X5 — XERIC
E	ENTISOLS	S	SPODOSOLS		X6 — ARDIC
	E1 — AQUENTS		S1 — UNDIFFER-		X7 — USTIC AND CRYIC
	E2 — ORTHENTS		ENTIATED		X8 — ARDIC AND CRYIC
	E3 — PSAMMENTS		S2 — AQUODS		
			S3 — HUMODS		MISCELLANEOUS
H	HISTOSOLS		S4 — ORTHODS	Z	Z1 — ICEFIELDS
	H1 — UNDIFFER-	U	ULTISOLS		
	ENTIATED		U1 — AQUULTS	Z	Z2 — RUGGED
			U2 — HUMULTS		MOUNTAINS
I	INCEPTISOLS		U3 — UDULTS		
	I1 — ANDEPTS		U4 — USTULTS	••••	SOUTHERN LIMIT OF
	I2 — AQUEPTS				CONTINUOUS
	I3 — OCHREPTS				PERMAFROST
	I4 — TROPEPTS				
	I5 — UMBREPTS			— —	SOUTHERN LIMIT OF
					DISCONTINUOUS
					PERMAFROST

For definitions of orders and suborders, see Soil Survey Staff, Soil Taxonomy,
A Basic System of Soil Classification for Making and Interpreting Soil Surveys,
Soil Conservation Service, U. S. Department of Agriculture (in press).

Soil Geography Unit, SCS, U.S. Department of Agriculture

AITOFF'S EQUAL AREA PROJECTION Adapted by V. C. Finch

The representation of international boundaries on this
map is not necessarily authoritative

FIGURE A.2 Generalized world soil map showing the probable occurrence of orders
and suborders according to *Soil Taxonomy*. See following pages for symbol explanations. [Courtesy USDA Soil Conservation Service, Soil Survey Staff]

Figure A.2 (continued)

A **ALFISOLS**—soils with argillic or natric horizon, medium-high bases; forest areas.
A1 BORALFS—cold.
A1a—with Histosols, cryic temperature regimes common.
A1b—with Spodosols, cryic temperature regimes.
A2 UDALFS—temperate to hot, usually moist.
A2a—with Aqualfs.
A2b—with Aquolls.
A2c—with Hapludults.
A2d—with Ochrepts.
A2e—with Troporthents.
A2f—with Udorthents.
A3 USTALFS—temperate to hot, dry more than 90 cumulative days during periods when temperature is suitable for plant growth.
A3a—with Tropepts.
A3b—with Troporthents.
A3c—with Tropustults.
A3d—with Usterts.
A3e—with Ustochrepts.
A3f—with Ustolls.
A3g—with Ustorthents.
A3h—with Ustox.
A3j—Plinthustalfs with Ustorthents.
A4 XERALFS—temperate or warm, moist in winter and dry more than 45 consecutive days in summer.
A4a—with Xerochrepts.
A4b—with Xerorthents.
A4c—with Xerults.

D **ARIDISOLS**—soils of dry areas.
D1 ARIDISOLS—undifferentiated.
D1a—with Orthents.
D1b—with Psamments.
D1c—with Ustalfs.
D2 ARGIDS—with horizons of clay accumulation.
D2a—with Fluvents
D2b—with Torriorthents.

E **ENTISOLS**—soils without pedogenic horizons.
E1 AQUENTS—seasonally or perennially wet.
E1a—Haplaquents with Udifluvents.
E1b—Psammaquents with Haplaquents.
E1c—Tropaquents with Hydraquents.
E2 ORTHENTS—loamy or clayey textures, many shallow to rock.
E2a—Cryorthents.
E2b—Cryorthents with Orthods.
E2c—Torriorthents with Aridisols.
E2d—Torriorthents with Ustalfs.
E2e—Xerorthents with Xeralfs.
E3 PSAMMENTS—sand or loamy sand textures.
E3a—with Aridisols.
E3b—with Orthox.
E3c—with Torriorthents
E3d—with Ustalfs.
E3e—with Ustox.
E3f—shifting sands.
E3g—Ustipsamments with Ustolls.

H **HISTOSOLS**—organic soils.
H1 HISTOSOLS—undifferentiated.
H1a—with Aquods.
H1b—with Boralfs.
H1c—with Cryaquepts.

I **INCEPTISOLS**—soils with few diagnostic features.
I1 ANDEPTS[1]—amorphous clay or vitric volcanic ash or pumice.
I1a—Dystrandepts with Ochrepts.
I2 AQUEPTS—seasonally wet.
I2a—Cryaquepts with Orthents.
I2b—Halaquepts with Salorthids.
I2c—Haplaquepts with Humaquepts.
I2d—Haplaquepts with Ochraqualfs.
I2e—Humaquepts with Psamments.
I2f—Tropaquepts with Hydraquents.
I2g—Tropaquepts with Plinthaquults.
I2h—Tropaquepts with Tropaquents.
I2j—Tropaquepts with Tropudults.

I3 OCHREPTS—thin, light-colored surface horizons and little organic matter.
I3a—Dystrochrepts with Gragiochrepts.
I3b—Dystrochrepts with Orthox.
I3c—Xerochrepts with Xerolls.
I4 TROPEPTS—continuously warm or hot.
I4a—with Ustalfs.
I4b—with Tropudults.
I4c—with Ustox.
I5 UMBREPTS—dark-colored surface horizons with medium to low base supply.
I5a—with Aqualfs.

M **MOLLISOLS**—soils with nearly black, organic-rich surface horizons and high base supply.
M1 ALBOLLS—light gray subsurface horizon over slowly permeable horizon; seasonally wet.
M1a—with Aquepts.
M2 BOROLLS—cold
M2a—with Aquolls.
M2b—with Orthids.
M2c—with Torriorthents.
M3 RENDOLLS—subsurface horizons with much calcium carbonate but no accumulation of clay.
M4 UDOLLS—temperate or warm, usually moist.
M4a—with Aquolls.
M4b—with Eutrochrepts
M4c—with Humaquepts.
M5 USTOLLS—temperate to hot, dry more than 90 cumulative days in year.
M5a—with Argialbolls.
M5b—with Ustalfs.
M5c—with Usterts.
M5d—with Ustochrepts.
M6 XEROLLS—cool to warm, moist in winter, and dry more than 45 consecutive days in summer.
M6a—with Xerorthents.

O **OXISOLS**—highly weathered soils with oxic horizon.
O1 ORTHOX—hot, nearly always moist.
O1a—with Plinthaquults.
O1b—with Tropudults.

[1]Many Andepts would now be included in the Andisols order.

O2 USTOX—warm or hot, dry for long periods but moist more than 90 consecutive days in the year.
    O2a—with Plinthaquults.
    O2b—with Tropustults.
    O2c—with Ustalfs.

S SPODOSOLS—soils of forest areas with spodic horizon, low bases.
S1 SPODOSOLS—undifferentiated.
    S1a—cryic temperature regimes; with Boralfs.
    S1b—cryic temperature regimes; with Histosols.
S2 AQUODS—seasonally wet.
    S2a—Haplaquods with Quartzipsamments.
S3 HUMODS—with accumulations of organic matter in subsurface horizons.
    S3a—with Hapludalfs.
S4 ORTHODS—with accumulations of organic matter, iron, and aluminum in subsurface horizons.
    S4a—Haplorthods with Boralfs.

U ULTISOLS—soils in forest areas with subsurface horizons of clay accumulation and low base supply.
U1 AQUULTS—seasonally wet.
    U1a—Ochraquults with Udults.
    U1b—Plinthaquults with Orthox.
    U1c—Plinthaquults with Plinthaquox.
    U1d—Plinthaquults with Tropaquepts.
U2 HUMULTS—temperate or warm and moist all of year; high content of organic matter.
    U2a—with Umbrepts.

U3 UDULTS—temperate to hot; never dry more than 90 cumulative days in the year.
    U3a—with Andepts.
    U3b—with Dystrochrepts.
    U3c—with Udalfs.
    U3d—Hapludults with Dystrochrepts.
    U3e—Rhodudults with Udalfs.
    U3f—Tropudults with Aquults.
    U3g—Tropudults with Hydraquents.
    U3h—Tropudults with Orthox.
    U3j—Tropudults with Tropepts.
    U3k—Tropudults with Tropudalfs.
U4 USTULTS—warm or hot; dry more than 90 cumulative days in the year.
    U4a—with Ustochrepts.
    U4b—Plinthustults with Ustorthents.
    U4c—Rhodustults with Ustalfs.
    U4d—Tropustults with Tropaquepts.
    U4e—Tropustults with Ustalfs.

V VERTISOLS—soils with high content of swelling clays; deep, wide cracks develop during dry periods.
V1 UDERTS—usually moist in some part of most years; cracks open less than 90 cumulative days in the year.
    V1a—with Usterts.
V2 USTERTS—cracks open more than 90 cumulative days in the year.
    V2a—with Tropaquepts.
    V2b—with Tropofluvents.
    V2c—with Ustalfs.

X **Soils in areas with mountains**—Soils with various moisture and temperature regimes; many steep slopes; relief and total elevation vary greatly within short distances and with changes in altitude; vertical zonation common.
X1 Cryic great groups of Entisols, Inceptisols, and Spodosols.
X2 Boralfs and cryic great groups of Entisols and Inceptisols.
X3 Udic great groups of Alfisols, Entisols, Inceptisols, and Ultisols.
X4 Ustic great groups of Alfisols, Inceptisols, Mollisols, and Ultisols.
X5 Xeric great groups of Alfisols, Entisols, Inceptisols, Mollisols, and Ultisols.
X6 Torric great groups of Aridisols and Entisols.
X7 Ustic and cryic great groups of Alfisols, Entisols, Inceptisols, and Mollisols; ustic great groups of Ultisols; cryic great groups of Spodosols.
X8 Aridisols, torric and cryic great groups of Entisols, and cryic great groups of Spodosols and Inceptisols.

Z **MISCELLANEOUS**
Z1 Icefields.
Z2 Rugged mountains—mostly devoid of soil (includes glaciers, permanent snowfields, and, in some places, areas of soil).

Major Soil Concept		Soil Order

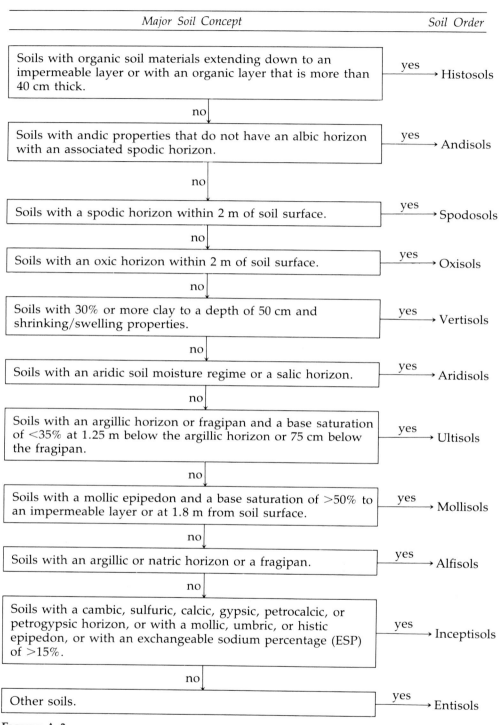

Soils with organic soil materials extending down to an impermeable layer or with an organic layer that is more than 40 cm thick. — yes → Histosols

no

Soils with andic properties that do not have an albic horizon with an associated spodic horizon. — yes → Andisols

no

Soils with a spodic horizon within 2 m of soil surface. — yes → Spodosols

no

Soils with an oxic horizon within 2 m of soil surface. — yes → Oxisols

no

Soils with 30% or more clay to a depth of 50 cm and shrinking/swelling properties. — yes → Vertisols

no

Soils with an aridic soil moisture regime or a salic horizon. — yes → Aridisols

no

Soils with an argillic horizon or fragipan and a base saturation of <35% at 1.25 m below the argillic horizon or 75 cm below the fragipan. — yes → Ultisols

no

Soils with a mollic epipedon and a base saturation of >50% to an impermeable layer or at 1.8 m from soil surface. — yes → Mollisols

no

Soils with an argillic or natric horizon or a fragipan. — yes → Alfisols

no

Soils with a cambic, sulfuric, calcic, gypsic, petrocalcic, or petrogypsic horizon, or with a mollic, umbric, or histic epipedon, or with an exchangeable sodium percentage (ESP) of >15%. — yes → Inceptisols

no

Other soils. — yes → Entisols

FIGURE A.3.
Simplified key showing the central concepts of the eleven orders in *Soil Taxonomy*.

Family Differentiae for *Soil Taxonomy*

The *Soil Taxonomy* system embodies six categories. From the highest to the lowest rank they are (1) orders, (2) suborders, (3) great groups, (4) subgroups, (5) families, and (6) series. The nomenclature for all categories except the family involves only one or two words. Since the family includes soils with common properties that affect the soil's response to management, the family name must include terms that describe these properties. This is accomplished by using nine different property classes, several of which may be used to differentiate the family, as follows.

1. Particle-size classes
2. Mineralogical classes
3. Soil temperature classes
4. Calcareous and reaction classes
5. Soil depth classes
6. Soil slope and shape classes
7. Soil consistence classes
8. Classes of coatings on sand
9. Permanent crack classes

Particle-Size Classes

Particle-size distribution is for the whole soil, not just the fine-earth fraction. Seven classes are used for most soils.

1. *Fragmental*—Stones, cobbles, gravel, and very coarse sand.

   Classes with less than 35% rock fragments (>2 mm in diameter):
2. *Sandy*—Fine earth is sand or loamy sand.
3. *Loamy*—Fine earth is loamy, very fine sand or finer, but less than 35% clay.
4. *Clayey*—Fine earth contains 35% or more clay.

   Classes with more than 35% rock fragments:
5. *Sandy-skeletal*—Fine earth is sand or loamy sand.
6. *Loamy-skeletal*—Fine earth is loamy, very fine sand, or finer, but less than 35% clay.
7. *Clayey-skeletal*—Fine earth is clay.

In soils developed in volcanic ejecta, these particle-size classes are replaced by terms that connote not only particle size but also mineralogy, such as *cindery, ashy, medial,* and *thixotropic.*

Mineralogical Classes

Mineralogical classes refer to the mineral or minerals that are most prominent in the soil. Two terms (vermiculitic/chloritic) may be used as well as mixed where several minerals are prominent. Common mineralogical classes include the following.

	Class	Dominant feature
	Applied to soils of any particle-size class	
1.	Carbonatic	Carbonates, some gypsum
2.	Ferritic	Iron oxides
3.	Gibbsitic	Aluminum oxides (gibbsite)
4.	Oxidic	High Fe, Al oxides in clay
5.	Gypsic	Gypsum
6.	Serpentinitic	Serpentine
7.	Glauconitic	Glauconite
	Applied to coarser classes (e.g., fragmental, sandy)	
8.	Micaceous	Micas
9.	Siliceous	Silica minerals
10.	Mixed	Mixtures
	Applied to finer classes (e.g., clayey)	
11.	Halloysitic	Halloysite
12.	Kaolinitic	Kaolinite
13.	Smectitic	Smectite
14.	Illitic[a]	Illite
15.	Vermiculitic	Vermiculite
16.	Chloritic	Chlorite
17.	Mixed	Mixtures of clays

[a] Refers to fine-grained micas.

SOIL TEMPERATURE CLASSES

These classes relate to the mean annual temperature at a standard depth (50 cm) and to differences between summer and winter temperatures, as follows.

Mean Annual Temperature (°C)	Differences Between Summer and Winter Temperatures	
	Greater than 5 °C	Less than 5 °C
<8	Frigid	Isofrigid
8–15	Mesic	Isomesic
15–22	Thermic	Isothermic
>22	Hyperthermic	Isohyperthermic

CALCAREOUS AND REACTION CLASSES

These classes refer to the presence or absence of carbonates and to the soil acidity/alkalinity. Terms such as *calcareous*, *noncalcareous*, *acid*, *nonacid*, and *allic* (high in Al^{3+}) are used.

SOIL DEPTH CLASSES

Depth of soil classes may be used in all orders of mineral soils, terms such as *micro* and *shallow* being used.

SLOPE AND SHAPE CLASSES

These classes are used to indicate slope and shape, particularly of soils in water-dominated (aquic) great groups.

SOIL CONSISTENCE CLASSES

Consistence classes are used to indicate cementation of some families, especially of Spodosols.

CLASSES OF COATINGS ON SAND

Coatings classes are used to indicate the coating of sand grains with silt and clay.

PERMANENT CRACK CLASSES

These classes are used to identify soils with permanent cracks.

Glossary
of Soil Science Terms[1]

A horizon The surface horizon of a mineral soil having maximum organic matter accumulation, maximum biological activity and/or eluviation of materials such as iron and aluminum oxides and silicate clays.

ABC soil A soil with a distinctly developed profile, including A, B, and C horizons.

absorption, active Movement of ions and water into the plant root as a result of metabolic processes by the root, frequently against an activity gradient.

absorption, passive Movement of ions and water into the plant root as a result of diffusion along a gradient.

AC soil A soil having a profile containing only A and C horizons with no clearly developed B horizon.

accelerated erosion *See* erosion.

acid rain Atmospheric precipitation with pH values less than about 5.6, the acidity being due to inorganic acids such as nitric and sulfuric that are formed when oxides of nitrogen and sulfur are emitted into the atmosphere.

acid soil A soil with a pH value <7.0. Usually applied to surface layer or root zone, but may be used to characterize any horizon. *See also* reaction, soil.

acidity, active The activity of hydrogen ion in the aqueous phase of a soil. It is measured and expressed as a pH value.

acidity, residual Soil acidity that can be neutralized by lime or other alkaline materials but cannot be replaced by an unbuffered salt solution.

acidity, salt replaceable Exchangeable hydrogen and aluminum that can be replaced from an acid soil by an unbuffered salt solution such as KCl or NaCl.

acidity, total The total acidity in a soil. It is approximated by the sum of the salt replaceable acidity plus the residual acidity.

actinomycetes A group of organisms intermediate between the bacteria and the true fungi that usually produce a characteristic branched mycelium. Includes many, but not all, organisms belonging to the order of Actinomycetales.

activated sludge Sludge that has been aerated and subjected to bacterial action.

adhesion Molecular attraction that holds the surfaces of two substances (e.g., water and sand particles) in contact.

adsorption The attraction of ions or compounds to the surface of a solid. Soil colloids adsorb large amounts of ions and water.

adsorption complex The group of organic and inorganic substances in soil capable of adsorbing ions and molecules.

[1] This glossary was compiled and modified from several sources including *Glossary of Soil Science Terms* (Madison, WI: Soil Sci. Soc. Amer., 1987), *Resource Conservation Glossary* (Anheny, IA: Soil Cons. Soc. Amer., 1982), and *Soil Taxonomy* (Washington, DC: U.S. Department of Agriculture, 1976).

aerate To impregnate with gas, usually air.

aeration, soil The process by which air in the soil is replaced by air from the atmosphere. In a well-aerated soil, the soil air is similar in composition to the atmosphere above the soil. Poorly aerated soils usually contain more carbon dioxide and correspondingly less oxygen than the atmosphere above the soil.

aerobic (1) Having molecular oxygen as a part of the environment. (2) Growing only in the presence of molecular oxygen, as aerobic organisms. (3) Occurring only in the presence of molecular oxygen (said of certain chemical or biochemical processes, such as aerobic decomposition).

aggregate (soil) Many soil particles held in a single mass or cluster such as a clod, crumb, block, or prism.

agric horizon *See* diagnostic subsurface horizons.

agronomy A specialization of agriculture concerned with the theory and practice of field-crop production and soil management. The scientific management of land.

air-dry (1) The state of dryness (of a soil) at equilibrium with the moisture content in the surrounding atmosphere. The actual moisture content will depend upon the relative humidity and the temperature of the surrounding atmosphere. (2) To allow to reach equilibrium in moisture content with the surrounding atmosphere.

air porosity The proportion of the bulk volume of soil that is filled with air at any given time or under a given condition, such as a specified moisture potential; usually the large pores.

albic horizon *See* diagnostic subsurface horizons.

Alfisols *See* soil classification.

alkali soil (Obsolete) A soil that contains sufficient alkali (sodium) to interfere with the growth of most crop plants. *See also* saline-sodic soil; sodic soil.

alkaline soil Any soil that has pH > 7. Usually applied to surface layer or root zone but may be used to characterize any horizon or a sample thereof. *See also* reaction, soil.

allophane An aluminosilicate mineral that has an amorphous or poorly crystalline structure and is commonly found in soils developed from volcanic ash.

alluvial soil (Obsolete) A soil developing from recently deposited alluvium and exhibiting essentially no horizon development or modification of the recently deposited materials.

alluvium A general term for all detrital material deposited or in transit by streams, including gravel, sand, silt, clay, and all variations and mixtures of these. Unless otherwise noted, alluvium is unconsolidated.

alpha particle A positively charged particle (consisting of two protons and two neutrons) that is emitted by certain radioactive compounds.

aluminosilicates Compounds containing aluminum, silicon, and oxygen as main constituents. An example is mocrocline, $KAlSi_3O_8$.

amendment, soil Any substance other than fertilizers, such as lime, sulfur, gypsum, and sawdust, used to alter the chemical or physical properties of a soil, generally to make it more productive.

amino acids Nitrogen-containing organic acids that couple together to form proteins. Each acid molecule contains one or more amino groups ($—NH_2$) and at least one carboxyl group ($—COOH$). In addition, some amino acids contain sulfur.

ammonification The biochemical process whereby ammoniacal nitrogen is released from nitrogen-containing organic compounds.

ammonium fixation The entrapment of ammonium ions by the mineral or organic fractions of the soil in forms that are insoluble in water and at least temporarily nonexchangeable.

amorphous material Noncrystalline constituents of soils.

anaerobic (1) Without molecular oxygen. (2) Living or functioning in the absence of air or free oxygen.

Andisols *See* soil classification.

anion Negatively charged ion; during electrolysis it is attracted to the positively charged anode.

anion exchange capacity The sum total of exchangeable anions that a soil can adsorb. Expressed as centimoles per kilogram (cmol/kg) of soil (or of other adsorbing material such as clay).

anthropic epipedon *See* diagnostic surface horizons.

antibiotic A substance produced by one species of organism that, in low concentrations, will kill or inhibit growth of certain other organisms.

Ap The surface layer of a soil disturbed by cultivation or pasturing.

apatite A naturally occurring complex calcium phosphate that is the original source of most of the phosphate fertilizers. Formulas such as $[3Ca_3(PO_4)_2] \cdot CaF_2$ illustrate the complex compounds that make up apatite.

argillic horizon *See* diagnostic subsurface horizons.

arid climate Climate in regions that lack sufficient moisture for crop production without irrigation. In cool regions annual precipitation is usually less than 25 cm. It may be as high as 50 cm in tropical regions. Natural vegetation is desert shrubs.

Aridisols *See* soil classification.

association, soil *See* soil association.

Atterberg limits Water contents of fine-grained soils at different states of consistency.

> **plastic limit (PL)** The water content corresponding to an arbitrary limit between the plastic and semi-solid states of consistency of a soil.

> **liquid limit (LL)** The water content corresponding to the arbitrary limit between the liquid and plastic states of consistency of a soil.

autotroph An organism capable of utilizing carbon dioxide or carbonates as the sole source of carbon and obtaining energy for life processes from the oxidation of inorganic elements or compounds such as iron, sulfur, hydrogen, ammonium, and nitrites, or from radiant energy. *Contrast with* heterotroph.

available nutrient That portion of any element or compound in the soil that can be readily absorbed and assimilated by growing plants. ("Available" should not be confused with "exchangeable.")

available water The portion of water in a soil that can be readily absorbed by plant roots. The amount of water released between the field capacity and the permanent wilting point.

B horizon A soil horizon usually beneath the A that is characterized by one or more of the following: (a) a concentration of silicate clays, iron and aluminum oxides, and humus, alone or in combination; (b) a blocky or prismatic structure, and (c) coatings of iron and aluminum oxides that give darker, stronger, or redder color.

bar A unit of pressure equal to one million dynes per square centimeter (10^6 dynes/cm^2).

base saturation percentage The extent to which the adsorption complex of a soil is saturated with exchangeable cations other than hydrogen and aluminum. It is expressed as a percentage of the total cation exchange capacity.

BC soil A soil profile with B and C horizons but with little or no A horizon. Most BC soils have lost their A horizons by erosion.

bedding (Engineering) Arranging the surface of fields by plowing and grading into a series of elevated beds separated by shallow depressions or ditches for drainage.

bedrock The solid rock underlying soils and the regolith in depths ranging from zero (where exposed by erosion) to several hundred feet.

bench terrace An embankment constructed across sloping fields with a steep drop on the downslope side.

beta particle A high-speed electron emitted in radioactive decay.

biodegradable Subject to degradation by biochemical processes.

biomass The amount of living matter in a given area.

bleicherde (Obsolete) The light-colored, leached A2 horizon of Spodosols.

blocky soil structure Soil aggregates with block-like shapes; common in B horizons of soils in humid regions.

blown-out land Areas from which all or almost all of the soil and soil material has been removed by wind erosion. Usually unfit for crop production. A miscellaneous land type.

border-strip irrigation *See* irrigation methods.

bottomland *See* floodplain.

breccia A rock composed of coarse angular fragments cemented together.

broad-base terrace A low embankment with such gentle slopes that it can be farmed, constructed across sloping fields to reduce erosion and runoff.

broadcast Scatter seed or fertilizer on the surface of the soil.

buffering capacity The ability of a soil to resist changes in pH. Commonly determined by presence of clay, humus, and other colloidal materials.

bulk blending Mixing dry individual granulated fertilizer materials to form a mixed fertilizer that is applied promptly to the soil.

bulk density, soil The mass of dry soil per unit of bulk volume, including the air space. The bulk volume is determined before drying to constant weight at 105 °C.

buried soil Soil covered by an alluvial, loessal, or other deposit, usually to a depth greater than the thickness of the solum.

C horizon A mineral horizon generally beneath the solum that is relatively unaffected by biological activity and pedogenesis and is lacking properties diagnostic of an A or B horizon. It may or may not be like the material from which the A and B have formed.

calcareous soil Soil containing sufficient calcium carbonate (often with magnesium carbonate) to effervesce visibly when treated with cold 0.1 N hydrochloric acid.

calcic horizon *See* diagnostic subsurface horizons.

caliche A layer near the surface, more or less cemented by secondary carbonates of calcium or magnesium precipitated from the soil solution. It may occur as a soft, thin soil horizon; as a hard, thick bed just beneath the solum; or as a surface layer exposed by erosion.

cambic horizon *See* diagnostic subsurface horizons.

capillary conductivity (Obsolete) *See* hydraulic conductivity.

capillary water (Obsolete) The water held in the "capillary" or *small* pores of a soil, usually with a tension >60 cm of water. *See also* moisture potential.

carbon cycle The sequence of transformations whereby carbon dioxide is fixed in living organisms by photosynthesis or by chemosynthesis, liberated by respiration and by the death and decomposition of the fixing organism, used by heterotrophic species, and ultimately returned to its original state.

carbon/nitrogen ratio The ratio of the weight of organic carbon (C) to the weight of total nitrogen (N) in a soil or in organic material.

carnivore An organism that feeds on animals.

"cat" clays Wet clay soils high in reduced forms of sulfur that upon being drained, become extremely acid because of the oxidation of the sulfur compounds and the formation of sulfuric acid.

catena A sequence of soils of about the same age, derived from similar parent material, and occurring under similar climatic conditions, but having different characteristics because of variation in *relief* and in *drainage*.

cation A positively charged ion; during electrolysis it is attracted to the negatively charged cathode.

cation exchange The interchange between a cation in solution and another cation on the surface of any surface-active material such as clay or organic matter.

cation exchange capacity The sum total of exchangeable cations that a soil can adsorb. Sometimes called "total-exchange capacity," "base-exchange capacity," or "cation-adsorption capacity." Expressed in centimoles per kilogram (cmol/kg) of soil (or of other adsorbing material such as clay).

cemented Indurated; having a hard, brittle consistency because the particles are held together by cementing substances such as humus, calcium carbonate, or the oxides of silicon, iron, and aluminum.

channery Thin, flat fragments of limestone, sandstone, or schist up to 15 cm (6 in.) in major diameter.

chelate (Greek, claw) A type of chemical compound in which a metallic ion is firmly combined with an organic molecule by means of multiple chemical bonds.

chert A structureless form of silica, closely related to flint, that breaks into angular fragments.

chisel, subsoil A tillage implement with one or more cultivator-type feet to which are attached strong knife-like units used to shatter or loosen hard, compact layers, usually in the subsoil, to depths below normal plow depth. *See also* subsoiling.

chlorite A 2:1:1-type layer-structured silicate mineral having 2:1 layers alternating with a magnesium-dominated octahedral sheet.

chlorosis A condition in plants relating to the failure of chlorophyll (the green coloring matter) to develop. Chlorotic leaves range from light green through yellow to almost white.

class, soil A group of soils having a definite range in a particular property such as acidity, degree of slope, texture, structure, land-use capability, degree of erosion, or drainage. *See also* soil structure; soil texture.

classification, soil *See* soil classification.

clastic Composed of broken fragments of rocks and minerals.

clay (1) A soil separate consisting of particles <0.002 mm in equivalent diameter. (2) A soil textural class containing >40% clay, <45% sand, and <40% silt.

clay mineral Naturally occurring inorganic material (usually crystalline) found in soils and other earthy deposits, the particles being of clay size, that is, <0.002 mm in diameter.

claypan A dense, compact, slowly permeable layer in the subsoil having a much higher clay content than the overlying material, from which it is separated by a sharply defined boundary. Claypans are usually hard when dry and plastic and sticky when wet. *See also* hardpan.

clod A compact, coherent mass of soil produced artificially, usually by such human activities as plowing and digging, especially when these operations are performed on soils that are either too wet or too dry for normal tillage operations.

coarse texture The texture exhibited by sands, loamy sands, and sandy loams except very fine sandy loam.

cobblestone Rounded or partially rounded rock or mineral fragments 7.5–25 cm (3–10 in.) in diameter.

cohesion Holding together: force holding a solid or liquid together, owing to attraction between like molecules. Decreases with rise in temperature.

colloid, soil (Greek, glue-like) Organic and inorganic matter with very small particle size and a correspondingly large surface area per unit of mass.

colluvium A deposit of rock fragments and soil material accumulated at the base of steep slopes as a result of gravitational action.

color The property of an object that depends on the wavelength of light it reflects or emits.

columnar soil structure *See* soil structure types.

compost Organic residues, or a mixture of organic residues and soil, that have been piled, moistened, and allowed to undergo biological decomposition. Mineral fertilizers are sometimes added. Often called "artificial manure" or "synthetic manure" if produced primarily from plant residues.

concretion A local concentration of a chemical compound, such as calcium carbonate or iron oxide, in the form of grains or nodules of varying size, shape, hardness, and color.

conduction The transfer of heat by physical contact between two or more objects.

conductivity, hydraulic *See* hydraulic conductivity.

conifer A tree belonging to the order Coniferae, usually evergreen, with cones and needle-shaped or scale-like leaves and producing wood known commercially as "softwood."

conservation tillage *See* tillage, conservation.

consistence The combination of properties of soil material that determine its resistance to crushing and its ability to be molded or changed in shape. Such terms as loose, friable, firm, soft, plastic, and sticky describe soil consistence.

constant charge The net surface charge of mineral particles, the magnitude of which depends only on the chemical and structural composition of the mineral. The charge arises from isomorphous substitution and is not affected by soil pH.

consumptive use The water used by plants in transpiration and growth, plus water vapor loss from adjacent soil or snow, or from intercepted precipitation in any specified time. Usually expressed as equivalent depth of free water per unit of time.

contour An imaginary line connecting points of equal elevation on the surface of the soil. A contour terrace is laid out on a sloping soil at right angles to the direction of the slope and nearly level throughout its course.

contour strip cropping Layout of crops in comparatively narrow strips in which the farming operations are performed approximately on the contour. Usually strips of grass, close-growing crops, or fallow are alternated with those of cultivated crops.

convection The transfer of heat through a gas or solution because of molecular movement.

corrugated irrigation *See* irrigation methods.

cover crop A close-growing crop grown primarily for the purpose of protecting and improving soil between periods of regular crop production or between trees and vines in orchards and vineyards.

creep Slow mass movement of soil and soil material down relatively steep slopes primarily under the influence of gravity, but facilitated by saturation with water and by alternate freezing and thawing.

crop rotation A planned sequence of crops growing in a regularly recurring succession on the same area of land, as contrasted to continuous culture of one crop or growing different crops in haphazard order.

crotovina A former animal burrow in one soil horizon that has been filled with organic matter or material from another horizon (also spelled "krotovina").

crumb A soft, porous, more or less rounded natural unit

of structure from 1 to 5 mm in diameter. *See also* soil structure types.

crushing strength The force required to crush a mass of dry soil or, conversely, the resistance of the dry soil mass to crushing. Expressed in units of force per unit area (pressure).

crust A surface layer on soils, ranging in thickness from a few millimeters to perhaps as much as 3 cm, that is much more compact, hard, and brittle, when dry, than the material immediately beneath it.

crystal A homogeneous inorganic substance of definite chemical composition bounded by plane surfaces that form definite angles with each other, thus giving the substance a regular geometrical form.

crystal structure The orderly arrangement of atoms in a crystalline material.

crystalline rock A rock consisting of various minerals that have crystallized in place from magma. *See also* igneous rock; sedimentary rock.

cultivation A tillage operation used in preparing land for seeding or transplanting or later for weed control and for loosening the soil.

deciduous plant A plant that sheds all its leaves every year at a certain season.

deflocculate (1) To separate the individual components of compound particles by chemical and/or physical means. (2) To cause the particles of the *disperse phase* of a colloidal system to become suspended in the *dispersion medium*.

delta An alluvial deposit formed where a stream or river drops its sediment load upon entering a quieter body of water.

denitrification The biochemical reduction of nitrate or nitrite to gaseous nitrogen, either as molecular nitrogen or as an oxide of nitrogen.

density *See* particle density; bulk density.

desalinization Removal of salts from saline soil, usually by leaching.

desert crust A hard layer, containing calcium carbonate, gypsum, or other binding material, exposed at the surface in desert regions.

detritus Debris from dead plants and animals.

desorption The removal of sorbed material from surfaces.

diagnostic horizons (As used in *Soil Taxonomy*): Combinations of specific soil characteristics that are indicative of certain classes of soils. Horizons that occur at the soil surface are called epipedons; those below the surface, diagnostic subsurface horizons.

diagnostic subsurface horizons The following diagnostic subsurface horizons are used in *Soil Taxonomy.*

    agric horizon A mineral soil horizon in which clay, silt, and humus derived from an overlying cultivated and fertilized layer have accumulated. The wormholes and illuvial clay, silt, and humus occupy at least 5% of the horizon by volume.

    albic horizon A mineral soil horizon from which clay and free iron oxides have been removed or in which the oxides have been segregated to the extent that the color of the horizon is determined primarily by the color of the primary sand and silt particles rather than by coatings on these particles.

    argillic horizon A mineral soil horizon characterized by the illuvial accumulation of layer-lattice silicate clays.

    calcic horizon A mineral soil horizon of secondary carbonate enrichment that is more than 15 cm thick, has a calcium carbonate equivalent of more than 15%, and has at least 5% more calcium carbonate equivalent than the underlying C horizon.

    cambic horizon A mineral soil horizon that has a texture of loamy very fine sand or finer, contains some weatherable minerals, and is characterized by the alteration or removal of mineral material. The cambic horizon lacks cementation or induration and has too few evidences of illuviation to meet the requirements of the argillic or spodic horizon.

    duripan A mineral soil horizon that is cemented by silica, to the point that air-dry fragments will not slake in water or HCl.

    gypsic horizon A mineral soil horizon of secondary calcium sulfate enrichment that is more than 15 cm thick.

    kandic horizon A horizon having a sharp clay increase relative to overlying horizons and having low-activity clays.

    natric horizon A mineral soil horizon that satisfies the requirements of an argillic horizon, but that also has prismatic, columnar, or blocky structure and a subhorizon having more than 15% saturation with exchangeable sodium.

    oxic horizon A mineral soil horizon that is at least 30 cm thick and characterized by the virtual *absence* of weatherable primary minerals or 2:1 lattice clays and the *presence* of 1:1 lattice clays and highly insoluble minerals such as quartz sand, hydrated oxides of iron and aluminum, low cation exchange capacity, and small amounts of exchangeable bases.

    petrocalcic horizon A continuous, indurated calcic horizon that is cemented by calcium carbonate and, in some places, with magnesium carbonate. It cannot be penetrated with a spade or auger when dry, dry fragments do not slake in water, and it is impenetrable to roots.

    petrogypsic horizon A continuous, strongly cemented, massive gypsic horizon that is cemented by calcium sulfate. It can be chipped with a spade when dry. Dry fragments do not slake in water and it is impenetrable to roots.

    placic horizon A black to dark reddish mineral soil horizon that is usually thin but that may range from 1 to 25 mm in thickness. The placic horizon is commonly cemented with iron and is slowly permeable or impenetrable to water and roots.

    salic horizon A mineral soil horizon of enrichment with secondary salts more soluble in cold water than gypsum. A salic horizon is 15 cm or more in thickness.

sombric horizon A mineral subsurface horizon that contains illuvial humus but has a low cation exchange capacity and low percentage base saturation. Mostly restricted to cool moist soils of high plateaus and mountainous areas of tropical and subtropical regions.

spodic horizon A mineral soil horizon characterized by the illuvial accumulation of amorphous materials composed of aluminum and organic carbon with or without iron.

sulfuric horizon A subsurface horizon in either mineral or organic soils that has a pH < 3.5, fresh straw-colored mottles (called jarosite mottles). Forms by oxidation of sulfide-rich materials and is highly toxic to plants.

diagnostic surface horizons The following diagnostic surface horizons are used in *Soil Taxonomy* and are called *epipedons*.

anthropic epipedon A surface layer of mineral soil that has the same requirements as the mollic epipedon but that has more than 250 mg/kg of P_2O_5 soluble in 1% citric acid, or is dry more than 10 months (cumulative) during the period when not irrigated. The anthropic epipedon forms under long-continued cultivation and fertilization.

histic epipedon A thin organic soil horizon that is saturated with water at some period of the year unless artificially drained and that is at or near the surface of a mineral soil.

mollic epipedon A surface horizon of mineral soil that is dark colored and relatively thick, contains at least 0.6% organic carbon, is not massive and hard when dry, has a base saturation of more than 50%, has less than 250 mg/kg P_2O_5 soluble in 1% citric acid, and is dominantly saturated with bivalent cations.

ochric epipedon A surface horizon of mineral soil that is too light in color, too high in chroma, too low in organic carbon, or too thin to be a plaggen, mollic, umbric, anthropic, or histic epipedon, or that is both hard and massive when dry.

plaggen epipedon A man-made surface horizon more than 50 cm thick that is formed by long-continued manuring and mixing.

umbric epipedon A surface layer of mineral soil that has the same requirements as the mollic epipedon with respect to color, thickness, organic carbon content, consistence, structure, and P_2O_5 content, but that has a base saturation of less than 50%.

diatoms Algae having siliceous cell walls that persist as a skeleton after death; any of the microscopic unicellular or colonial algae constituting the class Bacillariaceae. They occur abundantly in fresh and salt waters and their remains are widely distributed in soils.

diatomaceous earth A geologic deposit of fine, grayish, siliceous material composed chiefly or wholly of the remains of diatoms. It may occur as a powder or as a porous, rigid material.

diffusion The transport of matter as a result of the movement of the constituent particles. The intermingling of two gases or liquids in contact with each other takes place by diffusion.

dioctahedral An octahedral sheet, or a mineral containing such a sheet, that has two thirds of the octahedral sites filled with trivalent ions such as aluminum or ferric iron.

disintegration The breakdown of rock and mineral particles into smaller particles by physical forces such as frost action.

disperse (1) To break up compound particles, such as aggregates, into the individual component particles. (2) To distribute or suspend fine particles, such as clay, in or throughout a dispersion medium, such as water.

diversion dam A structure or barrier built to divert part or all of the water of a stream to a different course.

diversion terrace *See* terrace.

drain (1) To provide channels, such as open ditches or drain tile, so that excess water can be removed by surface or by internal flow. (2) To lose water (from the soil) by percolation.

drainage, soil The frequency and duration of periods when the soil is free from saturation with water.

drift Material of any sort deposited by geological processes in one place after having been removed from another. Glacial drift includes material moved by the glaciers and by the streams and lakes associated with them.

drumlin Long, smooth cigar-shaped low hills of glacial till, with their long axes parallel to the direction of ice movement.

dryland farming The practice of crop production in low-rainfall areas without irrigation.

duff The matted, partly decomposed organic surface layer of forest soils.

duripan *See* diagnostic subsurface horizons; hardpan.

dust mulch A loose, finely granular, or powdery condition on the surface of the soil, usually produced by shallow cultivation.

E horizon Horizon characterized by maximum illuviation (washing out) of silicate clays and iron and aluminum oxides; commonly occurs above the B horizon and below the A horizon.

earthworms Animals of the Lumbricidae family that burrow into and live in the soil. They mix plant residues into the soil and improve soil aeration.

ectotrophic mycorrhiza (ectomycorrhiza) A symbiotic association of the mycelium of fungi and the roots of certain plants in which the fungal hyphae form a compact mantle on the surface of the roots and extend into the surrounding soil and inward between cortical cells, but not into these cells. Associated primarily with certain trees. *See also* endotrophic mycorrhiza.

edaphology The science that deals with the influence of soils on living things, particularly plants, including man's use of land for plant growth.

electrokinetic potential In a colloidal systems, the difference in potential between the immovable layer attached to the surface of the dispersed phase and the dispersion medium.

eluviation The removal of soil material in suspension (or in solution) from a layer or layers of a soil. (Usually, the loss of material in *solution* is described by the term "leaching.") *See also* lelaching.

endotrophic mycorrhiza (endomycorrhiza) A symbiotic association of the mycelium of fungi and roots of a variety of plants in which the fungal hyphae penetrate directly into root hairs, other epidermal cells, and occasionally into cortical cells. Individual hyphae also extend from the root surface outward into the surrounding soil. *See also* vesicular arbuscular mycorrhiza.

Entisols *See* soil classification.

eolian soil material Soil material accumulated through wind action. The most extensive areas in the United States are silty deposits (loess), but large areas of sandy deposits also occur.

epipedon A diagnostic surface horizons that includes the upper part of the soil that is darkened by organic matter, or the upper eluvial horizons, or both. [*Soil Taxonomy.*]

erosion (1) The wearing away of the land surface by running water, wind, ice, or other geological agents, including such processes as gravitational creep. (2) Detachment and movement of soil or rock by water, wind, ice, or gravity. The following terms are used to describe different types of water erosion.

    accelerated erosion Erosion much more rapid than normal, natural, geological erosion; primarily as a result of the activities of humans or, in some cases, of animals.

    gully erosion The erosion process whereby water accumulates in narrow channels and, over short periods, removes the soil from this narrow area to considerable depths, ranging from 1–2 feet to as much as 23–30 m (75–100 ft).

    natural erosion Wearing away of the Earth's surface by water, ice, or other natural agents under natural environmental conditions of climate, vegetation, and so on, undisturbed by man. Synonymous with *geological erosion.*

    rill erosion An erosion process in which numerous small channels of only several centimeters in depth are formed; occurs mainly on recently cultivated soils. *See also* rill.

    sheet erosion The removal of a fairly uniform layer of soil from the land surface by runoff water.

    splash erosion The spattering of small soil particles caused by the impact of raindrops on very wet soils. The loosened and separated particles may or may not be subsequently removed by surface runoff.

esker A narrow ridge of gravelly or sandy glacial material deposited by a stream in an ice-walled valley or tunnel in a receding glacier.

essential element A chemical element required for the normal growth of plants.

eutrophic Having concentrations of nutrients optimal (or nearly so) for plant or animal growth. (Said of nutrient solutions or of soil solutions.)

eutrophication A process of aging of lakes whereby aquatic plants are abundant and waters are deficient in oxygen. The process is usually accelerated by enrichment of waters with surface runoff containing nitrogen and phosphorus.

evapotranspiration The combined loss of water from a given area, and during a specified period of time, by evaporation from the soil surface and by transpiration from plants.

exchange capacity The total ionic charge of the adsorption complex active in the adsorption of ions. *See also* anion exchange capacity; cation exchange capacity.

exchangeable sodium percentage The extent to which the adsorption complex of a soil is occupied by sodium. It is expressed as follows.

$$ESP = \frac{\text{exchangeable sodium (cmol/kg soil)}}{\text{cation exchange capacity (cmol/kg soil)}} \times 100$$

facultative organism An organism capable of both aerobic and anaerobic metabolism.

fallow Cropland left idle in order to restore productivity, mainly through accumulation of water, nutrients, or both. Summer fallow is a common stage before cereal grain in regions of limited rainfall. The soil is kept free of weeds and other vegetation, thereby conserving nutrients and water for the next year's crop.

family, soil In soil classification, one of the categories intermediate between the great group and the soil series. Families are defined largely on the basis of physical and mineralogical properties of importance to plant growth. [*Soil Taxonomy.*]

fauna The animal life of a region.

ferrihydrite, $Fe_5HO_8 \cdot 4H_2O$ A dark reddish brown poorly crystalline iron oxide that forms in wet soils.

fertility, soil The quality of a soil that enables it to provide essential chemical elements in quantities and proportions for the growth of specified plants.

fertilizer Any organic or inorganic material of natural or synthetic origin added to a soil to supply certain elements essential to the growth of plants. The major types of fertilizers include

    bulk blended fertilizers Solid fertilizer materials blended together in small blending plants, delivered to the farm in bulk, and usually spread directly on the fields by truck or other special applicator.

    granulated fertilizers Fertilizers that are present in the form of rather stable granules of uniform size, which facilitate ease of handling the materials and reduce undesirable dusts.

    liquid fertilizers Fluid fertilizers that contain essential elements in liquid forms either as soluble nutrients or as liquid suspensions or both.

    mixed fertilizers Two or more fertilizer materials mixed together. May be as dry powders, granules, pellets, bulk blends, or liquids.

fertilizer requirement The quantity of certain plant nutrient elements needed, in addition to the amount supplied by the soil, to increase plant growth to a designated optimum.

fibric materials *See* organic soil materials.

field capacity (field moisture capacity) The percentage of water remaining in a soil two or three days after its having been saturated and after free drainage has practically ceased.

fine-grained mica A silicate clay having a 2:1-type lattice structure with much of the silicon in the tetrahedral sheet having been replaced by aluminum and with considerable interlayer potassium, which binds the layers together and prevents interlayer expansion and swelling, and limits interlayer cation exchange capacity.

fine texture Consisting of or containing large quantities of the fine fractions, particularly of silt and clay. (Includes clay loam, sandy clay loam, silty clay loam, sandy clay, silty clay, and clay textural classes.)

first bottom The normal floodplain of a stream.

fixation (1) For other than elemental nitrogen: the process or processes in a soil by which certain chemical elements essential for plant growth are converted from a soluble or exchangeable form to a much less soluble or to a nonexchangeable form; for example, potassium, ammonium, and phosphate "''fixation.'' (2) For elemental nitrogen: process by which gaseous elemental nitrogen is chemically combined with hydrogen to form ammonia.

  biological nitrogen fixation Occurs at ordinary temperatures and pressures. It is commonly carried out by certain bacteria, algae, and actinomycetes, which may or may not be associated with higher plants.

  chemical nitrogen fixation Takes place at high temperatures and pressures in manufacturing plants; produces ammonia, which is used to manufacture most fertilizers.

flagstone A relatively thin rock or mineral fragment 15–38 cm in length commonly composed of shale, slate, limestone, or sandstone.

flocculate To aggregate or clump together individual, tiny soil particles, especially fine clay, into small clumps or floccules. Opposite of *deflocculate or disperse*.

floodplain The land bordering a stream, built up of sediments from overflow of the stream and subject to inundation when the stream is at flood stage. Sometimes called bottomland.

flora The sum total of the kinds of plants in an area at one time.

fluorapatite A member of the apatite group of minerals containing fluorine. Most common mineral in rock phosphate.

fluvial deposits Deposits of parent materials laid down by rivers or streams.

fluvioglacial *See* glaciofluvial deposits.

foliar diagnosis An estimation of mineral nutrient deficiencies (excesses) of plants based on examination of the chemical composition of selected plant parts, and the color and growth characteristics of the foliage of the plants.

fragipan Dense and brittle pan or subsurface layer in soils that owe their hardness mainly to extreme density or compactness rather than high clay content or cementation. Removed fragments are friable, but the material in place is so dense that roots penetrate and water moves through it very slowly.

friable A soil consistency term pertaining to the ease of crumbling of soils.

frigid *See* soil temperature classes.

fritted micronutrients Sintered silicates having total guaranteed analyses of micronutrients with controlled (relatively slow) release characteristics.

fulvic acid A term of varied usage but usually referring to the mixture of organic substances remaining in solution upon acidification of a dilute alkali extract from the soil.

furrow irrigation *See* irrigation methods.

fungi Simple plants that lack a photosynthetic pigment. The individual cells have a nucleus surrounded by a membrane, and they may be linked together in long filaments called *hyphae*, which may grow together to form a visible body.

gamma ray A high-energy ray (photon) emitted during radioactive decay of certain elements.

genesis, soil The mode of origin of the soil, with special reference to the processes responsible for the development of the solum, or true soil, from the unconsolidated parent material.

geological erosion *See* erosion, natural.

geothite, FeOOH A yellow-brown iron oxide mineral that accounts for the brown color in many soils.

gibbsite, Al(OH)$_3$ An aluminum trihydroxide mineral most common in highly weathered soils such as oxisols.

gilgai The microrelief of soils produced by expansion and contraction with changes in moisture. Found in soils that contain large amounts of clay that swells and shrinks considerably with wetting and drying. Usually a succession of microbasins and microknolls in nearly level areas or of microvalleys and microridges parallel to the direction of the slope.

glacial drift Rock debris that has been transported by glaciers and deposited, either directly from the ice or from the meltwater. The debris may or may not be heterogeneous.

glacial till *See* till.

glaciofluvial deposits Material moved by glaciers and subsequently sorted and deposited by streams flowing from the melting ice. The deposits are stratified and may occur in the form of outwash plains, deltas, kames, eskers, and kame terraces.

gley soil (Obsolete) Soil developed under conditions of poor drainage resulting in reduction of iron and other elements and in gray colors and mottles.

granular structure Soil structure in which the individual grains are grouped into spherical aggregates with indistinct sides. Highly porous granules are commonly called crumbs. A well-granulated soil has the best structure for most ordinary crop plants. *See also* soil structure types.

grassed waterway A natural or constructed waterway covered with erosion-resistant grasses that permits removal of runoff water without excessive erosion.

gravitational potential *See* soil water potential.

gravitational water Water that moves into, through, or out of the soil under the influence of gravity.

great group *See* soil classification.

greenhouse effect The entrapment of heat by upper atmosphere gases such as carbon dioxide, water vapor, and methane just as glass traps heat for a greenhouse. Increases in the quantities of these gases in the atmosphere will likely result in global warming that may have serious consequences for humankind.

green manure Plant material incorporated with the soil while green, or soon after maturity, for improving the soil.

groundwater Subsurface water in the zone of saturation that is free to move under the influence of gravity.

gully erosion *See* erosion.

gypsic horizon *See* diagnostic subsurface horizon.

halophyte A plant that requires or tolerates a saline (high salt) environment.

hardpan A hardened soil layer, in the lower A or in the B horizon, caused by cementation of soil particles with organic matter or with materials such as silica, sesquioxides, or calcium carbonate. The hardness does not change appreciably with changes in moisture content and pieces of the hard layer do not slake in water. *See also* caliche; claypan.

harrowing A secondary broadcast tillage operation that pulverizes, smooths, and firms the soil in seedbed preparation, controls weeds, or incorporates material spread on the surface.

heaving The partial lifting of plants, buildings, roadways, fenceposts, etc., out of the ground, as a result of freezing and thawing of the surface soil during the winter.

heavy metals Metals with particle densities >5.0 Mg/m^3.

heavy soil (Obsolete in scientific use) A soil with a high content of the fine separates, particularly clay, or one with a high drawbar pull, hence difficult to cultivate.

hematite, Fe_2O_3 A red iron oxide mineral that contributes red color to many soils.

herbicide A chemical that kills plants or inhibits their growth; intended for weed control.

herbivore A plant-eating animal.

hemic materials *See* organic soil materials.

heterotroph An organism capable of deriving energy for life processes only from the decomposition of organic compounds and incapable of using inorganic compounds as sole sources of energy or for organic synthesis. *Contrast with* autotroph.

histic epipedon *See* diagnostic surface horizons.

Hisotosols *See* soil classification.

horizon, soil A layer of soil, approximately parallel to the soil surface, differing in properties and characteristics from adjacent layers below or above it. *See also* diagnostic subsurface horizons; diagnostic surface horizons.

horticulture The art and science of growing fruits, vegetables, and ornamental plants.

humic acid A mixture of variable or indefinite composition of dark organic substances, precipitated upon acidification of a dilute alkali extract from soil.

humid climate Climate in regions where moisture, when distributed normally throughout the year, should not limit crop production. In cool climate annual precipitation may be as little as 25 cm; in hot climates, 150 cm or even more. Natural vegetation in uncultivated areas is forests.

humification The processes involved in the decomposition of organic matter and leading to the formation of humus.

humin The fraction of the soil organic matter that is not dissolved upon extraction of the soil with dilute alkali.

humus That more or less stable fraction of the soil organic matter remaining after the major portions of added plant and animal residues have decomposed. Usually it is dark in color.

hydraulic conductivity An expression of the readiness with which a liquid such as water flows through a solid such as soil in response to a given potential gradient.

hydrogen bond The chemical bond between a hydrogen atom in one molecule and a highly electronegative atom such as oxygen or nitrogen in another polar molecule.

hydrologic cycle The circuit of water movement from the atmosphere to the Earth and back to the atmosphere through various stages or processes, as precipitation, interception, runoff, infiltration, percolation, storage, evaporation, and transpiration.

hydrous mica *See* fine-grained mica.

hydroxyapatite A member of the apatite group of minerals rich in hydroxyl groups. A nearly insoluble calcium phosphate.

hygroscopic coefficient The amount of moisture in a dry soil when it is in equilibrium with some standard relative humidity near a saturated atmosphere (about 98%), expressed in terms of percentage on the basis of oven-dry soil.

hyperthermic *See* soil temperature classes.

igneous rock Rock formed from the cooling and solidification of magma and that has not been changed appreciably since its formation.

illite *See* fine-grained mica.

illuvial horizon A soil layer or horizon in which material carried from an overlying layer has been precipitated from solution or deposited from suspension. The layer of accumulation.

immature soil A soil with indistinct or only slightly developed horizons because of the relatively short time it has been subjected to the various soil-forming processes. A soil that has not reached equilibrium with its environment.

immobilization The conversion of an element from the inorganic to the organic form in microbial tissues or in plant tissues, thus rendering the element not readily available to other organisms or to plants.

imogolite A poorly crystalline aluminosilicate mineral with an approximate formula $SiO_2Al_2O_32.5H_2O$; occurs mostly in soils formed from volcanic ash.

impervious Resistant to penetration by fluids or by roots.

Inceptisols *See* soil classification.

indurated (soil) Soil material cemented into a hard mass that will not soften on wetting. *See also* consistence; hardpan.

infiltration The downward entry of water into the soil.

infiltration rate A soil characteristic determining or describing the *maximum* rate at which water *can* enter the soil under specified conditions, including the presence of an excess of water.

inoculation The process of introducing pure or mixed cultures of microorganisms into natural or artificial culture media.

inorganic compounds All chemical compounds in nature except compounds of carbon other than carbon monoxide, carbon dioxide, and carbonates.

insecticide A chemical that kills insects.

intergrade A soil that possesses moderately well-developed distinguishing characteristics of two or more genetically related great soil groups.

interlayer (mineralogy) Materials between layers within a given crystal, including cations, hydrated cations, organic molecules, and hydroxide groups or sheets.

ions Atoms, groups of atoms, or compounds that are electrically charged as a result of the loss of electrons (cations) or the gain of electrons (anions).

iron-pan An indurated soil horizon in which iron oxide is the principal cementing agent.

irrigation efficiency The ratio of the water actually consumed by crops on an irrigated area to the amount of water diverted from the source onto the area.

irrigation methods Methods by which water is artificially applied to an area. The methods and the manner of applying the water are as follows.

   border-strip The water is applied at the upper end of a strip with earth borders to confine the water to the strip.

   center pivot Automated sprinkler irrigation achieved by automatically rotating the sprinkler pipe or boom, supplying water to the sprinkler heads or nozzles, as a radius from the center of the field to be irrigated.

   check-basin The water is applied rapidly to relatively level plots surrounded by levees. The basin is a small check.

   corrugation The water is applied to small, closely spaced furrows, frequently in grain and forage crops, to confine the flow of irrigation water to one direction.

   drip A planned irrigation system where all necessary facilities have been installed for the efficient application of water directly to the root zone of plants by means of applicators (orifices, emitters, porous tubing, perforated pipe, etc.) operated under low pressure. The applicators may be placed on or below the surface of the ground.

   flooding The water is released from field ditches and allowed to flood over the land.

   furrow The water is applied to row crops in ditches made by tillage implements.

   sprinkler The water is sprayed over the soil surface through nozzles from a pressure system.

   subirrigation The water is applied in open ditches or tile lines until the water table is raised sufficiently to wet the soil.

   wild-flooding The water is released at high points in the field and distribution is uncontrolled.

interstratification Mixing of silicate layers within the structural framework of a given silicate clay.

isomorphous substitution The replacement of one atom by another of similar size in a crystal lattice without disrupting or changing the crystal structure of the mineral.

isotopes Two or more atoms of the same element that have different atomic masses because of different numbers of neutrons in the nucleus.

joule The SI energy unit defined as a force of one newton applied over a distance of one meter; 1 joule = 0.239 calorie.

kame A conical hill or ridge of sand or gravel deposited in contact with glacial ice.

kaolinite An aluminosilicate mineral of the 1:1 crystal lattice group; that is, consisting of single silicon tetrahedral sheets alternating with single aluminium octahedral sheets.

labile Descriptive of a substance in soil that readily undergoes transformation or is readily available to plants.

lacustrine deposit Material deposited in lake water and later exposed either by lowering of the water level or by the elevation of the land.

land A broad term embodying the total natural environmental of the areas of the Earth not covered by water. In addition to soil, its attributes include other physical conditions such as mineral deposits and water supply; location in relation to centers of commerce, populations, and other land; the size of the individual tracts or holdings; and existing plant cover, works of improvement, and the like.

land capability classification A grouping of kinds of soil into special units, subclasses, and classes according to their capability for intensive use and the treatments required for sustained use. One such system has been prepared by the USDA Soil Conservation Service.

land classification The arrangement of land units into various categories based upon the properties of the land or its suitability for some particular purpose.

land-use planning The development of plans for the uses of land that, over long periods, will best serve the general welfare, together with the formulation of ways and means for achieving such uses.

laterite An iron-rich subsoil layer found in some highly weathered humid tropical soils that, when exposed and allowed to dry, becomes very hard and will not soften when rewetted. When erosion removes the overlying layers, the laterite is exposed and a virtual pavement results. *See also* plinthite.

layer (clay mineralogy) A combination in silicate clays of (tetrahedral and octahedral) sheets in a 1:1, 2:1, or 2:1:1 combination.

leaching The removal of materials in solution from the soil by percolating waters. *See also* eluviation.

leaf area index (LAI) Area of leaves per unit of land on which the plants are growing.

legume A pod-bearing member of the Leguminosae family, one of the most important and widely distributed plant families. Includes many valuable food and forage species, such as peas, beans, peanuts, clovers, alfalfas, sweet clovers, lespedezas, vetches, and kudzu. Nearly all legumes are associated with nitrogen-fixing organisms.

Liebig's law The growth and reproduction of an organism are determined by the nutrient substance (oxygen, carbon dioxide, calcium, etc.) that is available in minimum quantity, the *limiting factor*.

light soil (Obsolete in scientific use) A coarse-textured soil; a soil with a low drawbar pull and hence easy to cultivate. *See also* coarse texture; soil texture.

lignin The complex organic constituent of woody fibers in plant tissue that, along with cellulose, cements the cells together and provides strength. Lignins resist microbial attack and after some modification may become part of the soil organic matter.

lime (agricultural) In strict chemical terms, calcium oxide. In practical terms, a material containing the carbonates, oxides and/or hydroxides of calcium and/or magnesium used to neutralize soil and acidity.

lime requirement The mass of agricultural limestone, or the equivalent of other specified liming material, required to raise the pH of the soil to a desired value under field conditions.

limestone A sedimentary rock composed primarily of calcite ($CaCO_3$). If dolomite ($CaCO_3 \cdot MgCO_3$) is present in appreciable quantities, it is called a dolomitic limestone.

limiting factor *See* Liebig's law.

liquid limit (LL) *See* Atterberg limits.

loam The textural class name for soil having a moderate amount of sand, silt, and clay. Loam soils contain 7–27% clay, 28–50% silt, and 23–52% sand.

loamy Intermediate in texture and properties between fine-textured and coarse-textured soils. Includes all textural classes with the words loam or loamy as a part of the class name, such as clay loam or loamy sand. *See also* loam; soil texture.

loess Material transported and deposited by wind and consisting of predominantly silt-sized particles.

luxury consumption The intake by a plant of an essential nutrient in amounts exceeding what it needs. For example, if potassium is abundant in the soil, alfalfa may take in more than it requires.

lysimeter A device for measuring percolation and leaching and evapotranspiration losses from a column of soil under controlled conditions.

macronutrient A chemical element necessary in large amounts (usually 50 mg/kg in the plant) for the growth of plants. Includes C, H, O, N, P, K, Ca, Mg, and S. ("Macro" refers to quantity and not to the essentiality of the element.) *See also* micronutrient.

marl Soft and unconsolidated calcium carbonate, usually mixed with varying amounts of clay or other impurities.

marsh Periodically wet or continually flooded area with the surface not deeply submerged. Covered dominantly with sedges, cattails, rushes, or other hydrophytic plants. Subclasses include freshwater and saltwater marshes.

matric potential *See* soil water potential.

mature soil A soil with well-developed soil horizons produced by the natural processes of soil formation and essentially in equilibrium with its present environment.

maximum water-holding capacity The average moisture content of a disturbed sample of soil, 1 cm high, which is at equilibrium with a water table at its lower surface.

mechanical analysis (Obsolete) *See* particle-size analysis; particle-size distribution.

medium texture Intermediate between fine-textured and coarse-textured (soils). It includes the following textural classes: very fine sandy loam, loam, silt loam, and silt.

mellow soil A very soft, very friable, porous soil without any tendency toward hardness or harshness. *See also* consistence.

mesic *See* soil temperature classes.

metamorphic rock A rock that has been greatly altered from its previous condition through the combined action of heat and pressure. For example, marble is a metamorphic rock produced from limestone, gneiss is produced from granite, and slate from shale.

methane, CH_4 An odorless, colorless gas commonly produced under anaerobic conditions. When released to the upper atmosphere, methane contributes to global warming. *See also* greenhouse effect.

micas Primary aluminosilicate minerals in which two silica tetrahedral sheets alternate with one alumina/magnesia octahedral sheet with entrapped potassium atoms fitting between sheets. They separate readily into thin sheets or flakes.

microfauna That part of the animal population which consists of individuals too small to be clearly distinguished without the use of a microscope. Includes protozoans and nematodes.

microflora That part of the plant population which consists of individuals too small to be clearly distinguished without the use of a microscope. Includes actinomycetes, algae, bacteria, and fungi.

micronutrient A chemical element necessary in only extremely small amounts (<50 mg/kg in the plant) for the growth of plants. Examples are B, Cl, Cu, Fe, Mn, and Zn. ("Micro" refers to the amount used rather than to its essentiality.) *See also* macronutrient.

microrelief Small-scale local differences in topography, including mounds, swales, or pits that are only a meter or so in diameter and with elevation differences of up to 2 m. *See also* gilgai.

mineralization The conversion of an element from an organic form to an inorganic state as a result of microbial decomposition.

mineral soil A soil consisting predominantly of, and having its properties determined predominantly by, mineral matter. Usually contains <20% organic matter, but may contain an organic surface layer up to 30 cm thick.

minimum tillage *See* tillage, conservation.

minor element (Obsolete) *See* micronutrient.

moderately coarse texture Consisting predominantly of coarse particles. In soil textural classification, it includes all the sandy loams except the very fine sandy loam. *See also* coarse texture.

moderately fine texture Consisting predominantly of intermediate-sized (soil) particles or with relatively small amounts of fine or coarse particles. In soil textural classification, it includes clay loam, sandy loam, sandy clay loam, and silty clay loam. *See also* fine texture.

moisture equivalent The weight percentage of water retained by a previously saturated sample of soil 1 cm in thickness after it has been subjected to a centrifugal force of 1000 times gravity for 30 min.

moisture potential *See* soil water potential.

mole drain Unlined drain formed by pulling a bullet-shaped cylinder through the soil.

mollic epipedon *See* diagnostic surface horizons.

Mollisols *See* soil classification.

montmorillonite An aluminosilicate clay mineral in the smectite group with a 2:1 expanding crystal lattice, with two silicon tetrahedral sheets enclosing an aluminum octahedral sheet. Isomorphous substitution of magnesium for some of the aluminum has occurred in the octahedral sheet. Considerable expansion may be caused by water moving between silica sheets of contiguous layers.

mor Raw humus; type of forest humus layer of unincorporated organic material, usually matted or compacted or both; distinct from the mineral soil, unless the latter has been blackened by washing in organic matter.

moraine An accumulation of drift, with an initial topographic expression of its own, built within a glaciated region chiefly by the direct action of glacial ice. Examples are ground, lateral, recessional, and terminal moraines.

morphology, soil The constitution of the soil including the texture, structure, consistence, color, and other physical, chemical, and biological properties of the various soil horizons that make up the soil profile.

mottling Spots or blotches of different color or shades of color interspersed with the dominant color.

mucigel The gelatinous material at the surface of roots grown in unsterilized soil.

muck Highly decomposed organic material in which the original plant parts are not recognizable. Contains more mineral matter and is usually darker in color than peat. *See also* muck soil; peat.

muck soil (1) A soil containing 20–50% organic matter. (2) An organic soil in which the organic matter is well decomposed.

mulch Any material such as straw, sawdust, leaves, plastic film, and loose soil that is spread upon the surface of the soil to protect the soil and plant roots from the effects of raindrops, soil crusting, freezing, evaporation, etc.

mulch tillage *See* tillage, conservation.

mull A humus-rich layer of forested soils consisting of mixed organic and mineral matter. A mull blends into the upper mineral layers without an abrupt change in soil characteristics.

mycorrhiza The association, usually symbiotic, of fungi with the roots of seed plants. *See also* ectotrophic mycorrhiza; endotrophic mycorrhiza; vesicular-arbuscular mycorrhiza.

natric horizon *See* diagnostic subsurface horizon.

necrosis Death associated with discoloration and dehydration of all or parts of plant organs, such as leaves.

nematodes Very small worms abundant in many soils and important because some of them attack and destroy plant roots.

neutral soil A soil in which the surface layer, at least to normal plow depth, is neither acid nor alkaline in reaction. In practice this means the soil is within the pH range of 6.6–7.3. *See also* acid soil; alkaline soil; pH; reaction, soil.

nitrification The biochemical oxidation of ammonium to nitrate, predominantly by autotrophic bacteria.

nitrogen assimilation The incorporation of nitrogen into organic cell substances by living organisms.

nitrogen cycle The sequence of chemical and biological changes undergone by nitrogen as it moves from the atmosphere into water, soil, and living organisms, and upon death of these organisms (plants and animals) is recycled through a part or all of the entire process.

nitrogen fixation The biological conversion of elemental nitrogen (N_2) to organic combinations or to forms readily utilize in biological processes.

nodule bacteria *See* rhizobia.

no tillage *See* tillage, conservation.

nucleic acids Complex compounds found in plant and animal cells; may be combined with proteins as nucleoproteins.

O horizon Organic horizon of mineral soils.

ochric epipedon *See* diagnostic surface horizons.

order, soil *See* soil classification.

organic soil A soil that contains at least 20% organic matter (by weight) if the clay content is low and at least 30% if the clay content is as high as 60%.

organic fertilizer By-product from the processing of animal or vegetable substances that contain sufficient plant nutrients to be of value as fertilizers.

organic soil materials (As used in *Soil Taxonomy* in the United States): (1) Saturated with water for prolonged periods unless artificially drained and having

18% or more organic carbon (by weight) if the mineral fraction is more than 60% clay, more than 12% organic carbon if the mineral fraction has no clay, or between 12 and 18% carbon if the clay content of the mineral fraction is between 0 and 60%. (2) Never saturated with water for more than a few days and having more than 20% organic carbon. Histosols develop on these organic soil materials. There are three kinds of organic materials.

fibric materials The least decomposed of all the organic soil materials, containing very high amounts of fiber that are well preserved and readily identifiable as to botanical origin.

hemic materials Intermediate in degree of decomposition of organic materials between the less decomposed fibric and the more decomposed sapric materials.

sapric materials The most highly decomposed of the organic materials, having the highest bulk density, least amount of plant fiber, and lowest water content at saturation.

ortstein An indurated layer in the B horizon of Spodosols in which the cementing material consists of illuviated sesquioxides (mostly iron) and organic matter.

osmotic pressure Pressure exerted in living bodies as a result of unequal concentrations of salts on both sides of a cell wall or membrane. Water moves from the area having the lower salt concentration through the membrane into the area having the higher salt concentration and, therefore, exerts additional pressure on the side with higher salt concentration.

osmotic potential *See* soil water potential.

outwash plain **A** deposit of coarse-textured materials (e.g., sands, gravels) left by streams of melt water flowing from receding glaciers.

oven-dry soil Soil that has been dried at 105 °C until it reaches constant weight.

oxic horizon *See* diagnostic subsurface horizon.

oxidation ditch An artificial open channel for partial digestion of liquid organic wastes in which the wastes are circulated and aerated by a mechanical device.

Oxisols *See* soil classification.

pans Horizons or layers, in soils, that are strongly compacted, indurated, or very high in clay content. *See also* caliche; claypan; fragipan; hardpan.

parent material The unconsolidated and more or less chemically weathered mineral or organic matter from which the solum of soils is developed by pedogenic processes.

particle density The mass per unit volume of the soil particles. In technical work, usually expressed as metric tons per cubic meter (Mg/m^3) grams per cubic centimeter (g/cm^3.)

particle size The effective diameter of a particle measured by sedimentation, sieving, or micro-metric methods.

particle-size analysis Determination of the various amounts of the different separates in a soil sample, usually by sedimentation, sieving, micrometry, or combinations of these methods.

particle-size distribution The amounts of the various soil separates in a soil sample, usually expressed as weight percentages.

pascal An SI unit of pressure equal to one newton per square meter.

peat Unconsolidated soil material consisting largely of undecomposed, or only slightly decomposed, organic matter accumulated under conditions of excessive moisture. *See also* organic soil materials; peat soil.

peat soil An organic soil containing more than 50% organic matter. Used in the United States to refer to the stage of decomposition of the organic matter, "peat" referring to the slightly decomposed or undecomposed deposits and "muck" to the highly decomposed materials. *See also* muck; muck soil; peat.

ped A unit of soil structure such as an aggregate, crumb, prism, block, or granule, formed by natural processes (in contrast to a clod, which is formed artificially).

pedon The smallest volume that can be called "a soil." It has three dimensions. It extends downward to the depth of plant roots or to the lower limit of the genetic soil horizons. Its lateral cross section is roughly hexagonal and ranges from 1 to 10 m^2 in size depending on the variability in the horizons.

peneplain A once high, rugged area that has been reduced by erosion to a lower, gently rolling surface resembling a plain.

penetrability The ease with which a probe can be pushed into the soil. May be expressed in units of distance, speed, force, or work depending on the type of penetrometer used.

percolation, soil water The downward movement of water through soil. Especially, the downward flow of water in saturated or nearly saturated soil at hydraulic gradients of the order of 1.0 or less.

permafrost (1) Permanently frozen material underlying the solum. (2) A perennially frozen soil horizon.

permanent charge *See* constant charge.

permanent wilting point *See* wilting point.

permeability, soil The ease with which gases, liquids, or plant roots penetrate or pass through a bulk mass of soil or a layer of soil.

petrocalcic horizon *See* diagnostic subsurface horizon.

petrogypsic horizon *See* diagnostic subsurface horizon.

pH, soil The negative logarithm of the hydrogen ion activity (concentration) of a soil. The degree of acidity (or alkalinity) of a soil as determined by means of a glass, quinhydrone, or other suitable electrode or indicator at a specified moisture content or soil/water ratio, and expressed in terms of the pH scale.

pH-dependent charge That portion of the total charge of the soil particles that is affected by, and varies with, changes in pH.

phase, soil A subdivision of a soil series or other unit of classification having characteristics that affect the use and management of the soil but do not vary sufficiently to differentiate it as a separate series. Included are such characteristics as degree of slope, degree of erosion, and content of stones.

photomap A mosaic map made from aerial photographs to which place names, marginal data, and other map information have been added.

phyllosphere The leaf surface.

physical properties (of soils) Those characteristics, processes, or reactions of a soil that are caused by physical forces and that can be described by, or expressed in, physical terms or equations. Examples of physical properties are bulk density, water-holding capacity, hydraulic conductivity, porosity, pore-size distribution, and so on.

placic horizon See diagnostic subsurface horizons.

plaggen epipedon See diagnostic surface horizons.

plant nutrients See essential elements.

plastic limit (PL) See Atterberg limits.

plastic soil A soil capable of being molded or deformed continuously and permanently, by relatively moderate pressure, into various shapes. See also consistence.

platy Consisting of soil aggregates that are developed predominantly along the horizontal axes; laminated; flaky.

plinthite (brick) A highly weathered mixture of sesquioxides or iron and aluminum with quartz and other diluents that occurs as red mottles and that changes irreversibly to hardpan upon alternate wetting and drying.

plow layer The soil ordinarily moved when land is plowed; equivalent to surface soil.

plow pan A subsurface soil layer having a higher bulk density and lower total porosity than layers above or below it, as a result of pressure applied by normal plowing and other tillage operations.

plow-plant See tillage, conservation.

plowing A primary broad-base tillage operation that is performed to shatter soil uniformly with partial to complete inversion.

polypedon (As used in Soil Taxonomy) Two or more contiguous pedons, all of which are within the defined limits of a single soil series; commonly referred to as a soil individual.

pore size distribution The volume of the various sizes of pores in a soil. Expressed as percentages of the bulk volume (soil plus pore space).

porosity, soil The volume percentage of the total soil bulk not occupied by solid particles.

primary mineral A mineral that has not been altered chemically since deposition and crystallization from molten lava.

primary tillage See tillage, primary.

prismatic soil structure A soil structure type with prism-like aggregates that have a vertical axis much longer than the horizontal axes.

productivity, soil The capacity of a soil for producing a specified plant or sequence of plants under a specified system of management. Productivity emphasizes the capacity of soil to produce crops and should be expressed in terms of yields.

profile, soil A vertical section of the soil through all its horizons and extending into the parent material.

protein Any of a group of nitrogen-containing organic compounds formed by the polymerization of a large number of amino acid molecules and that, upon hydrolysis, yield these amino acids. They are essential parts of living matter and are one of the essential food substances of animals.

puddled soil Dense, massive soil artificially compacted when wet and having no aggregated structure. The condition commonly results from the tillage of a clayey soil when it is wet.

rain, acid See acid rain.

reaction, soil The degree of acidity or alkalinity of a soil, usually expressed as a pH value.

Extremely acid	< 4.5
Very strongly acid	4.5–5.0
Strongly acid	5.1–5.5
Medium acid	5.6–6.0
Slightly acid	6.1–6.5
Neutral	6.6–7.3
Mildly alkaline	7.4–7.8
Moderately alkaline	7.9–8.4
Strongly alkaline	8.5–9.0
Very strongly alkaline	9.1 and higher

regolith The unconsolidated mantle of weathered rock and soil material on the earth's surface; loose earth materials above solid rock. (Approximately equivalent to the term "soil" as used by many engineers.)

residual material Unconsolidated and partly weathered mineral materials accumulated by disintegration of consolidated rock in place.

rhizobia Bacteria capable of living symbiotically with higher plants, usually in nodules on the roots of legumes, from which they receive their energy, and capable of converting atmospheric nitrogen to combined organic forms; hence, the term symbiotic nitrogen-fixing bacteria. (Derived from the generic name Rhizobium.)

rhizosphere That portion of the soil in the immediate vicinity of plant roots in which the abundance and composition of the microbial population are influenced by the presence of roots.

rill A small, intermittent water course with steep sides; usually only a few centimeters deep and hence no obstacle to tillage operations.

rill erosion See erosion.

riprap Broken rock, cobbles or boulders placed on earth surfaces, such as the face of a dam or the bank of a stream, for protection against the action of water (waves); also applied to brush or pole mattresses, or brush and stone, or other similar materials used for soil erosion control.

rock The material that forms the essential part of the Earth's solid crust, including loose incoherent masses such as sand and gravel, as well as solid masses of granite and limestone.

rotary tillage See tillage, rotary.

runoff The portion of the precipitation on an area that is discharged from the area through stream channels. That which is lost without entering the soil is called *surface runoff* and that which enters the soil before reaching the stream is called *groundwater runoff* or *seepage flow* from groundwater. (In soil science "runoff" usually refers to the water lost by surface flow; in geology and hydraulics "runoff" usually includes both surface and subsurface flow.)

salic horizon *See* diagnostic subsurface horizons.

saline-sodic soil A soil containing sufficient exchangeable sodium to interfere with the growth of most crop plants and containing appreciable quantities of soluble salts. The exchangeable sodium adsorption ratio is >13, the conductivity of the saturation extract is >4 dS/m (at 25°C), and the pH is usually 8.5 or less in the saturated soil.

saline soil A nonsodic soil containing sufficient soluble salts to impair its productivity. The conductivity of a saturated extract is >4 dS/m, the exchangeable sodium adsorption ratio is less than about 15, and the pH is <8.

salinization The process of accumulation of salts in soil.

saltation Particle movement in water or wind where particles skip or bounce along the stream bed or soil surface.

sand A soil particle between 0.05 and 2.0 mm in diameter; a soil textural class.

sapric materials *See* organic soil materials.

saturation extract The solution extracted from a saturated soil paste.

saturation percentage The water content of a saturated soil paste, expressed as a dry weight percentage.

savanna (savannah) A grassland with scattered trees, either as individuals or clumps. Often a transitional type between true grassland and forest.

secondary mineral A mineral resulting from the decomposition of a primary mineral or from the reprecipitation of the products of decomposition of a primary mineral. *See also* primary mineral.

second bottom The first terrace above the normal floodplain of a stream.

sedimentary rock A rock formed from materials deposited from suspension or precipitated from solution and usually being more or less consolidated. The principal sedimentary rocks are sandstones, shales, limestones, and conglomerates.

seedbed The soil prepared to promote the germination of seed and the growth of seedlings.

self-mulching soil A soil in which the surface layer becomes so well aggregated that it does not crust and seal under the impact of rain but instead serves as a surface mulch upon drying.

semiarid Term applied to regions or climates where moisture is more plentiful than in arid regions but still definitely limits the growth of most crop plants. Natural vegetation in uncultivated areas is short grasses.

separate, soil One of the individual-sized groups of mineral soil particles—sand, silt, or clay.

septic tank An underground tank used in the deposition of domestic wastes. Organic matter decomposes in the tank, and the effluent is drained into the surrounding soil.

series, soil *See* soil classification.

sewage sludge Settled sewage solids combined with varying amounts of water and dissolved materials, removed from sewage by screening, sedimentation, chemical precipitation, or bacterial digestion.

shear Force, as of a tillage implement, acting at right angles to the direction of movement.

sheet (mineralogy) A flat array of more than one atomic thickness and composed of one or more levels of linked coordination polyhedra. A sheet is thicker than a plane and thinner than a layer. Example: tetrahedral sheet, octahedral sheet.

sheet erosion *See* erosion.

shelterbelt A wind barrier of living trees and shrubs established and maintained for protection of farm fields. Syn. windbreak.

shifting cultivation A farming system in which land is cleared, the debris burned, and crops grown for 2–3 years. When the farmer moves on to another plot, the land is then left idle for 5–15 years; then the burning and planting process is repeated.

side dressing The application of fertilizer alongside row crop plants, usually on the soil surface. Nitrogen materials are most commonly side-dressed.

silica/alumina ratio The molecules of silicon dioxide (SiO_2) per molecule of aluminum oxide (Al_2O_3) in clay minerals or in soils.

silica/sesquioxide ratio The molecules of silicon dioxide (SiO_2) per molecule of aluminum oxide (Al_2O_3) plus ferric oxide (Fe_2O_3) in clay minerals or in soils.

silt (1) A soil separate consisting of particles between 0.05 and 0.002 mm in equivalent diameter. (2) A soil textural class.

silting The deposition of water-borne sediments in stream channels, lakes, reservoirs, or on floodplains, usually resulting from a decrease in the velocity of the water.

site index A quantitative evaluation of the productivity of a soil for forest growth under the existing or specified environment.

slag A product of smelting, containing mostly silicates; the substances not sought to be produced as matte or metal and having a lower specific gravity.

slash and burn *See* shifting cultivation.

slick spots Small areas in a field that are slick when wet because of a high content of alkali or exchangeable sodium.

slope The degree of deviation of a surface from horizontal, measured in a numerical ratio, percent, or degrees.

smectite A group of silicate clays having a 2:1 type lattice structure with sufficient isomorphous substitution in either or both the tetrahedral and octahedral sheets to give a high interlayer negative charge and high cation exchange capacity and to permit significant interlayer expansion and consequent shrinking and swelling of the clay. Montmorillonite, beidellite, and saponite are in the smectite group.

sodic soil A soil that contains sufficient sodium to interfere with the growth of most crop plants, and in which the sodium adsorption ratio is 13 or greater.

sodium adsorption ratio (SAR)

$$SAR = \frac{[Na^+]}{\sqrt{\frac{1}{2}([Ca^{2+}] + [Mg^{2+}])}}$$

where the cation concentrations are in millimoles per liter (mmole/L).

soil (1) A dynamic natural body composed of mineral and organic materials and living forms in which plants grow. (2) The collection of natural bodies occupying parts of the earth's surface that support plants and that have properties due to the integrated effect of climate and living matter acting upon parent material, as conditioned by relief, over periods of time.

soil air The soil atmosphere; the gaseous phase of the soil, being that volume not occupied by soil or liquid.

soil alkalinity The degree or intensity of alkalinity of a soil, expressed by a value >7.0 on the pH scale.

soil amendment Any material, such as lime, gypsum, sawdust, or synthetic conditioner, that is worked into the soil to make it more amenable to plant growth.

soil association A group of defined and named taxonomic soil units occurring together in an individual and characteristic pattern over a geographic region, comparable to plant associations in many ways.

soil classification (*Soil Taxonomy*) The systematic arrangement of soils into groups or categories on the basis of their characteristics.

  order The category at the highest level of generalization in the soil classification system. The properties selected to distinguish the orders are reflections of the degree of horizon development and the kinds of horizons present. The eleven orders are as follows.

    Andisols Soils developed from volcanic ejecta. The colloidal fraction is dominated by allophane and/or Al-humus compounds.

    Alfisols Soils with gray to brown surface horizons, medium to high supply of bases, and B horizons of illuvial clay accumulation. These soils form mostly under forest or savanna vegetation in climates with slight to pronounced seasonal moisture deficit.

    Aridisols Soils of dry climates. They have pedogenic horizons, low in organic matter, that are never moist as long as 3 consecutive months. They have an ochric epipedon and one or more of the following diagnostic horizons: argillic, natric, cambic, calcic, petrocalcic, gypsic, salic, or a duripan.

    Entisols Soils have no diagnostic pedogenic horizons. They may be found in virtually any climate on very recent geomorphic surfaces.

    Histosols Soils formed from materials high in organic matter. Histosols with essentially no clay must have at least 20% organic matter by weight (about 78% by volume). This minimum organic matter content rises with increasing clay content to 30% (85% by volume) in soils with at least 60% clay.

    Inceptisols Soils that are usually moist with pedogenic horizons of alteration of parent materials but not of illuviation. Generally, the direction of soil development is not yet evident from the marks left by various soil-forming processes or the marks are too weak to classify in another order.

    Mollisols Soils with nearly black, organic-rich surface horizons and high supply of bases. They have mollic epipedons and base saturation greater than 50% in any cambic or argillic horizon. They lack the characteristics of Vertisols and must not have oxic or spodic horizons.

    Oxisols Soils with residual accumulations of low-activity clays, free oxides, kaolin, and quartz. They are mostly in tropical climates.

    Spodosols Soils with subsurface illuvial accumulations of organic matter and compounds of aluminum and usually iron. These soils are formed in acid, mainly coarse-textured materials in humid and mostly cool or temperate climates.

    Ultisols Soils that are low in bases and have subsurface horizons of illuvial clay accumulations. They are usually moist, but during the warm season of the year some are dry part of the time.

    Vertisols Clayey soils with high shrink-swell potential that have wide, deep cracks when dry. Most of these soils have distinct wet and dry periods throughout the year.

  suborder This category narrows the ranges in soil moisture and temperature regimes, kinds of horizons, and composition, according to which of these is most important.

  great group The classes in this category contain soils that have the same kind of horizons in the same sequence and have similar moisture and temperature regimes.

  subgroup The great groups are subdivided into central concept subgroups that show the central properties of the great group, intergrade subgroups that show properties of more than one great group, and other subgroups for soils with atypical properties that are not characteristic of any great group.

  family Families are defined largely on the basis of physical and mineralogic properties of importance to plant growth.

  series The soil series is a subdivision of a family and consists of soils that are similar in all major profile characteristics.

soil complex A mapping unit used in detailed soil surveys where two or more defined taxonomic units are so intimately intermixed geographically that it is undesirable or impractical, because of the scale being used, to separate them. A more intimate mixing of smaller areas of individual taxonomic units than that described under *soil association*.

soil conditioner Any material added to a soil for the purpose of improving its physical condition.

soil conservation A combination of all management and land-use methods that safeguard the soil against depletion or deterioration caused by nature and/or humans.

soil correlation The process of defining, mapping, naming, and classifying the kinds of soils in a specific soil survey area, the purpose being to ensure that soils are adequately defined, accurately mapped, and uniformly named.

soil erosion *See* erosion.

soil fertility *See* fertility, soil.

soil genesis The mode of origin of the soil, with special reference to the processes or soil-forming factors responsible for the development of the solum, or true soil, from the unconsolidated parent material.

soil geography A subspecialization of physical geography concerned with the areal distributions of soil types.

soil horizon *See* horizon, soil.

soil management The sum total of all tillage operations, cropping practices, fertilizer, lime, and other treatments conducted on or applied to a soil for the production of plants.

soil map A map showing the distribution of soil types or other soil mapping units in relation to the prominent physical and cultural features of the Earth's surface.

soil mechanics and engineering A subspecialization of soil science concerned with the effect of forces on the soil and the application of engineering principles to problems involving the soil.

soil moisture potential *See* soil water potential.

soil monolith A vertical section of a soil profile removed from the soil and mounted for display or study.

soil morphology The physical constitution, particularly the structural properties, of a soil profile as exhibited by the kinds, thicknesses, and arrangement of the horizons in the profile, and by the texture, structure, consistence, and porosity of each horizon.

soil profile A vertical section of the soil from the surface through all its horizons, including C horizons. *See also* horizon, soil.

soil organic matter The organic fraction of the soil that includes plant and animal residues at various stages of decomposition, cells and tissues of soil organisms, and substances synthesized by the soil population. Commonly determined as the amount of organic material contained in a soil sample passed through a 2 mm sieve.

soil porosity *See* porosity, soil.

soil productivity *See* productivity, soil.

soil reaction *See* reaction, soil; pH, soil.

soil salinity The amount of soluble salts in a soil, expressed in terms of percentage, milligrams per kilogram, parts per million (ppm), or other convenient ratios.

soil separates *See* separate, soil.

soil series *See* soil classification.

soil solution The aqueous liquid phase of the soil and its solutes consisting of ions dissociated from the surfaces of the soil particles and of other soluble materials.

soil structure The combination or arrangement of primary soil particles into secondary particles, units, or peds. These secondary units may be, but usually are not, arranged in the profile in such a manner as to give a distinctive characteristic pattern. The secondary units are characterized and classified on the basis of size, shape, and degree of distinctness into classes, types, and grades, respectively.

soil structure classes A grouping of soil structural units or peds on the basis of size from the very fine to very coarse.

soil structure grades A grouping or classification of soil structure on the basis of inter- and intra-aggregate adhesion, cohesion, or stability within the profile. Four grades of structure, designated from 0 to 3, are recognized.
0: *structureless*—no observable aggregation.
1: *weakly* durable peds.
2: *moderately* durable peds.
3: *strong*, durable peds.

soil structure types A classification of soil structure based on the shape of the aggregates or peds and their arrangement in the profile, including platy, prismatic, columnar, blocky, subangular blocky, granulated, and crumb.

soil survey The systematic examination, description, classification, and mapping of soils in an area. Soil surveys are classified according to the kind and intensity of field examination.

soil temperature classes (*Soil Taxonomy*) Classes are based on mean annual soil temperature and on differences between summer and winter temperatures at a depth of 50 cm.
  1. Soils with 5 °C and greater difference between summer and winter temperatures are classed on the basis of mean annual temperatures.
     frigid: <8 °C mean annual temperature.
     mesic: 8–15 °C mean annual temperature.
     thermic: 15–22 °C mean annual temperature.
     hyperthermic: >22 °C mean annual temperature.
  2. Soils with <5 °C difference between summer and winter temperatures are classed on the basis of mean annual temperatures.
     isofrigid: <8 °C mean annual temperature.
     isomesic: 8–15 °C mean annual temperature.
     isothermic: 15–22 °C mean annual temperature.
     isohyperthermic: <22 °C mean annual temperature.

soil textural class A grouping of soil textural units based on the relative proportions of the various soil separates (sand, silt, and clay). These textural classes, listed from the coarsest to the finest in texture, are sand, loamy sand, sandy loam, loam, silt loam, silt, sandy clay loam, clay loam, silty clay loam, sandy clay, silty clay, and clay. There are several subclasses of the sand, loamy sand, and sandy loam classes based on the dominant particle size of the sand fraction (e.g., loamy fine sand, coarse sandy loam).

soil texture The relative proportions of the various soil separates in a soil.

soil water potential (total) A measure of the difference between the free energy state of soil water and that of pure water. Technically it is defined as "that amount of work that must be done per unit quantity of pure water in order to transport reversibly and isothermically an infinitesimal quantity of water from a pool of pure water, at a specified elevation and at atmospheric pressure to the soil water (at the point under consideration)." This *total* potential consists of the following potentials.

> matric potential That portion of the total soil water potential due to the attractive forces between water and soil solids as represented through adsorption and capillarity. It will always be negative.
>
> osmotic potential That portion of the total soil water potential due to the presence of solutes in soil water. It will generally be negative.
>
> gravitational potential That portion of the total soil water potential due to differences in elevation of the reference pool of pure water and that of the soil water. Since the soil water elevation is usually chosen to be higher than that of the reference pool, the gravitational potential is usually positive.

solum (plural sola) The upper and most weathered part of the soil profile; the A, E, and B horizons

specific surface The solid particle surface area per unit mass or volume of the solid particles.

splash erosion *See* erosion.

spodic horizon *See* diagnostic subsurface horizons.

Spodosols *See* soil classification.

sprinkler irrigation *See* irrigation methods.

stratified Arranged in or composed of strata or layers.

strip cropping The practice of growing crops that require different types of tillage, such as row and sod, in alternate strips along contours or across the prevailing direction of wind.

structure, soil *See* soil structure.

stubble mulch The stubble of crops or crop residues left essentially in place on the land as a surface cover before and during the preparation of the seedbed and at least partly during the growing of a succeeding crop.

subirrigation *See* irrigation methods.

subsoil That part of the soil below the plow layer.

subsoiling Breaking of compact subsoils, without inverting them, with a special knife-like instrument (chisel), which is pulled through the soil at depths usually of 30–60 cm and at spacings usually of 1–2 m.

summer fallow A cropping system that involves management of uncropped land during the summer to control weeds and store moisture in the soil for the growth of a later crop.

surface runoff *See* runoff.

surface soil The uppermost part of the soil, ordinarily moved in tillage, or its equivalent in uncultivated soils. Ranges in depth from 7 to 25 cm. Frequently designated as the "plow layer," the "Ap layer," or the "Ap horizon."

symbiosis The living together in intimate association of two dissimilar organisms, the cohabitation being mutually beneficial.

talus Fragments of rock and other soil material accumulated by gravity at the foot of cliffs or steep slopes.

taxonomy, soil The science of classification of soils; laws and principles governing the classifying the soil. *See also* soil classification.

tensiometer A device for measuring the negative pressure (or tension) of water in soil *in situ*; a porous, permeable ceramic cup connected through a tube to a manometer or vacuum gauge.

tension, soil-moisture *See* soil water potential.

terrace (1) A level, usually narrow, plain bordering a river, lake, or the sea. Rivers sometimes are bordered by terraces at different levels. (2) A raised, more or less level or horizontal strip of earth usually constructed on or nearly on a contour and designed to make the land suitable for tillage and to prevent accelerated erosion by diverting water from undesirable channels of concentration; sometimes called diversion terrace.

texture *See* soil texture.

thermal analysis (differential thermal analysis) A method of analyzing a soil sample for constituents, based on a differential rate of heating of the unknown and standard samples when a uniform source of heat is applied.

thermic *See* soil temperature classes.

thermophilic organisms Organisms that grow readily at temperatures above 45 °C.

tile, drain Pipe made of burned clay, concrete, or ceramic material, in short lengths, usually laid with open joints to collect and carry excess water from the soil.

till (1) Unstratified glacial drift deposited directly by the ice and consisting of clay, sand, gravel, and boulders intermingled in any proportion. (2) To plow and prepare for seeding; to seed or cultivate the soil.

tillage The mechanical manipulation of soil for any purpose; but in agriculture it is usually restricted to the modifying of soil conditions for crop production.

tillage, conservation Any tillage sequence that reduces loss of soil or water relative to conventional tillage, including the following systems.

> minimum tillage The minimum soil manipulation necessary for crop production or meeting tillage requirements under the existing soil and climatic conditions.
>
> mulch tillage Tillage or preparation of the soil in such a way that plant residues or other materials are left to cover the surface; also called *mulch farming, trash farming, stubble mulch tillage, plowless farming*.
>
> no-tillage system A procedure whereby a crop is planted directly into a seedbed not tilled since harvest of the previous crop; also zero tillage.
>
> plow-planting The plowing and planting of land in a single trip over the field by drawing both plowing and planting tools with the same power source.
>
> ridge till Planting on ridges formed by cultivation during the previous growing period.

sod planting A method of planting in sod with little or no tillage.

strip till Planting is done in a narrow strip that has been tilled and mixed, leaving the remainder of the soil surface undisturbed.

subsurface tillage Tillage with a special sweep-like plow or blade that is drawn beneath the surface, cutting plant roots and loosening the soil without inverting it or without incorporating residues of the surface cover.

wheel track planting A practice of planting in which the seed is planted in tracks formed by wheels rolling immediately ahead of the planter.

tillage, conventional The combined primary and secondary tillage operations normally performed in preparing a seedbed for a given crop grown in a given geographic area.

tillage, primary Tillage that contributes to the major soil manipulation, commonly with a plow.

tillage, rotary An operation using a power-driven rotary tillage tool to loosen and mix soil.

tillage, secondary Any tillage operations following primary tillage designed to prepare a satisfactory seedbed for planting.

tilth The physical condition of soil as related to its ease of tillage, fitness as a seedbed, and its impedance to seedling emergence and root penetration.

top dressing An application of fertilizer to a soil after the crop stand has been established.

toposequence A sequence of related soils that differ, one from the other, primarily because of *topography* as a soil formation factor.

topsoil (1) The layer of soil moved in cultivation. *See also* surface soil. (2) Presumably fertile soil material used to top-dress roadbanks, gardens, and lawns.

trace element (Obsolete) *See* micronutrient.

truncated Having lost all or part of the upper soil horizon or horizons.

tuff Volcanic ash usually more or less stratified and in various states of consolidation.

tundra A level or undulating treeless plain characteristic of arctic regions.

type, soil *See* soil type.

Ultisols *See* soil classification.

umbric epipedon *See* diagnostic surface horizons.

universal soil loss equation (USLE) An equation for predicting the average annual soil loss per unit area per year, $A = RKLSPC$, where R is the climatic erosivity factor (rainfall plus runoff), K is the soil erodibility factor, L is the length of slope, S is the percent slope, C is the cropping and management factor, and P is the soil erosion practice factor.

unsaturated flow The movement of water in a soil that is not filled to capacity with water.

varnish, desert A glossy sheen or coating on stones and gravel in arid regions.

vermiculite A 2:1-type silicate clay usually formed from mica that has a high net negative charge stemming mostly from extensive isomorphous substitution of aluminum for silicon in the tetrahedral sheet.

Vertisols *See* soil classification.

vesicular arbuscular mycorrhiza A common endomycorrhizal association produced by phycomycetous fungi of the genus *Endogone* and characterized by the development of two types of fungal structures: (a) within root cells small structures known as arbuscles and (b) between root cells storage organs known as vesicles. Host range includes many agricultural and horticultural crops. *See also* endomycorrhiza.

virgin soil A soil that has not been significantly disturbed from its natural environment.

waterlogged Saturated with water.

water potential, soil *See* soil water potential.

water-stable aggregate A soil aggregate stable to the action of water such as falling drops, or agitation as in wet-sieving analysis.

water table The upper surface of groundwater or that level below which the soil is saturated with water.

water table, perched The surface of a local zone of saturation held above the main body of groundwater by an impermeable layer of stratum, usually clay, and separated from the main body of groundwater by an unsaturated zone.

water use efficiency Dry matter or harvested portion of crop produced per unit of water consumed.

weathering All physical and chemical changes produced in rocks, at or near the Earth's surface, by atmospheric agents.

wilting point (permanent wilting point) The moisture content of soil, on an oven-dry basis, at which plants wilt and fail to recover their turgidity when placed in a dark humid atmosphere.

windbreak Planting of trees, shrubs, or other vegetation perpendicular, or nearly so, to the principal wind direction to protect soils, crops, homesteads, etc., from wind and snow.

xerophytes Plants that grow in or on extremely dry soils or soil materials.

zero tillage *See* tillage, conservation.

zeta potential *See* electrokinetic potential.

Index

O

Periodic Table of the Elements

Based on $^{12}_{6}C$. Numbers in parentheses are the mass numbers of the most stable isotopes of radioactive elements.

Group IA	Group IIA	Group IIIB	Group IVB	Group VB	Group VIB	Group VIIB	← Group	Group VIIIB →
1 H 1.01								
3 Li 6.94	4 Be 9.01							
11 Na 22.99	12 Mg 24.30							
19 K 39.10	20 Ca 40.08	21 Sc 44.96	22 Ti 47.88	23 V 50.94	24 Cr 52.00	25 Mn 54.94	26 Fe 55.85	27 Co 58.93
37 Rb 85.47	38 Sr 87.62	39 Y 88.91	40 Zr 91.22	41 Nb 92.91	42 Mo 95.94	43 Tc (98)	44 Ru 101.07	45 Rh 102.91
55 Cs 132.91	56 Ba 137.33	57 La 138.91	72 Hf 178.49	73 Ta 180.95	74 W 183.85	75 Re 186.21	76 Os 190.2	77 Ir 192.22
87 Fr (223)	88 Ra (226)	89 Ac (227)	104 Unq (261)	105 Unp (262)	106 Unh (263)	107 Uns (262)	108 Uno (265)	109 Une (266)

58 Ce 140.12	59 Pr 140.91	60 Nd 144.24	61 Pm (145)	62 Sm 150.36
90 Th (232)	91 Pa (231)	92 U (238)	93 Np (237)	94 Pu (244)